ELEMENTS OF PLANE SURVEYING

ELEMENTS OF PLANE SURVEYING

Arthur R. Benton, Jr.

Philip J. Taetz
Texas A&M University

McGraw-Hill, Inc.

New York St. Louis San Francisco Auckland Bogotá
Caracas Hamburg Lisbon London Madrid Mexico Milan Montreal
New Delhi Paris San Juan São Paulo Singapore Sydney Tokyo Toronto

This book was set in Times Roman by Publication Services.
The editors were B. J. Clark and James W. Bradley.
The cover was designed by Rafael Hernandez.
Arcata Graphics/Halliday was printer and binder.

Cover photo: Global Positioning System (GPS) receiving
antenna on tripod. *Courtesy of Wild-Leitz, Inc.*

ELEMENTS OF PLANE SURVEYING

1 2 3 4 5 6 7 8 9 0 HAL HAL 9 0 9 8 7 6 5 4 3 2 1

ISBN 0-07-004884-3

Library of Congress Cataloging-in-Publication Data

Benton, Arthur R.
Elements of plane surveying / Arthur R. Benton, Jr., Philip J.
Taetz.
p. cm.
Includes bibliographical references and index.
ISBN 0-07-004884-3 (hc)
1. Surveying. I. Taetz, Philip J. II. Title.
TA545.B48 1991
526.9—dc20 90-25340

ABOUT THE AUTHORS

Arthur R. Benton, Jr., enjoyed a 23-year initial career as a commissioned officer in the U.S. Coast and Geodetic Survey and NOAA on field and office assignments in geodetic, hydrographic, and photogrammetric surveying. After taking early retirement from NOAA, he earned a Ph.D. in civil engineering and for the next 11 years taught courses in plane surveying and remote-sensing analysis in the Civil Engineering Department at Texas A&M University, publishing almost 30 papers on his remote-sensing specialty. He retired from teaching in order to write this text, which is based on a plane surveying course he codeveloped at Texas A&M. At the time of his retirement he headed his department's Geodesy, Photogrammetry, and Surveying Division.

Philip J. Taetz is a lecturer in the Civil Engineering Department at Texas A&M University where, since 1979, he has taught undergraduate basic surveying courses to civil engineering and non-engineering students. Before coming to Texas A&M University he served for over 22 years with the National Oceanic and Atmospheric Administration (NOAA) and its predecessor organizations, which included the U.S. Coast and Geodetic Survey (now the National Geodetic Survey, a part of NOAA).

CONTENTS

PREFACE

Formal teaching of surveying was once the exclusive province of civil engineering education, and the most popular texts continue to maintain the engineering orientation. But surveying is now taught in such disciplines as surveying technology, landscape architecture, building construction, and forestry, which are significantly less math-oriented than civil engineering. For that matter, an alarming number of civil engineering students now enter surveying courses lacking the ability to apply readily the ordinary math they've been taught; thus, surveying texts should reflect the disparate levels of math preparedness among both engineers and nonengineers. Unfortunately, most of the widely used texts are written at a level of complexity that overwhelms a growing proportion of the students that use them.

This book seeks to compensate for less-than-optimal preparation by offering a review of algebra, geometry, and trigonometry, including examples of problems surveyors are likely to encounter. This first-chapter review may be included, in whole or in part, as an integral portion of the course content or it may be passed over entirely—it is up to the instructor. In any event, ample demonstration of surveying mathematics is incorporated throughout the balance of the text.

Subsequent to the brief math review, the full scope of plane surveying theory and application is presented with no corners cut. There is, for example, a chapter on the analysis and adjustment of surveying measurements errors—introduced *after* material on measurement of distances, elevations, and angles. Some texts introduce error analysis first, before the student has gained a true understanding of what is involved in the measurements whose adjustments are being discussed.

On the other hand, we have left out material on geodesy, boundary surveys, photogrammetry, and computation of state plane coordinates, topics clearly beyond the scope of a first course in plane surveying.

The time will come when surveying texts eliminate references to older instrumentation, or at least limit discussion thereof. Examples are engineering levels and vernier transits, which these days are rarely purchased by practicing surveyors, who prefer instead self-leveling levels and optical-reading theodolites. Similarly, today's surveyors seldom use steel tapes to measure appreciable distances. Mainstream practice now makes use of electronic distance measurement (EDM), reserving tapes for short lines for which current-model EDMs are less accurate and for which tape corrections may often be ignored. Nevertheless, older equipment is still around, not only to use when newer equipment breaks down in the field, but in many schools where surveying is taught. Until modern instruments become much more reliable, are universally used in the field, and are in plentiful supply in the schools that teach surveying, texts such as this will have to cover the old material along with the new.

Thus, we feel it necessary to discuss the theory and operation of modern EDMs as well as the theory and application of steel-tape corrections—the latter is tedious, but unavoidable at this writing. Similarly, engineering levels and vernier transits are discussed along with self-leveling levels and theodolites, the latter two having become the dominant instruments in regular field use.

With the advent of low-powered, low-cost microcircuitry, a new class of instruments is steadily evolving that is changing the way field surveys are carried out. The speed and accuracy of electronic distance measurement systems, along with the ability to use EDMs in conjunction with light, more-precise theodolites, means topographic and traverse operations can be carried out more quickly and precisely than was possible just a decade or two ago. Coming into common use at this writing is the "total station," an instrument that combines the functions of the theodolite and EDM, computes horizontal and vertical distances from vertical angles and EDM slope distances, and, in some units, provides automatic recording of survey data.

Obviously, not all those who practice surveying are converting to the newer instruments at the same time. Regrettably, many, if not most institutions that teach surveying simply do not have the resources to replace quickly their old vernier transits and engineer's levels with theodolites and self-leveling levels, nor can they purchase sufficient EDMs for all their labs, let alone provide enough total stations to permit their hands-on use by students in field labs. Since texts such as this are used primarily in teaching, we must take cognizance of what is going on in the actual world of education and cover the instructional bases as best we can.

Perhaps that is just as well. Changes in instruments and field procedures have not changed the physical and mathematical concepts on which surveying is based, nor are they likely to. The fundamental concept of traverse surveys—i.e., position location by measurement of directions and distances—is unaffected by how the measurements are made. Topographic surveys are still based on determination of directions, distances, and elevation differences from a known point, though newer equipment makes topographic operations less tedious. Similarly, concepts of differential leveling do not change because we now use self-leveling levels, or even automatic laser levels.

This text concentrates on the timeless, straightforward theories on which plane surveying is based and discusses, in what we hope is reasonable depth, the range of instruments and methods in use today.

Text organization is consistent with the foregoing. The text starts with surveying definitions, units, and dimensions. This is followed by discussion of survey precision and significant figures, an appropriate transition into a review of algebra, geometry, and trigonometry as they apply to surveying computations. Distances, leveling, and angle measurement are next, including instruments and equipment in current and emerging use, their operation, and their systematic errors. This is followed by a chapter on random errors as encountered in surveying measurements, along with common methods of error adjustment.

A brief treatment of transit field operations is followed by a chapter on meridians, bearings, and azimuths, including magnetics, then final chapters on traverse, stadia/prism-pole methods, topographic work, construction surveys, curves, and earthwork. The chapter on traverse includes a thorough demonstration of the compass-rule adjustment as well as brief discussion of other adjustment methods, including least squares. A separate chapter deals with use of coordinates, including explanation of state plane coordinates and discussions of geographic information systems (GIS) and current applications of the evolving Global Positioning System (GPS) technology.

A short chapter on stadia and prism-pole methods leads into a more in-depth chapter on topography that progresses into newer procedures involving the transit-EDM combination and total-station.

Construction surveying methods not covered elsewhere are discussed next, while right-of-way surveys, cross sections, and borrow pits are included in final chapters on curves and earthwork.

Derivations requiring the calculus, along with detailed examples of notekeeping, are subjects of separate appendixes. These are followed by a final section that contains answers to representative end-of-chapter problems.

McGraw-Hill and the authors would like to thank the following reviewers for their many helpful comments and suggestions: Al Barnes, Colorado State University; Joseph Gust, Northern Arizona University; Joseph O. Ramsay, Sumter Area Technical Education College; and Charles Styron, Texas State Technical Institute.

Arthur R. Benton, Jr.
Philip J. Taetz

ELEMENTS OF PLANE SURVEYING

CHAPTER 1

INTRODUCTION

1.1 OVERVIEW OF PLANE SURVEYING

Surveying is defined as measurement of the surface of the earth. Its three primary branches are **geodesy**, **photogrammetry**, and **plane surveying**. Geodesy involves precise, large-area surveys that take the actual shape of the earth into account in establishing **primary control**; that is, positions and elevations that are in turn used by other surveyors. Photogrammetry uses aerial photographs to map large areas and, with the techniques of analytical aerotriangulation, to establish control positions of somewhat lower precision within the primary geodetic control network. Plane surveying consists of ground measurement and layout within areas small enough that the effects of earth curvature may, for the most part, be neglected.

Plane surveying is by far the most widely practiced of the three. In ancient times, when mathematicians were beginning to grapple with the concepts of size and shape of a round earth, early engineers had already developed workable plane surveying techniques to lay out roads, aqueducts, and buildings and, in so doing, to advance the civilizations of their time. Today, plane surveying is seldom done by engineers; it is often handled by specialized companies that have evolved to satisfy the needs of the engineering and legal communities and the public at large. The words "surveying" and "surveyor" as used henceforth in this text will mean plane surveying and plane surveyors.

A great deal of unnecessary mystery surrounds surveying, even though it is not as complicated as many people think. Surveying deals with measurement of just three properties: distances, difference in elevation, and angles. The various surveying activities, for example, building layout, leveling, traverse, and topographic mapping of small areas, are simply different combinations of the same three basic measurements along with a few easy-to-learn computations.

Distances are measured with **steel tapes** and with **electronic distance measurement (EDM)** equipment. As EDMs have come into more general use in surveying, steel tapes are employed less and less. For reasons discussed in Chap. 2, most surveyors reserve tapes for very short lines, doing all their long-line measurement with EDMs.

Elevation surveys use **surveyors' levels** for **differential leveling**, a process that measures differences in elevation between consecutive points. Increasingly, though, difference in elevation is determined with vertical angles and slope distances by **trigonometric leveling**. Both methods are covered in Chap. 3.

Horizontal and vertical angles are measured with **transits** and **theodolites**, instruments that make it possible to read angles to very small fractions of a degree by use of either verniers or optical scales. Most surveyors call a vernier-reading instrument a transit and an optical-reading instrument a theodolite. The design and operation of vernier and optical instruments is covered in Chap. 4.

1.2 UNITS OF MEASUREMENT

Surveying applications involve measurement or determination of **length**, **elevation**, **area**, **volume**, and **angles**, all of which require use of agreed-upon units. Although the length and elevation units used in geodetic surveying are metric, plane surveyors in the United States measure lengths and elevations in **English units** of feet, tenths of feet, and hundredths of feet. Similarly, land areas, volumes, and even angles have both metric and nonmetric counterparts.

The metric system is formally known as the **International System of Units**, or, simply, **SI**. The meter was originally defined as 1/10,000,000th of the meridional distance from the Equator to the North or South Pole. This being an impractical definition for most users, metal bars were later used as standards. The most recent standard meter is the distance between two marks on a bar of 90% platinum, 10% iridium. Although the platinum–iridium international standard is a good deal more stable than the iron bar it replaced, it is still subject to change over time. In 1960, after it was discovered that spectroscopic wavelengths of certain gaseous elements were exceptionally stable, the meter was redefined as 1,650,763.73 wavelengths of the orange-red portion of the spectrum produced by light from krypton 86, a rare atmospheric gas. In 1983, the General Conference on Weights and Measures (CGPM) defined the meter as the length a path of light travels in a vacuum in 1/299,792,458 second.

The foot originally adopted in the United States was based on a meter exactly 39.37 inches (in) long, which meant that 1 foot (ft) was equal to 0.3048006096 meter (m). The **International Foot**, established by the United States in 1959, is based on an inch precisely 2.54 centimeters (cm) in length. It is thus 30.48 cm, or 0.3048 meter, exactly. Thus, the older foot differs from the International Foot by just two parts in one million. However, because the legal description of land, along with related standards for surveying equipment, is based on the pre-1959 value, the older unit, renamed the **U.S. Survey Foot**, was retained for land surveying purposes for the time being. A new U.S. surveying standard based on the International Foot was defined in 1986 but only a handful of states have adopted it at this writing.

Metric units are in many ways more convenient to use than English units because they are based on a decimal system (multiples of ten). For this reason the changeover from English to metric will come about eventually; however, despite the advantages of a decimal system, it is unlikely that plane surveyors in the United States will be able to convert completely to metric units in the near future. As indicated above, plane surveying is closely connected to the legal description of land, which in this country is recorded in feet. It will be an enormous and time-consuming task to convert from English units to metric units our millions of land descriptions, recorded in tens of thousands of volumes distributed among county courthouses nationwide.

Similarly, and for some of the same reasons, plane surveyors stick to units of feet, tenths of feet, and hundredths of feet for measuring differences in elevation and for computing elevations.

Land area is described in units of square feet (ft^2) and **acres** (1 acre = 43,560 ft^2), more often the latter. The acre (1/640th of a square mile) is often associated with an obsolete linear measurement called a **Gunter's chain** that is 66 ft long and divided into 100 links, each 0.66 ft, or 7.92 in. long. Some older property records are recorded in units of chains and links rather than feet. Even today, a few surveyors like to refer to tape measurement as "chaining." One of the chain's interesting qualities is its relationship to the mile and the acre: 80 chains equals one **statute mile** of 5280 ft and 10 square chains constitute 1 acre.

As long as land is bought and sold in units of acres, square feet, or, occasionally, in lineal feet of frontage, surveyors will have to deal with these units rather than metric units of area. Should the need arise, the metric counterpart to an acre is the **hectare** (ha), an area unit equal to a 100-m square. Thus, a hectare is 10,000 square meters (m^2), which is about 2.471 acres. The hectare is more conveniently related to the square kilometer than the acre is to the square mile. A hectare unit has 100-m sides; thus, there being 10 such sides in a kilometer, there are 100 ha in a square kilometer.

Surveyors frequently deal with earthwork volumes; that is, the amount of earth to be brought into a project, removed from it, or, most often, the volume to be moved from one part of the project to another. These volumes are normally measured in **cubic yards** (27 cubic feet = 1 cubic yard). The equipment used to transport earth is described by its capacity in cubic yards, expressed simply as "yards." For example, a "10-yard dump" is meant to carry up to 10 cubic yards of material. It would be possible, of course, to relabel such equipment in terms of cubic meters (1 cubic yard = 0.7646 m^3), but there would be no point in doing so at present.

The decimal system extends to angle measurement. Unlike our familiar system, with its 360-degree circle and 90-degree quarter-circle, European instruments divide circles into 400 units, so their quarter-circle contains 100 units. These units, called **grads** (^{g}), or, more recently, **gon**, are subdivided decimally; for example, 36.7954^g or 36.7954 gons. A grad or gon is exactly 0.9 degree, giving a conversion factor of 0.9 degree per gon ($0.9° \cdot \text{gon}^{-1}$), so an angle given in grads or gons is easily converted to degrees and decimal degrees by multiplying by $0.9° \cdot \text{gon}^{-1}$ or $0.9° \cdot \text{grad}^{-1}$.

Example 1.1. Convert 36.7954 gon to decimal degrees.

Solution.

$$36.7954\,\text{gon} \times 0.9^\circ \cdot \text{gon}^{-1} = 33.11586^\circ$$

Most surveyors in the United States are unlikely to run across grads or gon. United States instruments use the units **degrees** (°), **minutes** (′), and **seconds** (″), where a degree is 1/360th of a circumference, a minute is 1/60th of a degree, and a second is 1/60th of a minute. The degree-minute-second system used on surveying instruments is unfortunately at odds with the system built into most hand-held calculators. To obtain trigonometric functions on a calculator, the angle entered must first be converted to degrees and decimal degrees. Conversely, when using a calculator's inverse-function key to obtain the angle whose sine, cosine, or tangent equals some numerical value, the angle will appear on the calculator in degree–decimal degree notation. Most calculators that provide trigonometric and inverse trigonometric functions have keys that convert quickly from degrees, minutes, and seconds to decimal degrees and vice versa; the user's problem is remembering to do the conversion.

For calculators lacking the proper keys, the conversion process is simple enough, though rather time-consuming.

Example 1.2. Convert the angle 33.11586° from Example 1.1 to degrees, minutes, and seconds.

Solution. Think of the value as 33° + 0.11586°. Since there are 60 minutes per degree ($60' \cdot {}^{\circ -1}$), the computation is done as follows:

$$33^\circ + 0.11586^\circ \times 60' \cdot {}^{\circ -1} = 33^\circ + 6.9516' \quad \text{or} \quad 33^\circ + 6' + 0.9516'$$

Note that the ° symbol canceled the ${}^{\circ -1}$ symbol. Similarly:

$$33^\circ + 6' + 0.9516' \times 60'' \cdot {}'^{-1} = 33^\circ 6' 57.096''$$

Note, also, that the ′ symbol canceled the ${}'^{-1}$ symbol. For reasons that will be explained in the following section, this value should be rounded to 33°6′57.1″.

The procedure is reversed to convert from degrees, minutes, and seconds to decimal degrees.

Example 1.3. Convert 33°6′57.096″ to decimal degrees.

Solution.

$$33^\circ + 6' + \frac{57.096''}{60'' \cdot {}'^{-1}} = 33^\circ + 6' + 0.9516' \quad \text{or} \quad 33^\circ 6.9516'$$

Similarly:

$$33^\circ + \frac{6.9516'}{60' \cdot {}^{\circ -1}} = 33^\circ + 0.11586^\circ \quad \text{or} \quad 33.11586^\circ$$

Note that the units in Examples 1.2 and 1.3 (degrees, minutes, and seconds) were carried along with the numerical values. This helped in deciding whether the conversion factor $60' \cdot {}^{\circ -1}$ should be multiplied by decimal degrees or divided into decimal degrees in order to convert to minutes. Because the ° symbol was canceled by the ${}^{\circ -1}$ symbol, leaving just the minutes unit, it became obvious the procedure was the correct one. If, in attempting to convert from decimal degrees to degrees, minutes, and seconds, the decimal-degree value had been mistakenly *divided* by 60 minutes per degree, the resulting degrees-squared-per-minute combination would have been a tipoff.

Units also include feet, meters, miles, square feet, acres, cubic yards, degrees Fahrenheit, and so on. Because a hand-held calculator accepts only numerical values and not units, it is easy for the user to ignore the latter. It is best, when setting up a problem on paper for calculator solution, to carry units through all the steps taken in arriving at the final equation. In this way, inadvertent errors in algebra will give an inappropriate final combination of units, warning the person doing the computing that something is indeed amiss.

Another necessary conversion involves the angle notation **radians**. A radian is the angle subtended[1] at the center of a circle by a circular arc one radius in length. Thus, there are 2π radians in a circle. Surveying instruments are calibrated in degrees (or gon), but not radians, so conversion from radians to degrees or from degrees to radians is not ordinarily required in survey calculations. However, when dealing with very small angles or angular increments, it is sometimes necessary to convert from radians to seconds or from seconds to radians. For conversion of a very small angle α whose value is best expressed in seconds, it may be assumed that its radian value (arc α) is equal to its sine or its tangent. That is,

$$\text{arc } \alpha \cong \sin \alpha \cong \tan \alpha$$

Example 1.4. A traverse distance of 1697.03 ft was measured from point A to point B with electronic distance measurement equipment. It was later discovered that B', the point actually measured to, was offset 0.13 ft from intended line AB. What is the angular error in the direction of line AB' with respect to line AB?

Solution. The angle, in radians, subtended by the 0.13-ft offset is 0.13 ft/1697.03 ft = 0.000076604 radian. With that value still in the calculator window, pushing *either* the arcsin or arctan key produces an answer of 0.004389110°, which converts to 15.80080″. This result, as discussed in the next section, should be rounded to 16″. This value could have been obtained by dividing 0.000076604 radian by the conversion factor **0.00000484814 radian · second**$^{-1}$.

When solving a problem with a hand-held calculator, it is usually best to run the entire string of operations without pausing to write down intermediate values

---

[1] To visualize what "subtended" means, consider the circular arc lying between two points on the circumference of a circle. If lines are drawn to the center of the circle from each end of the arc, the angle thus formed at the center is the angle subtended by the arc.

and without turning off the calculator between operations. For one thing, writing down an intermediate value, then reentering it on the calculator, allows two unneeded opportunities to make a mistake. For another, most calculators carry additional digits beyond those displayed in the calculator window. In the above example, had the 0.000076604-radian value been written down, then later entered into the calculator and converted to decimal degrees, a very slightly different decimal-angle result would, in turn, have provided a slightly different final value of the angle in seconds.

1.3 SIGNIFICANT FIGURES

Measurement precision in surveying is limited to a somewhat fixed number of **significant figures**; that is, the number of digits in the measured value that can be stated with reasonable certainty. A taped distance, for example, may measure 614.09 ft (the tape is readable to the nearest 0.01 ft), for a total of five significant figures. Had a short line segment of, say, 5.63 ft been measured, there would still have been a precision of 0.01 ft, but this time the measurement would have been valid to just three significant figures. This illustrates the difference between significant figures and decimal places; although both measurements were taken to two decimal places, the number of significant figures is quite different.

Some rules on significant figures might be helpful:

Significant figures are counted from left to right, beginning with the first nonzero digit; thus, in a decimal figure smaller than 1.0, all digits between the decimal point and the first nonzero digit are not significant.

Zeroes at the end of a decimal number are significant.

Consider the following examples:

393.069 has six significant figures.

16.0008 has six significant figures.

0.0005300 has four significant figures.

9.0 has two significant figures.

A number such as 36,000 can be ambiguous. Was the number meant to be 36,000, as opposed to 35,000 or 37,000? Or was it meant to be 36,000, rather than 35,999 or 36,001? Had the number been 36.000 instead of 36,000, it could be assumed with certainty that there were five significant figures; that is, had the zeroes not been significant, there would have been no point in their being there. If a number such as 36,000 is meant to be valid only to two significant figures, it is better to write it in scientific notation, that is, 36×10^3.

Hand-held calculators don't deal in significant figures; some fill up the calculator window with digits (significant or not), some hold to a fixed, user-determined number of decimal places, some eliminate all trailing zeros to the right of the decimal.

The following example illustrates how to deal with significant figures while using a 10-window calculator to solve an everyday problem.

Example 1.5. The inside length and inside diameter of a cylindrical tank are measured with a surveyor's tape and found to be 27.42 ft and 6.37 ft, respectively. What is the tank's volume in cubic feet?

Solution. The volume of a cylinder is $\pi D^2 H/4$, where D is diameter and H is height. The calculator provides a final value of 873.8486035 ft^3 or perhaps 873.85 ft^3 depending on the make of calculator being used and how it is set. In terms of significant figures, neither value is correct. Consider:

π is 3.141592654 . . . , 10 significant figures on a calculator.

4 in $\pi D^2 H/4$ is an exact value, that is, 4.0000000000. . . .

D = 6.37 ft, which has three significant figures.

H = 27.42 ft, which has four significant figures.

To decide which significant-figure value governs the number of significant figures in the answer, apply the multiplication rule:

In a string of multiplication-division calculations, the precision of the end result cannot exceed the precision of the weakest number in the string.

The weakest number is 6.37; thus, the answer must be 874 ft^3.

Note, in the above example, that a somewhat different answer would have been obtained had all values been rounded to three significant figures—the same as the weakest number in the string—*before* entering those numbers in the calculator. This indicates the need for a precautionary rule on significant figures:

Never round to significant figures in the middle of a problem; instead, wait until the final value is obtained and round in accordance with the appropriate rule(s).

The next example illustrates how to handle significant figures in an addition problem in which the significant-figure precision of all numbers is known.

Example 1.6. Add the following values and express the answer to the correct number of significant figures.

Solution.

```
2742.8443
  13.8
 298.1988
1403.
---------
4457.843100   calculator result
```

The multiplication rule does not apply here. The new situation requires a new rule:

In addition or subtraction, do not carry the answer beyond the leftmost digit containing the final significant figure.

Applying the rule, the leftmost final digit is the 3 in 1403. That digit is in the units column; thus, the answer cannot extend beyond the units column and we must round to the final answer: 4458.

In mixed equations having both add-subtract and multiply-divide components, apply the rules separately, making sure to avoid rounding down to significant figures until the end of a calculation series.

Example 1.7. A concrete pipe has an outside diameter of 3.85 ft and a thickness of 0.29 ft. What is the volume of a 50.00-ft section of pipe, to the correct number of significant figures?

Solution. The volume of a hollow cylinder equals the volume of the outer cylinder less the volume of the inner cylinder; that is,

$$\text{Volume} = \frac{\pi D_o^2 H}{4} - \frac{\pi D_i^2 H}{4} \Rightarrow\Rightarrow \frac{\pi(D_o^2 - D_i^2)H}{4}$$

where D_o and D_i are the outside and inside diameters, respectively, of the cylinder. Note that $D_i = D_o - 2T$, where T is wall thickness.

Substituting known values of D_o, H, and T, an answer of 162.1690128 is obtained on the calculator; obviously, far too many significant figures. In this case, T is a two-significant-figure value. If the multiplication rule were mistakenly applied to the entire problem, the final answer would round to 1.6×10^2 ft^3. However, as a first step, the subtraction rule is applied to the equation $D_i = D_o - 2T$ to obtain a value of 3.27 for D_i; that is, a three-significant-figure value, the same as D_o. The correct result is thus 162 ft^3.

1.4 ROUNDING

The calculator results in Examples 1.5 through 1.7 were rounded to correct-significant-figure values using a rounding rule familiar to most of us: *In terms of the final digit to which the answer is rounded, if the following digit is greater than 5, round up; if it is less than 5, round down.*

A problem arises when the following digit is exactly 5. Most people are taught to round up in this case. Trouble is, surveyors must fairly often decide between instrument readings that lie halfway between two divisions on a scale. if the surveyor always rounds these halfway values *up*, a **systematic error** would occur; that is, if the system of reading and recording measured values results in adding a large number of values that are biased upward by the traditional rounding rule, an error accumulates that increases with the number of instrument observations made. To avoid this, a convention must be applied that makes it more likely that the errors will compensate rather than accumulate. Thus:

If the digit beyond the digit being rounded to is exactly 5, round to make the preceding digit even.

Examples. The value 0.023945, rounded to four significant figures, is 0.02394; the value 26.75, rounded to three significant figures, is 26.8. Note that the first case is rounded down, the second is rounded up.

Remember that this special rule refers to a value of *exactly* 5 in the digit beyond the digit being rounded to. The two-significant-figure value of 28.503 is still 29 and the three-significant-figure value of 162.500001 is still 163 because in neither case is the 5 an exact value.

1.5 ACCURACY AND PRECISION

Most people tend to use "accuracy" and "precision" interchangeably when, in reality, the words have particular meanings that are rather different. A very **precise** measurement of something is one whose value lies very close to the average of a large number of repeated measurements of that same thing. Put another way, an instrument is said to be precise if repeated measurements of the same thing lie within a close range of one another.

A very **accurate** measurement is one whose value is close to the true value. The problem is that the true value of any measurement cannot be known: Therefore, it cannot be known with certainty that a given measurement is truly accurate. At best, the *assumption* is made that the average of a large number of measurements of the same thing will be accurate, but only if steps are taken to eliminate mistakes and to reduce systematic measurement errors or compensate for them.

To illustrate the concepts of high or low accuracy or precision, consider a golfer on a practice putting green who has aimed a large number of balls toward the cup. If she putted all of her balls either into the cup or into a very small circle whose center lay at the center of the cup, her putting would be called highly accurate and highly precise. If she putted all of her balls into an equally small circle whose center was displaced some distance from the cup, she would have been highly precise, but not accurate. If all her balls ended up within a rather large circle centered on the cup, she would have putted the group of balls accurately but not precisely. If the balls had stopped within an equally large circle whose center lay some distance from the cup, she'd have putted neither accurately nor precisely.

Surveyors, in particular, often must differentiate between the two terms when describing an instrument or a measurement system. For the most part, the best instruments are highly precise but not necessarily accurate. Methods of improving the accuracy of such instruments will be discussed in later chapters.

1.6 MATHEMATICS FOR SURVEYORS

Many students find, on beginning their surveying courses, that they are inadequately prepared to cope with the necessary mathematics—and this is often as true of engineers as it is of nonengineers. To help overcome that lack of confidence, the balance of this chapter will review the math topics surveyors need. It is not intended to substitute for missing courses—it is assumed that any student taking a course in plane surveying will have passed a course in plane trigonometry, meaning that earlier courses in algebra and geometry will also have been successfully completed. Nevertheless, the authors have found that many students in surveying have need for a math refresher and for a focusing of math skills, with particular emphasis on areas relevant to solving surveying problems. The following sections are intended to provide that focus.

1.7 ALGEBRA

An algebraic expression involves three things: one or more known values, one or more unknown values, and an "equal" sign that defines the relationship between the known and unknown values. A simple example is $x = a + b$, where x is unknown and a and b designate known values, say 3 and 4. Substituting for a and b into the equation, it is quickly determined that x is 7. Although this example is trivial, most equations surveyors deal with are not much more difficult to handle, usually yielding to ordinary addition, subtraction, multiplication, and division, along with routine algebraic manipulation.

The algebra surveyors use is usually limited to **linear equations**, **simultaneous linear equations**, and **quadratic equations**. The balance of this section will review these three concepts.

LINEAR EQUATIONS. A linear equation is one in which no unknown value is taken to a power higher than one. A linear equation could also be defined as one whose plot on a two-dimensional coordinate system is a straight line (see Sec. 1.8). The following examples of linear equations are discussed below:

$$8x + 7 = 3x + 2 \qquad 25y - 17 = 6.4y + 11 \qquad x - 3 = y - 7x + 9$$

Example 1.8. In the equation $8x + 7 = 3x + 2$, solve for x.

Solution. Linear equations are solved by rearranging to place the unknown value or values on one side of the equal sign and the known values on the other side. Thus, $3x$ is subtracted from both sides to obtain $5x + 7 = 2$; then, 7 is subtracted from both sides to obtain $5x = -5$; finally, dividing both sides by 5 gives the answer: $x = -1$.

Although Example 1.8 is quite simple, it illustrates the method of solving algebraic equations in general: First, isolate the unknown part on one side of the equal sign (by convention, usually the left) with the rest of the equation on the other side, then do whatever is necessary to arrive at the final solution.

Example 1.9. Solve for y in the expression $25y - 17 = 6.4y + 11$.

Solution. Add 17 to both sides and subtract $6.4y$ from both sides, leaving

$$18.6y = 28$$

Now,

$$y = \frac{28}{18.6} = 1.505376344 \Rightarrow\Rightarrow 1.5$$

(Note the application of the appropriate significant-figure rule.)

Like most professionals, surveyors are subject to Murphy's laws,[2] which guarantee that mistakes are bound to happen. To minimize the effect of errors, surveyors are careful to check both their field measurements and the results of their computations.

Example 1.9 is checked by substituting the exact value of y into the original equation. Substituting on the left side of the equal sign:

$$25y - 17 = 25 \times 1.505376344 - 17 = 20.63440860$$

Substituting into the right side of the equal sign:

$$6.4y + 11 = 6.4 \times 1.505376344 + 11 = 20.63440860 \quad \checkmark$$

Checking in this manner would be a rather tedious procedure if done with pencil and paper; on a calculator it takes very little time. Furthermore, a 10-significant-digit check on a calculator is no more involved than one with only three significant figures, especially if the exact-solution value is transferred to the calculator memory at the time it is computed.

The third linear-equation example, $x - 3 = y - 7x + 9$, is simplified to $8x - y = 12$ by the methods already discussed. If x and y are both unknown, there is no unique solution. Another equation in x and y is needed, after which the values of the two unknowns can be found by the method discussed next.

SIMULTANEOUS LINEAR EQUATIONS. The equation $8x - y = 12$ is of the form $ax + by = c$, where, in this case, $a = 8$, $b = -1$, and $c = 12$. If there is available a second equation $a'x + b'y = c'$, where, a', b', and c' are constants at least one of which is different from its a, b, or c counterpart, there is bound to be a unique solution for x and y that satisfies both equations. It will be demonstrated in Sec. 1.8 that two linear equations of this sort may be plotted as separate lines on a common xy coordinate system, with the x and y coordinates of the single intersection of the two lines being the x and y values of the solution. The *arithmetic* solution, however, is derived as follows:

General solution of two simultaneous equations. First, the two equations are placed one over the other:

---

[2] Murphy's laws:

1. Nothing is as easy as it looks.
2. Everything takes longer than you think it will.
3. Anything that *can* happen, *will* happen.

The third law is further complicated by the Surveyor's corollary: The probability of instrument breakdown is directly proportional to the remoteness of the project area from civilization.

$$ax + by = c$$
$$a'x + b'y = c'$$

The unknown y is eliminated by multiplying the first equation by b' and the second equation by b to obtain

$$ab'x + bb'y = b'c$$
$$a'bx + bb'y = bc'$$

Subtracting, $$ab'x - a'bx = b'c - bc'$$

Rearrangement leads to the following solution for the value of x in simultaneous equations of the form: $ax + by = c; a'x + b'y = c'$:

$$\boldsymbol{x = \frac{b'c - bc'}{ab' - a'b}} \qquad \textbf{(1.1)}$$

To solve for y, simply insert the value of x obtained from Eq. (1.1) into *one* of the simultaneous equations, then treat it as a single linear equation in y. As a check, insert the solved x and y values into the *other* simultaneous equation. See Example 1.10.

Example 1.10. Solve the following for x and y using Eq. (1.1):

$$27.6x + 9.47y = 218.3 \qquad 22.9x - 25.6y = 128.6$$

Solution. In the above equations, a is 27.6, b is 9.47, c is 218.3, a' is 22.9, b' is −25.6, and c' is 128.6. Substituting into Eq. (1.1):

$$x = \frac{218.3(-25.6) - 9.47(128.6)}{27.6(-25.6) - 9.47(22.9)}$$
$$= 7.370752082 \Rightarrow\Rightarrow 7.37$$

Rearranging the first simultaneous equation to solve for y:

$$y = \frac{218.3 - 27.6x}{9.47}$$

Substituting the calculator value of x just obtained:

$$y = \frac{218.3 - 27.6(7.370752082)}{9.47} = 1.569930570 \Rightarrow\Rightarrow 1.57$$

As a check, the calculator values of x and y are substituted into the left side of the second simultaneous equation as follows:

$$22.9x - 25.6y = 22.9(7.370752082) - 25.6(1.569930570)$$
$$= 128.6000001 \checkmark$$

The difference between the check value of c and the original value of c is due to internal rounding in the calculator.

Complex values of a, b, c, a', b', and c' such as those in Example 1.10 require the use of Eq. (1.1) for solution. It is often the case, however, in simultaneous equations with simpler coefficient values, that an easier solution presents itself.

Example 1.11. Solve the following simultaneous equations:

$$3x + y = 11$$
$$x - y = 1$$

Solution. In this case, we can solve the equations easily by adding them together, thus eliminating y; that is,

$$3x + x + y - y = 11 + 1$$

Collecting and canceling:

$$4x = 12$$
$$x = 3$$

Substituting for x in the second equation:

$$y = x - 1 = 3 - 1 = 2$$

Substituting into the first equation as a check:

$$3x + y = 3(3) + 2 = 11 \quad \checkmark$$

QUADRATIC EQUATIONS. So far, only linear equations have been considered. A nonlinear equation is one whose xy plot would yield something other than a straight line. The expressions $y = ax^2 + bx + c$ and $y = \log x$ are just two examples of nonlinear equations. The first is known as a **quadratic**, an equation surveyors must deal with fairly often. Quadratics generally have two solutions; sometimes both are relevant and sometimes the solver must decide which is relevant and which is not. The solution of this equation, called the **quadratic equation theorem**, is fairly well known. Its derivation, shown below, makes an interesting algebraic exercise.

General solution of quadratic equations. Consider the equation $ax^2 + bx + c = 0$. The solution is derived by establishing an equation on one side of the equal sign that is the square of a linear expression of the form $a'x + b'$, placing the remainder on the opposite side of the equal sign, then taking the square root. The first step is to multiply the original equation by a in order to convert the first term to the square of ax. The equation becomes

$$a^2x^2 + abx + ac = 0$$

In order to make the first and second terms compatible with the first two terms of the square of an $a'x + b'$ expression, the equation is multiplied by 4 to produce

$$4a^2x^2 + 4abx + 4ac = 0$$

To complete the square $(2ax + b)^2$, add b^2 to both sides of the equation and subtract $4ac$ from both sides to produce

$$4a^2x^2 + 4abx + b^2 = b^2 - 4ac$$

Taking the square root of both sides:

$$2ax + b = \pm\sqrt{b^2 - 4ac}$$

Subtracting b from both sides and dividing both sides by $2a$, the final equation is:

$$\boldsymbol{x = \frac{-b \pm \sqrt{b^2 - 4ac}}{2a}} \tag{1.2}$$

Example 1.12. Solve for x in the equation $x^2 + 6x - 27 = 0$.

Solution. Substituting $a = 1$, $b = 6$, and $c = -27$ into Eq. (1.2):

$$x = \frac{-6 \pm \sqrt{6^2 - 4(1)(-27)}}{2(1)}$$

$$= -\frac{6}{2} \pm \frac{\sqrt{36 + 108}}{2}$$

$$= -3 \pm \frac{\sqrt{144}}{2}$$

$$= -3 \pm \frac{12}{2}$$

$$= -3 \pm 6$$

$$x = +3 \qquad \text{and} \qquad x = -9$$

With a bit of intuition, however, the equation in Example 1.12 might have been factored directly into $(x + 9)(x - 3) = 0$, yielding the two separate linear equations: $x + 9 = 0$ and $x - 3 = 0$, from which it is obvious that $x = -9$ and $x = +3$. Unfortunately, it rarely happens that quadratics yield to direct, simple factoring of this sort. Given a more complex equation—usually the case with surveying-related quadratics—Eq. (1.2) becomes essential to the solution.

Example 1.13. $2.01y^2 - 16.94y = 2.58 + 3.78y^2 + 18.17$. Solve for y.

Solution. Assembling the equation into the form $ay^2 + by + c = 0$ produces the expression:

$$1.77y^2 + 16.94y + 20.75 = 0$$

where $a = 1.77$, $b = 16.94$, and $c = 20.75$. Substituting into Eq. (1.2):

$$y = \frac{-16.94 \pm [(16.94)^2 - 4(1.77)(20.75)]^{1/2}}{2(1.77)}$$

$$= -4.785310734 \pm 3.343057729$$

$$= -8.128368463; -1.442253005 \Rightarrow\Rightarrow -8.13; -1.44$$

An excellent check of the computation is obtained by substituting either of the full calculator-derived values into both sides of the *original* equation, then comparing results:

$$2.01(-8.128368463)^2 - 16.94(-8.128368463) = 270.4960133$$

$$2.58 + 3.78(-8.128368463)^2 + 18.17 = 270.4960132 \checkmark$$

Again, the difference in the two check values is due to internal rounding in the calculator. The first check having been made, another check, using the second value of y, would serve only to verify the final arithmetic operation that produced the second value of y.

It might occur to the reader that $(b^2 - 4ac)$ could be a negative value, in which case its square root would be imaginary—a most awkward concept for surveyors to deal with. Fortunately, it is unlikely that quadratics encountered in surveying would result in imaginary values.

1.8 COORDINATE GEOMETRY

Few of the elaborate theorems and definitions involved in plane geometry are of importance in surveying. On the other hand, it will be increasingly necessary for surveyors to have a working knowledge of **coordinate geometry**: the mathematical description of geometric figures on an xy coordinate system. This is because many cities and towns are now placing coordinate data for their "infrastructure" (that is, transportation rights-of-way, utilities, and the like) and local property boundaries onto a computer format. A completed data base of this sort constitutes a **geographic information system**, or **GIS**, in which the locations of all components of the infrastructure will have a common geographic reference, a great advantage when new construction is contemplated or system maintenance is needed.

In coordinate geometry, the ordinate (y axis) is usually positioned vertically with the abscissa (x axis) horizontal. y values that lie above the x axis are positive; those that lie below the x axis are negative. x values that lie to the right of the y axis are positive; those to the left of the y axis are negative.

In the plan view in which the GIS is constructed, the y axis runs north-south and the x axis east-west. It is more convenient, for the GIS frame of reference, if the origin ($y = 0; x = 0$) is located to the west and south of any point in the system itself so that all x and y coordinate values are positive.

State plane coordinate systems, with which many surveyors become involved, work in the same way, all coordinates having been assigned fairly large positive values.

The coordinate geometry examples that follow will use conventional notation; that is, x and y can be either positive or negative. Once the conventional coordinate system is understood, it is easy enough to switch one's thinking to GIS or state plane coordinate notation.

LINE EQUATIONS. A line may be described mathematically in many ways. This portion of the text will discuss two: the **general form** and the **two-point form**. A general-form equation may be expressed in several ways. The following is probably the most convenient for surveyors:

General line equation: $$y = mx + b \tag{1.3}$$

where y and x are the coordinates of a point on a line of slope **m** that crosses the y axis a distance b from the x axis. Slope, or gradient, is further defined as the tangent

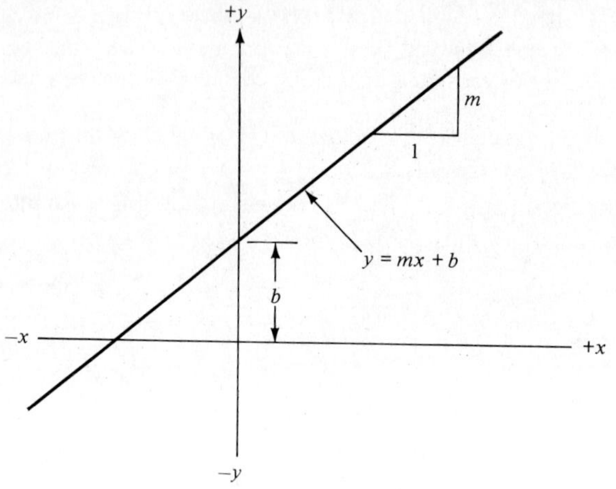

FIGURE 1.1
General form of the line equation.

of the angle of intersection of the line with the x axis. Figure 1.1 illustrates the line $y = mx + b$ on an xy coordinate system, with α the angle of intersection.

Example 1.14. Write the general equation of a line that crosses the y axis at $y = -3$ and rises 1 unit vertically for every 10 units of horizontal rightward run.

Solution. From the given data, the slope, m, is $+1/(+10) = 0.1$ and $b = -3$. Substituting, $y = 0.1x - 3$.

Example 1.15. A line whose slope is -0.28 passes through a point whose coordinates are $x = 10$, $y = -7$. What is the y coordinate of a point on the line whose x coordinate is 0?

Solution. The problem asks for the value of b. Manipulation of Eq. (1.3) produces

$$b = y - mx$$

Substituting for y, m, and x:

$$b = -7 - (-0.28)(10) = -4.2$$

Given the coordinates of two points on a line, the values of m and b are readily computed by the method of simultaneous equations. In Figure 1.2, the x- and y-coordinate values are given in parentheses using the standard convention wherein the x-coordinate is given first, followed by a comma, then the y coordinate.

Example 1.16. Compute m and b for the line in Fig. 1.2.

Solution. Substituting x and y coordinates of point 1 into Eq. (1.3):

$$4 = -6m + b$$

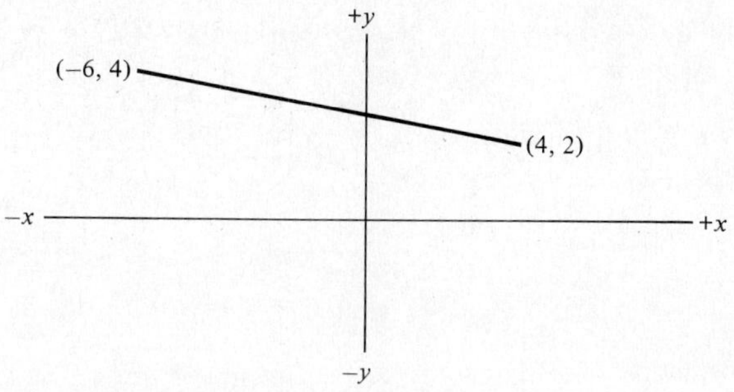

FIGURE 1.2
General-form line connecting two known points.

Substituting x and y coordinates of point 2 into Eq. (1.3):

$$2 = 4m + b$$

Subtracting,

$$2 = -10m + 0$$

From the remainder, $2 = -10m$, we find that $m = -0.2$. Substituting $m = -0.2$ into the *first* equation:

$$4 = -6(-0.2) + b$$

$$b = 4 - 1.2 = +2.8$$

Checking with computed values using the second equation:

$$2 = 4m + b = 4(-0.2) + 2.8 \equiv 2 \quad ✓$$

More often, however, simultaneous general-form line equations are used to determine the intersection of two lines. For the xy coordinates of a single point to satisfy the equations of two separate lines, the point must lie at the intersection of the lines.

Example 1.17. Compute the intersection of line $y = 2.3x + 1.3$ with the line specified in Example 1.16. See Fig. 1.3.

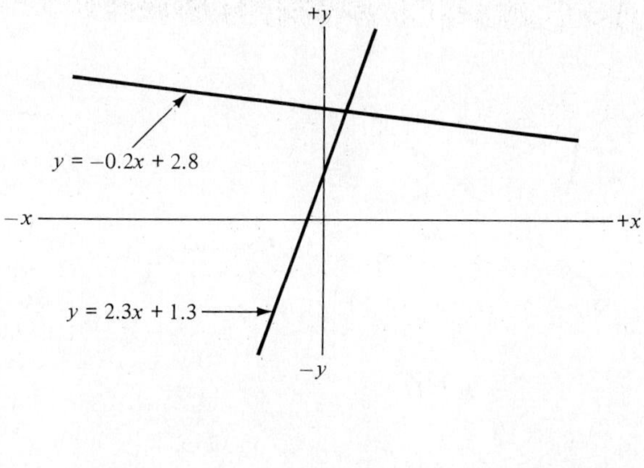

FIGURE 1.3
Intersecting general-form lines.

Solution. In Example 1.16 the line equation is

$$y = -0.2x + 2.8$$

The new line equation is $\quad y = 2.3x + 1.3$

Subtracting $\quad 0 = -2.5x + 1.5$

$$2.5x = 1.5$$

$$x = 0.6$$

Substituting $x = 0.6$ into the first equation:

$$y = -0.2(0.6) + 2.8 = 2.68$$

Check by substituting both x and y into the second equation:

$$2.68 = 2.3(0.6) + 1.3 \equiv 2.68 \checkmark$$

A surveyor usually will know the xy coordinates of two or more points on a line but not the values of m and b for that line. It is more straightforward to solve a line-intersection problem using the **two-point form**, in which two points on a line are known but m and b are neither known nor needed:

Two-point line equation: $\quad (y_2 - y_1)x - (x_2 - x_1)y = x_1y_2 - x_2y_1 \quad$ **(1.4)**

In Eq. (1.4), x and y are the coordinates of any point lying on the line, while x_1, y_1 and x_2, y_2 are the coordinates of the two specific points that define the line. The equation of a line defined by points 3 and 4 would be $(y_4 - y_3)x - (x_4 - x_3)y = x_3y_4 - x_4y_3$. Figure 1.4 shows two lines, one defined by points 1 and 2, the other defined by points 3 and 4.

Example 1.18. Compute the x and y coordinates of the intersection of the lines in Fig. 1.4 if coordinates of the four points are $x_1 = -6, y_1 = +3; x_2 = +4, y_2 = -1; x_3 = +3, y_3 = +5; x_4 = -5, y_4 = -2$.

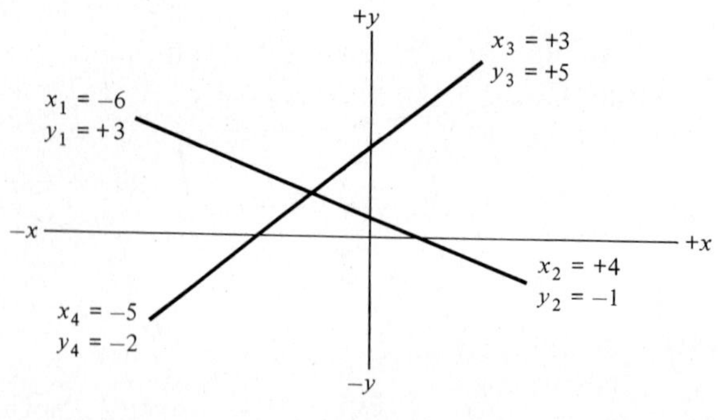

FIGURE 1.4
Two lines definable by the two-point line form.

Solution. The defining x- and y-coordinate values are substituted into the two-point equations as follows:

$$(-1-3)x - [4-(-6)]y = -6(-1) - 4(3) \Rightarrow\!\Rightarrow -4x - 10y = -6$$

$$(-2-5)x - (-5-3)y = 3(-2) - (-5)(5) \Rightarrow\!\Rightarrow -7x + 8y = 19$$

To solve, multiply the top equation by 4:

$$-16x - 40y = -24$$

and multiply the bottom equation by 5:

$$-35x + 40y = 95$$

Then add to obtain

$$-51x = 71$$

Therefore:

$$x = -\tfrac{71}{51} = -1.392156863$$

Rearranging the top equation:

$$10y = 6 - 4x$$

$$y = \frac{6-4x}{10}$$

Substituting:

$$y = \frac{6 - 4(-1.392156863)}{10} = 1.156862745$$

Substituting the computed x and y values into the bottom equation:

$$-7(-1.392156863) + 8(1.156862745) = 19.00000000 \;\checkmark$$

The computed x and y intersection values are rounded to whatever significant-figure values could be sustained by the precision of the known coordinate values for points 1, 2, 3, and 4. The values used in Example 1.18 were chosen for the sake of simplicity. For real-world coordinate systems, the defining-point coordinates would all be positive and probably at least five-digit numbers ending in tenths of feet or, more likely, hundredths of feet.

DISTANCE COMPUTATION. Distances between points on a coordinate system are easily computed from coordinate differences using the **theorem of Pythagoras**. Knowing the coordinates of points 1 and 2, the distance between them is given by

$$\textbf{Length}_{1-2} = \sqrt{(x_2 - x_1)^2 + (y_2 - y_1)^2} \qquad \textbf{(1.5)}$$

Example 1.19. Compute the distance between points 1 and 4 in Example 1.18.

Solution. From Example 1.18, $x_1 = -6, y_1 = +3, x_4 = -5, y_4 = -2$. Substituting into Eq. (1.5):

$$\text{Length}_{1-4} = \sqrt{[-5-(-6)]^2+(-2-3)^2} = \sqrt{(1)^2+(-5)^2}$$

$$= \sqrt{26} = 5.09902$$

CIRCLE EQUATIONS. The equation of a circle whose center is at the origin ($x = 0, y = 0$) is $x^2 + y^2 = r^2$, where r is the radius and x and y are the coordinates of any point on the circle. But circles are seldom centered on the origin. Therefore, use the general form:

$$(x-a)^2 + (y-b)^2 = r^2 \tag{1.6}$$

in which a and b are the x and y coordinates, respectively, of the center of the circle.

Example 1.20. What is the equation of a circle of radius equal to 12 units if its center is at $x = 3, y = -1$?

Solution. Here, $a = 3$, $b = -1$, and $r = 12$. Substituting, the circle equation must be

$$(x-3)^2 + (y+1)^2 = 144$$

Finding the intersections of a line with a circle is less difficult than it may seem, particularly if the line is in the general form.

Example 1.21. To four significant figures, compute the two intersections of a line whose equation is $y = x + 1$ with a circle whose equation is $(x-2)^2 + (y-4)^2 = 49$. See Fig. 1.5.

Solution. From the general equation, $y = x + 1$. Substituting $x + 1$ for y in the circle equation produces

$$(x-2)^2 + [(x+1)-4]^2 = 49$$

$$(x-2)^2 + (x-3)^2 = 49$$

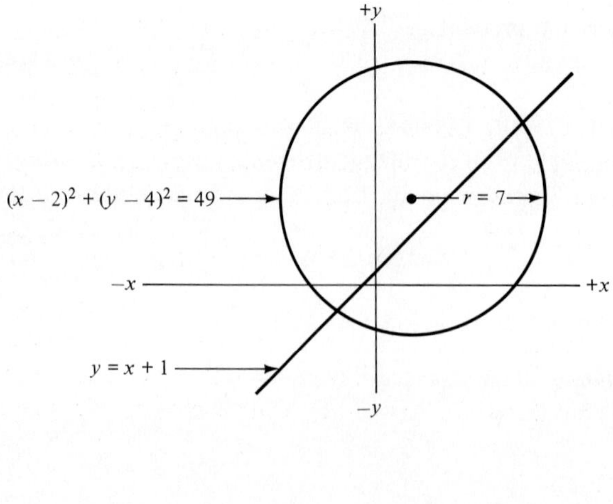

FIGURE 1.5
Intersection of a line with a circle.

Squaring the two values and collecting terms:

$$x^2 - 4x + 4 + x^2 - 6x + 9 = 49$$

$$2x^2 - 10x + 13 = 49$$

$$2x^2 - 10x - 36 = 0$$

Substituting into the quadratic equation theorem:

$$x = \frac{10 \pm \sqrt{100 + 289}}{4}$$

$$= 2.5 \pm 4.924428900$$

$$= 2.500 \pm 4.924$$

$$x = 7.424 \qquad x = -2.424$$

Substituting the two x values into the line equation $y = x + 1$ produces the coordinates of the two intersections: $x_1 = 7.424, y_1 = 8.424; x_2 = -2.424, y_2 = -1.424$.

Check by inserting the full calculator values of x_1 and y_1 into the circle equation:

$$(7.4244289 - 2)^2 + (8.4244289 - 4)^2 = 48.99999998 \quad ✓$$

The line in Example 1.21 was in the more convenient general form. Note that the two-point form, $(y_2 - y_1)x - (x_2 - x_1)y = x_1y_2 - x_2y_1$ can be converted by dividing by $x_2 - x_1$ and rearranging terms to obtain the general form $y = [(y_2 - y_1)/(x_2 - x_1)]x + (x_2y_1 - x_1y_2)/(x_2 - x_1)$, where m is $(y_2 - y_1)/(x_2 - x_1)$ and b is $(x_2y_1 - x_1y_2)/(x_2 - x_1)$. The end result isn't as complex as it looks because all of the subscripted terms are known numerical values; thus, expressions for m and b may be converted to decimal numbers quickly with a hand-held calculator.

1.9 TRIGONOMETRIC CONCEPTS

Surveyors use trigonometry primarily to deal with the relationships between angles and sides of triangles. For right triangles, the relationship is conveniently established using a circle marked at its north, south, east, and west points and having a line extending from O, the center of the circle, through the north mark (Fig. 1.6). The line is the origin, or zero line, for an angle α taken *clockwise* from the north origin.

The angle may be defined in degrees or radians. If in degrees, the circle is divided into 360 degrees and the clockwise angle from north to east is 90°, from north to south is 180°, from north to west is 270°, and from north back to north again is 360°.

As discussed earlier, there are 2π radians in a circle; thus, the clockwise angle from north to east is $\pi/2$ radians, from north to south is π radians, from north to west is $3\pi/2$ radians, and from north back to north again is 2π radians. Given that 360° is equal to 2π radians, the following conversion factors are readily obtained:

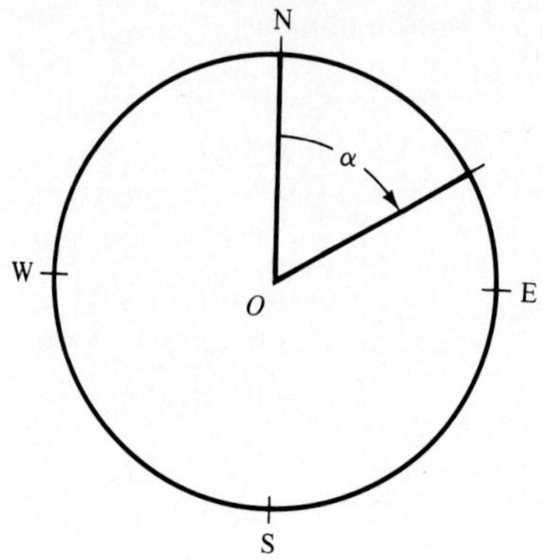

FIGURE 1.6
Circle with central angle clockwise from north.

$$\text{Degrees per radian} = \frac{360^\circ}{2\pi\,\text{rad}} = 57^\circ 17' 44.8'' \cdot \text{rad}^{-1}$$

$$\text{Minutes per radian} = \frac{360^\circ \times 60' \cdot {}^{\circ -1}}{2\pi\,\text{rad}}$$

$$= 3437.75' \cdot \text{rad}^{-1}$$

$$\text{Seconds per radian} = \frac{360^\circ \times 60' \cdot {}^{\circ -1} \times 60'' \cdot {}^{\circ -1}}{2\pi\,\text{rad}}$$

$$= 206,265'' \cdot \text{rad}^{-1}$$

$$\text{Radians per degree} = \frac{2\pi\ \text{radians}}{360\ \text{degrees}}$$

$$= 0.0174533\,\text{rad} \cdot {}^{\circ -1}$$

$$\text{Radians per minute} = \frac{2\pi\ \text{radians}}{360^\circ \times 60\ \text{min} \cdot {}^{\circ -1}}$$

$$= 0.00029088\,\text{rad} \cdot {}'^{-1}$$

$$\text{Radians per second} = \frac{2\pi\ \text{radians}}{360^\circ \times 60\ \text{min} \cdot {}^{\circ -1} \times 60'' \cdot {}'^{-1}}$$

$$= 0.00000484814\,\text{rad} \cdot {}''^{-1}$$

SINES AND COSINES. Returning to the circle, consider a line drawn from the center to point A, which lies along the arc from north to east (Fig. 1.7). The angle between the north origin line and the new line is α. Point B is at the intersection of the origin line with a perpendicular from the origin line to A. Note that OAB is a right triangle.

If OA (the radius of the circle) is one unit in length (thence the name **unit circle**), the **sine** of angle α is defined as the length of line AB and the **cosine** of α is defined as the length of line OB.

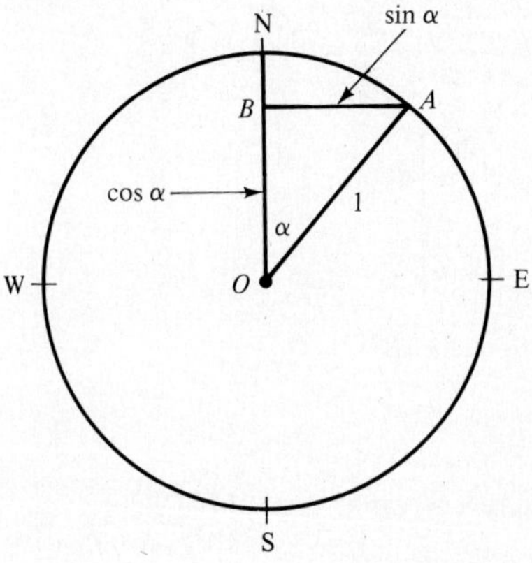

FIGURE 1.7
Unit circle with right triangle OAB.

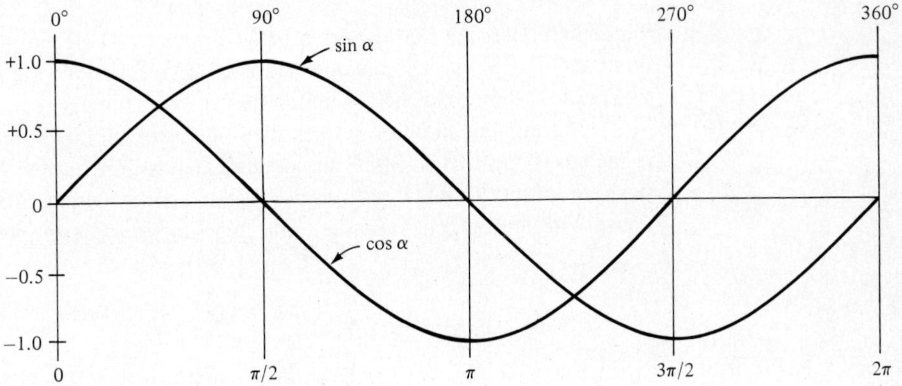

FIGURE 1.8
Plot of sine and cosine curves for a full circle.

Note that the hypotenuse OA of triangle OAB has a length of one unit. Sin $\alpha = AB = AB/1 = AB/\text{hypotenuse}$; thus, since AB is the side opposite angle α, sin α = side opposite α/hypotenuse. Similarly, cos α = side adjacent α/hypotenuse.

Line OA turns about point O so that α goes from 0° to 360°. As OA turns, and α increases from 0° to 90° (quadrant I), sin α changes value from 0 to +1; from 90° to 180° (quadrant II), sin α changes from +1 to 0; from 180° to 270° (quadrant III), sin α changes from 0 to −1; from 270° to 360° (quadrant IV), sin α changes from −1 to 0. Similarly, as α increases from 0° to 90°, cos α decreases from +1 to 0; from 90° to 180°, cos α decreases from 0 to −1; from 180° to 270°, cos α increases from −1 to 0; from 270° to 360°, cos α increases from 0 to +1. Figure 1.8 shows sine and cosine curves, or "waves," over 360°.

TANGENTS AND COTANGENTS. The **tangent** of α is defined as the side opposite α divided by the side adjacent to α ($\sin\alpha/\cos\alpha$). The **cotangent** of α is defined as 1/tan α; that is, the side adjacent to α divided by the side opposite α ($\cos\alpha/\sin\alpha$).

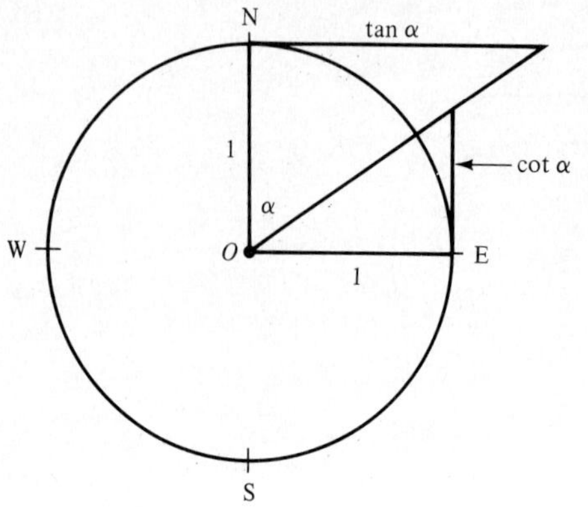

FIGURE 1.9
Unit circle showing tangent and cotangent lines.

Tan α and cot α are both positive in quadrants I and III and they are both negative in quadrants II and IV. The tangent and cotangent functions are shown on the unit circle in Fig. 1.9, including the algebraic sign of both functions by quadrant.

The tangent function passes from negative infinity to positive infinity twice as α increases from 0° to 360°; the cotangent function, on the other hand, passes from positive infinity to negative infinity twice over the same angular span, but 90° out of phase with the tangent function. The two functions are graphed in Fig. 1.10.

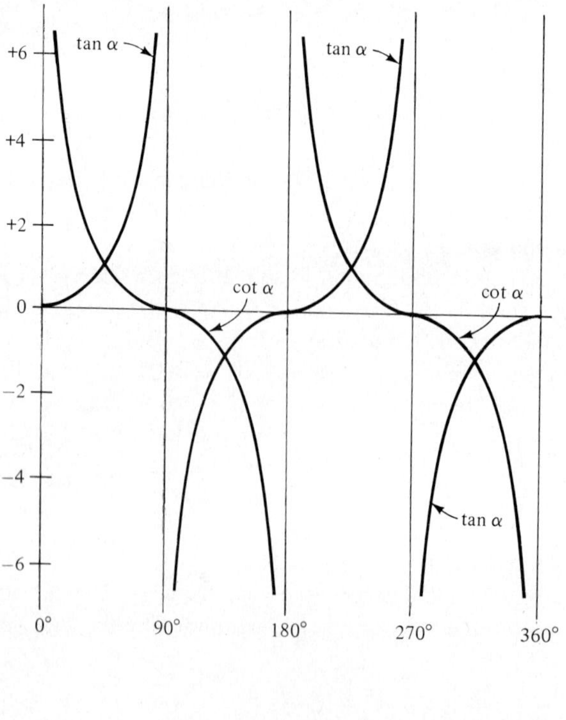

FIGURE 1.10
Plot of tangent and cotangent curves for a full circle.

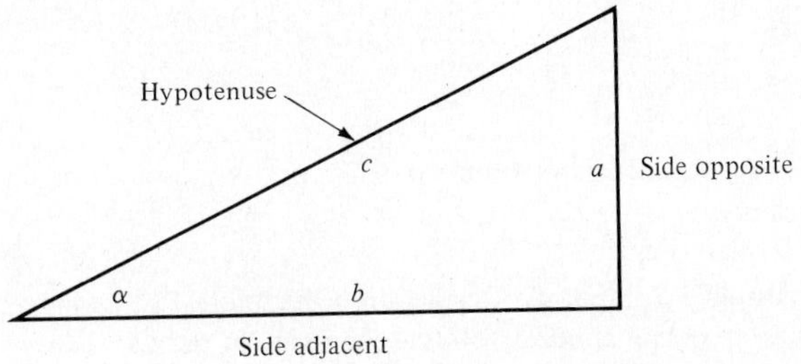

FIGURE 1.11
Right-triangle relationships.

The two remaining plane trigonometric functions are secant and cosecant, defined, respectively, as the reciprocal of the cosine function and the reciprocal of the sine function. With the increasing use of hand-held calculators, with which reciprocals of sine and cosine values are so quickly obtained, the secant and cosecant functions, as such, have becomes less important and their graphs are not particularly relevant to problems in plane surveying.

The foregoing discussion of sines, cosines, and tangents has developed three of the four key equations for solving right triangle problems. The fourth equation is the theorem of Pythagoras. The following equations relate to Fig. 1.11:

$$\sin \alpha = \frac{\text{side opposite } \alpha}{\text{hypotenuse}} \tag{1.7}$$

$$\cos \alpha = \frac{\text{side adjacent } \alpha}{\text{hypotenuse}} \tag{1.8}$$

$$\tan \alpha = \frac{\text{side opposite } \alpha}{\text{side adjacent } \alpha} \tag{1.9}$$

$$a^2 + b^2 = c^2 \tag{1.10}$$

1.10 RIGHT-TRIANGLE APPLICATIONS

Many of the important concepts in surveying may be described and understood using one or more of the four equations listed above. The following brief examples demonstrate a few applications; others will appear in later chapters. The following example is fairly common in construction surveying. It demonstrates the use of the Pythagorean theorem.

Example 1.22. A rectangular warehouse is to be 120.00 ft long and 50.00 ft wide. What diagonal measurement should the surveyor use to check form layout for the foundation?

Solution. The diagonal is the hypotenuse of a triangle whose sides are 50.00 ft and 120.00 ft. Thus,

$$\text{Diagonal}^2 = (50.00)^2 + (120.00)^2$$

$$\text{Diagonal} = \sqrt{(50.00)^2 + (120.00)^2}$$
$$= 130.00\text{ft}$$

The relationship between the sides and angles of a right triangle is shown in the next example.

Example 1.23. For the foundation of Example 1.22, what angle, to the nearest minute, does the diagonal make with the long side of the warehouse?

Solution. The solution triangle contains two known sides and an unknown included angle. In terms of the unknown angle, the long side of the warehouse is the adjacent side and the diagonal is the hypotenuse. Therefore,

$$\cos\alpha = \frac{\text{side adjacent}}{\text{hypotenuse}}$$
$$\alpha = \cos^{-1}\frac{120.00}{130.00}$$
$$= 22.6199° = 22°37'$$

Example 1.23 could have been solved by using only original data instead of computed data; that is, taking α as the angle whose tangent is the short side of the foundation divided by the long side. Thus, $\alpha = \tan^{-1}(50.00/120.00) \equiv 22.6199° = 22°37'$.

Example 1.24. The measured length of the shadow of a television tower is 358.9 ft at the instant of time when the sun's elevation angle above the horizon is known to be 64°15′. Assuming the ground is level, find the height of the tower to the nearest foot.

Solution. With respect to the solar elevation angle, the tower is the side opposite and the shadow is the side adjacent. Thus,

$$\tan\alpha = \frac{\text{height}}{\text{shadow}}$$
$$\text{Height} = \text{shadow} \times \tan\alpha$$
$$= 358.9(\tan 64°15')$$
$$= 744.0767 \Rightarrow\Rightarrow 744\text{ ft}$$

Example 1.25. The instrument glide path on a jet-aircraft approach to an airport is exactly −5°; that is, approaching aircraft descend on a 5° slope with respect to the horizontal. A tower whose elevation is 150 ft higher than the end of the runway is located directly under the glide path and 6400 ft back from the approach end of the runway. By how much will the tower be cleared by an approaching aircraft that is exactly in the glide path?

Solution. It is given that the top of the tower is 150 ft higher than the end of the runway. What must be determined is the height of the approaching aircraft with respect to the end of the runway when it is over the tower. The solution triangle includes the glide path as the hypotenuse and the glide path angle (α) as the included angle between the glide path and the 6400-ft horizontal distance to the tower. Thus,

$$\tan \alpha = \frac{\text{aircraft height}}{6400}$$

$$\text{Height} = 6400 \times \tan 5^\circ = 559.9 \text{ ft}$$

The vertical clearance is thus 560 ft − 150 ft = 410 ft.

Example 1.26. A 6.25% grade means 6.25 ft of rise in 100.00 ft of *horizontal* distance. How many lineal feet of paving must be used along a 6.25% grade as the highway advances a mile horizontally?

Solution. Paving length is the hypotenuse in a triangle whose side adjacent is 5280 ft and whose included angle must be obtained from the highway grade data. it is apparent the grade angle α is the angle whose tangent is 6.25/100.00; that is, $\alpha = \tan^{-1}(6.25/100.00)$. Rather than compute α first and then substitute into another equation, it is easier, when working with a calculator, to set up the entire problem and work it through all at once. Thus,

$$\cos \alpha = \frac{5280 \text{ ft}}{\text{paved distance}}$$

$$\text{Paved distance} = \frac{5280 \text{ ft}}{\cos \alpha}$$

Substituting for α,

$$\text{Paved distance} = \frac{5280}{\cos[\tan^{-1}(6.25/100.00)]}$$

$$= 5290.302449 \Rightarrow\Rightarrow 5290.30 \text{ ft}$$

This may also be solved by computing the hypotenuse of a triangle with a 5280-ft base and a height equal to the rise of a 6.25% grade in 1 mile. Thus,

$$\text{Rise} = (6.25 \text{ ft per } 100 \text{ ft}) \times (52.8 \text{ hundreds of ft}) = 330.00 \text{ ft}$$

Inserting the value for rise into Eq. 1.10:

$$\text{Paving (in ft)} = \sqrt{(330.00)^2 + (5280.00)^2} = 5290.302449 \text{ ft}$$

1.11 NON–RIGHT-TRIANGLE APPLICATIONS

It *must* be kept in mind that the basic trigonometric relationships for right triangles [Eqs. (1.7), (1.8), and (1.9)] along with the Pythagorean theorem [Eq. (1.10)] are valid *only* when dealing with right triangles. For the solution of triangles that are *not* right triangles, the **law of sines** or the **law of cosines** must be used.

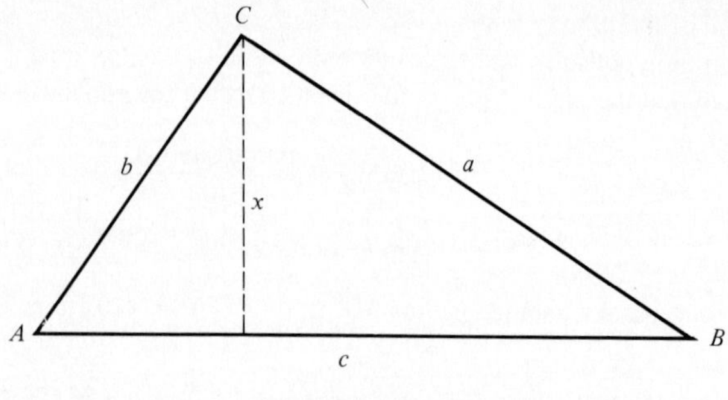

FIGURE 1.12
Triangle ABC modified to illustrate the law of sines.

GENERAL SOLUTION OF LAW OF SINES PROBLEMS. Figure 1.12 shows a triangle with angles A, B, and C and opposite sides a, b, and c. The perpendicular line dropped from C to line c has a length x and divides non-right triangle ABC into two right triangles, about which can be stated: $\sin A = x/b$ and $\sin B = x/a$. Equating values of x, it is seen that $x = b \sin A = a \sin B$. Transposing: $a/\sin A = b/\sin B$.

Had the perpendicular been dropped from A to the extension of line a and/or from B to the extension of line b, it may also be proved that $a/\sin A = c/\sin C$ or $b/\sin B = c/\sin C$. Thus,

Law of sines:

$$\frac{a}{\sin A} = \frac{b}{\sin B} = \frac{c}{\sin C} \tag{1.11}$$

Another convenient and necessary relationship is that the sum of the interior angles in any plane triangle must be 180°.

Sum of angles:

$$A + B + C = 180° \tag{1.12}$$

Knowing the value of any side and its opposite angle, if the value of either of the other sides or angles is known, all missing values may be computed using the law of sines and the fact that the sum of the angles in any plane triangle will be 180°.

Example 1.27. The courthouse record for a triangular lot is partially obliterated. From the plat and the legal description, the following data are available: the lengths of the sides corresponding to a and c in Fig. 1.12 are 138.09 ft and 273.44 ft, respectively, and the angle corresponding to C is 126°43′. From this limited data, compute the length of side b and the values of angles A and B.

Solution. Sides a and c are known, along with angle C. The relevant part of the law of sines is, therefore, $a/\sin A = c/\sin C$. Transposing,

$$\sin A = \frac{a \sin C}{c}$$

$$A = \sin^{-1} \frac{a \sin C}{c}$$

Substituting,

$$A = \sin^{-1} \frac{138.09 \sin 126°43'}{273.44}$$

$$= 23.87966626° \Rightarrow\Rightarrow 23°52'46.8''$$

Keeping in mind that rounding isn't done until the end of the problem, B is obtained from Eq. (1.12) as follows:

$$B = 180° - (A + C) = 180° - (23°52'46.8'' + 126°43')$$

$$= 180° - 150°35'46.8''$$

$$= 29°24'13.2''$$

The remaining value, b, is obtained from the relevant part of the law of sines:

$$\frac{b}{\sin B} = \frac{c}{\sin C}$$

$$b = \frac{c \sin B}{\sin C}$$

Substituting,

$$b = \frac{273.44 \sin 29°24'13.2''}{\sin 126°43'} = 167.4746 \text{ ft}$$

Rounding, final values are $A = 23°53'$, $B = 29°24'$, and $b = 167.47$ ft.

The above example is known as a "missing data problem," a category that will be discussed in detail in Chap. 8. Missing data problems assume that known values are exact values. For this reason, angle C was assumed to be $126°43'\,00.0''$ and angle A was carried to the nearest 0.1 second for the purposes of the computation.

Example 1.28. A rock, jutting up from the Niagara River just short of Niagara Falls, is to be located (Fig. 1.13). Triangle A–B–Rock is established by measuring a 2000.00-ft baseline AB along the river bank opposite the rock. Angles A and B are then measured and found to be 37°07.4′ and 39°22.0′, respectively. Compute sides A–Rock and B–Rock.

Solution. At first glance, there is no known side opposite a known angle, or vice versa. However, this is quickly rectified by using Eq. (1.12). Thus,

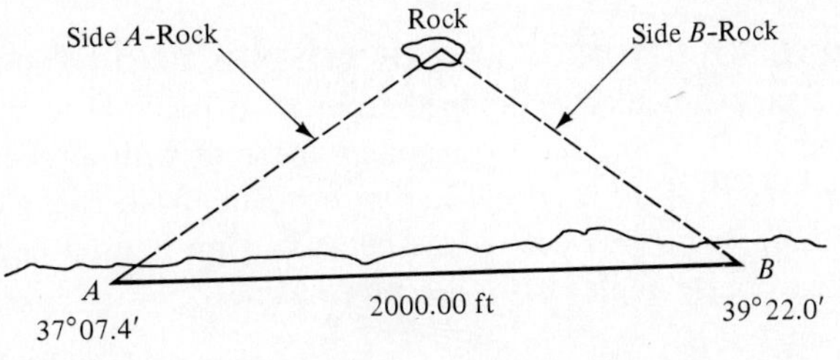

FIGURE 1.13
Rock in Niagara River offshore from base line AB.

$$\text{Angle at Rock} = 180° - (A + B)$$

$$= 180° - (37°07.4' + 39°22.0')$$

$$= 180°00.0' - 76°29.4' = 103°30.6'$$

Now, using Eq. (1.11):

$$\frac{A\text{–Rock}}{\sin B} = \frac{B\text{–Rock}}{\sin A} = \frac{\text{baseline}}{\sin \text{Rock}}$$

Transposing and substituting,

$$A\text{–Rock} = \frac{\text{baseline}}{\sin \text{Rock}} \times \sin B$$

and

$$B\text{–Rock} = \frac{\text{baseline}}{\sin \text{Rock}} \times \sin A$$

The expression "baseline divided by sine of angle at Rock" is used in both equations and its value should be stored in the calculator enroute to solving for *A*–Rock, thus simplifying the subsequent computation of *B*–Rock. Note that angles expressed in decimal-minute values don't directly convert to decimal degrees on the calculator. Decimal minutes must first be converted to seconds; for example, 103°30.6′ converts to 103°30′36″ by mentally multiplying 0.6 minute by 60 seconds per minute.

Substituting,

$$A\text{–Rock} = \frac{2000.00}{\sin 103°30.6' \times \sin 39°22.0'}$$

$$= 1304.6628 \Rightarrow\Rightarrow 1304.66 \text{ ft}$$

and

$$B\text{–Rock} = \frac{2000.00}{\sin 103°30.6' \times \sin 37°07.4'}$$

$$= 1241.4165 \Rightarrow\Rightarrow 1241.42\text{ft}$$

The law of sines has its limitations: The user must know either an angle and the *opposite* side, plus one of the other four values, or else the user must know *two* angles and *any* side. For situations where two angles and an included side are known, or the values of the three sides only are known, the **law of cosines** may be used.

GENERAL SOLUTION OF LAW OF COSINES PROBLEMS. In triangle *ABC*, (Fig. 1.14), a perpendicular is dropped from *A* to point *D* to make two right triangles, *ABD* and *ACD*. Using the Pythagorean theorem with respect to triangle *ABD*: $c^2 = h^2 + (p + a)^2 = h^2 + p^2 + 2ap + a^2$. But, from triangle *ACD*, $h^2 + p^2 = b^2$. Furthermore, $p = -b \cos C$ (the negative sign is used because p must be positive and C, an obtuse angle, has a negative cosine).

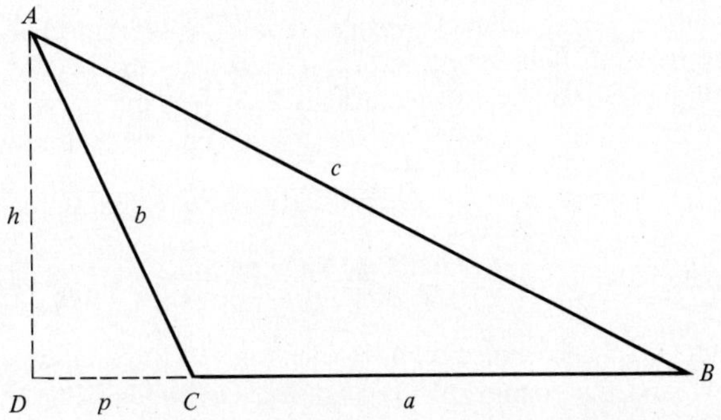

FIGURE 1.14
Triangle ABC modified to illustrate the law of cosines.

Substituting in the initial equation, $c^2 = b^2 - 2ab \cos C + a^2$. Transposing,

Law of cosines: $$c^2 = a^2 + b^2 - 2ab \cos C \quad \textbf{(1.13}a\textbf{)}$$

It can also be shown that the law of cosines can be expressed by

$$b^2 = a^2 + c^2 - 2ac \cos B \quad \textbf{(1.13}b\textbf{)}$$

$$a^2 = b^2 + c^2 - 2bc \cos A \quad \textbf{(1.13}c\textbf{)}$$

Knowing two sides and the included angle of any plane triangle, the other side can be computed.

Example 1.29. A power line is to be suspended between towers T_1 and T_2 across the mouth of a small embayment (Fig. 1.15). From a survey point at station X in the back of the embayment, distances XT_1 and XT_2 are measured by taping and found to be 874.68 ft and 907.41 ft, respectively. If angle X is 30°47′, compute distance T_1T_2.

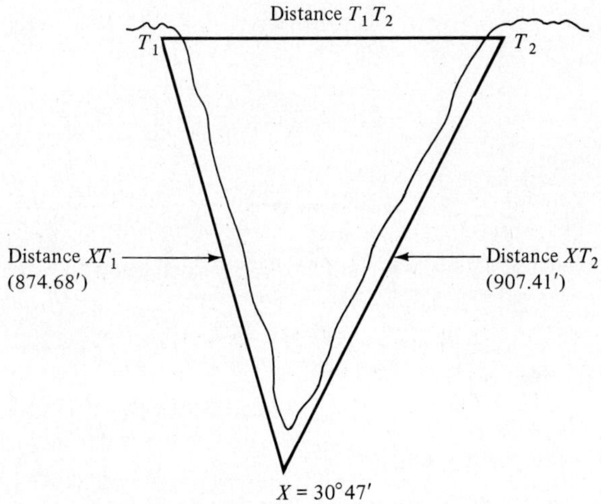

FIGURE 1.15
Power line at mouth of embayment.

Solution. Given the two known sides and the known included angle, the law of cosines can be used. Substituting in Eq. (1.13a), with the two known sides as a and b, and with the known angle C, the unknown side c is the square root of the right side of Eq. (1.13a). Thus,

$$T_1T_2 = \sqrt{(874.68)^2 + (907.41)^2 - 2(874.68)(907.41)\cos 30°47'}$$

$$= 474.0465 \Rightarrow\Rightarrow 474.05 \text{ ft}$$

Equations (1.13a, b, and c) can be transformed to equations in which the three sides are known and the cosines of the angles are unknown. For example, $c^2 = a^2 + b^2 - 2ab\cos C$; thus, $\cos C = (a^2 + b^2 - c^2)/2ab$. This, in turn, produces

$$\mathbf{C = \cos^{-1}\frac{a^2 + b^2 - c^2}{2ab}} \qquad \textbf{(1.14}\boldsymbol{a}\textbf{)}$$

Similarly,

$$\mathbf{B = \cos^{-1}\frac{a^2 + c^2 - b^2}{2ac}} \qquad \textbf{(1.14}\boldsymbol{b}\textbf{)}$$

$$\mathbf{A = \cos^{-1}\frac{b^2 + c^2 - a^2}{2bc}} \qquad \textbf{(1.14}\boldsymbol{c}\textbf{)}$$

Knowing all three sides of any plane triangle, any angle may be computed.

Example 1.30. Because a transit was damaged at the start of the survey of a triangular field, it was only possible to measure the horizontal lengths of the sides. Side a = 2038.94 ft, side b = 3324.99 ft, and side c = 2771.40 ft. Compute the three angles.

Solution. Compute angles by substituting into Eq. (1.14a, b, and c) as follows:

$$A = \cos^{-1}\frac{(3324.99)^2 + (2771.40)^2 - (2038.94)^2}{2 \times 3324.99 \times 2771.40}$$

$$B = \cos^{-1}\frac{(2038.94)^2 + (2771.40)^2 - (3324.99)^2}{2 \times 2038.94 \times 2771.40}$$

$$C = \cos^{-1}\frac{(2038.94)^2 + (3324.99)^2 - (2771.40)^2}{2 \times 2038.94 \times 3324.99}$$

Computing sequentially, the following are obtained:

$$A = 37°42'57.0''$$

$$B = 86°01'49.3''$$

$$C = 56°15'13.7''$$

Adding together for a check:

$$A + B + C = 180°00'00.0''$$

1.12 PROBLEMS

1.1. What is the basic difference between plane surveying and geodetic surveying?

Note: Answers to Probs. 1.2 through 1.10 are to be given to the proper number of significant figures.

1.2. Convert the following lengths to meters:
(*a*) 59 ft $9\frac{3}{4}$ in. (*b*) 1622.53 ft (*c*) 147.496 ft (*d*) 1 ft $8\frac{3}{16}$ in.

1.3. Convert the following lengths to feet:
(*a*) 463.983 m (*b*) 8.93 m (*c*) 26.4 m (*d*) 2836.147 m

1.4. How many acres in a rectangular field measuring 438.61 ft by 687.89 ft? How many hectares?

1.5. How many acres in a rectangular field measuring 35 chains 47 links by 82 chains 6 links?

1.6. How many hectares in a rectangular tract measuring 79.011 m by 126.883 m?

1.7. Convert the following angles from sexagesimal form (degrees, minutes, seconds) to decimal form:
(*a*) 89°15′32.1″ (*b*) 37°48′03.68″ (*c*) 123°06′47″ (*d*) 6°33′

1.8. Convert the following angles from decimal form to (1) degrees, minutes, and seconds and (2) gon:
(*a*) 43.8476° (*b*) 127.91° (*c*) 88.698° (*d*) 8.32°

1.9. Convert the following from gon to degrees, minutes, and seconds:
(*a*) 247.864 gon (*b*) 91.6947 gon (*c*) 135.22 gon (*d*) 48.916 gon

1.10. Convert the following angles from decimal degrees to radians:
(*a*) 342.586° (*b*) 12.91° (*c*) 156.9832° (*c*) 6.764°

1.11. Express the following numbers to three significant figures:
(*a*) 0.004659 (*b*) 18.8684 (*c*) 129.687 (*d*) 15159.2

1.12. Round the following numbers to four significant figures:
(*a*) 524.4502 (*b*) 1.8645 (*c*) 0.0411849 (*d*) 27.3276

1.13. State how many significant figures are in the numbers shown:
(*a*) 0.06920 (*b*) 4.980×10^4 (*c*) 6.477×10^{-3} (*d*) 87.00

1.14. State the answers to the following to the correct number of significant figures:
(*a*) $5321.694 - 829.0$ (*c*) 5866.9×47.2
(*b*) $3.6897 + 4322.46 + 677.8 + 44.936$ (*d*) $671.1 \div 10.968$

1.15. State the answers to the following problems to the correct number of significant figures:
(*a*) $\dfrac{32.1(563.92 + 88.479 - 211.1)}{79.478}$ (*b*) $\dfrac{(1328.6 - 47.911)(621.4 + 2.32)}{29.83 + 6.423 - 5.1}$

1.16. Solve the following linear equations:
(*a*) $4 + 9x = 2x + 32$ (*c*) $8y + 17 = 10y + 44$
(*b*) $6x - 13 = 12x - 66$ (*d*) $7.2y = 13y - 18.4$

1.17. Solve for the unknowns in the following simultaneous linear equations:
(*a*) $41x + 3 = 3y - 4x + 57$, $x + 14 = 8y - 16$
(*b*) $4.2y = 17.6z - 53.4$, $11.0y + 3.2z = 11.8$

1.18. Solve for the unknown in the following equations:
(*a*) $4x^2 - 13x = 44$ (*c*) $14x^2 + 12x = 10x + 40$
(*b*) $7.2x^2 - 13.1 = 16.8x$ (*d*) $22x^2 - 9x - 12 = 0$

1.19. Deciding to buy new rugs for your company office, you found a bargain roll of carpet containing 248 square yards. You had the carpet cut into 10 rugs, each the same width as the carpet-roll width and 1 ½ roll widths long, after which there were 6 ft left from the original roll. How wide was the roll?

1.20. A straight line on an xy coordinate system passes through (+8, −2) and (+13, +3), using standard coordinate-point designation where x is given first, y is second; that is, (x, y). What is the y value of the point where the line crosses the x axis?

1.21. A second straight line, whose equation is $y + x = 2$, crosses the line in Prob. 1.20. What are the coordinates of the intersection of the two lines?

1.22. The coordinates of two points are (+16, +4) and (−3, −6). What is the slope and the y intercept (at $x = 0$) of a straight line passing through these two points?

1.23. A line crosses the y axis at $y = -4.6$ and passes through a point whose coordinates are (+18, +4). What is the slope of the line? To the nearest minute, what is the acute angle of intersection of the line with the x axis?

1.24. Calculate the coordinates of the point of intersection of a line passing through points (−8, +7) and (+16, −3) with a second line that passes through point (−12, −3) and has a slope of +0.40.

1.25. Calculate the coordinates of the point of intersection of a line passing through points (−16, −12) and (+14, −2) with a second line that passes through points (−8, +4) and (+12, −6).

1.26. Calculate the coordinates of the point of intersection of a line passing through points (−22, +8) and (+10, −12) and a second line that is perpendicular to the first line and passes through (−13, −6) *Hint:* If m is the slope of a line, the slope of its perpendicular must be $-1/m$.

1.27. What is the equation of a circle with a radius of 15 units whose center is located at (+13, +4)?

1.28. Calculate both points of intersection of the line described in Prob. 1.22 and the circle described in Prob. 1.27.

1.29. Calculate both points of intersection of the line described in Prob. 1.23 and the circle described in Prob. 1.27.

1.30. To eight decimal places, determine the numerical values of the following trigonometric functions:

(*a*) sin 69°22′36.9″ (*c*) sec 87°12′48″ (*e*) csc 44°51′43.2″
(*b*) cot 123.4694° (*d*) cos 211.8879° (*f*) tan 304°08′29″

1.31. To the nearest integer second, determine the angles represented by the following trigonometric functions:

(*a*) $\csc^{-1}$ 1.367843 (*b*) $\cot^{-1}$ 7.669493 (*c*) $\cos^{-1}$ 0.894651 (*d*) $\tan^{-1}$ 1.444698

1.32. A rectangular foundation measures 29 ft $5\frac{3}{4}$ in. by 47 ft $10\frac{7}{8}$ in. In decimal feet, what is the length of the diagonal measurement (to the nearest 0.01 ft)?

1.33. A wheelchair-access ramp must rise 4.21 ft across existing level terrain. If the maximum permissible ramp slope is +5°00′, what is the minimum horizontal length of the ramp structure?

1.34. In Prob. 1.33, what is the length of the ramp measured along the slope?

1.35. A radio tower is to be erected on level ground. If the guy wire leading to the 500-ft point on the tower is to make an angle of 60°00′ with the ground, how far from the tower base should the guy anchor be set?

1.36. What is the horizontal distance and vertical rise for a stretch of highway that measures 2500.00 ft along a 4.75% grade?

1.37. The center of a circle of radius 8 units lies to the right of the y axis. The circle intersects the y axis at $y = +11$ and $y = -1$. Using trigonometry, compute the x coordinate of the center of the circle.

1.38. On a triangular lot, the side opposite a 63°12′ corner angle is 378.42 ft long. A second corner angle is 84°28′. What are the two missing side dimensions and the third angle?

1.39. The three sides of a triangular plot of land were measured as follows: a = 229.67 ft, b = 348.41 ft, c = 410.22 ft. To the nearest minute, determine the interior angles A, B, and C.

1.40. Two sides of a triangular-shaped building were measured as 214.63 ft and 188.19 ft. The angle between is 72°15′. Determine the length of the remaining side and the other two interior angles.

CHAPTER 2

DISTANCES

2.1 DISTANCE MEASUREMENT CONVENTIONS AND METHODS

HORIZONTAL DISTANCES. Distance measurement is an essential activity that enters into most operations carried out by surveyors. Its purpose is to establish the **horizontal distance** between two points, either by measuring between existing points or by setting a new point a required distance from an existing point. The word "horizontal" is important because property boundaries and other surveyed figures must be described in terms of the horizontal distances between the points that establish the figure. Horizontal distances between points at different elevations can be determined either by a procedure that allows the measuring device to be held horizontal or by directly measuring along a slope and computing horizontal distance from slope angles or from elevation differences. Procedures for obtaining horizontal distances from slope distances are discussed later in this chapter.

MEASUREMENT METHODS. Most people have measured distances by scaling from a map or by pacing between points. Pacing is adequate for secondary surveying operations that require only approximate distances. Most experienced surveyors are able to pace consistently within an accuracy range of about 2 percent by knowing the length of their own pace. Surveyors have frequent opportunities to check their pace length against measured distances, and it is well worth the small effort it takes to do so.

Pacing, no matter how carefully done, is far too inaccurate for most surveying measurement. Other methods in use, listed in order of increasing accuracy, are **tacheometry** (generally known as **stadia** in the United States), **taping**, and **electronic**

distance measurement (EDM). Stadia measurement is based on the fixed relationship between cross-hair spacing and focal length in the telescope of a transit, level, or theodolite. It provides precision on the order of 1 part in 500 if carefully done. The method will be discussed in Chap. 9.

Until fairly recently, reasonable accuracy in distance measurement was attainable only by careful use of steel surveying tapes. Today, however, increasing acceptance of electronic distance measurement has led to a decreasing use of the steel tape. The diminishing cost of EDM equipment, coupled with its precision and convenience, have made it increasingly popular for measuring distances longer than a tape length or so.

Over distances ranging from one or two tape lengths to a mile or more, electronic distance measurement should remain the method of choice for many years to come. However, a rapidly developing method, based on satellite positioning, could provide precision over long lines equivalent to that presently obtained with EDMs over lines many kilometers in length. The satellite method will be touched on briefly in Chap. 8. The balance of Chap. 2 will discuss the use of tapes and EDMs for measuring distances.

Although surveyors are using tapes less and less often these days, it is simpler to initiate the discussion of distance measurement procedures by referring to the familiar measuring tape than it is to embark on an explanation of a device as seemingly mysterious as an EDM. Besides, for the foreseeable future, the steel tape will be used for laying out small-area construction projects, for measuring short surveying lines for a variety of applications, and as an auxiliary measurement device when using the EDM. Then, too, all electronic and mechanical devices fail on occasion, making it necessary to fall back on the reliable steel tape. Thus, the steel tape continues to be important to the theory and practice of distance measurement.

2.2 STEEL TAPES AND TAPING EQUIPMENT

Hardware stores carry a variety of devices for the measurement of length and a number of these are found in the average home. Most people have used plastic and wooden rulers, along with relatively short cloth and metal tapes, for about as long as they can remember. These familiar devices, ordinarily graduated in feet and inches, are intended for convenience rather than precision and are mass-produced to satisfy a large group of casual users. Cloth tapes, even those that are metal-reinforced, are not stable enough dimensionally to provide the precision necessary for surveying. The thin, enamel-marked steel tapes most people are used to—even the 50-ft and 100-ft tapes used in construction—are also inadequate for surveying.

STEEL TAPES. Surveying tapes are precisely made of steel alloys chosen for strength, durability, and resistance to corrosion. They are generally either 100 ft or 30 m in length and are graduated decimally. Metric tapes are usually graduated in meters and tenths of meters (decimeters) over their entire length, but to centimeters and millimeters on the end decimeter. Foot tapes are ordinarily marked in whole feet over their length but to tenths and hundredths of feet on the end foot. This skimpy form of

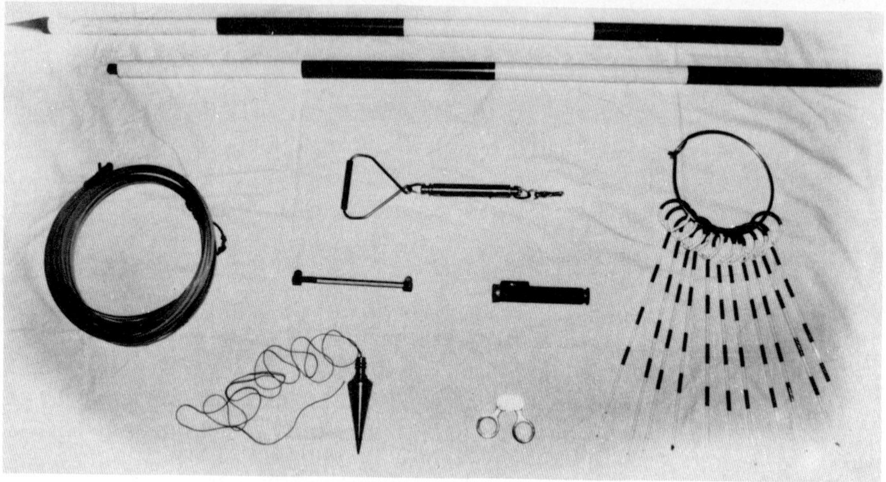

FIGURE 2.1
Equipment used by a taping party.

marking looks odd to people who are used to seeing a tape graduated to the smallest division over its entire length; however, this graduation style ensures that surveying tapes will be used in a particularly simple and straightforward manner that reduces the chance of measurement error.

TAPING EQUIPMENT. Although the basic concept of taping distances is not particularly complex, the proper and precise use of steel tapes in the field requires a number of accessories. Figure 2.1 illustrates a sampling of items of taping equipment, including a 100-ft steel tape, a two-section range pole, a set of taping pins on a carrying ring, a plumb bob, a hand level, a tension scale, a tape thermometer, and a tape clamp.

A steel surveying tape may be mounted on a reel or it may be gathered into 5-ft, figure-8 loops and twisted (or "thrown") into compact, $2\frac{1}{2}$-ft coils, as shown in Fig. 2.1. The latter method is easier said than done. For people who haven't been taught the double-coil technique, it is a much simpler matter to stretch out a reel-mounted tape, then detach it from its reel, than it is to uncoil and stretch out a double-coiled tape; similarly, it is simpler to reattach the tape to the reel and wind it in than it is to retrieve it back into the double-coiled configuration.

Whether on a reel or double-coiled, the tape is terminated at each end with a riveted loop through which a rawhide thong or braided-nylon strip is usually tied. The thongs at the tape ends allow the persons holding the tape to apply tension without kinking the tape. If, as often happens, a thong breaks in use in the field, a taping pin inserted through the loop will make a satisfactory temporary handle. A clamping handle, shown in use in Fig. 2.2, helps the surveyor avoid kinking the tape when holding it at a point other than an end. *Do not put sharp bends in a steel tape.* A straight tape will withstand about any tension two people can apply; if it is kinked or inadvertently looped, it will break easily under moderate tension.

FIGURE 2.2
Tape clamp in use.

Range poles have alternate red and white stripes and are usually 6 to 8 ft long and something over an inch in diameter. They are plunged into the ground at the two ends of a line that is to be taped. Taping party members sight on range poles during measurement to keep the tape on line. For convenience in transporting, most range poles are designed to come apart at the center, either at a ferrule or a threaded joint. Threaded-joint poles are preferable because they don't tend to stick together as they pick up dents and corrosion in use; in addition, the threads are compatible with those on most EDM reflector bases.

Taping pins are used to mark the positions of end points of the tape during the process of measuring distances. The set of 11 pins is carried in an openable ring as shown in Fig. 2.1. Pins are made of heavy steel wire that is formed into a wide loop at one end and sharply pointed at the other. Like range poles, they are alternately striped red and white so they may be easily seen, although well-worn taping pins tend to have lost most of their enamel coating strips of bright plastic tape can be tied into the loops to improve visibility.

A plumb bob is a conical brass device with a replaceable steel tip. It is suspended on a string to provide surveyors with a vertical reference. It is used on taping parties to establish a vertical line between a ground point and the tape. A plumb bob is needed whenever it is not feasible to hold the tape directly against the ground (see Fig. 2.2).

A surveyor's hand level is a simple instrument that allows the user to view simultaneously a level bubble, a horizontal cross hair, and a line of sight. When the

FIGURE 2.3
Using a surveyor's hand level.

level is held so the center of the bubble coincides with the cross hair, the line of sight is level. The hand level enables the taping party to hold the tape horizontal when measuring over uneven terrain. Figure 2.3 shows a hand level in use, steadied against a range pole to mark the height of the desired horizontal line.

For precise taping of base lines between monumented points—now rarely done—the taping party also uses multiple standardized tapes, a spring scale to maintain desired tape tension, a special thermometer to measure tape temperature, and either posts or taping tripods to support the tape during measurement. No particular effort is made to maintain the tape horizontal while doing high-precision taping; thus, tape ends will be at whatever elevations the successive posts or tripods happen to lie. Levels are then run between supports to determine elevation difference between tape ends.

2.3 USE OF STEEL TAPES IN THE FIELD

The basic concepts of taping are essentially unvarying no matter what the distance being measured. A consistent, orderly technique must be used to ensure reasonably precise results. For a distance shorter than a single tape length, the overall procedure is a bit less complex (range poles and taping pins need not be used); nevertheless, the same care must be taken to measure a short line as a long line.

TAPING ON LEVEL GROUND. It is quite easy for a taping party of two persons to measure horizontal distances on level ground. Sometimes the distance is measured between a starting point and an existing end point. Sometimes a predetermined distance is laid out between a starting point and a yet-to-be-established end point. In either case, the end points of the line are usually marked with surveyor's tacks set into the tops of 2 in. × 2 in. wooden hubs. For taped lines so long that the end point cannot be seen readily from the starting point, the line is marked with range poles set into the ground directly on line, but some 4 or 5 ft beyond either end so as not to interfere with the tape or the tapers. In the case of a line to be laid out, the far range pole is set somewhat beyond the estimated distance to the line end, the line direction being determined using a transit set up over the starting point.

The tape is stretched out with the zero mark at the forward end of the line and the 100-ft mark (or 30-m mark) bringing up the rear. The person at the zero end is called the "head taper" (designated HT in the field record book) and the person at the rear is called the "rear taper" (designated RT in the record book). Assuming a 100-ft tape is being used, the rear taper holds the 100-ft mark at the starting point (or the appropriate whole-foot mark if the distance to be measured is less than 100 ft). Using the far range pole as a guide, the rear taper motions the head taper onto the intended line by means of hand signals.

The head taper, holding the leather thong in one hand and a taping pin in the other, calls out "tension," and applies the appropriate tension, usually 10, 12, 15, or 20 pounds (lb). Holding the tape by the leather thong and taking care not to apply pressure on the sides of the hub, the rear taper positions the 100-ft mark against the top of the initial point and calls out "mark" at the instant the 100-ft mark lies directly on the center of the tack. The head taper, noting where the zero mark lies when the rear taper calls "mark," plunges the taping pin into the ground directly on that point, taking care that the pin is inserted perpendicular to the tape but at about 45° to the ground (see Fig. 2.4). The measurement is then repeated to ensure that the pin has been set at the proper point.

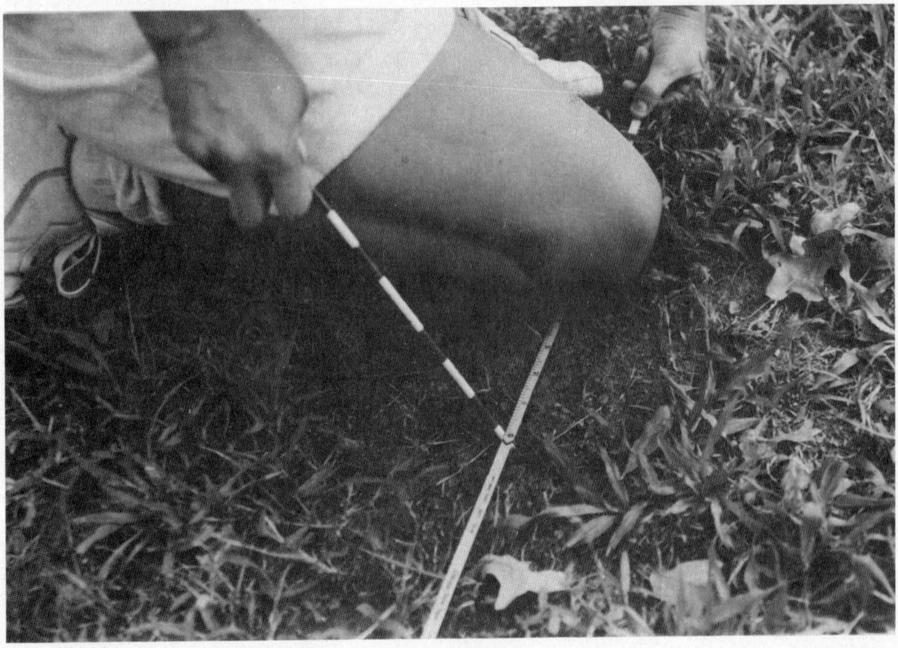

FIGURE 2.4
Position of taping pin with respect to tape.

After checking the measurement, the head taper calls "OK" and both tapers move forward along the line until the rear taper reaches the taping pin and calls "hold." The rear taper, holding the 100-ft mark at the taping pin, again motions the head taper on line. The head taper, now in position, again calls "tension" and the rear taper holds the 100-ft mark directly against the center of the taping pin shaft at the point where the taping pin enters the ground (Fig. 2.4). When the 100-ft measurement has been made and checked, the rear taper *removes the taping pin* from which the distance has just been measured and carries it along.

The line continues in this manner, the rear taper accumulating taping pins along the way, until reaching the point where the distance from the last taping pin to the end point is less than a full tape length. The procedure from here on depends on whether the tape in use is an **add tape** or, less commonly, a **cut tape**. The add tape is made so the tenth-foot and hundredth-foot graduations lie on an extra foot of tape beyond the zero mark; that is, an add tape is marked over a 101-ft length. A cut tape, or subtract tape, is marked over just a 100-ft distance, the tenth and hundredth graduations lying between zero and the 1-ft mark. On the add tape, the fine graduations on the extra foot begin at the zero mark and increase *outward* to the extra-foot mark. On the cut, or subtract tape, fine graduations increase from the zero mark *inward* to the 1-ft mark.

Figure 2.5*a* illustrates how the reading for the final taped segment is made with an add tape. The rear taper holds the particular whole-foot mark that puts the end point of line within the final graduated foot held by the front taper (57 ft in Fig. 2.5*a*). The rear taper calls "mark" and the front taper then reads the tenth-foot and hundredth-foot marks (0.37 ft in Fig. 2.5*a*). As before, the distance is remeasured for a check. The rear taper, who is then asked what foot mark is being held, calls "57." The front taper repeats "57" to the rear taper and records 57.37 ft in the record book for the tens, units, tenths, and hundredths digits. To determine the hundreds digit to be recorded, the rear taper counts the number of taping pins that have been picked up and carried forward (not including the last pin inserted, which will still be in the ground), say, 4 in this case. As a check, the front taper counts the number of taping pins that remain on the ring—this should be equal to 10 minus the number the rear taper is carrying, 6, in this case. The recorded line length is the distance measured in the final taped segment plus the number of pins the rear taper is carrying multiplied by 100; in this case, 457.37 ft.

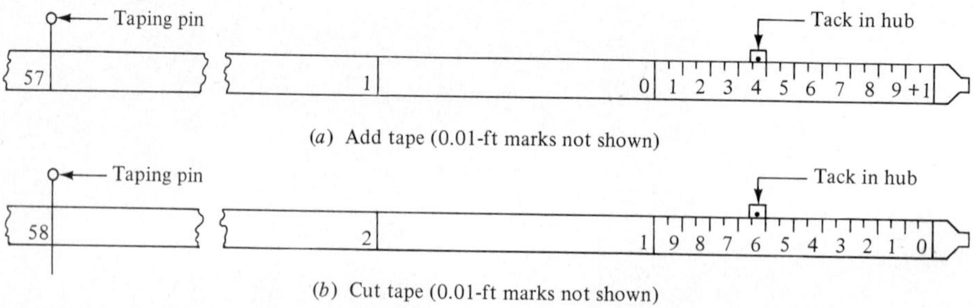

FIGURE 2.5
Reading tape-end graduations.

If the same distance had been measured with a cut tape, as in Fig.2.5*b*, the rear taper would have held 58, rather than 57, in order to place the end point in the range of the final graduated foot. In this case the front taper would have read 0.63 rather than 0.37. The final recorded reading is 400 plus 58.00 − 0.63 = 457.37 ft, as before.

The foregoing describes procedures at the end segment of a taped line *measured* between two existing points. In the case of a line being *laid out* between an existing point and an end point to be established, say 457.37 ft away, the head taper, using an add tape on the final segment, would have called out "hold 57" to the rear taper. With the rear taper holding the tape at the 57-ft mark, the front taper would then set a hub on line at the 0.37-ft mark. Measuring again, the tack position would be marked, on line, on the top of the hub. The check measurement is then made to the center of the newly set tack. With a cut tape, corresponding cut-tape procedures would be used.

Taped lines are ordinarily double-run to obtain a check on results and uncover significant errors in measurement. The check line is usually run in the opposite direction from the originally measured line.

HORIZONTAL MEASUREMENT ON UNEVEN GROUND. The procedure just described presupposes a level, relatively smooth measuring surface, but this does not occur often in real life. To measure horizontal distances directly on irregular ground, the taping party must use a hand level and a plumb bob. Figure 2.6 shows a profile of a 457.37-ft horizontal line measured downhill over unevenly sloping ground. In the downhill case, the rear taper holds the 100-ft mark alongside the taping pin (or on the beginning point if just starting out), while the head taper holds a plumb bob string on the zero mark of the horizontal tape. To level the tape, the head taper takes a position 100 feet along the line and sights the 100-ft mark through a horizontally held hand level (see Fig. 2.3). The zero end of the tape is then positioned at that

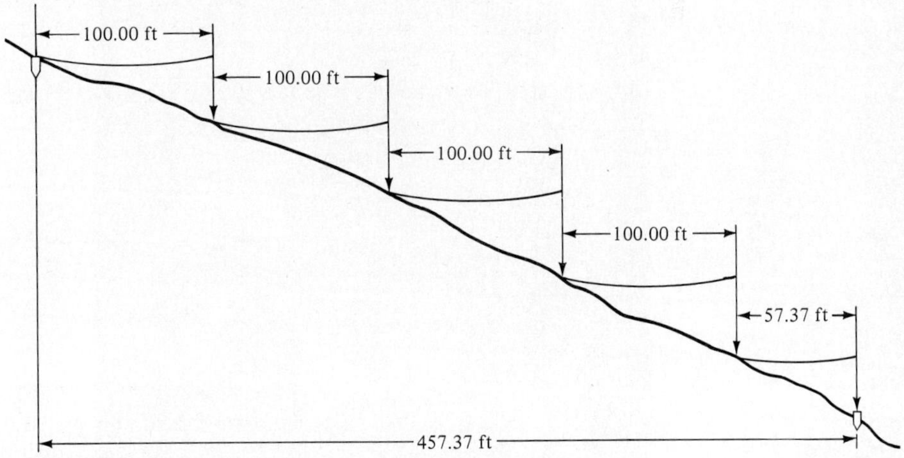

FIGURE 2.6
Tape measurement along unevenly sloping ground.

height above the ground, the plumb bob string is held at the zero mark, and string is let out until the plumb bob point is just above ground. On the call "mark," the taping pin is inserted into the ground at the point marked by the tip of the plumb bob.

This works almost as easily as described as long as the zero end of the tape is only a short distance above the ground and the wind is not blowing too hard. If the ground slopes relatively steeply or if the wind is blowing briskly, it is better to use three people in the taping party rather than two. Normally, in this case, the head taper will hold the plumb bob string in position and maintain tape tension while the other person steadies the plumb bob tip and inserts the taping pin into the ground. Note that the third person must take care to just *steady* the plumb bob rather than attempt to hold it in position. Under these circumstances, the head taper should stand sideways to the rear taper; that is, feet spread, with the tape crossing the front of the body, as in Fig. 2.7. Besides allowing a more effective brace against tape tension, this position usually affords a better windbreak for the plumb bob, reducing the degree of steadying the third taper is called on to provide.

The foregoing describes taping *downhill* over uneven terrain. For a check line measured in the opposite direction, the direction of progress must be *uphill*, a slightly

FIGURE 2.7
Breaking tape over steeply sloping ground.

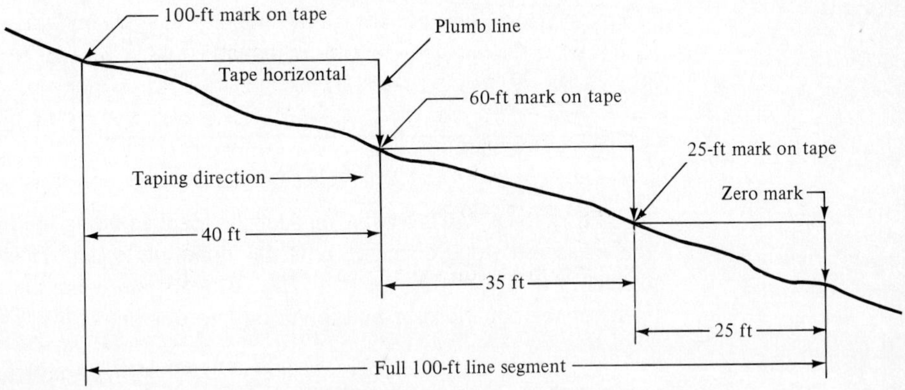

FIGURE 2.8
Procedure for breaking tape in full 100-ft increments.

more awkward procedure. In the latter case the head taper works alone while the rear taper uses the hand level to establish a horizontal tape line, then holds the plumb bob string on the 100-ft mark while the third taper steadies the plumb bob and calls "mark."

BREAKING TAPE. If the terrain along all or part of the measured line slopes more than about 5 ft in 100 ft, the taping party will need to "break tape"; that is, taping pins must be inserted, as needed, at intervals less than 100 ft. When doing so, special care must be taken to keep track of the lengths of the individual taped segments that have been measured over the entire line prior to arriving at the final segment. One method is to record each taped segment of the line as it is measured, depending on the check line to turn up instances of lost count or unrecorded segments. Another method forces the broken-tape segments to add to 100 ft. In Fig. 2.8, the taped line is proceeding from left to right and the first pin is inserted just 40 ft from the previous 100-ft pin; that is, at the 60-ft mark on the tape. For the next segment, the 60-ft mark on the tape is held against the pin just inserted and the next pin is inserted 35 ft away, at the 25-ft mark on the tape. For the final portion of this 100-ft section, the 25-ft mark is held at the pin just inserted and the last pin is placed at the zero mark on the tape. The main problem with this method is that the pin count no longer registers the number of full 100-ft sections in the line.

2.4 REDUCTION OF SLOPE DISTANCES

Surveyors usually prefer to measure distances as simply and easily as possible. Taping over uneven terrain, it is often easier to measure a slope distance. But whether working with tape or EDM, the need to convert a slope distance to a horizontal distance will often arise. There are two principal ways of doing this: (1) by measuring **slope angle** and solving a trigonometric computation, or (2) by measuring the **difference in elevation** between end points on the slope, then using the theorem of Pythagoras to find horizontal distance.

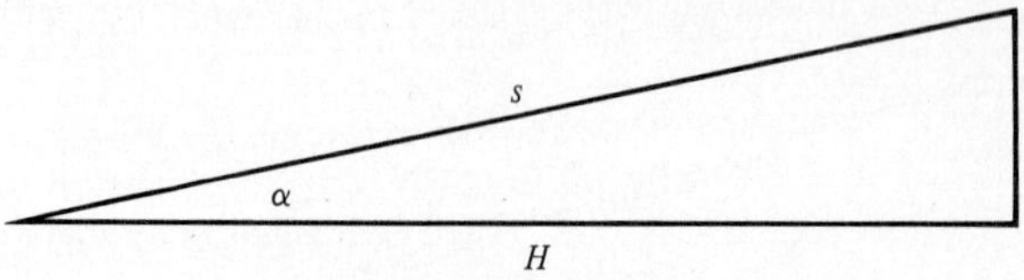

FIGURE 2.9
Horizontal distance from slope distance and slope angle.

SLOPE-ANGLE METHOD. Consider the idealized diagram in Fig. 2.9, in which s is the measured slope distance, α is the slope angle, and H is the required horizontal distance. It is seen that $\cos\alpha = H/s$. Since s and α are known while H is not, both sides of the equation are multiplied by s to obtain

$$H = s\cos\alpha \tag{2.1}$$

Example 2.1. A slope distance of 174.68 ft is measured between points A and B along a 3°38′ slope. Compute the horizontal distance.

Solution.

$$\begin{aligned} H = s\cos\alpha &= 174.68\cos 3°38' \\ &= 174.68\cos 3.63333333\ldots° \\ &= 174.3289 \Rightarrow\Rightarrow 174.33 \text{ ft} \end{aligned}$$

Although this example calls for determining horizontal distance between two *existing* points, the surveyor must often establish a *new point* a predetermined horizontal distance from one existing point. In the latter case, slope angle α and horizontal distance H are known, while s, the slope distance to be measured, must be computed. The original equation, $\cos\alpha = H/s$, is manipulated to obtain

$$s = \frac{H}{\cos\alpha} \tag{2.2}$$

Example 2.2. Point B, to be set along a 6°12′ slope from point A, is to lie 240.00 ft horizontally from A. Compute the slope distance to be laid out from A to establish B.

Solution.

$$\begin{aligned} s = \frac{H}{\cos\alpha} &= \frac{240.00}{\cos 6°12'} \\ &= \frac{240.00}{\cos 6.20000°} \\ &= 241.4120 \Rightarrow\Rightarrow 241.41 \text{ ft} \end{aligned}$$

EFFECTS OF SLOPE-ANGLE MEASUREMENT ERROR. In the examples just given, the answers were taken to five significant figures (hundredths of a foot) despite the fact that the angles involved were only stated to *about* three significant figures ("about," because portions of degrees, when given in minutes and seconds, have somewhat less significant-figure validity than if they were stated in decimals using the same number of digits).

When a multiplication-division equation includes trigonometric functions, the usual rule for significant figures does not strictly apply. One method of determining the number of valid significant figures is to test the effect on the numerical solution of varying the angle slightly. The slope angle in the first example is given as $3°38'$; that is, in whole minutes; thus, the angle read on the instrument must have been closer to $3°38'$ than it was to $3°37'$ or $3°39'$. From this it can be assumed that the reading precision of the transit was plus or minus *half* a minute; thus, the given angle may be expressed as $3°38' \pm \frac{1}{2}'$ or $3°38' \pm 30''$. The question now arises as to how large an error in computed horizontal distance comes about from an error equal to the reading precision of the transit.

Example 2.3. Given the data of Example 2.1, what difference in the final answer would occur if α were $30''$ larger or smaller?

Solution. The trial computation will be done using $3°38'30''$ (if $3°37'30''$ were used, the answer would not be significantly different). From Example 2.1, $H = 174.68 \cos 3°38' = 174.3289$ ft and $H' = 174.68 \cos 3°38'30'' = 174.3273$ ft. Therefore, the difference between H and $H' = 0.0016$ ft.

This shows that a half-minute error in angle measurement produces an error in H of less than half a hundredth of a foot; thus, it was valid to take the final answer to five significant figures in Example 2.1. In general, when dealing with shallow slope angles, it would take a fairly large error in slope-angle measurement to have an appreciable impact on computed H or computed s.

It is obviously tedious to recompute a slope-angle problem just to test the validity of a significant-figure assumption. It is simpler to use the next equation, developed from the calculus in App. Sec. A-1:

$$\Delta H = -s \sin \alpha \times \Delta\alpha \tag{2.3}$$

where ΔH is the resultant error in H from a possible slope-angle error $\Delta\alpha$. (The minus sign simply means H decreases as α increases, and vice versa.) *It is important that $\Delta\alpha$ be in radians*, a natural result of equations developed from the calculus. This is easily done using the conversion factor 0.00000484814 radian (rad) = 1 second (sec).

Example 2.4. Do Example 2.3 using Eq. (2.3).

Solution.

$$\begin{aligned}\Delta H &= -174.68 \sin 3°38' \times 30'' \times 0.00000484814 \text{ rad} \cdot \text{sec}^{-1} \\ &= -0.0016 \text{ ft}\end{aligned}$$

It is often helpful to consider horizontal-distance errors of this sort in terms of **relative accuracy**. Relative accuracy, in this case, is defined as the ratio of ΔH

to H, that is, $\Delta H/H$. This is expressed numerically as "1 : x" or, in words, as "1 part in x" (parts). To obtain $\Delta H/H$, divide Eq. (2.3) by H as follows: $\Delta H/H = (s \sin\alpha \times \Delta\alpha)/H$. Substituting $s\cos\alpha$ for H in the right-side denominator produces the value: $(s \sin\alpha \times \Delta\alpha)/(s\cos\alpha)$. Canceling and converting $\sin\alpha/\cos\alpha$ to $\tan\alpha$ provides the end result:

$$\textbf{Accuracy} = \tan\boldsymbol{\alpha} \times \boldsymbol{\Delta\alpha} \tag{2.4}$$

Example 2.5. In Example 2.1, find the relative accuracy if the slope angle is measured to $\pm 30''$.

Solution. Substituting into Eq. (2.4):

$$\text{Accuracy} = \tan\alpha \times \Delta\alpha = \tan 3°38' \times 30'' \times 0.00000484814 \text{ rad}\cdot\text{sec}^{-1}$$

$$= 0.000009235$$

This is converted to 1 part in x parts by depressing the $1/x$ key on the calculator, which puts the value of "x" in the calculator window. In this case, $1/0.000009235 = 108,281$. The final result:

$$\text{Accuracy} = 1{:}108,281 \Rightarrow\Rightarrow 1{:}108,000 \quad \text{approximately}$$

A relative accuracy of better than 1:100,000 is very good indeed, an indication that ordinary transit accuracy, when used to measure shallow slope angles, is sufficient to provide quite precise results in computed horizontal distances. Now consider the next example.

Example 2.6. For an accuracy of 1:50,000, how much error in α can be tolerated if α is 4°00′?

Solution. Equation (2.4), accuracy $= \tan\alpha \times \Delta\alpha$, is manipulated algebraically to obtain

$$\Delta\alpha(\text{radians}) = \frac{\text{accuracy}}{\tan\alpha}$$

and

$$\Delta\alpha(\text{seconds}) = \Delta\alpha(\text{radians}) \times \text{seconds per radian}$$

Substituting,

$$\Delta\alpha = \frac{1}{50,000 \tan 4°00'} \text{radians} \times 206,265 \text{ sec}\cdot\text{rad}^{-1}$$

$$= \pm 59.0'' \quad \text{or about} \pm 1 \text{ minute}$$

In the examples just discussed, the slope was defined in terms of slope angle. Slope may also be defined in terms of **grade**, that is, the number of feet of vertical change in elevation per hundred feet of *horizontal* distance, expressed as percent.

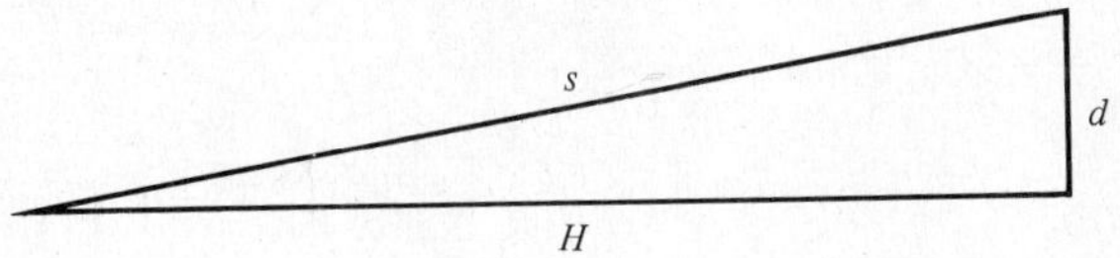

FIGURE 2.10
Horizontal distance from slope distance and elevation difference.

Thus, if a grade rises 4.00 ft in 100 ft of horizontal run, it is called a 4.00% grade. From Fig. 2.10, percent grade is equal to $100d/H$, or 100 tan α. Now, because percent grade is equal to 100 tan α, and because α is the angle whose sine is d/s [that is, $\alpha = \sin^{-1}(d/s)$], it is often convenient to use the following equation to compute percent grade from known values of d and s:

$$\textbf{Percent grade} = \mathbf{100\ \tan}\left(\mathbf{\sin}^{-1}\frac{\boldsymbol{d}}{\boldsymbol{s}}\right) \tag{2.5}$$

ELEVATION-DIFFERENCE METHOD. Conversion of slope distance to horizontal distance can also be done if the change in elevation over the length of the line is known. In Fig. 2.10, s and H are the same as before. The dimension d, discussed earlier, is the difference in elevation between line ends.

From the theorem of Pythagoras, $s^2 = H^2 + d^2$. This may be transposed to $H^2 = s^2 - d^2$. Taking the square root,

$$\boldsymbol{H} = \sqrt{\boldsymbol{s}^2 - \boldsymbol{d}^2} \tag{2.6}$$

Although this equation appears complicated, it is actually quite easy to solve with a hand-held calculator.

Example 2.7. Slope distance between points A and B is 278.47 ft and elevation difference is 19.37 ft. Compute the horizontal distance.

Solution.

$$H = \sqrt{s^2 - d^2} = \sqrt{(278.47)^2 - (19.37)^2}$$

$$= 277.7955 \Rightarrow\Rightarrow 277.80 \text{ ft}$$

Effects of elevation-difference measurement error. As is the case with the slope-angle method, computed horizontal distance is also subject to the effects of errors in measuring difference in elevation between the ends of a line. The amount of error may be found by recomputing the horizontal distance, this time using an elevation difference to which the anticipated error has been added or subtracted. In the following example, the error is added.

Example 2.8. Using Example 2.7 data, what is the maximum error in H if d is known only to ± 0.05 ft?

Solution. Horizontal distance will be computed twice, once with the original value of d and once using $d' = d + 0.05$ ft.

$$H = \sqrt{(278.47)^2 - (19.37)^2} = 277.7955 \text{ ft}$$

and

$$H' = \sqrt{(278.47)^2 - (19.42)^2} = 277.7920 \text{ ft}$$

Therefore, the difference between H and $H' = 0.0035$ ft.

The error in horizontal distance due to a given error in elevation difference may be computed somewhat more quickly using the following equation, derived from the calculus in App. Sec. A-2:

$$\Delta H = -\frac{d}{H} \times \Delta d \tag{2.7}$$

If d is quite small compared to H (or s), always the case when dealing with shallow slopes, Eq. (2.7) may be approximated to the more convenient expression:

$$\Delta H \cong -\frac{d}{s} \times \Delta d \tag{2.8}$$

In Eq. (2.8), the symbol $\cong$ means "approximately equal to."

Example 2.9. Solve Example 2.7 using Eq. (2.8).

Solution.

$$\text{Difference} = \Delta H \cong -\frac{19.37}{278.47} \times 0.05$$

$$\cong -0.0035 \text{ ft}$$

As was the case with Eq. (2.3), the minus sign may be ignored.

The fact that the results of Examples 2.8 and 2.9 are the same illustrates two important points about shallow-slope situations: (1) Eq. (2.6) provides five-significant-figure results even though elevation difference is known to less than five significant figures, and (2) a relatively large error in elevation-difference measurement will not produce a significant error in computed horizontal distance. Besides that, the expected error in elevation difference ordinarily will be much less than that given in Example 2.8; thus, error in computed horizontal distance from an expected error in measuring elevation difference will in all probability be negligible.

As was the case with the slope-angle method, elevation-difference-measurement error may also be considered in terms of error in relative accuracy. Relative accuracy, defined as $\Delta H/H$, is obtained by dividing both sides of Eq. (2.7) by H. Ignoring the minus sign,

$$\textbf{Accuracy} = \frac{\Delta H}{H} = \frac{d}{H^2} \times \Delta d \tag{2.9}$$

It is often more convenient to compute relative accuracy using the measured slope-distance data rather than the not-yet-computed horizontal distance. As before, when the value of d is much less than the value of s (or H), Eq. (2.9) may be approximated as follows:

$$\textbf{Accuracy} = \frac{\Delta H}{H} \cong \frac{d}{s^2} \times \Delta d \qquad \textbf{(2.10)}$$

Example 2.10. Using the data from Example 2.7, what maximum error in measurement of d may be tolerated if a relative accuracy of 1 part in 25,000 is to be maintained?

Solution. From Eq. (2.10), accuracy $\cong (d/s^2) \times \Delta d$. Performing the necessary algebraic manipulation isolates Δd to the left of the equal sign and results in $\Delta d \cong \text{accuracy} \times s^2/d$. Substituting,

$$\Delta d \cong \frac{(1/25{,}000)(278.47^2)}{19.47} = 0.16 \text{ ft}$$

As the reader will learn in the next chapter, it is relatively easy to maintain a difference-in-elevation measurement tolerance well under 0.16 ft over a slope distance as short as 278.47 ft.

2.5 CORRECTIONS FOR LENGTH, TENSION, TEMPERATURE, AND SAG

Until the advent of EDMs, precise field mearurement required that calibrated tapes be used and that measured values be corrected as necessary to compensate for differences in length, temperature, tension, and support. Electronic-distance-measurement equipment is self-compensating for the most part, with only a minimum of field calibration needed. Calibration of EDMs is explained further on in this chapter, for now, it is appropriate to look into tape correction.

All tapes have a nominal length; that is, the length designated by the markings on the tape. The purchaser of a calibrated surveying tape, however, is provided with a certificate of calibration giving the actual length of the tape under a particular set of conditions as to tension, temperature, and method of support. Thus, a steel tape whose **nominal length** is 100.000 ft could have a **calibrated length** of, say, 99.986 ft at the usual calibration temperature of 68°F (20°C), under tension of, say, 15 lb, with the tape lying on a flat surface. The surveyor may also have a tape calibrated under the same set of tension and support conditions anticipated in the field. If there is no testing lab available locally, tape calibration may be arranged at the National Institute of Standards and Technology (formerly National Bureau of Standards), Gaithersburg, Maryland 20899.

BASIC TAPE-CORRECTION EQUATION. If corrections are to be properly applied (to a measured distance to find true distance, or to a desired true distance in order to find the tape distance to be laid out), the correct algebraic sign must be used. By definition, for any and all corrections applied to a distance measured with a particular tape,

$$\textbf{True distance} = \textbf{measured distance} + \Sigma C \qquad (2.11a)$$

where measured distance is the length of the line as measured with a particular tape and ΣC designates the sum of all corrections for that distance with that tape. Each C must be designated by either a positive or negative sign. In addition, since there are different corrections for length, tension, and so forth, each C is identified by its own subscript; C_l, for example, is the correction for length.

For correcting a single tape length, Eq. 2.11*a* becomes

$$\textbf{True length} = \textbf{nominal length} + \Sigma C \qquad (2.11b)$$

where ΣC, this time, denotes the sum of the various corrections for a single tape length.

LENGTH CORRECTION. If the actual length of a tape under standard conditions (that is, the conditions of temperature, tension, and support under which the tape was calibrated) is greater or less than the tape's nominal, or marked length, the error is assumed to be proportional over the full length of the tape. That is, if a 100-ft tape is 99.976 ft long under standard conditions, the 50-ft marking is assumed to be half that distance, or 49.988 ft from the zero mark, the 25-ft mark is assumed half *that* distance, or 24.994 ft from zero, and so forth.

If the above tape were used in the field under standard conditions, 0.024 ft would be subtracted from each measured tape length to arrive at a true distance; thus C_l for that tape is -0.024 ft. In the case of a single length of the above tape, using Eq. (2.11*b*),

$$\begin{aligned}\text{True tape length} &= \text{nominal length} + C_l \\ &= 100.000 \text{ ft} + (-0.024 \text{ ft}) = 99.976 \text{ ft}\end{aligned}$$

For distances longer than a single tape length, the correction for one tape length must be multiplied by the number of tape lengths involved in the measurement.

Example 2.11. A distance of 756.94 ft is measured with the above tape under standard conditions. What is the true distance?

Solution. The total $C_l = -0.024$ ft per tape length $\times$ 7.5694 tape lengths, or -0.182 ft $\Rightarrow\Rightarrow$ -0.18 ft. Since there are no corrections other than for length, Eq. (2.11*a*) becomes

$$\begin{aligned}\text{True distance} &= \text{measured distance} + \text{total } C_l \\ &= 756.94 \text{ ft} + (-0.18 \text{ ft}) = 756.76 \text{ ft}\end{aligned}$$

At times it is simpler to treat length correction differently. For example, it can be stated that

$$\textbf{True line length} = \textbf{true tape length} \times \textbf{no. of lengths measured} \qquad (2.12)$$

Example 2.12. Solve Example 2.11 using Eq. (2.12).

Solution.

$$\text{True length} = 99.976 \text{ ft per tape length} \times 7.5694 \text{ tape lengths} = 756.76 \text{ ft}$$

Equation (2.12) is also convenient for computing the distance to be measured when *laying out* a line of a desired length.

Example 2.13. What distance should be measured in laying out a 756.94-ft line using the tape from Examples 2.11 and 2.12?

Solution. Equation (2.12) is manipulated to obtain

$$\begin{aligned}\text{Number of lengths measured} &= \frac{\text{true line length}}{\text{true tape length}} \\ &= \frac{756.94 \text{ ft}}{99.976 \text{ ft per tape length}} \\ &= 7.5712 \text{ tape lengths}\end{aligned}$$

Since the tape is nominally 100.00 ft long, the total measured length must be 7.5712 tape lengths × 100.000 ft per tape length = 757.12 ft.

Note that the same result would have been obtained by manipulating Eq. (2.11*a*) to obtain

$$\begin{aligned}\text{Measured distance} &= \text{true distance} - C_l \\ &= 756.94 \text{ ft} - (-0.18 \text{ ft}) = 757.12 \text{ ft}\end{aligned}$$

TEMPERATURE CORRECTION. Steel expands when heated and contracts when cooled. Expansion or contraction of a tape is directly proportional to tape length and to the difference between the temperature in the field and the temperature at standardization. That is,

$$\text{Change in length} \approx \text{original length} \times \text{change in temperature}$$

where the ≈ symbol means "is proportional to."

To replace the ≈ sign with the = sign—and thus make a numerical computation possible—a "constant of proportionality" is needed that reconciles both dimensions and units. In the present application, the constant of proportionality is called the **coefficient of thermal expansion** and its numerical value is a function of the length and temperature units being used as well as the material the tape is made of. The constant of proportionality for tape steel, designated α, is 0.0000065 $\text{ft}\cdot\text{ft}^{-1}\cdot{}^\circ\text{F}^{-1}$. Inserting α into the above equation, the correction for different-from-standard temperature, C_t, becomes

$$C_t = L\alpha(T - T_s) \qquad \textbf{(2.13)}$$

where T is the field temperature, in °F, at time of measurement, T_s is the temperature at time of standardization in °F, and L is the length of the measured line. Note that C_t for a single tape length is obtained by substituting tape length for L.

Example 2.14. A 100-ft tape is calibrated at 68°F but used in the field at 100°F. What is C_t for a single tape length?

Solution. Substituting into Eq. (2.13):

$$C_t = 100.00 \text{ ft} \times 0.0000065 \text{ ft}\cdot\text{ft}^{-1}\cdot{}^\circ\text{F}^{-1} \times (100^\circ\text{F} - 68^\circ\text{F})$$
$$= +0.0208 \Rightarrow\Rightarrow +0.021 \text{ ft}$$

For a long taped line, C_t may be obtained *either* by computing directly with Eq. (2.13) or by multiplying C_t for a single tape length by the number of tape lengths in the line.

Example 2.15. The tape from Example 2.14 is used to measure a line 1 mile long when the field temperature is 100°F. What is C_t for the 1-mile line?

Solution. Using Eq. (2.13),

$$C_t = 5280 \times 0.0000065 \times (100 - 68) = +1.098 \Rightarrow\Rightarrow +1.10 \text{ ft}$$

The same result is obtained by multiplying C_t for one tape length by the number of tape lengths; that is,

$$C_t \text{ (total)} = +0.0208 \text{ ft per tape length} \times 52.8 \text{ tape lengths}$$
$$= +1.098 \Rightarrow\Rightarrow +1.10 \text{ ft}$$

TENSION CORRECTION. Tape steel is elastic and stretches under tension. Within the limits of pull that surveyors are physically able to apply, the stretching is directly proportional to tension and to the length of the line being measured, and inversely proportional to the cross section of the tape. The constant of proportionality, called Young's modulus, or E, is a function of the material, the units of cross-sectional area, and the units of tension. For tape steel, with tape tension stated in pounds and tape cross section given in square inches, E is 29×10^6 lb·in^{-2}, on the average.

If a tape is standardized at one tension and used in the field at another, a tension correction, C_p, is applied as follows:

$$C_p = \frac{(P - P_s)L}{AE} \tag{2.14}$$

where P = tension in pounds at time of measurement
P_s = tension at time of calibration
L = line length *or* tape length in feet
A = tape cross section in in^2
E = Young's modulus, 29×10^6 lb·in^{-2}

Example 2.16. A 100-ft steel tape has a cross section of 0.0050 square inches (in^2) and a calibration tension of 20 lb. What is C_p if the tape is used in the field at 24-lb tension?

Solution. Substituting into Eq. (2.14)

$$C_p = \frac{(24 - 20) \times 100.00}{0.0050(29 \times 10^6)} = +0.0028 \Rightarrow\Rightarrow +0.003 \text{ ft}$$

This is a negligible value when dealing with a single tape length but it becomes significant over a longer line.

Example 2.17. For the tape and the field tension in Example 2.16, what is C_p for a 2500.00-ft measured line?

Solution. Substituting into Eq. (2.14),

$$C_p = \frac{(24 - 20) \times 2500.00}{0.0050(29 \times 10^{-6})} = +0.069 \Rightarrow\Rightarrow +0.07 \text{ ft}$$

Note that the answer would have been the same if the result of Example 2.16 (+0.0028 ft per tape length) had been multiplied by the number of 100-ft tape lengths in 2500 ft (25).

In the foregoing examples involving tension correction, the cross-sectional area of the tape was given. When tape cross section is not known, it may be computed from tape weight and the unit weight of steel, which is 490 lb per cubic foot.

Example 2.18. A particular 100-ft steel tape is actually 103.5 ft long between its extreme ends. If the entire tape weighs 1.72 lb, what is its cross section in square inches?

Solution. Tape weight must be equal to tape volume (V) multiplied by weight per unit volume (ρ); that is, $W = V\rho$. But $V = LA$, where L is length and A is cross section. Thus, $W = LA\rho$. Transposing, $A = W/L\rho$. Now, since W is in pounds, L is in feet, and ρ is in pounds per cubic foot, A must be in square feet. To convert A to square inches, its value in square feet must be multiplied by 144 in^2·ft^{-2}, that is,

$$A, \text{in}^2 = \left(\frac{W \text{ lb}}{L \text{ ft} \times \rho \text{ lb·ft}^{-3}}\right)(144 \text{ in}^2\text{·ft}^{-2})$$

Substituting,

$$A = \left(\frac{1.72}{103.5 \times 490}\right)(144) = 0.0049 \text{ in}^2$$

SAG CORRECTION. Experience tells us that a tape held at its ends will sag in the middle. Because of the sag, the *horizontal* distance between the support points will be less for a tape held at each end than it will be for a tape that is fully supported along a plane surface. The correction for this loss of length, designated C_s, is

$$C_s = \frac{-nw^2L^3}{24P^2} \tag{2.15}$$

where L = distance between supports
n = number of unsupported lengths of length L
w = weight of the tape *per unit length* in pounds per foot
P = tape tension in pounds

The negative sign simply means that a tape held ends-only, or supported at intervals, is shorter than a tape laid flat.

Example 2.19. The weight of a 100-ft steel tape between the zero and 100-ft marks is 2.0 lb. It is held at 15-lb tension, supported at the two ends only. What is C_s for a single tape length?

Solution. In terms of Eq. (2.15): $n = 1$ and $w = 2.0$ lb/100 ft, or 0.020 lb·ft^{-1}. Substituting into Eq. (2.15),

$$C_s = \frac{-(0.020)^2(100)^3}{24(15)^2} = -0.074 \text{ ft}$$

Suppose, though, that the 100-ft tape in the above example had been held at the zero and 100-ft points and supported at the 50-ft point as well.

Example 2.20. Solve Example 2.19 if the tape is supported at the two ends and the middle.

Solution. This time, $n = 2$ and $L = 50$. Substituting into Eq. (2.15),

$$C_s = \frac{-(2)(0.020)^2(50)^3}{24(15)^2} = -0.019 \text{ ft}$$

Examples 2.19 and 2.20, taken together, show that sag correction is not proportional to length. Thus, for a line that is not a whole number of tape lengths long, C_s cannot be computed by simply multiplying the correction for one tape length by the decimal number of tape lengths in the line. The proper procedure is to compute C_s separately for each different L involved in measurement of the line. Treating Eq. (2.15) in this manner, then collecting terms, results in the following general expression for C_s, for *any* combination of L's and n's:

$$C_s = \frac{-(n_1L_1^3 + n_2L_2^3 + \cdots + n_nL_n^3)w^2}{24P^2} \tag{2.16}$$

where n_i is the number of tape sections of unsupported length L_i. For the most part, Eq. (2.16) will involve only two sets of n's and L's, n_1 designating the number of full tape lengths and n_2 designating the single less-than-100-ft length at the end of the line.

Example 2.21. A 100-ft tape weighing 0.015 lb·ft^{-1} was used at a tension of 20 lb, supported at the ends only, to measure a 278.35-ft distance between two points. What is C_s for the line?

Solution. If the tape was held ends-only, two 100-ft segments and 78.35-ft segment must have been measured. Returning to the Eq. (2.16) terminology, $n_1 = 2$, $L_1 = 100$ ft, $n_2 = 1$, and $L_2 = 78.35$ ft. Substituting into Eq. (2.16),

$$C_s = \frac{-[2(100)^3 + (78.35)^3](0.015)^2}{24(20)^2} = -0.058 \Rightarrow\Rightarrow -0.06 \text{ ft}$$

Over undulating terrain, tape measurement of a line can involve a number of segments whose lengths are each less than 100 ft between supports. In that case, Eq. (2.16) must be used in its entirety.

The three foregoing examples provide the value of the sag correction resulting from holding the tape at intervals rather than having the tape fully supported, *but at the same tension as the fully supported tape*. Consider, though, a tape that is calibrated at one tension, either fully supported or supported at ends only, and used in the field at a *different* tension, supported at the two ends only or supported at intervals. In that case, determining the sag correction is at least a two-step process. The following example demonstrates what is probably the most complicated case: a tape that is calibrated at one tension, unsupported, then used in the field at a different tension, also unsupported.

Example 2.22. The tape in Example 2.18 was calibrated at 68°F under a tension of 12 lb, supported at the ends only, and found to be 99.992 ft long. What is the length of the tape if used in the field at 68°F, supported at the ends only, but at a tension of 20 lb?

Solution. The first step is to find the length of the tape when fully supported at the calibration tension of 12 lb; that is, *subtract* the sag correction at 12-lb tension. The next step is to increase the tension to 20 lb, still fully supported, and again find the tape length; that is, *add* the correction for an increase in tension from 12 lb to 20 lb. The final step is to pick up the tape by its two ends, holding 20-lb tension; that is, *add* the sag correction at 20-lb tension.

C_s at 12 lb is computed by substituting into Eq. (2.15), recalling that the Example 2.18 tape weighs 1.72 lb:

$$C_s \text{ at } 12 \text{ lb} = -\left[\frac{(0.0172)^2}{24(12)^2}\right](100^3) = -0.0856 \text{ ft}$$

Now, to find C_p at 20 lb, laid flat, substitute into Eq. (2.14), recalling that the Example 2.18 tape cross section is 0.0049 in^2:

$$C_p \text{ at } 20 \text{ lb} = \frac{(20-12)(100)}{0.0049(29\times 10^6)} = +0.0056 \text{ ft}$$

To find C_s at 20 lb, substitute again into Eq. (2.15):

$$C_s \text{ at } 20 \text{ lb} = \left[\frac{(0.0172)^2}{24(20)^2}\right](100)^3 = -0.0308 \text{ ft}$$

From the basic tape-correction equation, the corrected length is the nominal length plus the sum of the corrections. Substituting,

$$\text{Corrected length} = 99.992 - (-0.0856) + 0.0056 + (-0.0308)$$
$$= 100.0524 \Rightarrow\Rightarrow 100.052 \text{ ft}$$

In the above example, tape length was taken to just three significant figures (100 ft) in the correction equations. This simplification introduces a negligible error in terms of C_s and C_p, which contain three significant figures at worst.

2.6 CORRECTIONS FOR OFFSET AND GRADE

During field measurement it frequently occurs that the tape is offset horizontally or vertically from its intended alignment. In either case, the measured length is the hypotenuse of a right triangle in which the offset is the short leg and the corrected length is the long leg. Thus,

$$\textbf{Corrected length} = \sqrt{L^2 - d^2} \tag{2.17}$$

where L is the measured line length and d is *either* the horizontal or the vertical offset from the intended line.

Equation (2.11*a*), manipulated to place correction left of the equal sign and the corrected and measured lengths on the right, results in Correction = true distance − measured distance. Substituting Eq. (2.17) for corrected length produces

Offset correction:

$$C_o = \sqrt{L^2 - d^2} - L \tag{2.18a}$$

Grade correction:

$$C_g = \sqrt{L^2 - d^2} - L \tag{2.18b}$$

Example 2.23. A 154.23-ft line is measured from A to B. It was then discovered that the taping pin was offset 1.38 ft perpendicular to the intended line AB. What is the correct length of AB?

Solution. The offset produces an error in the measured distance from point A to the taping pin and another error in the measured distance from the taping pin to B. Thus, the corrected value for line AB is the sum of the two corrected distances. Applying Eq. (2.17) to both segments:

$$\text{Corrected distance } AB = \sqrt{(100.00)^2 - (1.38)^2} + \sqrt{(54.23)^2 - (1.38)^2}$$
$$= 99.9905 + 54.2124 = 154.203 \Rightarrow\Rightarrow 154.20 \text{ ft}$$

Note that the large offset in the foregoing example produced rather a small error in measured distance. Consider now a more realistic offset, of the magnitude one would expect in the normal course of tape measurement between two points.

Example 2.24. After completing tape measurement of a 644.39-ft distance between P and Q, the rear taper recalled that the first taping pin was offset 0.4 ft from the intended line. What correction should be applied to line PQ?

Solution. After the first taping pin was incorrectly set 100.00 ft from P, the rear taper lined up the head taper properly with the range pole at Q when setting the remaining five pins along the line. Thus, the taping party actually measured one 100.00-ft hypotenuse from P to the first taping pin and one 544.39-ft hypotenuse from the first taping pin to Q. The total correction must then be the sum of the two corrections.

Substituting into Eq. (2.18*a*),

$$C_o \text{ (total)} = \sqrt{(100.00)^2 - (0.4)^2} - 100.00 + \sqrt{(544.39)^2 - (0.4)^2} - 544.39$$
$$= 0.0009 \text{ ft}$$

The above correction is negligible with respect to the hundredth-of-a-foot value to which the measured distance was recorded. In the normal case, the rear taper will be able to line the head taper onto the intended line within one or two tenths of a foot, an indication that naked-eye alignment is quite adequate for ordinary taping, with no additional correction needed.

2.7 APPLICATION OF COMBINED TAPE CORRECTIONS

Generally, the surveyor must deal with a number of corrections simultaneously because the calibrated length of the tape, along with field conditions of temperature, tension, and support, are going to be different from the ideal. The proper approach is to calibrate the correction for each factor separately, determine the sum of the corrections, then apply them to the measured line length as appropriate. As discussed earlier, there are liable to be pitfalls when computing C_s.

Example 2.25. A 100-ft steel tape weighs 1.2 lb and has a cross section of 0.0034 in^2. Its calibrated length is 99.950 ft at 68°F, 15-lb tension, supported throughout. A line was measured with this tape at 4°F, 12-lb tension, supported at the two ends and the 50-ft mark, and with a difference in elevation of 2.8 ft between the two ends of the measured line. If the measured distance was 188.84 ft, what is the total correction and the corrected distance?

Solution. The corrections involved here are C_l, C_t, C_p, C_s, and C_g. Substituting into the appropriate equations,

$$C_l \text{ (one tape length)} = \text{true} - \text{nominal} = 99.950 - 100.000 = -0.050 \text{ ft}$$

$$C_l \text{ (total)} = -0.050 \text{ ft}\cdot\text{length}^{-1} \times 1.8884 \text{ tape lengths} = -0.0944 \text{ ft}$$

$$C_t \text{ (total)} = 188.84 \times 0.0000065 \times (4 - 68) = -0.0786 \text{ ft}$$

$$C_p \text{ (total)} = \frac{(12 - 15)188.84}{0.0034(29 \times 10^6)} = -0.0057 \text{ ft}$$

$$C_s\text{(total)} = \frac{-[3 \times (50)^3 + (38.84)^3]0.012^2}{24 \times (12)^2} = -0.0181 \text{ ft}$$

$$C_g \text{ (total)} = \sqrt{(188.84)^2 - (2.8)^2} - 188.84 = \underline{-0.0208 \text{ ft}}$$

$$\Sigma C = -0.2176 \text{ ft}$$

Total correction, to hundredths of a foot = −0.22 ft.

Recalling Eq. (2.11*a*):

$$\text{True distance} = \text{measured distance} + \Sigma C$$

$$= 188.84 - 0.22 = 188.62 \text{ ft}$$

The above example deals with measured distance between two existing points. Had the example called for finding the distance to be measured with that tape to *lay out* a desired horizontal distance, the procedure would have been the same except for the final step.

Example 2.26. Given the same tape and field conditions as in Example 2.25, what distance should be measured in order to lay out a true horizontal distance of 188.84 ft?

Solution. The sum of the corrections, computed in Example 2.25, is still −0.22 ft. Now, however, Eq. (2.11*a*) is rearranged to the form:

$$\text{Measured distance} = \text{true distance} - \Sigma C$$

$$= 188.84 - (-0.22) = 189.06 \text{ ft}$$

2.8 TAPING ACCURACY EXPECTATIONS

Along with errors arising from discrepancies in tape length, tension, temperature, conditions of support, and offset, there is also the possibility of errors due to such factors as incorrectly set taping pins, from the rear taper's being unable to hold zero precisely, from the head taper's inability to read precisely between tape graduations or being unable to properly position the plumb bob, especially when it is windy. Considering all the potential errors that may occur, it would appear that taping is a chancy proposition at best. In reality, it is unlikely that simultaneous maximum errors will occur, or, even if they do, that they will be cumulative rather than compensating.

Measuring taped lines in both directions will not only eliminate major errors due to misreading the tape or transposing numbers, it will also reduce the possibility of accumulating a large number of small errors. Small random errors can be held to a minimum by simply exercising care in the way the taping is carried out. Even so, a few random errors are unavoidable. Consider now the cumulative impact of a number of small, random taping errors on a measurement of approximately a full 100-ft tape length.

The length of the tape itself will ordinarily be known to within ± 0.01 ft. If tape temperature is known within 10°F, temperature correction error will not exceed ±0.006 ft. For a tape of average weight, if tension is known to within 2 lb, combined error in tension and sag correction should not exceed ±0.02 ft. Errors in holding zero and reading the tape should not exceed ±0.005 ft each and errors due to faulty positioning of the plumb bob should not exceed ±0.02 ft.

Average errors will probably be on the order of half the above-listed maxima. It is explained in Chap. 5 that the total anticipated or "standard" error will not be additive; rather, it will be equal to the square root of the sum of the squares of the individual errors. That being the case, given the possible errors above, standard error in measuring one tape length should be on the order of ±0.016 ft, or about 1 part in 6000.

Not only is 1:6000 a favorable expectation for single-tape-length measurement, it may also be anticipated that accuracy will *improve* for measured lines longer than a single tape length. This is because total standard error for a line is equal to the square root of the sum of the squares of the standard errors in the individual taped segments. Consider a 10-tape-length line, each tape length having an additional taping-pin-setting error of ±0.01 ft; the total standard error would be about ±0.059 ft, or 1 part in about 17,000.

The foregoing discussion of tapes and tape corrections explains the rather tedious distance-measurement process that was carried out routinely by most surveyors as recently as the 1970s. The advent and current near-universal use of low-cost, precise, electronic distance measurement equipment has made tape measurement obsolescent. Nowadays, most surveyors reserve tapes for measuring rather short lines—normally less than a tape length—on which tapes are generally more accurate and more convenient to use. At the same time, the surveyor must keep in mind that tapes have one enormous advantage over EDMs: Tapes are not subject to the occasional circuit breakdown and ultimate component aging that are inherent in any complex device that is dependent on microcircuitry, particularly one that is used in an outdoor environment. For that reason, it is much the wiser course for surveyors to keep their taping equipment handy and to maintain at least minimal skills in the use of tapes for long-line measurement.

2.9 ELECTRONIC DISTANCE MEASUREMENT (EDM)

Tape measurement is painstakingly slow over appreciable distances and is subject to a large number of errors. An EDM instrument, on the other hand, provides quite accurate distance readings in the twinkling of an eye. Wide use of EDMs has come about through the development of low-cost, reliable microcircuitry—the same technology that has given us quartz-crystal-regulated watches and TV remote controls that transmit coded pulses of infrared light. EDMs now enable surveyors to quickly measure relatively long line-of-sight distances to very close tolerances. Further improvements in EDM accuracy, range, ease of use, and reliability should be expected as a result of competition among the manufacturers of the many instruments now on the market.

All electronic distance measurement equipment operates under the same basic principle: determining the round-trip distance between two points in terms of an exact number of wavelengths of modulated electromagnetic energy. The first such surveying instrument was the Geodimeter, a Swedish development introduced in the late 1940s, which transmitted visible light and provided exceedingly accurate measurement with a range up to 40 kilometers (km) (25 miles) at night. The Tellurometer, introduced in South Africa in the late 1950s, used microwaves to provide a range up to 50 miles, day or night. An added advantage of microwave transmission is a capability for operating in haze or light rain. However, microwave instruments require a station at both ends of a line: the *master* station transmits a signal that is received by the *remote* station, which, after a brief, precise time delay, retransmits the signal to the master station.

Geodimeters and Tellurometers are used primarily for geodetic measurement over long distances. Compared to today's short-range instruments, they are bulky, expensive, and somewhat awkward to operate. Since their introduction, though, modern microprocessors have evolved rapidly and brought about the development of new, smaller EDMs that use coherent light, that is, light generated in a narrow band of the light spectrum rather than broad-band visible light. Operation of these newer, shorter-range EDMs is simplicity itself and most professional surveyors find them quite affordable because of the ease, speed, and efficiency with which they can be used for measuring distances up to a mile or more.

Figures 2.11 through 2.16 illustrate a representative sample of the popular, modulated-light EDMs available at this writing. They transmit coherent light in the near-infrared band and are designed to measure relatively short distances. Their ranges under average atmosphere conditions vary from about 1.3 km to 4.0 km. All but one are designed to be mounted either on top of a theodolite or on a tribrach, the latter being a device that adapts the EDM directly to a tripod and that has its own optical plummet for centering the EDM over a survey point. Most EDMs in this group have an accuracy of ±(5 mm + 5 ppm), where ppm stands for "parts per million." The exception is the Wild Distomat (Figure 2.16), whose advertised range and accuracy are 4.0 km and ±(1 mm + 1 ppm) but which is designed for direct mounting on the telescopes of Wild theodolites.

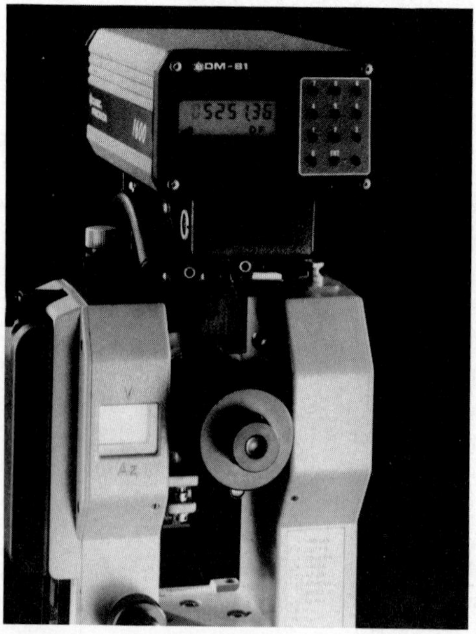

FIGURE 2.11
Cubic Precision DM-81 EDM mounted on a theodolite. (*Courtesy of Cubic Corporation.*)

FIGURE 2.12
Lietz REDMINI2 EDM. (*Couretsy of The Lietz Company.*)

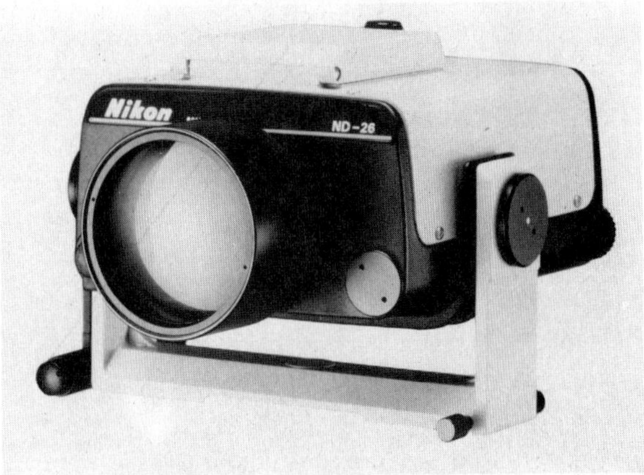

FIGURE 2.13
Nikon ND-26 EDM with theodolite mount. (*Courtesy of Nikon, Inc.*)

FIGURE 2.14
IRI Stinger II EDM on tribrach. (*Courtesy of IRI, Inc.*)

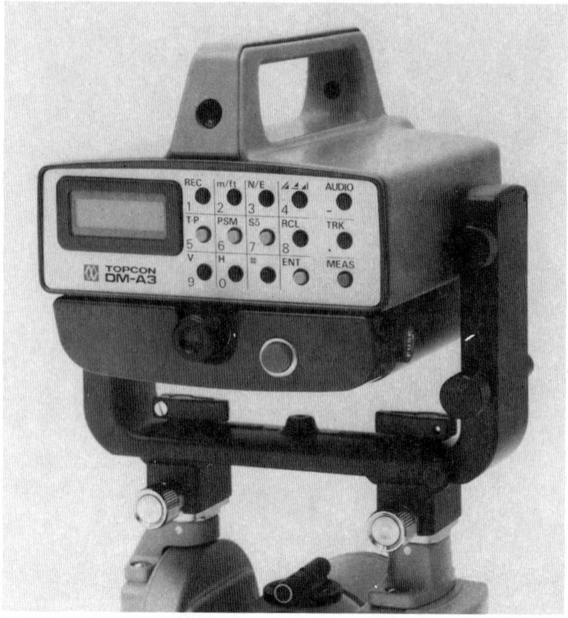

FIGURE 2.15
Topcon DM-A3 mounted on theodolite. (*Courtesy of Topcon Instrument Corporation of America*.)

FIGURE 2.16
Wild Distomat DI 2000 EDM mounted on telescopes of three theodolites. (*Courtesy of Wild Leitz USA, Inc*.)

Most of the figures show the EDMs from the back side, on which the controls are located. Figures 2.13 and 2.16 show the front side, on which are located the lens or lenses through which the infrared light is transmitted and received.

Although the instruments shown here have significantly shorter range than microwave instruments or the newer Geodimeters (which use coherent light), they are quite appropriate for the shorter distances with which most surveyors are almost exclusively concerned.

2.10 EDM THEORY

All EDMs, whether they transmit light or microwaves, incorporate a **carrier frequency** and two or more **modulation frequencies**. The carrier frequency is the electromagnetic radiation frequency that is actually transmitted. A flashlight, for example, transmits energy at light frequencies. An FM radio station transmits a signal in the 88- to 108-MHz FM band, where MHz stands for megahertz, or millions of cycles per second.

The modulation frequency, on the other hand, is a lower frequency that actually carries information. Consider an FM station with an assigned carrier frequency of, say, 88.5 MHz. To broadcast music, the station must modulate, or superimpose, audio frequencies in the range of 60 Hz to 16,000 Hz onto its carrier frequency. A listener's FM radio tuned to the 88.5-MHz carrier frequency would receive the signal and demodulate it in order to reproduce the music being broadcast.

Similarly, an EDM generates a transmitted carrier frequency and modulates the carrier at a much lower frequency. EDM distances are determined in terms of the number of modulation-frequency wavelengths that stretch from the EDM to the reflector (for a light instrument) or to the remote station (for a microwave instrument) and back to the EDM.

The wavelength of the modulation frequency is determined from

$$\lambda = \frac{V}{f} \tag{2.19}$$

where λ is modulation wavelength in meters, V is the velocity of electromagnetic radiation in the atmosphere in meters per second, and f is modulation frequency in hertz.

The term "electromagnetic radiation" simply refers to any radiation in the entire electromagnetic spectrum, from the extremely short-wavelength radiation of cosmic rays up through the progressively longer wavelengths of gamma rays, x-rays, ultraviolet light, visible light, infrared light, microwaves, and common broadcast frequencies. EDMs transmit on discrete, narrow-band segments of the electromagnetic spectrum. As will be discussed further along, the velocity of electromagnetic radiation in air, though relatively constant, does have some variation as a result of differences in air temperature, air density, and carrier-frequency wavelength.

Consider an EDM that transmits electromagnetic energy on a carrier frequency that is modulated at a precisely known f. If the velocity of the electromagnetic radiation

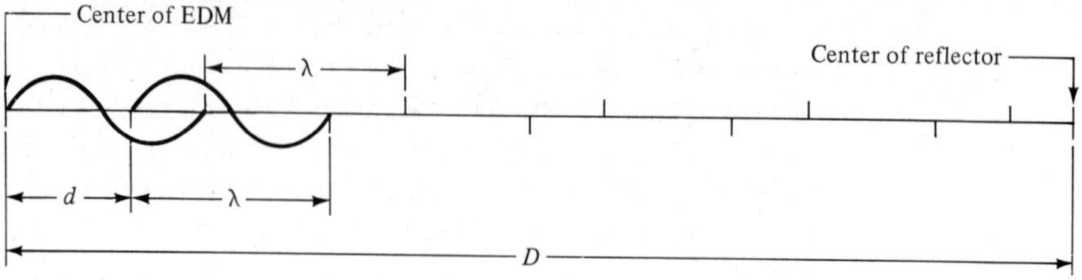

FIGURE 2.17
EDM transmission path along a measured line.

in air is just as precisely known, the wavelength λ must be known with the same precision as f or V. Figure 2.17 shows a light-transmitting EDM on the left and its reflector on the right, with outgoing whole wavelengths marked along the top of the horizontal line and return whole wavelengths marked along the underside. For simplicity, only the initial outgoing wavelength and the final returning wavelength are shown in their true sine-wave form. The distance from the EDM to the reflector and back is some whole number of modulation wavelengths, n, plus a fraction of a wavelength known as a **phase difference** and designated d. The measured distance D is the one-way distance between the EDM and the reflector, or

$$D = \frac{n\lambda + d}{2} \tag{2.20}$$

D could also be expressed as $n\lambda/2 + d/2$; that is, some integer number of half-wavelengths plus half of the phase difference value.

PHASE-DIFFERENCE MEASUREMENT. Consider the "phase angle" between outgoing and incoming waves, which varies from 0° to 360°. That is, if the outgoing and incoming waves were exactly in phase, the phase *angle* would be 0° and the phase *difference* would also be zero; if the phase *angle* were 180°, the phase *difference* would be $\lambda/2$, and so forth. The EDM interprets phase angle in terms of phase difference and it is simple enough, using a single modulated wavelength, to measure the phase difference portion of Eq. (2.20); however, the wavelength count n is ambiguous. Thus, in terms of a single modulation frequency, the distance D is a fairly exact value of $d/2$ plus some still-unknown whole number of half-wavelengths, $\lambda/2$.

Consider a light-actuated EDM with its reflector set up 2429.382 m away. Consider further that the EDM's modulated wavelength is precisely 20 m, and thus its half-wavelength is precisely 10 m. The EDM would determine a phase difference of 18.764 m, or a $d/2$ value of 9.382 m. Thus, the single modulation frequency could only establish that the measured distance is 9.382 m plus *any* multiple of 10 m; that is, 9.382 or 19.382 or 29.382 and so on, up to the maximum range of the EDM. Obviously, determination of the correct distance requires that the EDM also establish the value of the whole number n.

Solving for *n*. Older EDM instruments resolved the n-ambiguity by using as many as four modulation wavelengths that differed from one another by factors of 10, a system known as **decade modulation**. Consider an older EDM with a precise, crystal-controlled oscillator that produced a primary wavelength of exactly 20 m and a primary half-wavelength of exactly 10 m. Assume that phase-angle measurement determined $d/2$ to be the 9.382-m value discussed above. By synchronizing a second modulation frequency so it is exactly one-tenth the frequency of the primary oscillator, a new modulation-frequency wavelength of 200 m and half-wavelength of 100 m are established. With the second modulation frequency the EDM would measure a phase difference to the same number of significant figures as the first phase difference and find the half-phase difference to be 29.38 m. The last three digits would be ignored and the tens-of-meters digit, 2, would be inserted before the previous value, 9.382, to give 29.382 m.

In the next step, the EDM would generate a new modulation frequency, with a half-wavelength of 1000 m, that would measure a phase difference of 429.4 m and provide the hundreds-of-meters digit, which is 4. In the same manner, a final modulation frequency would provide a thousands-of-meters digit of 2.

Assembling the thousands, hundreds and tens digits, in order, and adding in the initial $d/2$ value, a measurement reading of 2429.382 m would be provided. If the EDM were set to display distance in feet rather than meters, the value 2429.382 m would be multiplied internally by a conversion factor of 3937 ft per 1200 m to obtain 7970.40 ft.

For the most part, newer EDMs have dropped the decade-modulation technique in favor of a faster system that establishes the value of n in an elegantly simple manner, using just two very slightly different wavelengths. If the wavelengths were close enough, the number of whole wavelengths from the EDM to the reflector and back would be the same for the two wavelengths (or, at most, different by one whole wavelength), but the phase differences obtained from the two separate wavelengths would be significantly different. With the first modulation frequency, the EDM-measured round-trip distance, using Eq. (2.20), would be $n\lambda_1 + d_1$. With the second frequency, the round-trip distance would be $n\lambda_2 + d_2$. Since round-trip distance is the same with either equation, it can be said that

$$n\lambda_1 + d_1 = n\lambda_2 + d_2$$

Rearranging to solve for n:

$$\boldsymbol{n = \frac{d_1 - d_2}{\lambda_2 - \lambda_1}} \qquad \textbf{(2.21)}$$

Note that the two wavelengths, λ_1 and λ_2, are precisely established and the two phase differences, d_1 and d_2, are precisely measured. The whole number n is thus readily determined. Note also that the phase difference d can be any value between zero and λ; thus, the maximum range of $d_1 - d_2$ must also be equal to the numerical value of λ. The hypothetical maximum value of n that can be determined is thus equal to λ_1 (the shorter of the two wavelengths) divided by $(\lambda_2 - \lambda_1)$.

Consider, for example, a two-frequency, modulated-light EDM whose primary modulation-frequency wavelength is exactly 40 m and whose secondary modulation-frequency wavelength is exactly 40.16 m. The wavelength difference $\lambda_2 - \lambda_1$ is exactly 0.16 m. Thus, the largest value of n that can be determined from Eq. (2.21) is $\frac{40}{0.16}$, or 250. The maximum measurable round-trip distance is $n_{max}\lambda_1$, or, 250 × 40 m = 10,000 m, a one-way hypothetical maximum range of 5 km, which is somewhat longer that the maximum range of any of the EDMs shown in Fig. 2.11 through 2.16. The range of the EDM is governed by the power output of the carrier and the efficiency of the reflector array. The design wavelength difference must be small enough to permit a maximum n value that accommodates the EDM's maximum range. Consider what happens if the measured distance were such that n from the second, longer wavelength were one less than n from the primary wavelength. Expressed algebraically, the following situation would occur:

$$(n-1)\lambda_2 + d_2 = n\lambda_1 + d_1$$

$$n\lambda_2 - \lambda_2 + d_2 = n\lambda_1 + d_1$$

Rearranging to solve for n,

$$n = \frac{d_1 + \lambda_2 - d_2}{\lambda_2 - \lambda_1}$$

In this case, the EDM's initial value of $d_1 - d_2$ would be *negative*, requiring that λ_2 be added to d_1 and activating the above equation in solving for n.

MICROWAVE EDMS. The foregoing discussion has concentrated on the shorter-range versions of modulated-light EDMs that are more than adequate for almost all plane surveying measurements. Long-range, modulated-light EDMs and microwave EDMs, used mostly for geodetic measurements, are available for the relatively few plane surveying applications that require measurement of distances longer than a mile or so. Besides having longer range, microwave EDMs are also usable under marginal weather conditions that limit modulated-light EDMs. However, microwave EDMs have a number of disadvantages: the need for two operators, limitation in applications brought about by the relatively large size of the remote unit, problems from unwanted reflections due to the broad cone of microwave transmission, and greater variation in carrier-frequency velocity resulting from atmospheric changes.

Relative advantages and disadvantages aside, microwave instruments function very much like light instruments. The microwave carrier frequency is modulated in the same manner as the light carrier. The remote station serves the same function as the reflector. Thus, the instrument measures round-trip distance between master station and remote station as a whole number of modulated wavelengths plus the phase difference. Microwave instruments also rely on multiple modulation frequencies to determine number of whole wavelengths.

Because most measurement applications in today's plane surveying operations are easily handled by small, inexpensive, and convenient modulated-light EDMs, further discussion of EDMs will be limited to those instruments.

2.11 ATMOSPHERIC EFFECTS

Light travels faster in a vacuum than in air; generally speaking, the heavier the air, the lower the velocity. Velocity in air, v, is related to velocity in a vacuum by the equation:

$$v = \frac{c}{n_a} \tag{2.22}$$

where c, velocity of light in a vacuum, is 299,792.5 km per second and n_a is the **refractive index** of ambient air (that is, the air through which the light passes). The refractive index of ambient air depends on air temperature, air pressure, and water vapor pressure. The effect of vapor-pressure changes is significant for microwave EDMs but not for any but the most precise uses of modulated-light instruments. The equation for n_a, with vapor pressure neglected, is

$$n_a = 1 + \frac{0.359474(n_g - 1)p}{273.2 + t} \tag{2.23}$$

where t is air temperature in C°; p is air pressure in millimeters of mercury, or **torr**; and n_g is the refractive index in a **standard atmosphere** of 0°C and 760 torr. The value of n_g, in turn, is

$$n_g = 1 + \left(287.604 + \frac{4.8864}{\lambda_c^2} + \frac{0.068}{\lambda_c^4}\right) \times 10^{-6} \tag{2.24}$$

where λ_c is wavelength, in micrometers (μm), of the EDM light source. The two most prevalent EDM light sources at this writing are the infrared light-emitting diode (LED), with design wavelengths that vary between 0.850 and 0.930 μm, depending on manufacturer, and the 0.6328-μm helium-neon laser. A less-used EDM light source is the mercury-vapor lamp, whose wavelength is 0.5500 μm.

Example 2.27. For a mercury-vapor light source, what is the refractive index in a standard atmosphere?

Solution. Substituting in Eq. (2.24),

$$n_{g,0.5500} = 1 + \left[287.604 + \frac{4.8864}{(0.5500)^2} + \frac{0.068}{(0.5500)^4}\right] \times 10^{-6}$$

$$= 1 + (287.604 + 16.153 + 0.743) \times 10^{-6} = 1.0003045$$

Example 2.28. An EDM with an infrared LED source wavelength of 0.865 μm is used on a day when air temperature is 77°F and air pressure is 30.15 in. of mercury. If the EDM modulation frequency is exactly 15 MHz, what is the modulated wavelength λ_m?

Solution. Several steps are required. After calculating n_g, air temperature and pressure must be converted to °C and torr so that n_a may be computed in order to obtain v_a and, finally, λ_m.

Substituting into Eq. (2.24),

$$n_{g,0.865} = 1 + \left[287.604 + \frac{4.8864}{(0.865)^2} + \frac{0.068}{(0.865)^4}\right] \times 10^{-6}$$

$$= 1 + (287.604 + 6.531 + 0.121) \times 10^{-6} = 1.0002943$$

Converting °F to °C by substituting into $C = \frac{5}{9}(F - 32)$,

$$\text{Temperature (°C)} = \frac{5(77 - 32)}{9} = 25.0\text{°C}$$

Converting inHg to mmHg (torr) using the factor 25.4 mm per inch,

$$\text{Pressure (torr)} = 30.15 \text{ inHg} \times 25.4 \text{ mm·in}^{-1} = 765.8\text{torr}$$

Substituting into Eq. (2.23),

$$n_a = 1 + \frac{0.359474(1.0002943 - 1)(765.8)}{273.2 + 25.0}$$

$$= 1.0002717$$

Substituting into Eq. (2.22),

$$v = \frac{c}{n_a} = \frac{299,792.5}{1.0002717} = 299,711.1 \text{ km per second}$$

Substituting into Eq. (2.19),

$$\lambda_m = \frac{v}{f} = \frac{299.7111 \times 10^6 \text{ m·sec}^{-1}}{15.00000 \times 10^6 \text{ cycles·sec}^{-1}}$$

$$= 19.98074 \text{ m}$$

Examination of the solution to Eq. (2.23) in the above problem shows that errors in air pressure and air temperature on the order of 10 torr or 3°C will only change λ_m—and therefore EDM-measured distance—about 3 parts per million. But 3 ppm is better than the accuracy specifications of all but one of the newer-model EDMs described earlier. Surveyors can readily determine pressure and temperature well within 10 torr and 3°C using relatively inexpensive aneroid barometers and mercury thermometers.

EDM users don't really need to compute atmospheric corrections. Correction functions are built into the EDM's microprocessors, so the operator need only key in values of temperature and pressure, or, with some instruments, correction factors that substitute for the function $p/(273.2 + C)$. In the latter case, the surveyor uses ambient temperature and pressure to enter a nomogram and obtain a correction factor that is dialed or keyed into the EDM control panel.

It must be kept in mind that air pressures at higher altitudes are well under sea-level norms. The following table shows differences in standard air pressure with change in elevation:

	Pressure of standard air	
Elevation, ft	mmHg, torr	inHg
Sea level	760	29.9
1,000	731	28.8
2,000	705	27.8
3,000	679	26.7
4,000	655	25.8
6,000	607	23.9
8,000	563	22.2
10,000	522	20.6
12,000	485	19.1

When relatively long EDM distances are measured along steep slopes, it is well to determine pressure at both ends of the line and average the two values.

2.12 EDM ERRORS

Advertised accuracy specifications for most of the EDM's discussed earlier (Sec. 2.9) are listed as ±(5 mm + 5 ppm). The first term is independent of distance and the second is proportional to distance. The two terms are usually characterized as "constant" error and proportional error, respectively. However, the distance-independent first term is not truly constant; it is cyclic and it arises primarily from uncertainty in measuring phase difference. Error in phase-difference measurement is sinusoidal, occurring twice per modulation-frequency wavelength and, for most instruments now on the market, having an *amplitude* of about 5 mm with respect to the average phase-distance-measurement error. Figure 2.18 shows a typical graph of this particular error for a present-generation EDM.

Some currently available instruments advertise accuracies with range-independent components significantly less that 5 mm, an indication that the technology is available to compensate or overcome the relatively large-amplitude error characterized in Fig. 2.18.

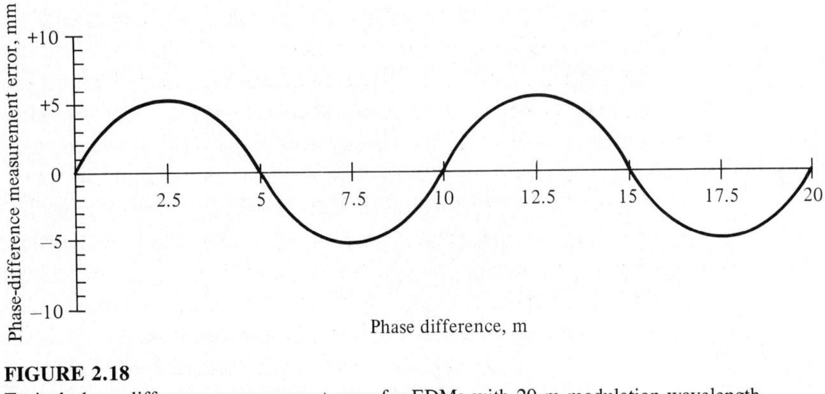

FIGURE 2.18
Typical phase-difference-measurement error for EDMs with 20-m modulation wavelength (10-m error cycle).

The proportional component of error results from uncertainties in both modulation-frequency wavelength and light velocity. Modern EDMs, using temperature-compensated crystals to generate the primary modulation frequency, can maintain modulation-frequency wavelength to better than one part per million. Keying the proper atmospheric corrector into the EDM can hold the uncertainty in light velocity to somewhat better than ± 5 ppm.

The errors just discussed are not only being reduced significantly in the latest instruments coming into the market, they are already essentially negligible for most plane surveying applications. There is one additional error, however, that is significant and is inherent in the design of EDM systems. It has to do with uncertainties in the length of the light path within the prism reflectors and, to a certain extent, within the EDM itself.

The EDM reflector uses the principle of the **corner reflector**, which consists of three mirrors on the inside corner of a cube, each one at precise right angles to the other two. Light entering the reflector along a particular ray path will hit one mirror, reflect to the second, then to the third, and return along a path exactly parallel to the entry path. The effective location of an *equivalent* reflecting surface that is perpendicular to the light path is the same no matter what part of the reflector the entering path hits. For a variety of reasons, the equivalent reflecting surface will be offset to some extent from the mounting center of the reflector array. Similarly, the effective location of the light source and the light detector within the EDM tends to be offset from the mounting center of the EDM.

The combined light-path error due to the two offsets is compensated by the **instrument correction**, designated C_I. This error can be readily determined by the user and adjusted out of the instrument. Consider points A, B, and C on a common straight line, as shown in Fig. 2.19. With the EDM set up over A, distances AB and AC are measured. Then distance BC is measured with the EDM set up over B. For any EDM measurement, it can be said that the measured distance plus C_I is equal to the actual distance. Thus:

$$\text{EDM-measured } AB + C_I = \text{actual } AB$$

and

$$\text{EDM-measured } BC + C_I = \text{actual } BC$$

Adding the two equations,

$$\begin{aligned}\text{EDM-measured } AB + \text{EDM-measured } BC + 2C_I &= \text{actual } AB + \text{actual } BC \\ &= \text{actual } AC\end{aligned}$$

But EDM-measured $AC + C_I$ = actual AC; therefore, equating AC:

$$\text{EDM-measured } AB + \text{EDM-measured } BC + 2C_I = \text{EDM-measured } AC + C_I$$

from which is obtained the following equation for C_I:

$$\mathbf{C_I = \textbf{EDM-measured } AC - (\textbf{EDM-measured } AB + \textbf{EDM-measured } BC)} \quad \mathbf{(2.25)}$$

FIGURE 2.19
Field layout for determining C_I.

Note that determination of instrument correction does not require that any of the *true* distances be known—Eq. (2.25) uses only measured distances.

Example 2.29. The three EDM distances, *AB*, *AC*, and *BC*, were found to be 75.692 m, 188.090 m, and 112.393 m, respectively. Compute C_I.

Solution. Substituting into Eq. (2.25),

$$C_I = 188.060 - (75.692 + 112.393) = -0.025 \text{ m, or } -25 \text{ mm}$$

Once C_I is determined, as in Example 2.29, its value in millimeters (centimeters on some instruments) may be keyed into the EDM. Thereafter, when using the same reflector array, the appropriate instrument correction is automatically applied to any measured distance.

2.13 USE OF THE EDM IN THE FIELD

In order to measure the distance between two points, the EDM is set vertically over one point and the reflector array is set vertically over the other. The EDM is most often mounted on a theodolite, as in Figs. 2.11, 2.15, and 2.16. This allows slope angle to be measured concurrently with slope distance. The theodolite is mounted on a tripod and centered over the point with an optical plummet. The EDM, through its theodolite mount, is aligned vertically with the geometric center of the theodolite.

Occasionally an EDM is used as a stand-alone instrument and the slope distance is measured independent of slope-angle measurement. This is done by mounting the EDM on a tribrach that is in turn mounted on a tripod. The tribrach has its own optical plummet that allows precise positioning over a point. Tribrachs are often provided as part of the standard EDM equipment set; if not, they are available as an extra-cost accessory.

Figure 2.20 shows several styles of reflectors and reflector arrays, three of them mounted on tribrachs. Single reflectors are usually adequate for distances up to half a mile under average atmospheric conditions. Measurements of distances beyond that requires three or more reflectors. A listing of reflector requirements at different ranges and under different atmospheric conditions is provided with the EDM operating manual.

EDMs are operated in accordance with the design characteristics of individual models. Many modern EDMs can perform a number of other functions beyond simple measurement and their control panels reflect this added complexity. However, when used solely for measuring distances, they require very little of the operator.

In a typical field operation, the EDM and reflector are set up on their respective tripods, taking care to center both over their stations, the power is turned on to check battery charge, and the instrument constant is set in along with atmospheric correction(s). The EDM is then sighted on the reflector array, using horizontal and vertical tangent screws to center the cross hairs of the sighting telescope on the center of the reflector—or on the center of the array cluster if multiple reflectors are used. If the sighting telescope is not correctly boresighted—that is, with its optical axis parallel to the light-beam axis—pointing may be refined using the EDM's signal-strength meter

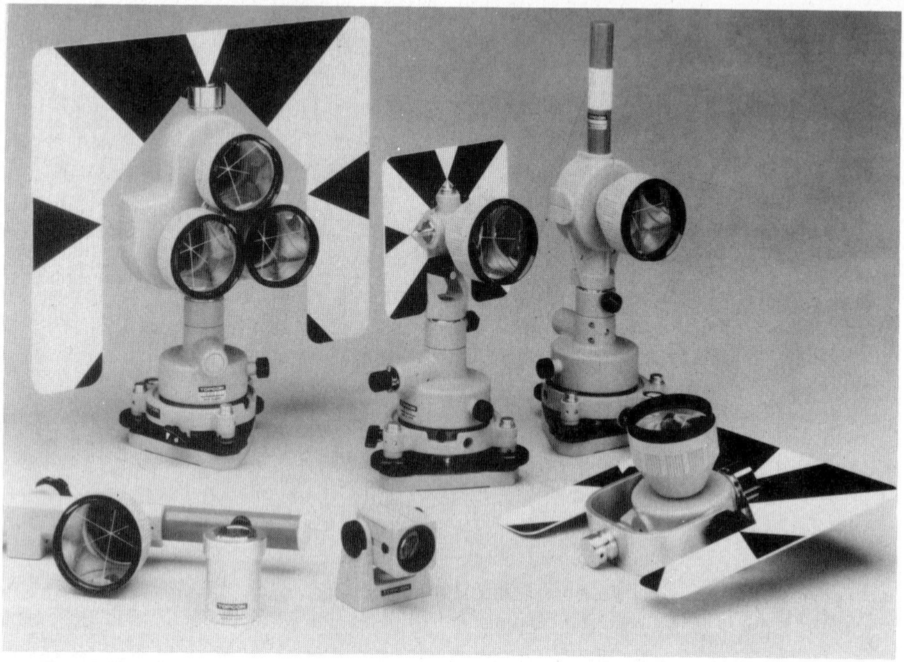

FIGURE 2.20
Examples of EDM reflectors. (*Courtesy of Topcon Instrument Corporation of America.*)

(or audio "chirp"), which tells the operator when reflected light is being received by the instrument.

Once properly sighted on the reflector array, distance measurement is usually accomplished by pressing a single button, after which the distance will appear on the EDM display, either in feet or in meters, depending on how the operator has set the feet-meters switch.

Many newer EDMs incorporate a range of computational capabilities, including reduction of slope distance to horizontal and vertical distance, along with a few of the exotic features usually associated with **total-station** instruments. A total station, covered in some detail in Chap. 4, is a combination theodolite-EDM that incorporates a fairly elaborate microprocessor and its own recording system.

2.14 EDM SLOPE REDUCTION

Section 2.4 of this chapter discussed the computation of horizontal distances using slope distances and slope angles, the assumption being that the slope-angle axis coincided with the line along which the slope distance was measured. In the real world of slope-distance measurement with EDMs, however, that simplifying assumption does not necessarily hold. Of necessity, slope distance is measured along a line between the EDM center and the reflector-array center. Slope angle, on the other hand, is measured along a line between the theodolite's optical center and the center of the target the theodolite is sighting on. For obvious reasons, the two lines cannot coincide and are rarely parallel.

Significant divergence of the EDM light path with respect to the line of sight of the theodolite can occur when slope angle is measured at a different time, and thus at a different setup height from the slope-distance measurement. That is, even though the theodolite, the EDM, the theodolite target, and the reflector array are centered vertically over their respective points, the slope-angle and slope-distance axes diverge because the theodolite and EDM are at differing heights above one point while the reflector array and theodolite target are at differing heights above the other point.

When the slope-distance and slope-angle measurement are taken from a common setup, there is still the problem of relative displacement of the EDM from the theodolite and, therefore, divergent lines of sight. In addition, because an EDM reflector makes a poor theodolite target at anything beyond fairly close range, their centers are usually somewhat displaced from one another, coincident reflector–target combinations (as shown in Fig. 2.20) are fairly recent developments.

Figure 2.21 is a generalized diagram of an EDM slope distance and a theodolite slope angle measured between points A and B. The EDM, designated E, is located in its usual position on top of the theodolite, designated T. The EDM reflector E' is on top of the theodolite target T'. Thus, the slope distance is measured along EE' and the slope angle α is measured along TT'. Dashed line EE'' is parallel to TT'; thus, the angle by which the slope-distance line differs from the slope-angle line, $\Delta\alpha$, is the angle measured at E between EE' and EE''.

The heights of the EDM and theodolite above point A are AE and AT, respectively. The heights of the reflector and theodolite target above point B are BE' and

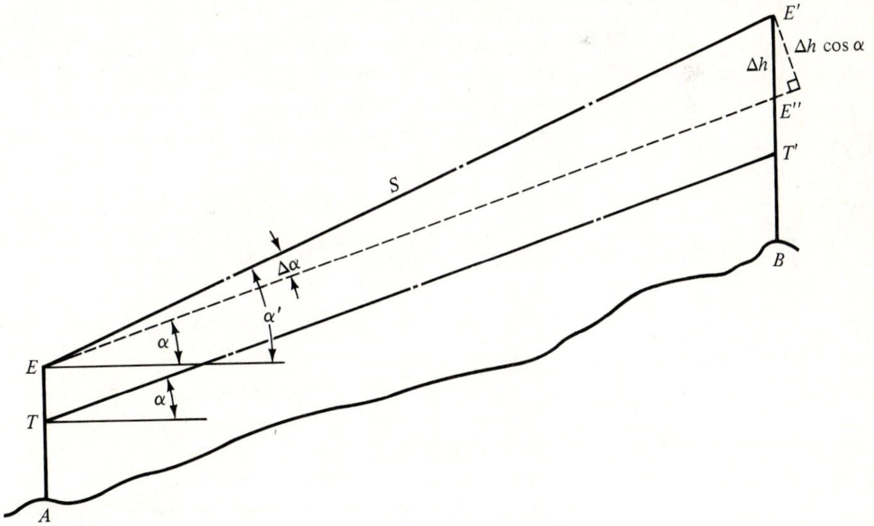

FIGURE 2.21
Divergence of slope distance and slope angle.

BT', respectively. The height of the EDM above the theodolite must then be $AE - AT$, while the height of the reflector above the theodolite target must be $BE' - BT'$. The height difference between the two, designated Δh, is equal to $E''E'$, defined *specifically* as $(BE' - BT') - (AE - AT)$. *Note that the position of the terms in this expression must be maintained as shown—exchanging their position can result in a Δh whose value and algebraic sign are both incorrect.*

In Fig. 2.21, it is seen that $\Delta\alpha$ is the angle subtended by the line $E'E''$ $\cos\alpha$, or $\Delta h \cos\alpha$. The value of $\Delta\alpha$, in radians, is very closely approximated by $(\Delta h \cos\alpha) \div s$. In turn, the value of $\Delta\alpha$, in seconds, must be $\Delta\alpha$ in radians divided by 0.000004848 radian per second of arc. Substituting,

$$\Delta\alpha\ (\text{seconds}) = \frac{[(BE' - BT') - (AE - AT)]\cos\alpha}{0.000004848s} \tag{2.26}$$

The horizontal distance H is equal to the slope distance multiplied by the cosine of the slope angle α' of the EDM-measured line EE'. From Fig. 2.21, $\alpha' = \alpha + \Delta\alpha$. Therefore,

$$H = s\cos(\alpha + \Delta\alpha) \tag{2.27}$$

Note the importance of algebraic sign to the above equations. A positive α or α' denotes an upward sloping **elevation angle**, while a negative α or α' denotes a downward sloping **depression angle**. Similarly, a positive $\Delta\alpha$ denotes a slope-distance line (EE') that rises with respect to the slope-angle line (TT'), while a negative $\Delta\alpha$ defines a line EE' that falls with respect to line TT'.

Consider the following situation where α is a depression angle rather than an elevation angle and where TE is greater than $T'E'$, so line EE' falls with respect to TT'.

Example 2.30. A theodolite-EDM setup is used to measure slope angle and slope distance between points *A* and *B*. The theodolite is 5.08 ft above *A* and the EDM is 5.80 ft above *A*. The theodolite target is 5.19 ft above *B* but the reflector is *below* the target and 4.83 ft above *B*. The slope angle (a depression angle) is −9°15′45″ and the EDM slope distance is 762.44 ft. Compute horizontal distance *AB*.

Solution. Substituting into Eq. (2.26),

$$\Delta\alpha = \frac{[(4.83 - 5.19) - (5.80 - 5.08)]\cos(-9°15'45'')}{0.000004848(762.44)}$$

$$= \frac{-1.08\cos(-9°15'45'')}{0.000004848(762.44)}$$

$$= -288 \text{ seconds} \Rightarrow\Rightarrow -4'48''$$

Substituting into Eq. (2.27),

$$H = 762.44\ \cos[-9°15'45'' + (-4'48'')] = 762.44\ \cos(-9°20'33'')$$

$$= 752.33\text{ft}$$

Examination of Eq. (2.26) shows that $\Delta\alpha$ decreases as slope distance increases and as Δh decreases. Consider what happens when one of the newer, more precise instruments with very small constant error is used on very short lines, say, on the order of a quarter of a tape length. With appreciable slope angles, which is often the case on very short lines, $\Delta\alpha$ becomes a very significant factor that must always be taken into account.

2.15 PROBLEMS

2.1. A surveyor with a pace length of 2.86 ft counted 285 paces between two property corners. The distance was later measured accurately with an EDM and found to be 829.77 ft. By what percentage was the paced distance in error?

2.2. A line measured 1233.95 ft with a 100-ft steel tape, but later calculations involving this measurement did not work out. The surveyor, whose pace length was 2.79 ft, paced the line as 475 paces. What conclusion, if any, might she draw about the distance measured with the steel tape?

2.3. For each set of conditions shown below, calculate the corrected horizontal distance.

Measured slope distance	Slope	Elevation difference
(*a*) 569.32 ft	+5°32′	
(*b*) 422.86 ft	+6.80%	
(*c*) 813.98 ft	−3°49′	
(*d*) 351.44 ft		−2.50 ft per 100 ft along slope
(*e*) 498.74 ft		+4.70 ft per 100 ft horizontally
(*f*) 763.68 ft		−26.4 ft total elevation difference
(*g*) 637.19 ft		+13.8 ft total elevation difference

2.4. For each set of conditions shown below, calculate the distance that must be measured on the designated slope to establish the specified horizontal distance.

Required horizontal distance	Slope	Elevation difference
(*a*) 632.12 ft	−4.64%	
(*b*) 466.61 ft	+6°24′	
(*c*) 713.37 ft		+3.19 ft per 100 ft horizontally
(*d*) 582.73 ft		−8.63 ft per 100 ft along slope
(*e*) 219.42 ft		+4.22 ft total elevation difference
(*f*) 372.55 ft	+7.17%	

2.5. In Prob. 2.3(*a*), what maximum error is introduced into the calculation of the horizontal distance by measuring the slope angle to just the nearest whole minute (that is, ±30″)?

2.6. Same as Prob. 2.5 except use the data from Prob. 2.3(*c*).

2.7. In Prob. 2.3(*f*), what maximum error is introduced into the calculation of the horizontal distance by measuring the difference in elevation to just three significant figures rather than four?

2.8. Same as Prob. 2.7 except use the data from Prob. 2.3(*g*).

2.9. In Prob 2.3(*a*), what minimum angular accuracy in measuring slope angle is required to assure a relative accuracy of 1:40,000 or better?

2.10. In Prob. 2.3(*a*), what relative accuracy was obtained if it is assumed that the recorded slope angle was measured to $\pm 30''$?

2.11. Same as Prob. 2.9 except use the data from Prob. 2.3(*c*).

2.12. Same as Prob. 2.10 except use the data from Prob. 2.3(*c*).

2.13. In Prob. 2.4(*d*), what maximum error in measurement of the elevation difference would be tolerated to achieve a relative accuracy of 1:30,000 or better?

2.14. In Prob. 2.4(*e*) what relative accuracy was achieved if the elevation difference was determined to no better than ±0.005 ft?

2.15. How much error was introduced in laying out a 200.00-ft line if the tape was out of vertical alignment by 1.50 ft *and* out of horizontal alignment by 1.00 ft at the 100-ft point?

2.16. After laying out a 100.00-ft distance with a 100-ft steel tape, it was noticed that the tape had become snagged at the 30-ft point, resulting in a 2.05-ft horizontal offset and a 1.20-ft vertical offset. What was the actual distance laid out?

2.17. In taping a distance between two points, an obstruction necessitated setting an intermediate point 7.2 ft to the side of the line. The distance from the beginning point to the intermediate point, as measured along horizontal ground, was 432.86 ft. The distance from the intermediate point to the end point, measured along a 4.80% slope, was 289.31 ft. What is the correct horizontal distance between the two points?

Note: In Probs. 2.18 through 2.23, the coefficient of thermal expansion α is 0.0000065 ft·ft·°F^{-1} and Young's modulus E is 29×10^6 lb·in^{-2}.

2.18. Determine the correct distances for lines measured under the indicated field conditions. The tape weighs 1.50 lb, it has a cross section of 0.0044 in^2, it is 100.004 ft between end points when supported throughout at a tension of 16 lb and a temperature of 68° F, but only 99.967 ft long when supported ends-only under the same conditions of tension and temperature.

Measured distance, ft	Support	Tension, lb	Temp, °F	Slope or elevation difference
(*a*) 535.69	Throughout	12	90	4.00%
(*b*) 246.71	Ends-only	16	50	Tape horizontal
(*c*) 413.66	Ends and 50-ft point	18	44	6°30′
(*d*) 689.43	Throughout	18	85	4.00 ft/100 ft horizontally
(*e*) 323.82	Ends-only	14	80	3.50 ft/100 ft along slope
(*f*) 776.44	Throughout	20	68	25.5 ft total elev. diff.
(*g*) 469.22	Ends, 25-, 50-, and 75-ft points	16	75	Tape horizontal

2.19. Determine the correct measurements to be made under the indicated field conditions to establish the desired true distances. The tape to be used weighs 1.80 lb and has a cross section of 0.0053 in^2. Its calibrated length is 100.012 ft between end points supported throughout at 14-lb tension, 68°F, but only 99.943 ft long supported ends-only under the same tension and temperature conditions.

	Desired distance distance, ft	Support	Tension lb	Temp °F	Slope or elevation difference
(*a*)	339.65	Ends-only	18	47	Tape horizontal
(*b*)	500.00	Throughout	12	55	4°45′
(*c*)	700.00	Ends-only	14	79	2.80 ft/100 ft horizontally
(*d*)	482.39	Ends-only	16	68	Tape horizontal
(*e*)	650.00	Ends and 50-ft point	14	95	3.50%
(*f*)	247.85	Throughout	10	87	4.40 ft/100 ft along slope
(*g*)	519.27	Throughout	20	32	16.8 ft total elev. diff.

2.20. A 100-ft steel tape weighs 2.05 lb, has a cross section of 0.0060 in^2, and is 99.974 ft long when supported ends-only at 15-lb tension and 68°F. A line was measured as 532.69 ft with this tape supported ends-only under 20-lb tension and a field temperature of 85°F. What is the corrected length of the line?

2.21. Determine the cross section of a steel tape whose weight between the zero and 100-ft points is 2.20 lb.

2.22. If the computed sag correction for a 100-ft steel tape was −0.064 ft per tape length with the tape supported ends-only, what would the sag correction be for one tape length supported at the 50-ft point as well as the end points, with all other conditions the same?

2.23. A tape with a 0.0053-in^2 cross section is 100.000 ft long when supported throughout at a tension of 15.0 lb and a temperature of 68°F. To the nearest 0.1 lb, what field tension must be used to compensate for a field temperature of only 58°F if the tape length is to remain 100.000 ft when supported throughout?

2.24. Explain why a distance cannot be measured with an EDM that has just one modulation frequency.

2.25. What atmospheric conditions should be measured and factored in when using EDMs operating in the infrared wavelength band?

2.26. For a helium-neon laser source, what is the refractive index in a standard atmosphere?

2.27. For the light source in Prob. 2.26, what is the index of refraction on a day when the field temperature is 55°F and the air pressure 29.99 inHg?

2.28. For an EDM with a helium-neon laser source and the field conditions in Prob. 2.27, what must the modulation frequency be if a modulation wavelength of exactly 20.00000 m is required?

2.29. Do the same as Prob. 2.28 except the EDM is used at an elevation of 10,000 ft with normal atmospheric pressure for that altitude.

2.30. An EDM measures with an accuracy of ±(5 mm + 5 ppm).

(*a*) What is the maximum error for a measured distance of 4562.32 ft?

(*b*) What is the maximum error for a measured distance of 30.00 ft?

2.31. In Prob. 2.30, determine the relative accuracy of each line and discuss the significance of these results.

2.32. For a particular EDM-reflector combination used at a test setup similar to that diagrammed in Fig. 2.19, distances *AB*, *AC*, and *BC* were measured as 246.32 ft, 510.41 ft, and 264.24 ft, respectively. Compute the instrument correction C_I.

2.33. Using the C_I determined in Prob. 2.32, what is the corrected length of a line measured as 842.63 ft?

2.34. The slope angle and distance between points *A* and *B* were measured with a theodolite-EDM combination with the EDM optical center 0.95 ft above the theodolite optical center and the theodolite optical center 4.99 ft above point *A*. A vertical angle of $+7°23'16''$ was read to a theodolite target set 5.19 ft above point *B*. An EDM distance of 2187.88 ft was read to a reflector mounted 1.50 ft above the theodolite target. What is the horizontal distance *AB*?

2.35. Do the same as Prob. 2.34 except the EDM reflector was located 1.50 ft *below* the theodolite target.

2.36. A slope distance of 1622.17 ft was measured from point *A*, elevation 465.32 ft, to point *B*, elevation 212.89 ft. The height-of-instrument of the EDM was 5.23 ft above point *A* and the height of the reflector was 4.67 ft above point *B*. What is the horizontal distance between *A* and *B*?

2.37. A slope distance of 409.72 ft was measured with an EDM mounted on top of a theodolite. The slope angle measured $+4°49'18''$ and $\Delta\alpha$ was found to be $-9'14''$. Compute the horizontal distance.

2.38. A slope distance of 344.08 ft was measured with an EDM mounted 0.49 ft above a theodolite set over point *A*. The theodolite measured a slope angle of $+6°29'45''$ to the center of a single EDM reflector set over point *B*. Compute horizontal distance *AB*.

CHAPTER 3

LEVELING

3.1 LEVELING DEFINITIONS

Leveling is done to establish the elevations of survey points or, more often, the *differences* in elevation between survey points. Contours cannot be drawn on maps without determining beforehand the elevations of a reasonably dense array of such points within the area being mapped. Similarly, layout and construction of water lines, sewer lines, highways, aqueducts, and navigation channels are dependent on knowing the elevation differences between points along the project right of way. In building construction, the tops of the forms around the perimeter of a foundation must be held as closely as possible to a constant elevation to assure that floors will be level.

Because leveling deals with vertical distances, it is necessary to define what "vertical" really means on a round earth. A **vertical line** is defined as a line pointing in the direction of gravity—a plumb line is therefore a vertical line. A plumb line will almost never point toward the center of the earth—on the one hand, it is deflected toward the further pole by the effect of the earth's rotation; on the other hand, since the earth is made up of a variety of materials of varying density, it is deflected toward denser masses of crustal matter that lie in the vicinity. A plumb bob in the foothills of the Himalayas, for example, will deflect toward the mountain range.

A **level surface** is everywhere perpendicular to the direction of gravity. Thus, since the direction of gravity is locally variable, a level surface is an undulating surface, not a regular, mathematically predictable surface of revolution about the earth's polar axis. In this respect, a **horizontal line** is not a line on a level surface; it is a line perpendicular to the direction of gravity. Strictly speaking, a horizontal line coincides with a level surface at a single point. Thus, a **horizontal plane** (a plane perpendicular to a vertical line) is tangent to a level surface at a single point. In relatively small areas, and over relatively small distances, it may be said that a horizontal plane and a level surface coincide. How small those areas and distances are, however, will be discussed further along.

Differences in elevation are the distances between level surfaces. The **elevation** of a particular point is defined as the vertical distance between the level surface on which the point lies and a level surface of reference called a **datum**. The most familiar datum, and the one most meaningful to the average surveyor, is **mean sea level**, or **MSL**. Mean sea level is defined as average sea level measured over a 19-year period, the length of one repeating lunar–solar cycle with respect to earth.

Mean sea level, as such, has meaning only to those specific coastal locations where 19-year measurements have been made. The National Geodetic Survey has established the **National Geodetic Vertical Datum of 1929 (NGVD29)**, a reference for the United States and Canada that is tied to mean sea level as measured at 26 primary tide gages, 21 in the United States and 5 in Canada. The elevations of vertical survey control stations in the United States and Canada are given as heights above—or, in a very few locations, below—the NGVD. These survey stations, called **bench marks**, are marked with permanent metal discs set into what are assumed to be nonmovable locations such as rock outcrops, concrete monuments, or building foundations.

A newer datum has recently been established, based on an adjustment of all existing data and designated the **North American Vertical Datum of 1988 (NAVD88)**. Bench-mark elevation values on the new datum are scheduled to be available in 1991.

Surveyors may often obtain locations and elevations of nearby bench marks by inquiring at the local county engineer's office. The same information is also available from the National Geodetic Information Center, National Oceanic and Atmospheric Administration, Rockville, Maryland, 20852.

3.2 EARTH CURVATURE AND REFRACTION

It was stated in an earlier chapter that plane surveyors may neglect the effects of earth curvature—for the most part. An exception occurs in the case of vertical distances, which are taken between level surfaces, but which are measured by means of a horizontal reference or a horizontal line of sight. Because a horizontal line diverges from a level surface, the surveyor must either compensate for the divergence or apply a correction, as necessary. Consider the primary cause of the divergence: the curvature of the earth.

CURVATURE. The earth curves out from under an observer's horizontal line of sight. To illustrate, Fig. 3.1 shows a horizontal line tangent to the earth at A. D is the distance from A along the horizontal line to point B', a vertical distance c above point B. If the earth's radius is E, it is seen that there is a right triangle whose legs are E and D and whose hypotenuse is $E + c$. Applying the theorem of Pythagoras:

$$E^2 + D^2 = (E + c)^2 = E^2 + 2Ec + c^2$$

For a distance D that is equivalent to even the longest line of sight on the earth, it is clear that the relative length of D in Fig. 3.1 is greatly exaggerated. For plane

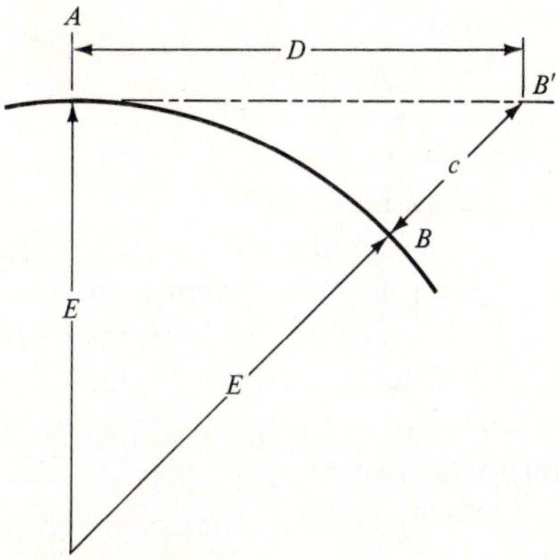

FIGURE 3.1
Curvature of earth with respect to a horizontal line.

surveyors, a line-of-sight distance D would be a few miles at the most, while the mean radius of the earth is about 3959 statute miles. Similarly, c would be a very small value with respect to D or E. That being the case, the c^2 term in the above equation may be neglected entirely, leaving

$$E^2 + D^2 = E^2 + 2Ec$$

$$D^2 = 2Ec$$

$$c = \frac{D^2}{2E}$$

Substituting for the 3959-statute-mile radius of the earth,

$$c_{\text{miles}} = \frac{D^2_{\text{miles}}}{2 \times 3959 \text{ miles}}$$

But c in miles is equal to c in feet divided by 5280 ft per mile. Substituting,

$$\frac{c_{\text{ft}}}{5280 \text{ ft}\cdot\text{mi}^{-1}} = \frac{D^2_{\text{miles}}}{2 \times 3959 \text{ miles}}$$

Multiplying and dividing as indicated,

$$\boldsymbol{c_{\text{ft}} = 0.667\, D^2_{\text{miles}}} \qquad \textbf{(3.1)}$$

REFRACTION. The term c in Eq. (3.1) is of little practical use on the earth because a line of sight in the atmosphere is not straight; it bends toward denser air. A beam of light that passes through air that is incrementally more dense on one side of the beam than the other will be incrementally slowed on the denser side and therefore be deflected to that side. In a normal, stable atmosphere, a line of sight will curve toward earth. Figure 3.2 shows a line of sight that is horizontal at A and curves downward toward B. In this case, the line of sight intersects a vertical line above point B at B''. The distance from B to B'' is $(c \,\&\, r)$, so designated because the distance the earth

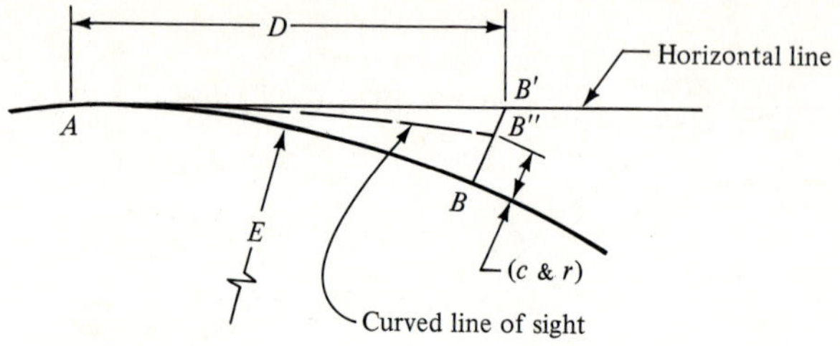

FIGURE 3.2
Curved line of sight in earth's atmosphere.

falls out from under a horizontal line of sight is a function of both curvature and refraction. Thus, the coefficient 0.667 in Eq. (3.1) must be reduced to account for atmospheric effects. The following equation may be used under average atmospheric conditions.

$$(\boldsymbol{c} \ \& \ \boldsymbol{r})_{\text{ft}} = \mathbf{0.574}\boldsymbol{D}^2 \tag{3.2}$$

where D is taken as horizontal distance in miles.

Example 3.1. Under average atmospheric conditions, how far will the earth fall from under an observer's line of sight at a horizontal distance of 5.25 miles?

Solution. Substituting into Eq. (3.2),

$$(c \ \& \ r)_{\text{ft}} = 0.574(5.25)^2 = 15.821 \Rightarrow\Rightarrow 15.8 \text{ ft}$$

More frequently, however, survey distances are measured or given in feet rather than miles. In this case, use

$$(\boldsymbol{c} \ \& \ \boldsymbol{r})_{\text{ft}} = \mathbf{0.0206}\boldsymbol{T}^2 \tag{3.3}$$

where T is horizontal distance in *thousands* of feet. Similarly, when using metric values

$$(\boldsymbol{c} \ \& \ \boldsymbol{r})_{\text{m}} = \mathbf{0.0675}\boldsymbol{K}^2 \tag{3.4}$$

where the subscript m designates $(c \ \& \ r)$ in meters and K is horizontal distance in *kilometers*.

Example 3.2. An observer used a long-range EDM to measure across a large lake. The instrument was set up on one shoreline 1.85 m above the waterline and the reflector was set up on the opposite shoreline. It took several attempts to get the reflector high enough to provide a clear EDM line of sight over the calm lake waters. If, after the final, successful attempt, the EDM-measured distance was 11,417.16 m, how high must the reflector have been above the opposite shore, to the nearest 0.01 m?

Solution. Note that the lake itself interfered with initial attempts to measure across it. In the final, successful attempt, the EDM line of sight was approximately tangent to the lake's surface. Figure 3.3 shows there are two distances involved in this problem: K_1 from

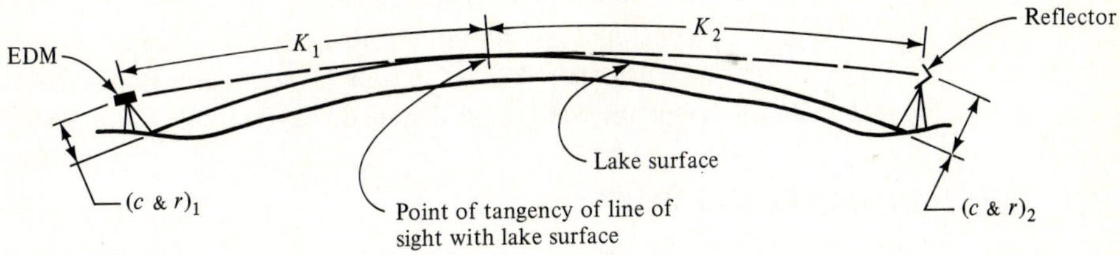

FIGURE 3.3
Diagram of Example 3.2.

the EDM to the point of tangency; K_2 from the point of tangency to the reflector. Together, the two distances add to 11,417.16 m. The first distance is obtained from Eq. (3.4), with the 1.85-m instrument height above water being the (c & r) value corresponding to K_1. Thus

$$(c \,\&\, r)_{\mathrm{m}} = 0.0675K^2$$

$$K^2 = \frac{(c \,\&\, r)_{\mathrm{m}}}{0.0675}$$

$$K = \sqrt{\frac{(c \,\&\, r)_{\mathrm{m}}}{0.0675}}$$

Substituting to obtain K_1,

$$K_1 = \sqrt{\frac{1.85}{0.0675}} = 5.23521 \text{ km} = 5235.21 \text{ m}$$

K_2 must be the total distance less K_1, or 11,417.16 m − 5235.21 m = 6181.95 m, which converts to 6.18195 km. Substituting K_2 into Eq. (3.4) to obtain the (c & r) value corresponding to K_2, which is also the height of the reflector above water,

$$(c \,\&\, r)_2 = 0.0675(6.18195)^2 = 2.5796 \Rightarrow\Rightarrow 2.58 \text{ m}$$

Example 3.3. Compute the maximum length of a horizontal line of sight such that the combined curvature and refraction effect may be neglected.

Solution. Vertical distances, like horizontal distances, are normally taken to the nearest 0.01 ft. A curvature-and-refraction value of 0.005 ft or more would therefore round to 0.01 ft or more and be of significance in vertical distance measurement. Converting Eq. (3.3),

$$(c \,\&\, r)_{\mathrm{ft}} = 0.0206T^2$$

$$T = \sqrt{\frac{(c \,\&\, r)_{\mathrm{ft}}}{0.0206}}$$

Substituting,

$$T = \sqrt{\frac{0.005}{0.0206}} = 0.493 \text{ thousand feet, or about 500 ft}$$

The result of Example 3.3 shows that the effects of curvature and refraction may be neglected for horizontal distances less than 500 ft. By the same token, the effects of curvature and refraction on vertical-distance computations *must* be taken into account for horizontal distances beyond 500 ft, or about a tenth of a mile.

3.3 TRIGONOMETRIC LEVELING

Recall, from Sec. 2.4, that horizontal distances may be obtained from slope distances and slope angles using the equation $H = s\cos\alpha$, with s and α as shown in Fig. 3.4. Similarly, vertical distance h may be obtained from

$$\mathbf{h = s \sin \alpha} \qquad (3.5)$$

It sometimes happens that the horizontal distance H will be known, rather than the slope distance s. In this case:

$$\mathbf{h = H \tan \alpha} \qquad (3.6)$$

Equations (3.5) and (3.6) are valid as shown only for horizontal distances of about 500 ft or less, as discussed earlier. For distances greater than 500 ft, the $(c \ \& \ r)$ correction must be added, resulting in

$$\mathbf{h = s \sin \alpha + (c \ \& \ r)} \qquad (3.7)$$

$$\mathbf{h = H \tan \alpha + (c \ \& \ r)} \qquad (3.8)$$

Example 3.4. In Fig. 3.4, the slope distance is 11,481.92 ft and the slope angle α is $+4°37'25''$. Assuming normal atmospheric conditions, find the difference in elevation between the EDM at A and its reflector at B.

Solution. From Eq. (3.3), $(c \ \& \ r) = 0.0206T^2$, where T is *horizontal* distance in thousands of feet. But $H = s\cos\alpha$ and $T = H/1000$. Therefore, $T = (s \ \cos\alpha)/1000$, or $(s/1000)\cos\alpha$. Substituting into Eq. (3.7),

$$h = 11{,}481.91 \ \sin(+4°37'25'') + 0.0206[11.48192 \ \cos(+4°37'25'')]^2$$

$$= 925.554 + 2.698 \Rightarrow\Rightarrow 928.25 \text{ ft}$$

In Example 3.4, the slope distance is measured between the EDM and the reflector; thus, the vertical distance h from Eq. (3.7) is also the vertical distance from the EDM to the reflector. In the usual situation, though, the surveyor needs the difference in elevation between the point over which the EDM is set and the point over which the reflector is set; that is, taking into account the EDM and reflector heights above their setup points. Figure 3.5 illustrates a typical elevation-difference measurement, with

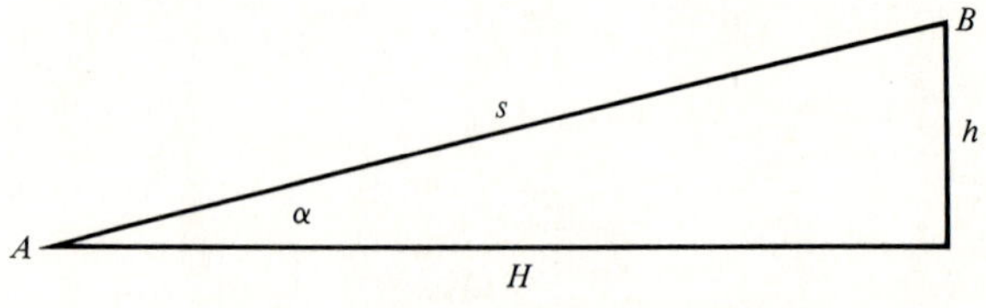

FIGURE 3.4
Profile view of trigonometric leveling.

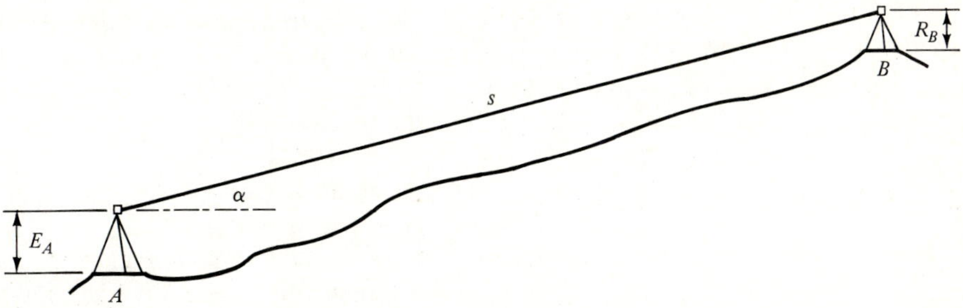

FIGURE 3.5
Elevation difference using slope angle and slope distance.

the EDM height at A and the reflector height at B designated E_A and R_B, respectively. From Fig. 3.5, with the elevation of the EDM setup point known, the elevation of the reflector setup point can be expressed as

$$\mathbf{Elev}_B = \mathbf{Elev}_A + \boldsymbol{E}_A + \boldsymbol{h} - \boldsymbol{R}_B \tag{3.9}$$

with Elev_A the known elevation of the EDM setup point and h the EDM-to-reflector elevation difference as defined by Eq. (3.7) or (3.8).

Example 3.5. Slope distance and slope angle are the same as in Example 3.4, the EDM optical center is 5.21 ft above point A, and the reflector is 3.72 ft above point B. If the elevation of point A is 644.79 ft above mean sea level (MSL), find the elevation above MSL of point B.

Solution. Substituting into Eq. (3.9), including the value of h from Example 3.4:

$$\text{Elev}_B = 644.79 + 5.24 + 928.25 - 3.72 = 1574.56 \text{ ft}$$

Figure 3.6 illustrates a common method of trigonometric leveling that requires two measurements of both vertical angle and slope distance from a single setup: (1) from the instrument to a reflector set over known elevation point A and (2) from the instrument to a reflector set over point B, whose elevation is to be determined. In this case it is not necessary to know either the elevation of the instrument or its height above ground. From Fig. 3.6 it is seen that the elevation of point B is determined from

$$\mathbf{Elev}_B = \mathbf{Elev}_A + \boldsymbol{R}_A - \boldsymbol{h}_A + \boldsymbol{h}_B - \boldsymbol{R}_B \tag{3.10}$$

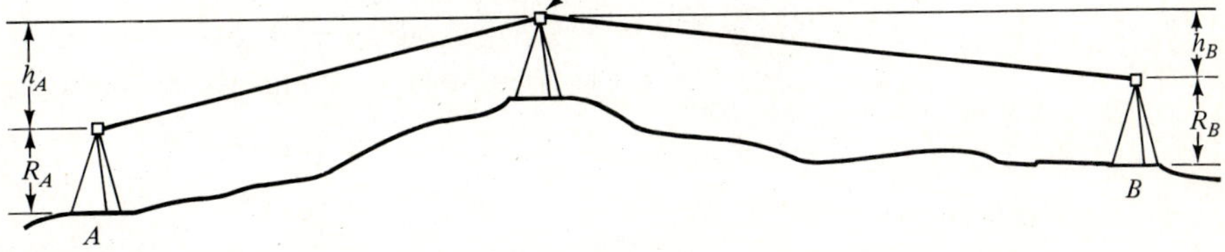

FIGURE 3.6
Elevation difference with theodolite-EDM between points.

where R_A and R_B are the reflector heights above points A and B, respectively, while h_A and h_B are respective elevation differences, from Eq. (3.7), from instrument to A and instrument to B.

Example 3.6. In Fig. 3.6, the elevation of A is 1462.34 above MSL and the reflector height above A is 5.35 ft. Slope distance and slope angle to the reflector at A are 8463.22 ft and $-10°26.3'$, respectively. Slope distance and slope angle to the reflector at B are 11,877.69 ft and $-6°42.9'$, respectively. The height of the reflector above point B is 8.73 ft. Compute the elevation of B.

Solution. Computing h_A and h_B from Eq. (3.7),

$$h = s \sin\alpha + (c \,\&\, r) = s \sin\alpha + 0.0206\left[\left(\frac{s}{1000}\right)\cos\alpha\right]^2$$

Substituting to find h_A and h_B,

$$\begin{aligned} h_A &= 8463.22 \sin(-10°26.3') + 0.0206[8.46322 \cos(-10°26.3')]^2 \\ &= -1533.342 + 1.427 = -1531.92 \text{ ft} \\ h_B &= 11,877.69 \sin(-6°42.9') + 0.0206[11.87769 \cos(-6°42.9')]^2 \\ &= -1388.867 + 2.867 = -1386.00 \text{ ft} \end{aligned}$$

Substituting into Eq. (3.10),

$$\text{Elev}_B = 1462.34 + 5.35 - (-1531.92) + (-1386.00) - 8.73 = 1604.88 \text{ ft}$$

The $(c \,\&\, r)$ corrections in the solutions of Examples 3.4, 3.5, and 3.6 are relatively large; thus, use of Eqs. (3.2) through (3.4) under other-than-normal conditions of atmospheric temperature and pressure could introduce significantly erroneous corrections. The problem of an imperfectly known $(c \,\&\, r)$ value is avoided by measuring slope angle and distance from both ends of the line simultaneously. From Fig. 3.5, difference in elevation from point A to point B is

$$\Delta\text{Elev}_{A\text{-to-}B} = E_A + s_{A\text{-to-}B} \sin\alpha_A + (c \,\&\, r)_A - R_B$$

Similarly, with the EDM at B and the reflector at A, the difference in elevation from point B to point A is expressed as

$$\Delta\text{Elev}_{B\text{-to-}A} = E_B + s_{B\text{-to-}A} \sin\alpha_B + (c \,\&\, r)_B - R_A$$

But the difference in elevation from A to B can also be expressed as the negative value of the difference in elevation from B to A, or

$$\Delta\text{Elev}_{A\text{-to-}B} = -E_B - s_{B\text{-to-}A} \sin\alpha_B - (c \,\&\, r)_B + R_A$$

Taking the correct A-to-B elevation difference as the average of the two A-to-B values produces

$$\Delta\text{Elev}_{A\text{-to-}B} = \frac{E_A - E_B + s_{A\text{-to-}B} \sin\alpha_A - s_{B\text{-to-}A} \sin\alpha_B + (c \,\&\, r)_A - (c \,\&\, r)_B - R_B + R_A}{2}$$

Given simultaneous or near-simultaneous observations at points A and B, and given the number of significant figures in the final $(c \,\&\, r)$ values, $(c \,\&\, r)_A$ and $(c \,\&\, r)_B$ will be the same; thus, they will cancel. This leaves the expression

$$\Delta \mathbf{Elev}_{A\text{-to-}B} = \frac{\boldsymbol{E_A - E_B + s_{A\text{-to-}B}\sin\alpha_A - s_{B\text{-to-}A}\sin\alpha_B - R_B + R_A}}{2} \tag{3.11}$$

Example 3.7. Recall, from Examples 3.4 and 3.5, that the following measurements were made from an EDM set over point A (Elev. 644.79 ft) on a reflector over point B: $s_{A\text{-to-}B} = 11{,}481.92$ ft; $\alpha_A = +4°37'35''$, $E_A = 5.24$ ft; $R_B = 3.72$ ft. Very soon after measurements were taken from A to B, the following set of measurements was made from B to A: $s_{B\text{-to-}A} = 11{,}482.20$ ft; $\alpha_B = -4°40'07''$; $E_B = 5.63$ ft, and $R_A = 3.24$ ft. Compute the elevation of B based on the combined data.

Solution. Because the measurement sets were nearly simultaneous, the $(c \,\&\, r)$ correction is ignored. Substituting into Eq. (3.11),

$$\text{Elev}_{A\text{-to-}B}$$

$$= \frac{5.24 - 5.63 + 11{,}481.92\sin(+4°37'25'') - 11{,}482.20\sin(-4°40'07'') - 3.72 + 3.24}{2}$$

$$= 929.62 \text{ ft}$$

$$\text{Elev}_B = \text{Elev}_A + \text{Elev}_{A\text{-to-}B} = 644.79 + 929.62 = 1574.41 \text{ ft}$$

Comparing elevation values obtained in Examples 3.5 and 3.7 shows a fairly significant difference. The Example 3.5 result includes a $(c \,\&\, r)$ value based on average atmospheric conditions. In Example 3.7, curvature and refraction is fully compensated. It may be assumed that the Example 3.7 result is more nearly the correct elevation.

3.4 DIFFERENTIAL LEVELING: CONCEPT AND TERMINOLOGY

Differential leveling measures the difference in height of a horizontal line of sight above two ground points; that is, the difference in elevation between the two points. Figure 3.7 is a profile diagram showing a **level** set up between two ground points

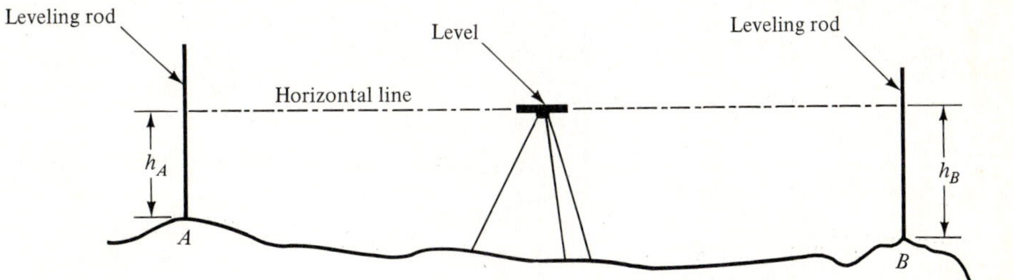

FIGURE 3.7
Principle of differential leveling.

over which a **leveling rod** is held. In surveying terminology, a level is an instrument that provides a horizontal line of sight at a fixed height above ground. A leveling rod is a rigid scale that is held vertically and is used to measure heights above ground points—in Fig. 3.7, the rod is held over points *A* and *B*. The height of the line of sight above point *A* is h_A and its height above point *B* is h_B.

The difference in elevation between *A* and *B* is therefore $h_A - h_b$. If the elevation of *A* is known, the elevation at *B* must be the elevation of *A* plus the algebraic difference in the *A*-to-*B* elevation. In Fig. 3.8, the elevation at point *A* is 248.45 ft and the heights of the line of sight above points *A* and *B* are 4.28 ft and 5.41 ft, respectively. Therefore, $\text{Elev}_B = \text{Elev}_A + (h_A - h_B) = 248.45 + (4.28 - 5.41) = 248.45 + (-1.13) = 247.32$ ft.

Differential leveling is every bit as simple as the above diagrams illustrate. Methods for recording and computing leveling data are demonstrated further along in this chapter. First, though, it would be well to consider the basic differential leveling terminology as well as the instrumentation involved in the differential leveling operation.

A marked, permanent point of known elevation, as mentioned earlier, is called a **bench mark**, or **BM**. Point *A* in Figs. 3.7 and 3.8 could well be a bench mark.

A **turning point**, or **TP**, is an elevation point *temporarily* used during the differential leveling operation. Point *B*, in Figs. 3.7 and 3.8, is an example of a turning point.

The elevation of the horizontal line of sight in Figs. 3.7 and 3.8 is called the **height of instrument**, or **HI**. In Fig. 3.8, HI = 248.45 + 4.28 = 247.32 ft.

A reading taken over a point of known elevation is a **backsight**, or **BS**. In Fig. 3.8, the 4.28-ft reading over point *A* is a backsight. A reading taken over a new point to establish its elevation is called a **foresight**, or **FS**. in Fig. 3.8, the 5.41-ft sight over *B* is a foresight.

Some texts designate a backsight as a plus (+) sight and a foresight as a minus (−) sight. This can be confusing, however, in those cases where the backsight point and/or the foresight point lies above the HI rather than below. For example, in leveling during construction of a tunnel, the tunnel floor is often covered with rubble while the tunnel ceiling has a number of clean, convenient rock outcrops that would serve as ideal points against which the base of the leveling rod may be set. Similarly, the underside of a beam on an overpass can be a convenient location for a turning point

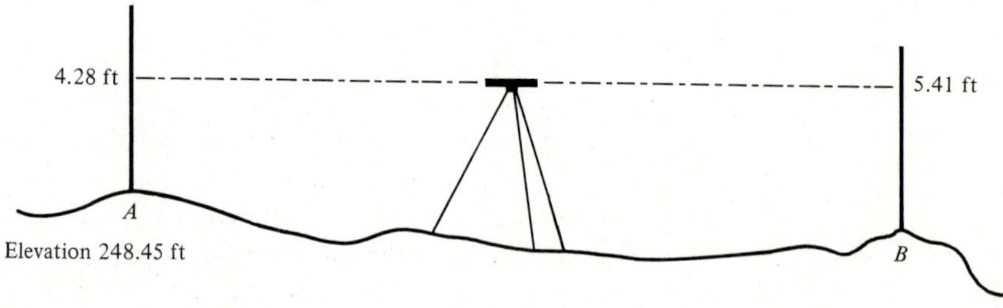

FIGURE 3.8
Carrying elevations between ground points.

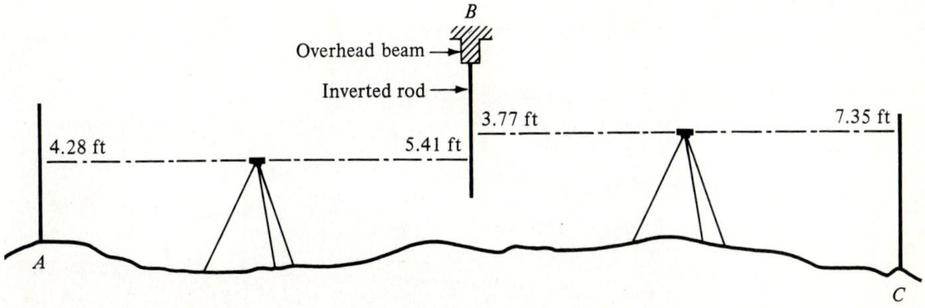

FIGURE 3.9
Differential leveling with an elevation point lying above two horizontal lines of sight.

or a bench mark. In all such cases, the rod is held inverted; that is, with the rod zero against the overhead point of elevation as diagrammed in Fig. 3.9.

Figure 3.9 shows point *A* as before, point *B* lying above the HI, and an additional elevation point *C*, which lies below the HI. The reading on the inverted rod held on *B* is still 5.41 ft but in this case, since the point lies above the HI, the elevation at *B* is equal to the HI elevation *plus* the rod reading on *B*; that is, 247.32 ft + 5.41 ft = 252.73 ft. Similarly, the 3.77-ft backsight reading on point *B*, taken from the second instrument position, must be *subtracted* from the elevation of *B* to obtain the second HI; that is, 252.73 − 3.77 = 248.96 ft. The elevation of point *C* is then obtained in the conventional way; that is, $\text{Elev}_C = \text{Elev}_{\text{HI}} - \text{FS} = 248.96 - 7.35 = 241.61$ ft.

3.5 DIFFERENTIAL LEVELING INSTRUMENTS AND FIELD METHODS

A level is a tripod-mounted telescope whose optical axis is perpendicular to a vertical reference, the latter being either a level vial or, in the case of a self-leveling instrument, a pendulum-mounted compensator. The telescope m has two basic features: (1) it contains a mechanism for bringing the telescope axis to horizontal and (2) it permits rotation about the vertical axis so that the line of sight may be pointed in the desired horizontal direction.

ENGINEER'S LEVEL. The simplest form of optical level is the engineer's level, which combines a telescope with a long, sensitive level vial. The vial and telescope axes are maintained parallel to one another so that the act of bringing the level bubble to center also brings the telescope optical axis to the horizontal. The two best-known engineer's levels are the wye, now rarely seen, and the dumpy, which is still made and sold in the United States. Figure 3.10 shows a surveying student peering through the telescope eyepiece of a dumpy level set up on a leveling tripod.

INSTRUMENT TELESCOPES. Whether on a level, a transit, a theodolite, or a total station, instrument telescopes are similar in design and function. Each has a set of internal cross hairs (Fig. 3.11) and two separate focusing mechanisms, one on the eyepiece and one on the telescope barrel. The first brings the cross hairs into sharp focus. The second focuses the field of view of the observer—the level rod, for example.

FIGURE 3.10
Engineer's level in use.

Focusing of the telescope should be done in the order indicated above: cross hairs first, then field of view. After initial focusing, the observer should always check by moving the eye slightly from side to side or up and down. If the cross hairs remain stationary on the field of view, the instrument is in correct focus. If the cross hairs move—a condition known as **parallax**—refine the focus, cross hairs first, until the cross hairs remain stationary when the eye is moved.

Assuming the telescope in Fig. 3.10 was properly focused and pointed, the pictured instrument observer saw sharply defined cross hairs superimposed on the face of a level rod.

LEVELING TRIPODS. Tripods are made with leg length either fixed or adjustable. Adjustable-leg tripods are convenient for a number of surveying uses, but fixed-leg tripods are more rigid and stable. An engineer's level is threaded directly to the tripod

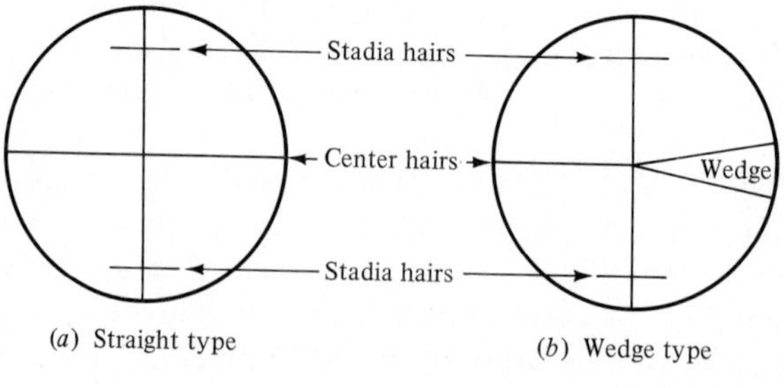

FIGURE 3.11
Common cross-hair styles.

head, the level having a female thread and the tripod having a male thread. Older engineer's levels and tripods have a variety of thread pitches. When working with such instruments, care must be taken that the tripod head and the level base match one another.

LEVELING RODS. There are a number of leveling rods available. The Philadelphia rod, in almost universal use today, is shown in Figs. 3.12 through 3.15. Figure 3.12 shows the Philadelphia rod by itself.

In Fig. 3.13, the rodwoman is facing the rod toward the instrument observer, keeping it upright by lining it up with whatever vertical structure may be visible to her in the background behind the observer; for example, a building, tower, or something of that nature. If there is no such vertical reference available, the observer will use hand signals to line the rod up with the level's vertical cross hair.

Although the above methods will work to make the rod vertical with respect to the telescope field of view, it is more difficult to hold the rod perpendicular to the

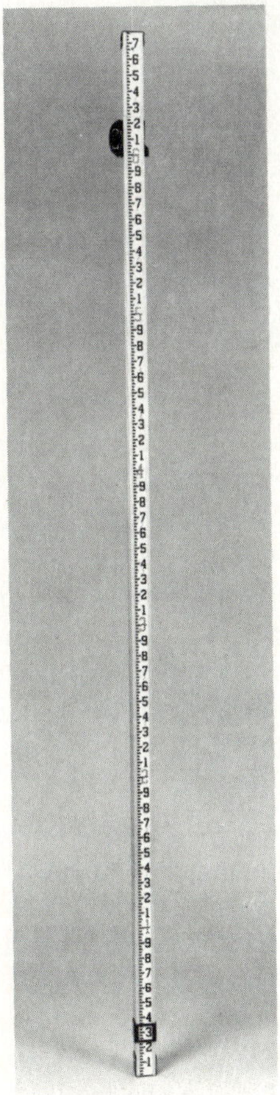

FIGURE 3.12
Philadelphia rod. (*Courtesy of Chicago Steel Tape Co.*)

FIGURE 3.13
Vertically held Philadelphia rod.

horizontal line of sight at the same time. An older method involves the observer's directing the rodperson to tilt the rod slowly toward, then away from the instrument. The level observer continuously monitors the rod reading at the point where the instrument's center horizontal cross hair intersects the face of the rod; when that reading is a minimum, the rod face is vertical. This tedious, time-consuming process can be avoided by using a rod level, seen in Fig. 3.14. The level bracket is held firmly against the rod, as shown, and the rod is manipulated to center the bubble within the bull's-eye, which brings the rod to vertical. However, since a bull's-eye level breaks readily when dropped on a hard surface, the need for rocking the rod can arise unexpectedly.

Unlike a surveyor's tape, a level rod face is marked continuously along its full length. On a Philadelphia rod, large, red numerals designate whole feet while small, black numerals designate tenths of feet. The alternate black bars and white spaces between tenth-of-a-foot marks are each one hundredth of a foot wide. Figure 3.15 (left) illustrates the manner in which a Philadelphia rod face is marked. Tenth-of-a-foot points and odd five-hundredths-of-a-foot points are indicated by projecting angled cuts on somewhat longer black bars. The exact tenth-of-a-foot point is designated by a projection on *top* of the adjacent black bar while the exact five-hundredths-of-a-

FIGURE 3.14
Holding a rod vertical using a rod level.

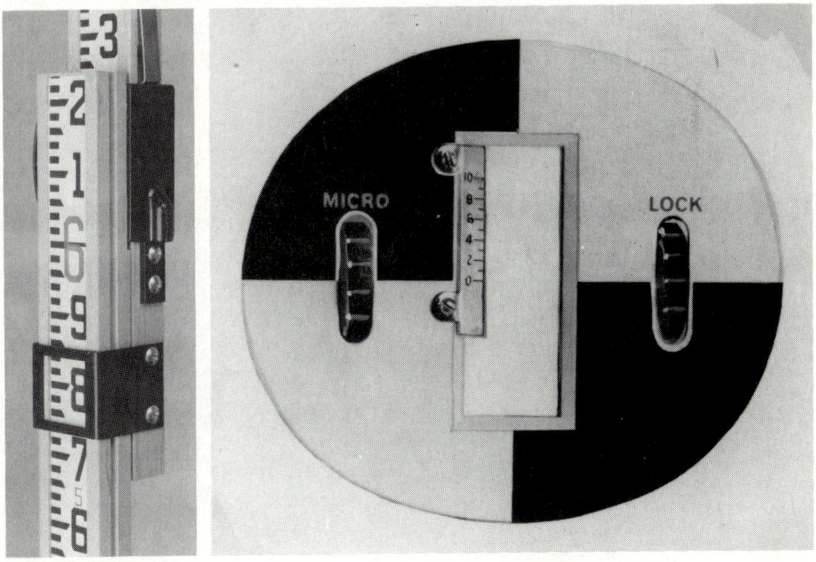

FIGURE 3.15
Close-up of Philadelphia rod and rod target.

foot point is designated by a projection on the bottom of its corresponding adjacent black bar. Other exact hundredth-of-a-foot marks are designated by either the top or bottom edge of shorter, squared-off black bars. This complex-seeming reading method is quickly learned. If the rod is close enough to the observer—say, within 50 ft or so—the rod reading may be interpolated to thousandths of a foot.

DUMPY LEVEL. Figure 3.16 shows a dumpy level by itself, close up. The upwardly convex level vial contains an air bubble that moves as the vial is tilted. The vial's top surface is scribed with lines spaced a constant distance apart. The sensitivity of the vial, defined as the angular distance between lines, is dependent on vial radius as well as the linear spacing between lines; that is, the larger the vial radius, the greater the movement of the bubble for a given amount of tilt, and, therefore, the greater the sensitivity. The dumpy level in Fig. 3.16 has a bubble sensitivity of 20 seconds, which means that tilting the telescope 20 seconds of arc will move the bubble a full division between graduations. Consider another vial with exactly eight graduations per inch (0.125 in. per graduation) and a radius of curvature of 72 ft (864 in.). The angle, in radians, subtended by a single division on the vial is equal to the arc distance between graduations divided by the radius, or

$$\text{Sensitivity} = \frac{0.125 \text{ in.}}{864 \text{ in.}} = 0.0001447 \text{ rad}$$

Dividing by 0.000004848 rad per second:

$$\frac{0.0001447 \text{ rad}}{0.000004848 \text{ rad}\cdot\text{sec}^{-1}} = 29.84 \text{ seconds} \cong 30 \text{ seconds}$$

The bubble sensitivity may be found easily by another method if the vial radius is not known. Simply set up and level the instrument, turn the telescope until it lines up with an opposing pair of leveling screws, then position a level rod at a measured

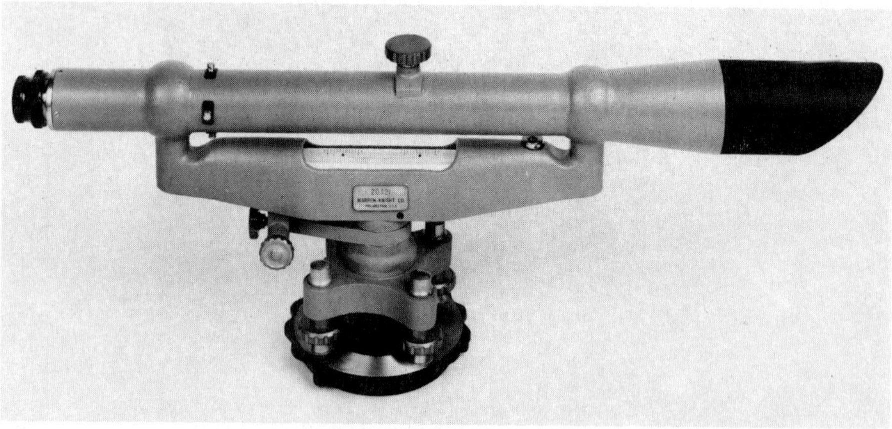

FIGURE 3.16
Dumpy Level. (*Courtesy of Warren-Knight Instrument Co.*)

distance from the instrument center and on line with the telescope axis. Using the pair of leveling screws on line with the telescope axis, move the bubble until one end exactly coincides with the furthest line on one side of the bubble vial. Read the rod. Again using the same pair of leveling screws, move the bubble to the nearest line on the same side of the bubble vial as the first line. Read the rod again. Compute as follows:

Example 3.8. Sighting on a rod 62.00 ft away, with the edge of the level bubble coincident with a vial marking, the rod reading is 4.687 ft. After tilting the level through exactly three vial divisions, the rod reading is 4.705 ft. What is the bubble sensitivity?

Solution. Sensitivity is the angular arc for *one* division. Angular arc in radians for the *three* divisions in the example is the difference in rod reading divided by the distance to the rod from the telescope center, or

$$\text{Angular arc (radians)} = \frac{4.705 - 4.687}{62.00} = \frac{0.018}{62.00} = 0.0002903 \text{ rad}$$

$$= \frac{0.0002903 \text{ rad}}{0.000004848 \text{ rad}\cdot\text{sec}^{-1}} = 59.89 \text{ seconds}$$

For one division,

$$\text{Angular arc} = 59.89/3 \text{ divisions}$$

$$\cong 20 \text{ seconds sensitivity}$$

The dumpy level (Fig. 3.16) is adjusted with the single capstan screw to the right of the bubble vial and four smaller capstan screws next to the left telescope support. The large capstan adjusts the bubble to center; the top and bottom screws in the set of four smaller capstans adjust the telescope cross-hair reticle up and down, and thus adjust the telescope optical axis parallel to the vial axis.

Unfortunately, few levels ever approach perfect adjustment; that is, (1) when the bubble is perfectly centered, the vertical axis of the instrument is unlikely to be perfectly vertical, and (2) when the vertical axis of the instrument is truly vertical, the telescope optical axis is unlikely to be perfectly horizontal. The first problem can be compensated for when the instrument is being leveled, in the manner discussed below. The second is compensated during the differential leveling process by properly positioning the level setup between rod setup points—the procedure is discussed at the end of this section.

There is a simple way of bringing the vertical axis of the engineer's level to vertical even though the bubble is not in perfect adjustment. To do this, simply mount the level on a tripod and make sure that the leveling screws are snug, but not tight. This maintains a desirably moderate tension on the ball-socket mount, which is located in the center of the four leveling screws. Unclamp the telescope and turn it so the bubble axis is parallel to an opposite pair of leveling screws. Facing the side of the telescope, center the bubble by turning the left-hand and right-hand leveling screws. To maintain steady tension on the ballsocket, *it is essential that the leveling screws be turned equal amounts, but in opposite directions*. The bubble moves in the same direction as the *left* thumb. When the bubble is centered with the first pair of leveling screws, turn the telescope clockwise 90° and bring the bubble to center in the same manner with the other pair of leveling screws. Return the telescope to its original position, readjusting the bubble to center, as necessary. Once the bubble is centered with the instrument turned to both positions, turn the telescope 180° from the original orientation and, using the original pair of leveling screws, bring the bubble exactly halfway to the center position. Turn the telescope 90° clockwise again and, using the second pair of leveling screws, repeat the procedure to bring the bubble halfway to center. Check by turning the telescope in 90° increments to make sure the bubble remains in the same offset position—when it does, the vertical axis will be actually vertical.

Even with the vertical axis vertical, the telescope axis will probably not be truly horizontal. However, it *will* be tilted a constant amount no matter what its directional orientation, a very desirable condition that can be readily compensated, as discussed further along.

Since the reason we go to all this trouble of leveling the instrument is to insure a constant tilt of the telescope axis for any direction the telescope may be pointed, it would seem there should be a better method for doing this. There is. A tilting dumpy level is leveled in a simpler manner that allows centering the bubble at each directional orientation of the telescope, thereby accomplishing the same end.

TILTING DUMPY LEVEL. A tilting level is given that name because the telescope and bubble vial may be tilted, teeter-totter fashion, on a horizontal axis perpendicular to the line of sight. Figure 3.17 shows a precise version of this instrument. In use, a tilting level is pointed in the direction of the rod and approximately leveled by using the three leveling screws to center a bull's-eye bubble (left center, Fig. 3.17). Using an auxiliary eyepiece, both ends of the bubble in the main level vial are viewed simultaneously, as seen in Fig. 3.18*a*. The adjusting knob (adjacent to the bull's-eye level in Fig. 3.17) is then used to bring the two ends of the bubble together, as seen in Fig. 3.18*b*, after which the rod reading is taken through the telescope optics in the

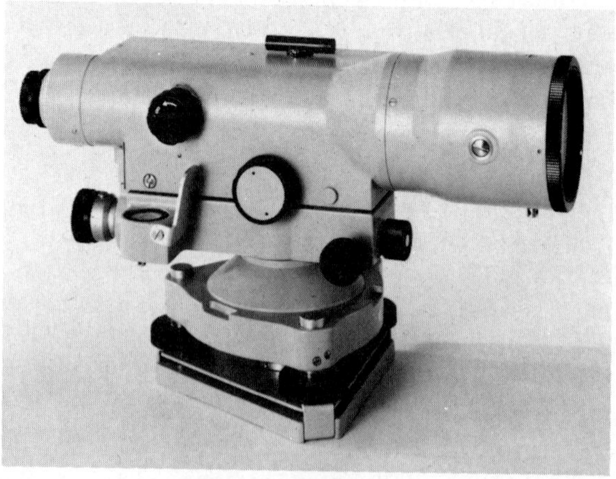

FIGURE 3.17
Sokkisha PL-1. (*Courtesy of The Lietz Co.*)

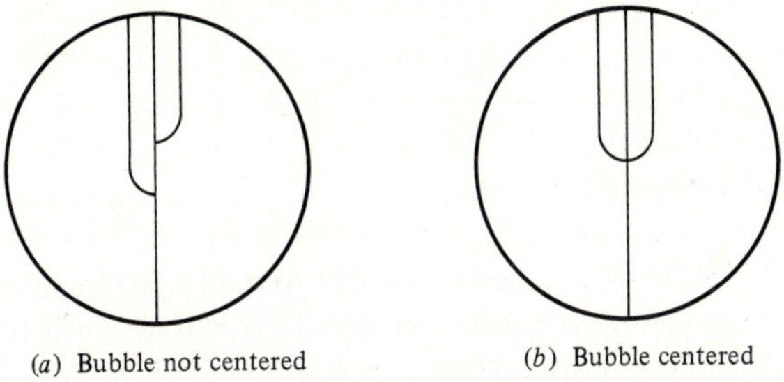

FIGURE 3.18
Tilting level coincidence bubble.

normal manner. Because the instrument's vertical axis is only approximately vertical, it is necessary for the telescope to be releveled for each rod reading from every instrument position. Even so, the single adjusting knob on a tilting dumpy affords a quicker and more reliable method of leveling the telescope optical axis than does the set of four leveling screws on the ordinary dumpy.

SELF-LEVELING LEVELS. The premise of leveling is that the telescope optical axis is horizontal. Because the telescope on a dumpy level is perpendicular to the telescope turning axis, the leveling procedure concentrates on making the turning axis vertical prior to taking rod readings. The tilting level takes a more direct approach in that the telescope itself is leveled just before a rod reading is taken. In both cases, however, the user must perform the final leveling manually.

The self-leveling level, on the other hand, is just that: An internal, pendulum-mounted mirror or prism compensates for a modest amount of tilt in the telescope, bringing the line of sight to horizontal when the instrument is approximately leveled. Since it takes time and a certain amount of expertise to do the leveling procedure on either an ordinary dumpy level or a tilting dumpy level, having the leveling done automatically can speed the surveyor's job significantly and/or allow the use of less-skilled (lower-paid) people on leveling jobs.

Figure 3.19 is a simplified schematic illustrating how the pendulum compensator works, the tilt angle in the figure being considerably exaggerated to demonstrate the

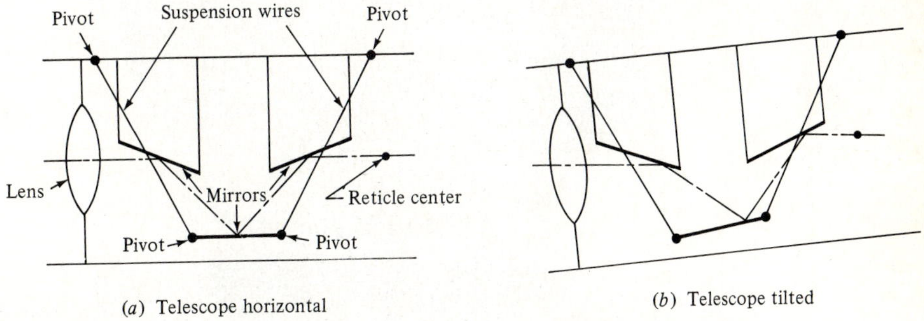

FIGURE 3.19
Self-leveling compensator.

principle. The support linkage for the pendulum mirror is designed to tilt the mirror surface exactly twice the amount the telescope barrel is tilted. The end result is a line of sight through the objective lens that is horizontal when the telescope barrel itself is tilted.

The self-leveling level is far and away the most popular instrument in current use. Since its widest use is for construction work, most models sold today incorporate a horizontal circle that makes it possible to read horizontal directions from the setup position at the same time that differential elevations are being determined with significantly greater precision. Figures 3.20 through 3.23 show four such levels that are relatively inexpensive, reasonably precise, and have the directional feature as well. Their advertised accuracies range from ±1.5 mm to ±5 mm in 1 km of double-run leveling and all four provide one-degree divisions on the horizontal circle.

To operate the self-leveling level, position the tripod, set up the instrument, and approximately level it by bringing the bubble of the circular level (below and left of the instrument nametag in Fig. 3.20) within the bull's-eye. By so doing, the compensator pendulum—whose working range is greater than that of the bull's-eye—is freed to swing and thus bring the line of sight horizontal. Most of the instruments shown in Figs. 3.20 through 3.23 also incorporate mirrors that permit the user to view the circular level without having to move away from the eyepiece end of the telescope.

Figure 3.24 shows a similar automatic level with somewhat greater precision. It provides a 1-km, double-run leveling accuracy of ±1.0 mm (±0.4 mm with a micrometer reading attachment) and scale markings at 10-minute intervals on the horizontal circle.

Although self-leveling levels aren't yet up to the precision of tilting levels, the best are fairly close. Figure 3.25 is just one example of a high-precision automatic level. Its advertised accuracy is ±0.3 mm in 1 km of double-run leveling when used with a parallel-plate micrometer attachment. This compares quite favorably with the

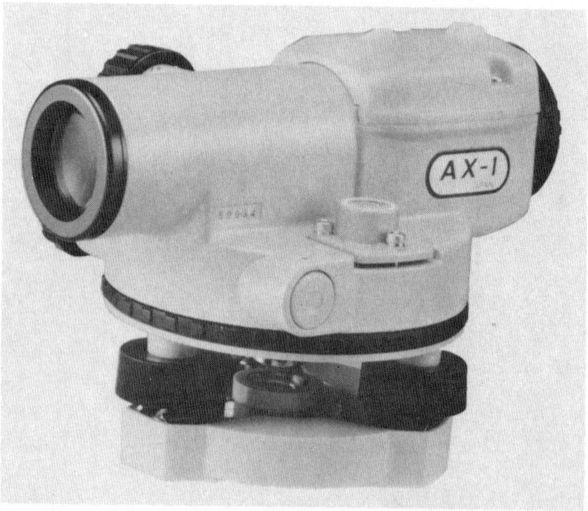

FIGURE 3.20
Nikon AX-1 automatic level. (*Courtesy of Nikon, Inc., Surveying Instrument Group*).

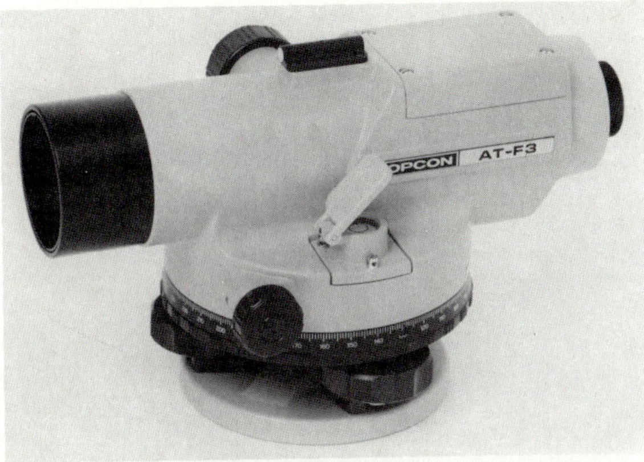

FIGURE 3.21
Topcon AT-F3 automatic level. (*Courtesy of Topcon Instrument Corporation of America*).

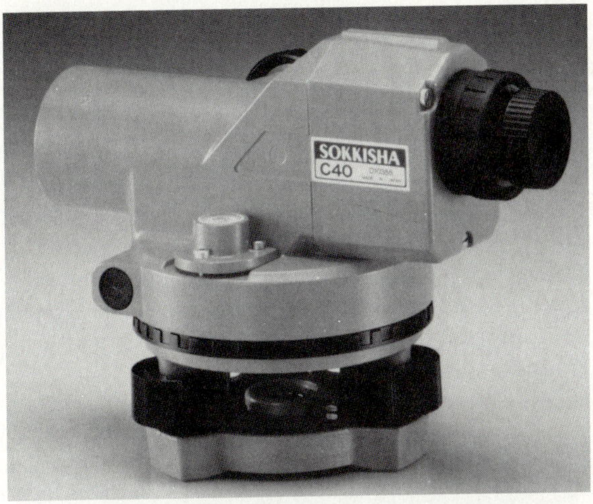

FIGURE 3.22
Sokkisha C40 automatic level. (*Courtesy of The Lietz Co.*)

FIGURE 3.23
Wild NA28 automatic level. (*Courtesy of Wild Leitz, Ltd.*)

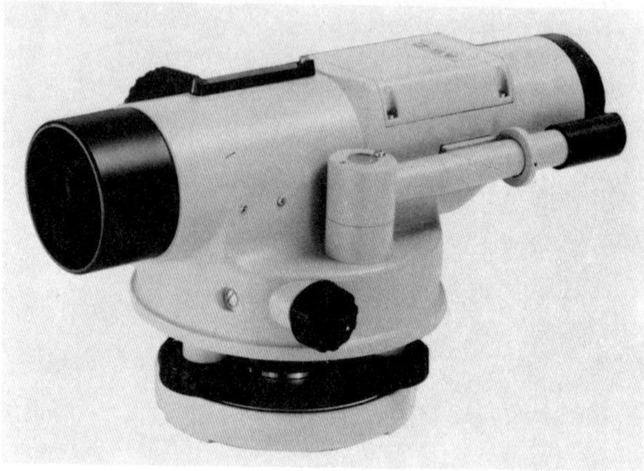

FIGURE 3.24
Nikon AS-C automatic level. *(Courtesy of Nikon, Inc., Surveying Instrument Division.)*

tilting level shown in Fig. 3.17, a first-order instrument with a built-in parallel-plate micrometer that has an advertised accuracy of ±0.2 mm for 1 km of double-run leveling.

A parallel-plate micrometer eliminates interpolation in rod reading by allowing the user to raise or lower the instrument's line of sight (while maintaining it at the horizontal) thus moving the image of the horizontal cross hair a closely measured distance up or down the rod face in order to align it with the sharp edge of an adjacent rod marking. Wedge-type cross hairs similar to Fig. 3.11*b* allow exceedingly precise alignment. The result is a very significant increase in rod-reading accuracy when using geodetic level rods—to a tenth of a millimeter, equivalent to about $\frac{1}{30}$ of a 0.01-ft division on an ordinary Philadelphia rod.

BALANCING BACKSIGHT AND FORESIGHT DISTANCES. As discussed earlier, there's no guarantee that the optical axis of a level telescope will be truly horizontal just because the dumpy level's vertical axis is truly vertical, the tilting level's main bubble is truly centered, or the self-leveling level's pendulum compensator is working

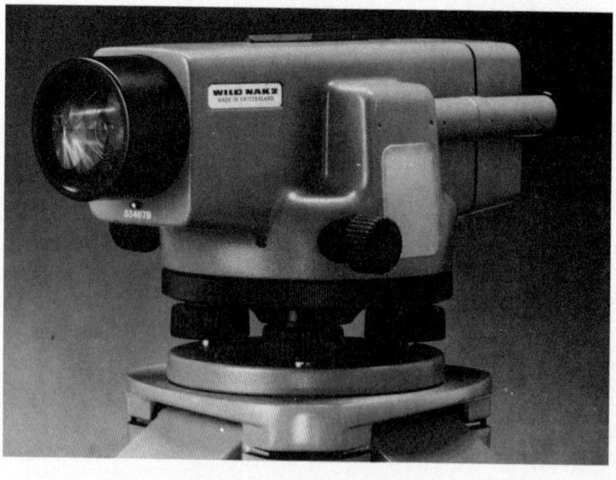

FIGURE 3.25
Wild NAK2 universal automatic level. *(Courtesy of Wild Leitz, Ltd.)*

properly. To some extent, the telescope axis will always deviate from a true horizontal to a lesser or greater degree. Errors due to an inclined telescope are compensated by balancing the horizontal backsight distance with the horizontal foresight distance. Assume, for the moment, that a level telescope axis tilts upward 30 seconds when the instrument is otherwise properly leveled. For a 150-ft backsight distance, error in cross-hair position is 150 tan $00°00'30''$, or a bit over 0.02 ft, so the rod reading would be 0.02 ft higher than the correct reading for a horizontal line of sight. For a 150-ft foresight distance with this instrument, properly leveled, the rod reading would also be 0.02 ft higher because of the upward tilt of the optical axis. Since backsights are additive and foresights subtractive, the 0.02-ft error would cancel out in the computations.

On the other hand, if the backsight distances were always 150 ft and the foresight distances were always 75 ft when using the above instrument, the backsight error of 0.02 ft would *not* be compensated because in this case the foresight error would be just 0.01 ft. There would thus be a cumulative positive error of 0.01 ft per level setup.

For ordinary leveling, even when the surveyor has reason to believe the instrument is in good adjustment, it is always good practice to maintain backsight and foresight distances about the same. Pacing is sufficiently accurate for this measurement, even though it involves the pace lengths of two different people, the instrument observer and the rodperson. Another advantage of this procedure is that it avoids cumulative curvature and refraction errors as well.

For precise leveling, backsight and foresight distances must be measured and leveling computations must be adjusted for cumulative backsight-foresight imbalance. Actually, distances may be quickly and easily measured using the **stadia hairs** on the level's cross-hair reticle, as seen in Fig. 3.11*a* and *b*. The procedure for doing this is discussed in Sec. 3.8.

3.6 DIFFERENTIAL LEVELING PROCEDURE AND COMPUTATIONS

POINT-TO-POINT LEVELING PROCEDURE. Figure 3.26 shows a line of levels that starts from existing bench mark *A* and progresses to new bench mark *B* for the purpose of establishing the elevation of *B*. The distance and the elevation difference

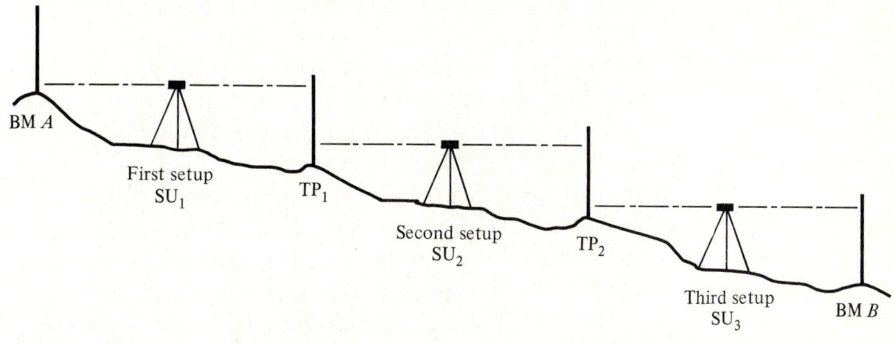

FIGURE 3.26
Differential-level line from BM *A* to BM *B*.

between A and B necessitate three instrument setups (SU_1, SU_2, and SU_3) and two turning points (TP_1 and TP_2).

In Sec. 3.4, a turning point was defined as a temporary elevation used during differential leveling. A TP must be a stable object with a definable high point on which the rod base is placed; for example, a rock outcrop, the top of a hydrant, a high point on a curbing, or a firmly fixed sprinkler head. Sometimes the area being surveyed affords a wide selection of good potential TP sites; sometimes there are few or none. The rodperson must be prepared to establish a TP in a place where no candidate TPs exist. The usual method is to carry a large spike or a hatchet and drive it firmly into the ground to use as a TP. The drawback is that the spike, or hatchet, which must be used again, is knocked out of the ground when the rodperson moves to the next TP; thus, it will no longer be possible to return to that TP.

To begin the line of levels shown in Fig. 3.26, the location for TP_1 is chosen at a convenient point along the intended route from BM A to BM B and the instrument is set up about halfway between at SU_1. With the rod held over BM A, the first backsight is taken and the backsight rod reading is recorded. The instrument person then releases the rodperson to move to TP_1, after which the first foresight reading is taken and recorded. Once the foresight reading is safely in the field book—which establishes the elevation of TP_1—the level is moved to SU_2, halfway between TP_1 and TP_2, where it is set up as before. A backsight rod reading is then taken on TP_1 and recorded. The rodperson having been directed to go to TP_2, a foresight is taken on TP_2 and recorded. With the foresight reading in the field book, the instrument is moved to SU_3, halfway between TP_2 and BM B. The process is repeated, taking and recording, in order, the backsight on TP_2 and the concluding foresight on BM B.

If the instrument person should inadvertently move the instrument before a foresight reading is taken, the original height-of-instrument for the setup is lost and the backsight reading just taken on the previous TP is no longer valid. The instrument *must* be set up again, the rodperson *must* return to the previous TP, and the process *must* begin again for that setup. Moral of the story: Be absolutely certain that a proper foresight reading has been made and recorded before picking up the instrument and moving to the next setup position. If the previous TP was a spike or hatchet head that had already been moved, the level party must return to the last reusable TP (a rock outcrop, say) and begin again, rejecting all subsequent rod readings, including the backsight taken on the rod held over the rock outcrop.

NOTEKEEPING FOR POINT-TO-POINT LEVELS. The standard field book, opened up, displays a left-hand horizontal-ruled page that is divided into six columns and a right-hand horizontal-ruled page that is divided into two columns. Field observations are recorded on the left-hand page while the right-hand page is used to record pertinent data about the area being surveyed, date, weather conditions, names and duties of survey party members, and a listing of the instruments being used. With ordinary differential levels, the left-hand page is used to record, in order, the following six columns of data: names of elevation points (such as BM A, TP_1), designated station or STA; backsight rod readings (BS); heights of instrument (HI); foresight rod readings (FS); elevations of the elevation points (ELEV); and backsight and foresight *distances*

(DIST), listed in feet or paces. For the short line of levels shown in Fig. 3.26, the left-hand page recording begins with the columns headed as shown, BM *A* entered as the first station, and its known elevation entered on the same line as shown:

STA	BS, ft	HI, ft	FS, ft	ELEV, ft	DIST, ft
BM *A*				**438.62**	

With the level set up at SU_1, the 0.65-ft backsight reading on BM *A* is recorded under BS and the paced distance from BM *A* to the instrument, translated into feet, is recorded under DIST. The HI elevation (HI = BM *A* + BS = 438.62 + 0.65 = 439.27) is recorded under HI as shown:

STA	BS, ft	HI, ft	FS, ft	ELEV, ft	DIST, ft
BM *A*	**0.65**	**439.27**		438.62	**150**

The foresight is now taken on TP_1, the 6.43-ft rod reading is recorded *on the next line* under FS, and distance from instrument to TP_1 is entered after a slash. The elevation of TP_1 (TP_1 = HI − FS, = 439.27 − 6.43 = 432.84) is entered on the same line as shown:

STA	BS, ft	HI, ft	FS, ft	ELEV, ft	DIST, ft
BM *A*	0.65	439.27		438.62	150
TP_1			**6.43**	**432.84**	**/153**

With the recording completed for the first setup, the instrument is moved to the second setup and the 1.66-ft backsight rod reading from the second setup is recorded as shown below, along with the distance from TP_1 to the second setup. The HI of the second setup is computed as before (TP_1 + BS = HI = 432.84 + 1.66 = 434.50).

STA	BS, ft	HI, ft	FS, ft	ELEV, ft	DIST, ft
BM *A*	0.65	439.27		438.62	150
TP_1	**1.66**	**434.50**	6.43	432.84	**100**/153

The foresight rod reading over TP_2 and the instrument-to-TP_2 distance are entered on the next line as shown below, along with the TP_2 elevation (TP_2 = HI − FS = 434.50 − 5.95 = 428.55).

STA	BS, ft	HI, ft	FS, ft	ELEV, ft	DIST, ft
BM *A*	0.65	439.27		438.62	150
TP_1	1.66	434.50	6.43	432.84	100/153
TP_2			**5.95**	**428.55**	**/97**

Entries for the third instrument setup are continued in the same manner. The completed page is shown below in Fig. 3.27.

The elevation of BM *B* is determined twice: the first as a result of the sequential procedure; the second as a check, shown on the last three lines of Fig. 3.27. The second computation of BM *B* was obtained by adding the sum of the backsights and the sum of the foresights to the elevation of BM *A*. The backsight sum is $(+0.65)+(+1.66)+(+0.83) = +3.14$ ft; the foresight sum is $(-6.43)+(-5.95)+(-6.22) = -18.60$ ft. Now, BM B = BM $A+\Sigma$ BS $+\Sigma$ FS $= 438.62+3.14-18.60 = 423.16$ ft, checking the first value. Note that all backsights are positive and all foresights negative, because elevation points in the example are normal ground points. Keeping track of backsight and foresight distances in the last column helps to maintain their cumulative sums in reasonable balance, a necessity if leveling errors are to be minimized. These errors are discussed in Sec. 3.7.

Consider now a situation with an elevation point overhead rather than on the ground. A foresight on an overhead point would be positive, rather than negative, and a backsight on an overhead point would be negative, rather than positive. Figure 3.28 shows a single-run line of levels with one intermediate point overhead. The rod is held inverted against the overhead point for each sight taken on that point. The foresight rod reading on PT 1 (Fig. 3.28) is designated (+)3.81 rather than simply 3.81, to indicate that the value must be treated differently than in the usual case with a ground point. The backsight on PT 1 is designated (−)6.92 for the same reason. The overall treatment of the computation is shown in Fig. 3.29.

STA	BS, ft	HI, ft	FS, ft	ELEV, ft	DIST, ft
BM *A*	0.65	439.27		438.62	150
TP_1	1.66	434.50	6.43	432.84	100/153
TP_2	0.83	429.28	5.95	428.55	124/97
BM B			6.22	423.16	/122
ΣBS	+3.14	ΣFS	−18.60		374/372
	−18.60				
	−15.46 +438.62 = 423.16 ✓				

FIGURE 3.27
Differential-level notes–single run.

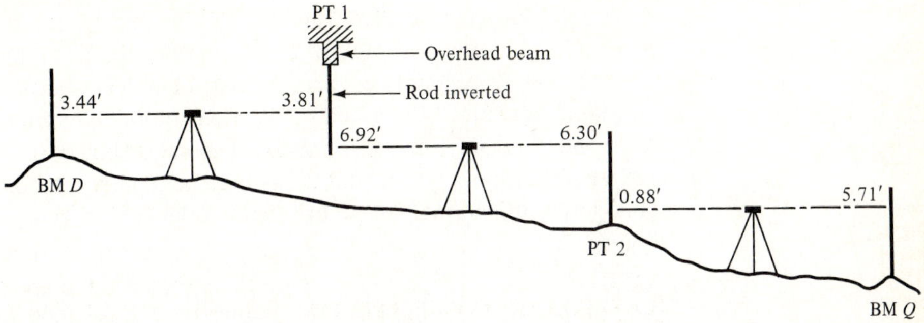

FIGURE 3.28
Differential-level line with one overhead point.

Note that the elevation of PT 1 was computed by *adding* the foresight on PT 1 to the HI of the first instrument setup; that is, PT 1 = 219.51 + 3.81 = 223.32 ft. Similarly, the HI of the second setup was obtained by *subtracting* the backsight on PT 1 from the elevation of PT 1; that is, HI = 223.32 − 6.92 = 216.40 ft. Except for the check, the remainder of the computation was carried out in the usual manner.

In computing the check value for the BM *Q* elevation, the sums of the foresights and backsights take the parenthesized algebraic signs into account. Thus, Σ BS = (+3.44) + (−6.92) + (+0.88) = −2.60 ft and Σ FS = (+3.81) + (−6.30) + (−5.71) = −8.20 ft. The check value for BM *Q* is still equal to BM *D* + Σ BS + Σ FS = 216.07 − 2.60 − 8.20 = 205.27 ft, which checks the initial computation.

LOOP LEVELING PROCEDURE. Although it is gratifying to find the check values in Figs. 3.27 and 3.29 in agreement with the initial values for the final elevations, the checks only verify the arithmetic. Any backsight or foresight reading can be in error by a significant amount and that error would go undetected if a line of levels is run in just one direction. To obtain a true check on the accuracy of the work, the line must be run in both directions, returning to the point of origin. The cumulative error,

STA	BS, ft	HI, ft	FS, ft	ELEV, ft	DIST, ft
BM D	3.44	219.51		216.07	122
PT 1	(−)6.92	216.40	(+)3.81	223.32	104/124
PT 2	0.88	210.98	6.30	210.10	110/99
BM Q			5.71	205.27	/114
ΣBS	−2.60	ΣFS	−8.20		336/337

−8.20
−10.80 +216.07 = 205.27 ✓

FIGURE 3.29
Differential-level notes, single run, one overhead point.

known as **closure**, will then equal the difference between the actual elevation of the point of origin and its computed elevation after the level line is run out and back.

Rather than simply retrace the line of levels—particularly when intermediate elevations are being established—it is often helpful to return to the origin by a different route and thus establish another series of elevation points at convenient locations. With proper planning, auxiliary elevations may be readily established throughout a project area during the normal course of running a level loop between two primary elevation points.

LEVEL-LOOP COMPUTATIONS. Consider a level loop that starts from a known elevation at BM 1, establishes two intermediate points enroute to new BM 2, then establishes three more intermediate points on the return to BM 1 along a different path. All elevation points are normal ground points; thus, no special notation is needed for backsight and foresight rod readings, all backsight readings are treated as positive values while all foresight rod readings are treated as negative values. The computational method shown in Fig. 3.30 is basically similar to those shown in Figs. 3.27 and 3.29. Note that horizontal sight distances are given in paces in this example.

As before, the arithmetic check simply shows that the sequence of addition and subtraction was correctly done or that two or more compensating errors have been made. It is far more likely, however, that the arithmetic is correct—two or more compensating errors that add to zero is a most unlikely occurrence. The fact that the out-and-back elevation difference is the same in the two computations is an indication that the arithmetic is correct. The fact that that difference is only 0.01 ft is an indication that the random errors involved in the leveling procedure were quite small or there were compensating large errors. Again, the former is significantly more likely than the latter.

STA	BS, ft	HI, ft	FS, ft	ELEV, ft	DIST, paces
BM 1	5.41	253.86		248.45	35
PT A	9.37	259.95	3.28	250.58	33/34
PT B	8.93	266.67	2.21	257.74	30/34
BM 2	9.79	275.02	1.44	265.23	38/29
PT C	6.10	279.54	1.58	273.44	33/40
PT D	3.43	270.95	12.02	267.52	35/34
PT E	1.03	259.50	12.48	258.47	33/33
BM 1			11.04	248.46	/34
ΣBS	+44.06	ΣFS	−44.05		237/238

$$\begin{array}{l} -44.05 \\ \hline +\ 0.01 + 248.45 = 248.46\ \checkmark \end{array}$$

FIGURE 3.30
Differential-level notes for level-loop circuit.

Taking the 0.01-ft difference at face value, the question then arises as to how good that result might be. To rate a 0.01-ft result, though, one needs to evaluate the complexity of the survey. It is one thing to obtain a 0.01-ft out-and-back difference for four rod readings—one pair out, one pair back; it is quite another to close that well when a large number of setups are involved. In "scoring" a survey closure, two factors are involved: the way the score is determined and the standards used in evaluating that score.

CLOSURE ERROR. Every sight taken during a level survey provides an opportunity for a small, random error of some sort. Some errors cancel and some don't; to some extent, though, errors will tend to accumulate. The amount of cumulative random error likely to occur on any survey depends on the inherent precision of the instruments used, the care with which observations are made, and the total number of observations taken. For a survey involving repeated observations of a single type, the anticipated cumulative error is proportional to the square root of the number of observations. In a level survey, with two sights per setup, the cumulative error is also proportional to the square root of the number of setups. Finally, since sight distances tend to average out, cumulative error for a level survey should be proportional to the square root of the distance run. This can be expressed mathematically as $C \approx \sqrt{d}$, where C is **closure error**; that is, the expected difference between the elevation of the origin and its computed elevation at the end of an out-and-back level survey. The value d is the round-trip distance and $\approx$ means "is proportional to." If the $\approx$ sign is to be legitimately replaced by an equal sign, a constant of proportionality must be inserted to the right of the equal sign, that constant being a function of the distance units. This leads to separate constants and separate notations for metric and English units as follows:

$$C_{\text{mm}} = m\sqrt{K} \tag{3.12a}$$

$$C_{\text{ft}} = f\sqrt{M} \tag{3.12b}$$

where C_{mm} is maximum allowable closure error in millimeters for a double-run level section whose round-trip distance is K kilometers and C_{ft} is maximum allowable closure error in feet for a similar section M miles long. Values for m and f are a function of the instruments and procedures used as well as the units of distance; that is, for precise surveys run to close tolerances, m and f would have smaller values than would be the case for ordinary surveys.

STANDARDS OF ACCURACY FOR CLOSURES. The National Geodetic Survey (NGS) of the U.S. Department of Commerce has published national standards for maximum allowable closures on surveys classified to three different categories of accuracy: first order, second order, and third order. In NGS terminology, first-order surveys are made with very precise instruments that incorporate parallel-plate micrometers observing on metric rods with Invar scales. First-order work is done almost exclusively by government survey parties for the purpose of establishing and extending the primary vertical control network in the United States. Second-order surveys use the same leveling equipment as first-order. They are mostly government surveys as well, and are usually run to establish secondary control within the primary network, thus providing convenient bench marks at reasonable intervals to support local surveys of lower-order accuracy. Second-order surveys are run for "densification" within large metropolitan areas—that is, to increase the number of bench marks within a given area—and on major engineering projects to support studies of crustal movement and subsidence. Discussion of first- and second-order surveying procedures is not within the scope of this text. Third-order standards, which may be achieved by local surveyors if reasonably precise leveling equipment and methods are used, is done mostly for municipal control and for small engineering and mapping projects.

Table 3.1 is a condensation of the National Geodetic Survey's published standards of accuracy for all three orders of level surveys.

TABLE 3.1
National Geodetic Survey standards for *m* and *f* Used for determining maximum allowable closure in first-, second-, and third-order surveys in the United States

	First order		Second order		Third order
	Class I	Class II	Class I	Class II	
m	3 mm	4 mm	6 mm	8 mm	12 mm
f		0.017 ft	0.025 ft	0.035 ft	0.050 ft

Example 3.9. What is the maximum allowable closure for a first-order, class II, double-run section 10 km long, one way? For a third-order, double-run section 10 miles long, one way?

Solution. In the first part, the value of m for a first-order, class II line (4 mm) is inserted into Eq. (3.12a) to produce

$$C_{\text{mm}} = m\sqrt{K} = 4\sqrt{20} = 17.9 \text{ mm}$$

To solve the second part, insert the third-order f value (0.050 ft) into Eq. (3.12b):

$$C_{\text{ft}} = f\sqrt{M} = 0.050\sqrt{20} = 0.224 \text{ ft}$$

The three-significant-figure answer for C_{mm} is valid for two reasons: First, first-order levels are run to an integer number of kilometers; second, the 4-mm value for m is an exact number. Besides, first-order rod readings are taken to better than millimeter precision, so maximum closure error should reflect this capability. By the same token, third-order rod readings to a thousandth of a foot are feasible with relatively short sights and reasonable care in reading the rod.

Example 3.10. An 8-km line of levels is double-run using first-order equipment and procedures. If the closure was 17 mm, to what order and class of accuracy was this line of levels run?

Solution. To determine order and class, compute the m value for this closure and distance, then compare with Table 3.1 values. Manipulating Eq. (3.12a),

$$C_{\text{mm}} = m\sqrt{K}$$

$$m = \frac{C_{\text{mm}}}{\sqrt{K}} = \frac{17}{\sqrt{16}} = 4.25 \text{ mm}$$

Considering the m values listed in Table 3.1, the above answer falls between the value for first-order, class II and second-order, class I. Therefore, since the value 4.25 mm is not up to the first-order, class II standard, the survey must have been done to the standards for second-order, class I.

Example 3.11. An 8.6-mile line of levels is double-run using equipment and methods just adequate for third-order standards, with rod readings taken to the nearest thousandth of a foot. If closure was 0.054 ft, to what order and class was the line of levels run?

Solution. To check the resultant f value:

$$f = \frac{c_f}{\sqrt{M}} = \frac{0.054}{\sqrt{(2)(8.6)}} = 0.013 \text{ ft}$$

Although the 0.013-ft is well up in the first-order range, the levels were run using third-order equipment and techniques; thus, the survey must be considered third order.

The fact that the value of f in the above problem seems to indicate first-order, class II accuracy or better should be considered a lucky accident resulting from compensating errors. *Third-order methods and equipment can produce no better than third-order results!*

ADJUSTMENT OF A LEVEL CIRCUIT. When a level loop is run, or when a line of levels originates at one point of known elevation and ends at a different point of known elevation, there is an automatic check on the amount of closure for that circuit or line. In either event, the closure thus determined may be adjusted by apportioning the closure value through a multisection line of levels in some logical manner.

As already discussed, cumulative leveling error varies as the square root of the number of rod readings; thus, average error per setup decreases as number of setups increases. However, once actual cumulative error in a line of levels has been determined, closure correction for each level section should be directly proportional to the number of setups between sections. Thus, the proportion of the cumulative error used to adjust the elevation of each elevation point along the circuit must be proportional to the cumulative number of setups to each point from the start of the circuit.

Example 3.12. A level loop is run from BM 44, of known elevation 382.194 ft, to TBMs A, B, and C and back to BM 44 with results as shown below. Find the adjusted elevations of the TBMs.

Station	Observed elev., ft	Setups
BM 44	382.194 (fixed)	
TBM *A*	416.208	11
TBM *B*	422.633	4
TBM *C*	397.045	8
BM 44	382.213	5

Solution. Total correction is found by subtracting the elevation of BM 44 at the end of the circuit from its fixed starting elevation. Because errors in observed elevation are assumed proportional to the cumulative number of setups along the level circuit, the corrections to be applied to the observed elevations are assumed proportional to the cumulative number of setups along the circuit. The cumulative number of setups to each station is entered in the fourth column (below), divided by the number of setups in the loop (fifth column, left) and multiplied by the total correction to determine the correction for each station (fifth column, right), with the adjusted elevations then entered in the sixth column.

Station	Observed elev., ft	Setups	Cumulative	Correction computations	Adj. elev.
BM 44	382.194				382.194
TBM *A*	416.208	11	11	$\frac{11}{28}(-0.019) = -0.007'$	416.201
TBM *B*	422.633	4	15	$\frac{15}{28}(-0.019) = -0.010'$	422.623
TBM *C*	397.045	8	23	$\frac{23}{28}(-0.019) = -0.016'$	397.029
BM 44	382.213 Corr. −0.019	5	28	$\frac{28}{28}(-0.019) = -0.019'$	382.194

3.7 LEVELING ERRORS

Surveying is done in a nonideal environment by ordinary people using imperfect equipment; thus, survey errors may be categorized as natural errors, personal errors, and instrument errors.

NATURAL ERRORS. Curvature and refraction, as discussed earlier, can cause errors of 0.005 ft over sight distances of 500 ft. For ordinary leveling, however, maximum sight distances are normally no greater than 150 ft, in which case the curvature and refraction effect for a single sight would be just under 0.0005 ft, which is negligible when reading to 0.01 ft. What errors might arise or accumulate from this factor are compensated by the simple expedient of keeping backsight and foresight distances reasonably balanced for each setup.

Circumstances sometimes require close-tolerance rod readings, very long sights, and significant, unavoidable imbalance between backsight and foresight distances. In such cases, corrections for curvature and refraction must be applied to individual rod readings. Because curvature and refraction brings the cross hair too high, rod readings must be *reduced* by the appropriate $(c \& r)$ correction value. Thus, $\text{Elev}_B = \text{Elev}_A + [h_A - (c \& r)_A] - [h_B - (c \& r)_B]$.

For first-order work, with rod readings taken to closer tolerances than ordinary leveling, sight distances may not exceed 60 m, backsight-foresight imbalance may be no greater than 5 m for a single setup, and cumulative imbalance may not exceed 10 m for a level section (between bench marks). These specifications hold the worst-case curvature and refraction error for a section of first-order levels to a bit under 0.1 mm.

On bright days, the sun's heat on one side of the telescope barrel can warp the optical line of sight. This is corrected by shading the instrument. Casual passersby are surprised to see an umbrella over the observer while other party members are out in the hot sun. Actually, the umbrella is there to shade the instrument rather than the observer. At the same time, that is an added benefit of being an observer rather than a rodperson or recorder.

Ground settlement and instability are two separate problems that create slightly different errors. Settlement error results from the ground's being pushed down more or less uniformly by the weight of the observer or the observing party. In general, the ground will settle slowly over the entire period observations are made from the setup; thus, the HI will be higher when the backsight is taken than when the foresight reading is made. The end result is a positive (uphill) bias to all pairs of readings in which settlement is a factor.

Instability results from a much more localized settlement, in which the ground will sink more under the weight of the observer than it will under the instrument; that is, it will behave to a certain extent like a trampoline. It becomes difficult to use an ordinary dumpy level where there is instability because the observer stands in a different position when sighting through the telescope than when leveling the instrument, causing the telescope to tilt while that sight is taken. Instability error occurs in thawing permafrost in polar latitudes but less frequently elsewhere. Instability is easily detected because the main bubble will move noticeably as the observer walks around the instrument. To counter this, the level party should attempt a different route where the soil may be firmer.

PERSONAL ERRORS. Rod-reading errors are often a factor, particularly with inexperienced instrument observers. Small reading errors tend to compensate. Large errors, including transposition of numbers as well as inadvertent reading of the wrong foot, tenth, or hundredth, do not compensate. Another reading error, which may compensate in the long run, is a personal bias on the part of some observers in favor of certain numbers or number combinations.

Parallax—movement of the cross hairs with respect to the field of view as the eye is moved up or down—is a common personal error usually resulting from the observer's being in too great a hurry. The error is easily eliminated by careful refocus of the cross hairs.

Errors caused by the rod being out of plumb are ordinarily the fault of the observer rather than the rodperson. When a properly functioning bull's-eye rod level is used, this error won't occur. With no rod level available, the observer must direct that the rod be waved slowly toward and away from the instrument, then take the minimum rod reading as the true reading.

Overconfidence on the part of the observer sometimes results in too-long sight distances with respect to telescope optics or weather conditions. The resultant blurring of the image of the rod face will make it very difficult to determine which rod marking is adjacent to the center horizontal cross hair.

INSTRUMENT ERROR. Leveling instruments, depending on their design and overall quality, are quite likely to be out of perfect adjustment to some extent. Such misadjustment makes the telescope optical axis inclined rather than horizontal. Figure 3.31 exaggerates the slope of the telescope optical axis to better illustrate the problem. Assuming the instrument is otherwise in adjustment, that is, the bubble of the dumpy or tilting level is centered or, for a self-leveling level, the compensator is functioning, the actual line of sight is tilted an angle α with respect to the horizontal. At the sight distance D, the error in sight reading due to the tilt of the telescope is $D \tan \alpha$.

Figure 3.32 shows a level setup with backsight rod position on the left and two foresight rod positions at the right; one the same distance, D, from the instrument as the backsight rod position, the other at a somewhat greater distance, D'. For the situation with foresight distance the same as backsight distance, the error is $D \tan \alpha$ in both cases; thus, errors cancel. Where the foresight distance is D', the error is $D' \tan \alpha$ and the *uncompensated* error difference in this case is $(D' - D) \tan \alpha$. Figure 3.32 also illustrates the effect of *cumulative* imbalance, where D and D' denote sums of backsight and foresight distances, respectively. Rather than compensate for a misadjusted bubble by balancing backsight and foresight distances at each setup, the same result is achieved by balancing backsight and foresight distance sums over an entire line, as illustrated earlier.

With ordinary levels, the error due to a modest amount of backsight-foresight imbalance may be tolerated unless the instrument is a good deal out of adjustment. In first-order work, a telescope axis tilted 10 seconds can cause an error of about 0.2 mm in the 10-m backsight-foresight imbalance permitted over a section of levels—twice the maximum error due to curvature and refraction. To minimize this problem, high-accuracy level procedures utilize a "collimation correction," in which the tilt of the telescope axis is determined by the observer and the resultant error compensated mathematically. Field notes for determining collimation correction are illustrated in Appendix B, Example 5. However, as already shown, when backsight and foresight distances are maintained in reasonable balance, collimation error is not a problem.

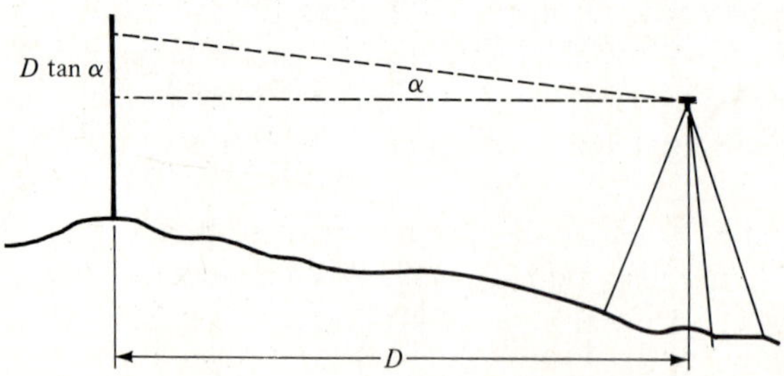

FIGURE 3.31
Inclined line of sight.

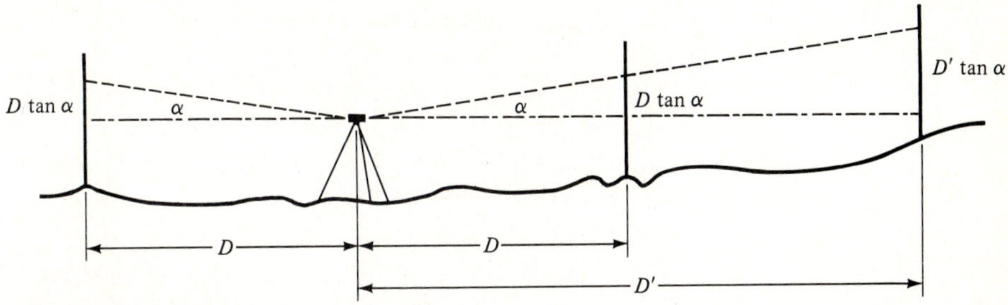

FIGURE 3.32
Imbalance in backsight and foresight distances.

For low-precision leveling operations such as profiles and cross sections, balanced sights are not feasible and minor errors may be overlooked. However, setting foundation heights is an unbalanced-sight activity deserving closer scrutiny. Elevation errors that exceed $\frac{1}{4}$ in. in 100 ft across a finished foundation should not be permitted in quality construction; this translates to a slope angle of $\pm 43''$ or a collimation correction of 0.00021 ft per foot. If a level is used where collimation error might be a factor, the error should be determined as illustrated in Appendix B, Example 5, and, if necessary, the level should be adjusted by a qualified instrument technician.

3.8 THREE-WIRE LEVELS

Three-wire leveling offers three distinct advantages over ordinary differential leveling: (1) internal checks on rod readings, (2) more refined rod-reading values through averaging, and (3) better measurement of backsight and foresight distances by use of the **stadia principle**, discussed next. Although most three-wire leveling is done by geodetic parties using precise levels, Invar-faced metric rods, and double-run techniques, the method can be adapted to engineering and construction leveling using a dumpy or self-leveling level and a Philadelphia rod on single-run work.

THE STADIA PRINCIPLE. The reticles of modern leveling instruments have three horizontal cross hairs, the top and bottom hairs being shorter than, and equally spaced from, the center hair. Figure 3.11 shows two different styles of cross-hair reticles with stadia hairs. Figure 3.33, an idealized diagram of the optics on a level telescope, shows the effective location of the cross-hair reticle with respect to the rod and to the observer's eye. The distance between the top and bottom hairs on the reticle is i and the distance from eye to reticle is the instrument focal length f. The distance from eye to rod is D and the distance on the rod between top and bottom hair readings, called the **stadia interval**, or **stadia intercept**, is I. The small isosceles triangle (in which f is the height and i forms the base) is similar to the large isosceles triangle (in which D is the height and I forms the base). The ratio of $f{:}i$, which is fixed on any instrument telescope, is almost invariably 100:1. Thus, the ratio of the sight distance to the rod intercept is also fixed at the same value.

Put another way, sight distance to a rod is equal to 100 multiplied by the difference between the readings on the top and bottom cross hairs for that sight. Philadelphia

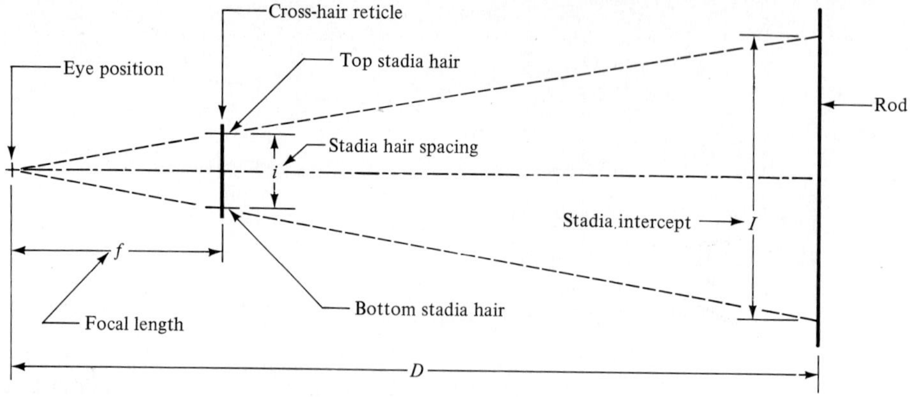

FIGURE 3.33
Idealized optical diagram showing stadia principle.

rods are marked in feet; therefore, sight distances calculated from the rod intercept will also be in feet.

THREE-WIRE OBSERVATIONS AND COMPUTATIONS. Because one of the distinct advantages of three-wire leveling is improved accuracy, it is much better to interpolate rod readings to the nearest 0.001 ft rather than read the rod to only 0.01 ft. To do this requires sight distances of around 50 ft or less, necessitating more frequent setups. For the occasional application requiring higher accuracy than that provided by ordinary differential leveling, this is a small enough price to pay.

Consider a line of levels that runs between bench marks *A* and *B* using three-wire techniques. The examples that follow demonstrate how rod readings are recorded and computed using the standard six-column format for ordinary field book use. The first illustration shows the field book set out with the proper column headings, BM *A* entered in the first column, its elevation entered in the last column, and the three backsight cross-hair readings on BM *A* entered, top to bottom, in the second column on the next three lines.

Three-wire leveling

	Backsight data		**Foresight data**		
Sta.	**Rdgs.**	**Stadia**	**Rdgs.**	**Stadia**	**Elev., ft**
BM *A*					**340.393**
	2.326				
	2.079				
	1.834				

While the rodperson is moving to the foresight rod point, the observer (or recorder if there is one) should compute the stadia entries and the mean of the three backsight cross-hair readings as shown below. The top stadia entry is the top hair minus the middle hair; the bottom stadia entry is the middle hair minus the bottom hair.

Three-wire leveling

	Backsight data		Foresight data		
Sta.	**Rdgs.**	**Stadia**	**Rdgs.**	**Stadia**	**Elev., ft**
BM *A*					**340.693**
	2.326				
	2.079	**24.7**			**+2.0797**
	1.834	**24.5**			**342.7727 (HI)**
Sums	**6.239**	**49.2**			
Mean	+2.0797				

Stadia distances equal 100 multiplied by rod-reading differences; for example, (2.326 − 2.079) 100 = 24.7. The two stadia values add to 49.2, the backsight distance in feet. The mean of the three backsight cross-hair readings is added to the elevation of BM *A* in the last column to give the elevation of the HI. The next illustration shows the three foresight rod readings being treated in a similar manner.

Three-wire leveling

	Backsight data		Foresight data		
Sta.	**Rdgs.**	**Stadia**	**Rdgs.**	**Stadia**	**Elev., ft**
BM *A*					340.693
	2.326		**3.711**		
	2.079	24.7	**3.448**	**26.3**	+2.0797
	1.834	24.5	**3.188**	**26.0**	342.7727
Sums	6.239	49.2	**10.347**	**52.3**	**−3.4490**
Mean	2.0797		**−3.4490**		
TP 1					**339.3237**

Note that TP 1 has been entered in the first column and its computed elevation—the sum of the HI and the mean of the three foresight rod readings—has been entered in the last column.

The next illustration shows additional entries and computations for the backsight on TP 1 and the foresight on TP 2. Two new entries—the cumulative sums of the stadia sight distances—are shown on the same line as TP 2.

Three-wire leveling

	Backsight data		Foresight data		
Sta.	**Rdgs.**	**Stadia**	**Rdgs.**	**Stadia**	**Elev., ft**
BM *A*					340.693
	2.326		3.711		
	2.079	24.7	3.448	26.3	+2.0797
	1.834	24.5	3.188	26.0	342.7727
Sums	6.239	49.2	10.347	52.3	−3.4490
Mean	+2.0797		3.4490		

(continued)

Three-wire leveling (Cont.)

	Backsight data		Foresight data		
Sta.	**Rdgs.**	**Stadia**	**Rdgs.**	**Stadia**	**Elev., ft**
TP 1					339.3237
	3.756		3.415		
	3.550	20.6	3.179	23.6	+3.5497
	3.343	20.7	2.943	23.6	342.8734
Sums	10.649	41.3	9.537	47.2	−3.1790
Mean	3.5497		3.1790		
TP 2		90.5		99.5	339.6944

The cumulative backsight sum is seen to be 9 ft less than the cumulative foresight sum. This tells the observer that the next instrument setup should be placed so the backsight distance is about 9 ft greater than the foresight distance in order to balance the backsight and foresight distances. Figure 3.34 shows the completed page of three-wire leveling notes for the BM *A*–to–BM *B* level line.

Page checks are important for first- and second-order level surveys because they provide frequent tests of the validity of the ongoing computation. For the same reasons, the page check is helpful to surveyors incorporating three-wire techniques into level surveys using ordinary equipment.

Three-wire leveling

	Backsight data		Foresight data		
Sta.	**Rdgs.**	**Stadia**	**Rdgs.**	**Stadia**	**Elev., ft**
BM *A*					340.693
	2.326		3.711		
	2.079	24.7	3.448	26.3	+2.0797
	1.834	24.5	3.188	26.0	342.7727
Sums	6.239	49.2	10.347	52.3	−3.4490
Mean	2.0797		3.4490		
TP 1					339.3237
	3.756		3.415		
	3.550	20.6	3.179	23.6	+3.5497
	3.343	20.7	2.943	23.6	342.8734
Sums	10.649	41.3	9.537	47.2	−3.1790
Mean	3.5497		3.1790		
TP 2		90.5		99.5	339.6944
	6.318		4.147		
	6.040	27.8	3.917	23.0	+6.0397
	5.761	27.9	3.684	23.3	345.7341
Sums	18.119	55.7	11.748	46.3	−3.9160
Mean	6.0397		3.9160		
BM *B*		146.2		145.8	341.8181

ΣBS + 11.6691 ΣFS − 10.5440

Page Check: 340.693 + 11.6691 − 10.5440 = 341.818 ✓

Stadia Imbalance: BS − FS = 146.2 − 145.8 = +0.6ft

FIGURE 3.34

Three-wire recording format for standard field book.

Note also the value of maintaining a record of the cumulative sums of sight distances. By so doing, sight-distance adjustments may be made at any point during a level survey, thus assuring a reasonable balance and eliminating a major source of error.

Probably the best reason for taking the extra care involved in the three-wire procedure is that it provides an otherwise unavailable check on individual rod readings by affording a comparison of the top and bottom halves of the stadia intercepts. An appreciable difference between these two values is a clear indication that at least one rod reading is incorrect. Generally, all three rod readings should be reobserved if the difference between top- and bottom-half intercepts exceeds 0.004 ft.

3.9 RECIPROCAL LEVELS

In emphasizing the importance of balancing backsight and foresight distances, it is assumed that the observer usually has that option. A different situation arises when elevations must be carried between points on opposite sides of a river or chasm and no adequate crossing structure is available. In this case, not only will it be impossible to balance backsights and foresights, the observer must also take longer sights than are desirable. The resultant errors are very significantly reduced by the method of **reciprocal levels**, in which multiple long sights are taken from *both* sides of the crossing and averaged. This compensates for curvature and refraction and for misadjustment of the bubble. Figure 3.35 shows a river in the path of a line of levels that progresses from left to right. The elevation of BM A, on the left bank, has been established and an elevation is now to be determined for BM B. This is done by erecting rods over BMs A and B, then observing from setup points A' and B', adjacent to BM A and BM B, respectively. From setup A', backsight $A'A$ and foresight $A'B$ are taken. From setup B', backsight $B'A$ and foresight $B'B$ are taken.

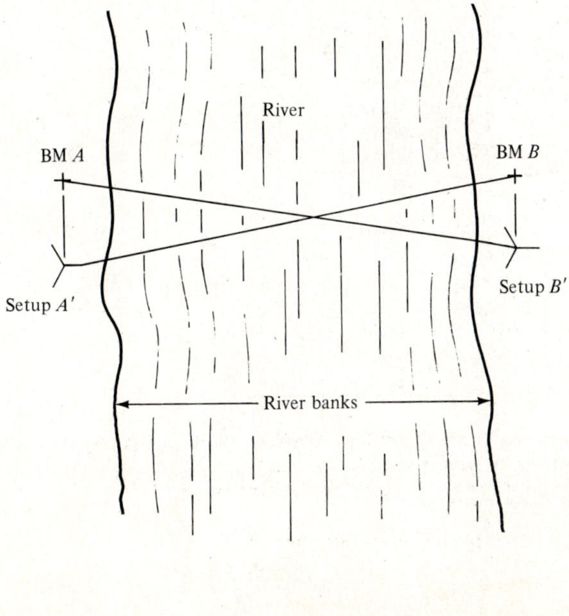

FIGURE 3.35
Reciprocal leveling across a river.

River crossings involve sight distances too great for the rod to be read directly. To get around this problem, a target similar to the one in Fig. 3.15 is placed on the rod. As directed by hand signals, the rodperson moves the target up or down until the center cross hair of the telescope coincides exactly with the center of the pattern on the rod target. Using the target's **vernier** scale, the rodperson reads and records the target reading to the nearest 0.001 ft. The rodperson then displaces the target up or down on the rod and another sight is taken, read and recorded in the same manner. Three or more long-sight readings of this sort are normally taken from each instrument setup point. The two sets of readings are taken close together in time to eliminate significant changes in refraction. Once all readings have been taken from both setups, the elevation difference is computed as shown in the following example.

Example 3.13. In Fig. 3.35, an elevation of 377.298 ft has been established at BM *A*, a concrete monument on one bank of a river. A similar monument, BM *B*, is on the opposite bank, about 500 ft from BM *A*. Setting up at *A*′, a few feet from BM *A*, the observer takes a backsight reading on *A* of 5.318 ft, followed by four foresights on *B* of 6.249 ft, 6.252 ft, 6.257 ft, and 6.251 ft, then finishes with a closing backsight on *A* of 5.317 ft. Setting up on *B*′, a few feet from *B*, the observer takes one foresight on *B* of 5.049 ft, followed by four backsights on *A* of 4.136 ft, 4.130 ft, 4.128 ft, and 4.135 ft, before taking a closing foresight on *B* of 5.049 ft. Record the readings in a field record book and compute the elevation of BM *B*.

Solution. A five-column recording is shown below, the readings from setup *A*′ recorded first, the readings from setup *B*′ second, and leaving spaces between data entries for computation in the places indicated.

Reciprocal leveling

Sta.	BS	FS	ΔElev.	Elev., ft
BM *A*	5.318			377.298
	5.317			
	———			
BM *B*		6.249		
		6.252		
		6.257		
		6.251		
		———		
BM *A*	4.136			
	4.130			
	4.128			
	4.135			
	———			
BM *B*		5.049		
		5.049		
		———		

Starting with the data from setup *A*′, the backsight and foresight averages are taken and both the *A*-to-*B* difference in elevation and the initial elevation of *B* are computed as shown below.

Reciprocal leveling

Sta.	BS	FS	ΔElev.	Elev., ft
BM *A*	5.318			377.298
	5.317			
	Avg 5.3175			
BM *B*		6.249		
		6.252		
		6.257		
		6.251		
		Avg 6.2522	−0.9347	376.3633
BM *A*	4.136			
	4.130			
	4.128			
	4.135			
BM *B*		5.049		
		5.049		

Setup *B'* computations are completed in a similar manner, then the average of the two elevation differences is applied to the elevation of BM *A* to obtain the mean value of the elevation of BM *B*. A minor check is obtained by comparing the value so obtained to the average of the two computed elevations for BM *B*.

Reciprocal leveling

Sta.	BS	FS	ΔElev.	Elev., ft
BM *A*	5.318			377.298
	5.317			
	Avg 5.3175			
BM *B*		6.249		
		6.252		
		6.257		
		6.251		
		Avg 6.2522	−0.9347	376.3633
BM *A*	4.136			
	4.130			
	4.128			
	4.135			
	Avg 4.1322			
BM *B*		5.049		
		5.049		
		Avg 5.0490	−0.9168	376.3812
		Mean	−0.9258	
BM *B*				376.372 ✓
		Δ Elev misclosure =	0.0179 *ft*	

Note that the value of the misclosure in the above example, 0.0179 ft, comes to about 0.009 ft for each setup. A brief study of the data will show that this is probably because the values of the two long-sight data sets are too large by that amount. Because the long sights in the example were given to be on the order of 500 ft, about 0.005 ft of the too-high readings can be attributed to curvature and refraction. The balance, 0.004 ft, could be attributed to an upward tilt of the telescope axis due to bubble misadjustment (or misadjustment of the self-leveling pendulum in the case of an automatic level). If this is so, the angular tilt would be $\tan^{-1}$ 0.004/500, or roughly $1\frac{1}{2}$ seconds, about what may be expected from ordinary self-leveling levels. As already noted, the reciprocal method compensates errors due to curvature and refraction and to a misadjusted bubble.

3.10 PROFILES

Profile leveling is a specialized form of differential leveling in which elevations are taken at specific points along the intended route of a highway, railroad, aqueduct, transmission line, drainage, or sewer line. The purpose is to establish the profile and cross section along the right-of-way. Profile data are essential for calculating the volume of earthwork on a route construction project; that is, the amount of earth that must be brought in or carted away.

Definition of the right-of-way profile requires elevations at **full stations** (that is, at integer hundreds of feet from the starting point of the project), at high and low points, at directional-change points, and where the grade changes appreciably. Because the locations of such points are defined by their position along the centerline of the right-of-way, they probably won't meet normal criteria for selecting rod points; that is, they will usually not have a distinct high point and will seldom provide firm support under the weight of the rod. The rod base will probably land on sloping ground and, to some extent, depress the natural surface. For this reason, elevations of most points along the right-of-way cannot be established to the 0.01-ft-or-better precision expected on a line of ordinary differential levels.

Because rod-point locations on the right-of-way are almost always inadequate for normal elevation points, a dual system must be used when doing profile leveling. One system, based on the use of carefully selected rod points and 0.01-ft-or-better precision, is run to maintain the overall integrity of the right-of-way survey; the other system provides right-of-way elevations where needed to about 0.1-ft precision, which is adequate considering overall requirements for profiling. Profile leveling is the name given to the specialized surveying procedure that runs these two systems concurrently and provides elevations of two distinctly different orders of accuracy.

Profile levels, which are run initially to satisfy reconnaissance and design requirements before any earth is moved on the project, often need to be repeated during the construction phase. For this reason, bench marks established to support the project must be located far enough off the right-of-way to remain undisturbed during the course of construction. Turning points, being temporary in nature, may or may not fall within the right-of-way boundaries. Right-of-way points, because they are literally in the middle of construction, must frequently be reestablished during the various project stages.

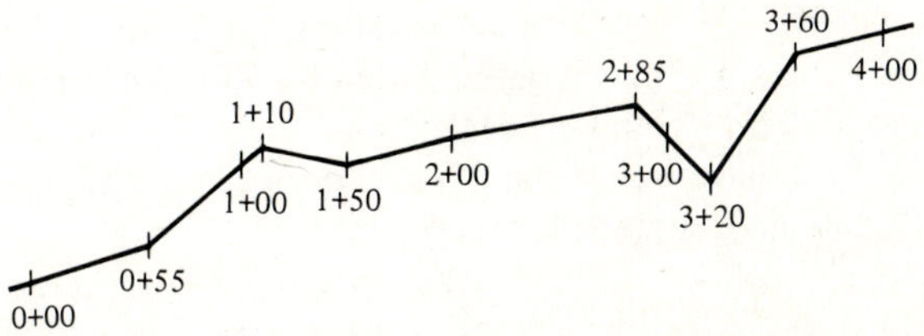

FIGURE 3.36
Right-of-way profile.

Figure 3.36 shows a right-of-way profile along the first four stations of a highway project, with elevations taken at full stations, high points, low points, and points of sudden change in grade. The initial full station is designated 0+00, meaning it is located at zero horizontal distance from the project's beginning. The first change in grade is 55 ft horizontally from the beginning point and is thus designated 0+55, in keeping with the same convention. Station 1+00 is exactly one station, or 100 ft horizontally, from the beginning point, and so on.

Profile leveling in this right-of-way segment involves two bench marks (*A* and *B*) and one turning point (TP 1) but, because the bench marks and the turning point lie *off* the right-of-way, they are not shown in Fig. 3.36.

To keep the dual system of elevations separate, sights on right-of-way points (called "side shots" by many surveyors) are listed in a separate column in the field record book as **i**ntermediate **f**ore**s**ights or **IFS**s. Intermediate foresights are usually taken on multiple rod points from a single setup. When lengths of intermediate foresights become too long for the rod to be read with reasonable assurance, the instrument is moved to the next setup point. Referring again to Fig. 3.36, the IFS on station 2+00 was the last that could be read easily from the first setup, so a foresight was made on TP 1 and the instrument was advanced to a convenient point from which the next series of IFSs could be taken.

The illustration below shows the record-book headings for profile levels and the conventional manner in which the initial HI is computed using a backsight on the bench mark that was established near the beginning point of the project.

Profile levels

Sta.	BS	HI	FS	IFS	Elev., ft
BM *A*	**6.68**	**925.32**			**918.64**

With the HI of the first setup established, intermediate foresights may be taken to every right-of-way rod point within telescope range. The next illustration shows how the first IFS is recorded in the field record and how the initial right-of-way elevation is computed.

Profile levels

Sta.	BS	HI	FS	IFS	Elev., ft
BM *A*	6.68	925.32			918.64
0+00				**9.3**	**916.0**

As indicated earlier, IFSs are usually valid to only 0.1 ft. Following the rule for significant figures, the elevation of station 0+00 is taken to no better than the nearest 0.1 ft; that is, 925.32−9.3 = 916.0.

The illustration below continues with the balance of the intermediate foresights taken from the first setup.

Profile levels

Sta.	BS	HI	FS	IFS	Elev., ft
BM *A*	6.68	925.32			918.64
0+00				9.3	916.0
0+55				**7.6**	**917.7**
1+00				**3.9**	**921.4**
1+10				**3.01**	**922.31**
1+50				**4.9**	**920.4**
2+00				**3.5**	**921.8**

Values of the IFS and computed elevation at station 1+10 show that the rod point was firm enough and well enough defined to be valid to 0.01 ft. As indicated earlier, it was not feasible to take IFSs beyond station 2+00 from the first setup. Therefore, a foresight was taken on a normal TP, the instrument moved, and a new HI established. The field book layout for doing this is illustrated below.

Profile levels

Sta.	BS	HI	FS	IFS	Elev., ft
BM *A*	6.68	925.32			918.64
0+00				9.3	916.0
0+55				7.6	917.7
1+00				3.9	921.4
1+10				3.01	922.31
1+50				4.9	920.4
2+00				3.5	921.8
TP 1	**6.12**	**930.29**	**1.15**		**924.17**

With the HI established on the new instrument position, additional IFSs are taken to the end of this right-of-way segment and a closing foresight is taken to the next bench mark. The completed computation for this entire profiling segment is shown in Fig. 3.37. Note that the check computation is done in the usual way, using data from the BS and FS columns but ignoring the IFS values.

Profile levels

Sta.	BS	HI	FS	IFS	Elev., ft
BM *A*	6.68	925.32			918.64
0+00				9.3	916.0
0+55				7.6	917.7
1+00				3.9	921.4
1+10				3.01	922.31
1+50				4.9	920.4
2+00				3.5	921.8
TP 1	6.12	930.29	1.15		924.17
2+85				6.1	924.2
3+00				7.6	922.7
3+20				9.7	920.6
3+60				3.8	926.5
4+00				2.7	927.6
BM *B*			0.41		929.88
Σ	+12.80		−1.56		
	−1.56				
	+11.24 + 918.64 = 929.88 ft				

FIGURE 3.37
Notekeeping for profile levels.

3.11 LASER LEVELING: CONCEPT AND INSTRUMENTATION

In running an ordinary line of levels, elevations are carried forward using the differential leveling procedure, the observer and rodperson leapfrogging along their intended route, generally taking just one backsight and one foresight from each setup. In profile leveling, on the other hand, ordinary differential leveling is carried out only to support multiple, lower-order, profile shots on points along a right-of-way centerline; in this case, elevations are less important than the relative elevations of sequential points along the centerline.

Yet another form of leveling is carried out to support more localized construction work. Here, the emphasis is on determining the relative elevations of a number of points within a fairly small project area, using, when feasible, a single instrument location chosen because it "sees" all the other points to which elevation differences are to be taken. Rather obviously, this sort of leveling can be carried out by an observer operating a conventional optical instrument while directing one or more rodpersons to all the elevation points deemed necessary to satisfy project requirements.

Recently, however, an unconventional instrument has come into use, one which eliminates the need for the instrument observer as such. A **laser level**, or **electronic level**, stands unattended on its tripod, transmitting a reasonably horizontal laser beam that is rotated to form a reasonably horizontal height-of-instrument reference plane. The rod counterpart is a laser sensor called a **detector**. The detector is raised or lowered through the plane of the transmitted laser beam until a maximum signal is received, at which point, neglecting certain small, inherent errors, the detector is at the HI of the transmitting instrument.

FIGURE 3.38
Spectra-Physics EL-1 level and detector. (*Courtesy of Spectra-Physics, Construction and Agriculture Division.*)

LASER LEVELING INSTRUMENTS. Electronic leveling technology is still undergoing fairly rapid change with respect to accuracy, size, range, cost, and user-application. Figures 3.38 and 3.39 show two currently available levels and their detectors. Note that both levels incorporate standard tripod mounts. A bull's-eye level—used for rough leveling to bring the more precise self-leveling mechanism within range—is seen below the manufacturer's logo on the tripod mount of the instrument in Fig. 3.38.

The detector may be attached to a Philadelphia rod by means of a movable bracket, as illustrated by Fig. 3.40, or used for a hand-held application, as shown in Fig. 3.41.

FIGURE 3.39
Topcon RL-20 level and detector. (*Courtesy of Topcon Instrument Corporation of America.*)

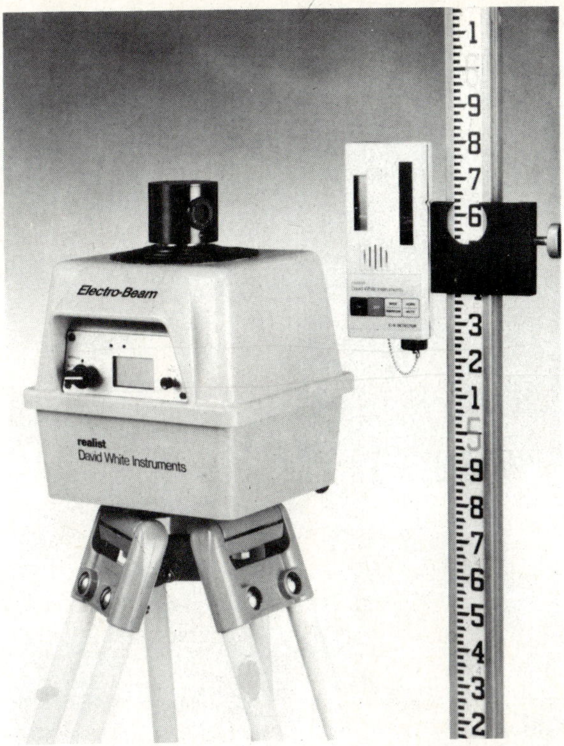

FIGURE 3.40
Realist Electro-Beam level and detector. (*Courtesy of David White Instruments.*)

FIGURE 3.41
Electronic level applications using hand-held detector. (*Courtesy of David White Instruments.*)

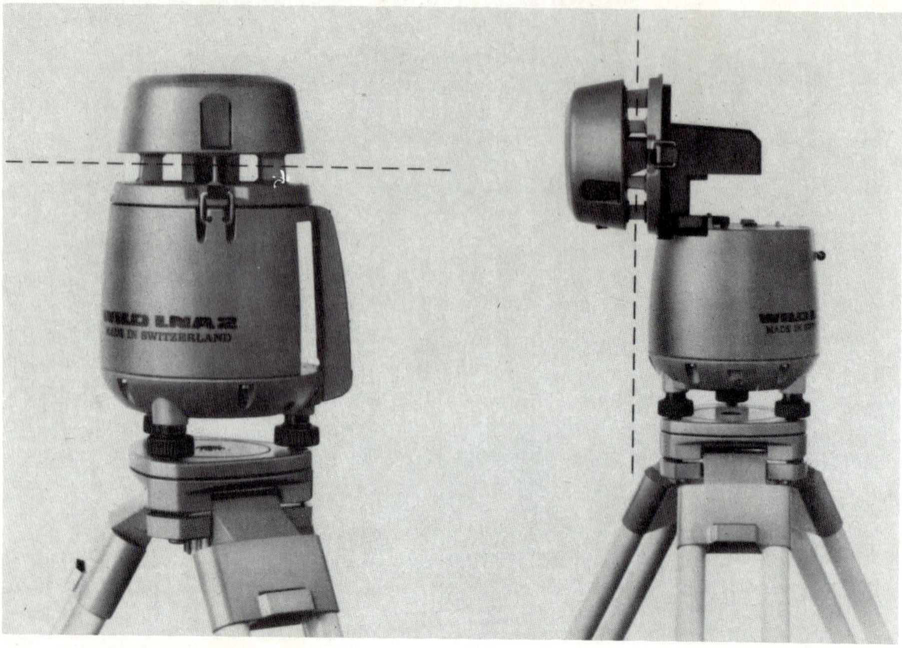

FIGURE 3.42
Wild LNA2 level; horizontal mode at left, vertical mode at right. (*Courtesy of Wild Leitz USA, Inc.*)

Most laser-level instruments currently on the market can be adapted to provide a vertical reference plane—useful for a number of construction applications—as well as a horizontal plane. Figure 3.42 shows another instrument in both its horizontal and vertical configuration.

LASER-LEVELING ACCURACY. The laser level generates a beam with a laser diode in much the same way as an EDM. Using a self-leveling device for a vertical reference, the beam axis is transmitted upward, then deflected to the horizontal by a prism that is rotated to form the horizontal plane. As is the case with any other leveling instrument, the laser level suffers some inherent deviation between the assumed horizontal and the true horizontal. Together, the self-leveling device and the rotating prism mechanism will provide a probable cumulative error on the order of 5 to 15 arc seconds; that is, anywhere from about 0.005 ft to 0.015 ft at a range of 200 ft. In addition, the laser beam is not a cylindrical beam at all, it is a very narrow conical beam. The result is a laser "plane" whose cross section (1) is tilted slightly with respect to the horizontal and (2) thickens with increasing distance from the center. The effect of the conical beam is to make the center of the beam less well defined with increasing distance from the laser level. This, in turn, makes for increasing uncertainty in positioning the detector as range increases.

In ordinary leveling, the level rod really does nothing on its own. It just stands there being read by the observer as it is held vertical by the rodperson; thus, rod-reading errors can hardly be blamed on the rod. The laser detector, on the other hand, is an electronic sensing device that must be switched on and must work properly if it is to do its job. As such, it has accuracy limitations that result from the size and sensitivity of the individual sensor cells, the configuration of the cell matrix, and how well defined the laser beam is when it strikes the matrix. These combined factors can produce errors in the range of 0.0025 ft to 0.01 ft, depending on the make and model of the detector as well as its distance from the laser beam source.

It may also occur to the reader that curvature and refraction errors are not compensated when using a laser level. As determined earlier, (c & r) error is about 0.005 ft at 500 ft and 0.02 ft at the extreme range of 1000 ft.

Present laser-leveling technology does not provide the accuracy the surveyor might hope for—error in the horizontal reference is about 10 times that of a self-leveling level of the same relative quality. Even so, the rather rough accuracy that makes this instrument less than ideal for most surveying tasks is more than adequate for a broad range of construction activities. Too, it takes fewer people to operate and may be used by less-skilled people. These combined advantages can result in early payoff of its purchase cost.

LASER LEVEL OPERATION AND APPLICATIONS. The instrument, once rough-leveled and turned on, will self-level and transmit a horizontal reference until turned off—some models will turn off automatically if disturbed sufficiently to knock the compensator out of self-leveling range. For most instruments, a fresh battery lasts 10 hours or more—the detector battery lasts even longer. The chief advantage of the laser level is unattended operation, allowing use of one or more detectors that are taken wherever elevations are needed within range of the level—from 300 ft to 1000 ft, depending on make, model, and mode of operation of the instrument. Most construction applications involve an operating range well within the maximum range.

A laser level is ideal for outdoor, early-construction tasks such as setting forms for foundations, grading, and trenching. It also satisfies a range of needs for indoor, late-construction work such as setting building modules, marking sheetrock battens at constant height above floor level, shimming baseplates, and setting raised flooring for computer rooms. The instrument is still coming into general use; thus, new applications are still emerging. Surveyors, who tend to be conservative and wary of new technology, are starting to operate laser levels in conjunction with conventional levels. At this writing, an efficient surveyor would use a conventional leveling instrument to establish one or more elevation points within a project area, then take advantage of the great speed of the laser level for in-project applications requiring less-critical accuracy.

3.12 PROBLEMS

3.1. Leveling establishes elevations of survey points or differences in elevation between survey points, usually by either the differential or trigonometric leveling process. Describe other methods that might be used to establish elevations or differences in elevation.

3.2. A high-intensity flashing light is located on top of an offshore oil-drilling platform, 185 ft above water level. At what maximum distance from the platform should the light be visible on a calm night to a shipboard observer whose height-of-eye above water is 18 ft?

3.3. Standing on her observing platform, a lifeguard with binoculars was able to see the roof of a concession stand located at the water's edge on the opposite side of a lake. She noticed, as she sat down, that the roof disappeared from view just as she came to a seated position. If her height-of-eye, seated, was 10.2 ft above water, and if the lake was 8.8 miles wide at that point, what was the height of the roof above lake level, to the nearest 0.1 ft?

3.4. Microwave transmission towers of equal height are to be built on opposite shores of a 15.0-mile-wide lake. To minimize reflection from the water surface, the direct-line transmission path must clear the water surface a minimum of 15.0 ft. What is the minimum height of the towers above the lake surface?

3.5. An EDM slope distance and theodolite slope angle are measured to point Low from instruments set over BM Mesa (elev. 1988.86 ft), with recorded results as follows: the slope distance is 4672.13 ft and the slope angle is $-7°53.2'$; the height above BM Mesa of *both* the EDM and the theodolite is 5.15 ft, while the height above point Low of *both* the EDM reflector and the rod target is 4.93 ft. With normal atmospheric conditions, what is the elevation of point Low, to the nearest 0.01 ft?

3.6. A theodolite-EDM combination was set over hilltop BM Q, whose elevation is 1422.67 ft; the theodolite center was 5.04 ft above BM Q and the EDM center was 0.48 ft above the center of the theodolite. A vertical angle of $-18°18'50''$ was taken to a target 4.39 ft above point X in the valley. An EDM slope distance of 2817.63 ft was measured to a reflector 0.77 ft *below* the center of the theodolite target. What is the elevation of point X, to the neatest 0.01 ft, assuming normal atmospheric refraction conditions?

3.7. A theodolite-EDM combination was set over a bench mark and a vertical angle of $+3°42'18''$ was read to a target on a level rod held on the high point of a hill. The EDM slope distance to the reflector was 10,582.12 ft. The elevation of the bench mark was 1469.81 ft above MSL. Heights of the EDM, theodolite, target, and reflector were all 5.08 ft. Assuming normal refraction conditions, what is the elevation of the high point on the hill, to the nearest 0.01 ft?

3.8. Using a theodolite-EDM combination, a vertical angle of $+5°18'52''$ was taken to a target set 5.83 ft above a BM of elevation 563.69 ft. The EDM slope distance to a reflector set over the BM was 1421.50 ft. From the same setup, a vertical angle of $-11°42'12''$ was then taken to a target set 3.29 ft over a point at the bottom of the hill and an EDM slope distance of 710.89 ft was taken to a reflector set over that same point. In both cases, the height of the reflector above the target was the same as the height of the EDM center above the theodolite. With normal atmospheric conditions, what is the elevation of the point at the bottom of the hill, to the nearest 0.01 ft?

3.9. In Prob. 3.8, what is the elevation of the ground beneath the theodolite-EDM combination if the theodolite HI was 5.06 ft?

3.10. Referring to Prob. 3.5, nearly simultaneous reciprocal observations were taken from point Low to BM Mesa. Compute the elevation of Low using the Prob. 3.5 data *and* the following Low-to-Mesa results: slope distance is 4672.12 ft, slope angle is $7°52.6'$; height of EDM *and* theodolite over point Low is 5.09 ft, height of EDM reflector *and* theodolite target over BM Mesa is 5.46 ft?

3.11. Explain a likely cause of the small difference in elevation of point Low as calculated in Prob. 3.5 (nonreciprocal observations) and Prob. 3.10 (reciprocal observations), and state which result is theoretically better.

3.12. How are bench marks and turning points different?

3.13. In viewing a leveling rod through your instrument telescope, you notice that by moving your eye up and down, the horizontal cross hairs appear to move up and down on the rod face. What is the problem and how is it corrected?

3.14. The bull's-eye bubble on a rod level was out of adjustment, causing the fully extended rod to be 6 in. out of plumb at the 13-ft mark, the rod being inclined toward the observer. If the reading on the inclined rod was 7.832 ft, what would the correct reading have been for a rod held vertically on the same elevation point?

3.15. The baseplate on the level rod used in Prob. 3.14 was $1\frac{5}{8}$ in. square. If the rod was sitting on a perfectly flat surface and was 6 in. out of plumb at the 13-ft mark, with the rod inclined directly away from the observer, what would have been the correct reading on a vertically held rod if the rod reading for the conditions described above was 7.832 ft?

3.16. A dumpy level had a bubble-vial radius of 68.8 ft. A second dumpy level had a bubble sensitivity of 20 seconds per division at 0.01-ft spacing per division. Which instrument, the first or the second, had the more sensitive bubble?

3.17. You wish to determine the bubble sensitivity of the leveling instrument you are using. You set a vertical rod 200.00 ft from the telescope optical center and take a rod reading of 6.392 ft on the center cross hair with one end of the level bubble in coincidence with a division marker. By turning one level screw to tilt the telescope upward, you move the bubble exactly five divisions and read 6.513 ft on the center hair. What is the bubble sensitivity? If the division spacing is exactly 2 mm, what is the radius of curvature of the level vial in meters?

3.18. Compensators in self-leveling levels have only a limited range through which they can swing freely. Describe a simple observing practice to use in the field to find out if the compensator is working.

3.19. Differential-leveling observations were taken as shown below. Complete and check the level notes, and identify closure error.

STA	BS, ft	HI, ft	FS, ft	ELEV, ft	DIST, ft
BM *A*	4.73			369.76	103
TP 1	6.19		7.68		151/97
TP 2	8.81		10.32		122/155
BM *B*	4.10		3.66		167/130
TP 3	7.07		9.33		91/159
TP 4	9.28		2.69		133/96
TP 5	1.92		4.74		144/126
BM *A*			3.65		/150

3.20. Differential levels were run in the normal backsight-then-foresight sequence from BM *R* (elevation 379.43 ft above MSL) to BM *S* as follows: 6.31, 0.47; 9.22, 3.88; 7.71, 4.92; 3.19, 11.16. Then, from BM *S* to BM *T* the sequence was as follows: 8.33, 2.97;

7.93, 9.41; 5.72, 8.68. The final sequence, from BM T back to BM R, was as follows: 1.91, 7.40; 10.02, 6.96; 4.79, 9.25. Prepare a complete set of level notes, including check, and identify the closure error.

3.21. Do the same as Prob. 3.20 except use 468.10 ft for the elevation of BM R and a final foresight rod reading of 9.30 ft.

3.22. Do the same as Prob. 3.20 except use 219.67 ft for the elevation of BM R and a final foresight rod reading of 9.32 ft.

3.23. You were asked to determine the elevation above MSL of a point on the ceiling of a cave near the cave entrance. You set up your level midway between the entrance and an existing bench mark outside whose elevation was 400.394 ft above MSL. Your backsight to the BM was 8.117 ft. For the foresight reading of 5.276 ft, the rod was held upside down, with zero against the elevation point on the cave ceiling. Compute the elevation of the ceiling point.

3.24. In Prob. 3.20, assuming that appropriate instruments and methods were used, what order of accuracy was obtained if the total loop distance was 4000 ft?

3.25. In Prob. 3.21, assuming that appropriate instruments and methods were used, what order of accuracy was obtained if the total loop distance was 3600 ft?

3.26. In Prob. 3.22, assuming that appropriate instruments and methods were used, what order of accuracy was obtained if the total loop distance was 3800 ft?

3.27. Using Prob. 3.19 data, adjust the BM B elevation.

3.28. Using Prob. 3.20 data, adjust the elevations of BMs S and T.

3.29. Using Prob. 3.21 data, adjust the elevations of BMs S and T.

3.30. Using Prob. 3.22 data, adjust the elevations of BMs S and T.

3.31. As a prank, a friend sawed 0.03 ft off the bottom of your Philadelphia rod and replaced the brass base to hide the saw marks. Before you discovered this, you ran a level line from A to B, obtaining an A-to-B difference in elevation of -4.97 ft. If you used 24 setups, with one backsight and one foresight at each setup, what correction should be applied to the A-to-B elevation difference?

3.32. What would be the Prob. 3.23 result if the Prob. 3.31 rod had been used instead of a normal-length rod?

3.33. A rod made from unseasoned wood shrank to just 12.990 ft between the zero and 13-ft marks. Levels run between two points with this shrunken rod required 14 setups, with total backsight and foresight sums of 68.132 ft and 107.162 ft, respectively. What was the correct elevation difference between the points?

3.34. A leveling instrument that had just been precisely adjusted was carefully leveled and a backsight rod reading of 8.932 ft obtained to a vernier target on a rod 520 ft away. A foresight reading of 6.363 ft was then obtained from the same setup to the vernier target on a rod 297 ft away. Assuming normal atmospheric conditions, compute the difference in elevation between the two points.

3.35. A level line was run from BM R (elevation 681.371 ft) to BM S over rugged terrain where it wasn't feasible to balance backsight and foresight distances. The level was in precise adjustment; that is, the telescope was truly level for all rod readings. The observed data, listed below, were obtained using a vernier-reading rod target. Assuming normal atmospheric conditions, compute the elevation of BM S, adjusting as necessary for backsight-foresight imbalance.

Section	BS, ft	BS distance, ft	FS, ft	FS distance, ft
BM *R*–TP 1	4.832	257	10.671	633
TP 1–TP 2	1.543	91	12.220	488
TP 2–TP 3	3.109	704	8.907	566
TP 3–BM *S*	5.666	361	11.618	197

3.36. A line of three-wire levels was run from BM *J* (elev. 687.432 ft) to new BMs *K* and *L*, then to existing BM *H*, elevation 685.882 ft. BMs *J* and *H* had been established to second order, class I standards. Using the following data, prepare the proper three-wire level notes, including page check. Determine closure error and the order of leveling obtained between BM *J* and BM *H*. Rod reading data, in feet, are as follows:

BM *J* to TP 1:	BS 5.67, 5.451, 5.2777; FS 10.961, 10.693, 10.422
TP 1 to BM *K*:	BS 4.178, 3.922, 3.670; FS 8.881, 8.677, 8.471
BM *K* to BM *L*:	BS 7.469, 7.191, 6.910; FS 2.633, 2.381, 2.096
BM *L* to TP 2:	BS 11.433, 11.171, 10.913; FS 3.917, 3.681, 3.441
TP 2 to TP 3:	BS 8.709, 8.486, 8.259; FS 6.392, 6.135, 5.881
TP 3 to BM *H*:	BS 1.777, 1.504, 1.235; FS 7.987, 7.724, 7.460

3.37. Do the same as Prob. 3.36 except BM *J* elevation is 443.291 ft and the elevation of BM *H* is 441.723 ft.

3.38. Do the same as Prob. 3.36 except BM *J* elevation is 386.449 ft and the elevation of BM *H* is 384.903 ft.

3.39. Using the data from Prob. 3.36, compute the adjusted elevations at bench marks *K* and *L*.

3.40. Using the data from Prob. 3.37, compute the adjusted elevations at bench marks *K* and *L*.

3.41. Using the data from Prob. 3.38, compute the adjusted elevations at bench marks *K* and *L*.

3.42. Differential levels were run from BM *A* to BMs *B*, *C*, *D*, and *E*, then back to BM *A*. Elevation differences and distances run were as follows: BM *A* to BM *B*, +37.131 ft, 0.5 mile; BM *B* to BM *C*, +12.896 ft, 0.6 mile; BM *C* to BM *D*, −3.626 ft, 0.4 mile; BM *D* to BM *E*, −17.367 ft, 0.4 mile; BM *E* to BM *A*, −29.078 ft, 0.3 mile. The fixed elevation of BM *A* is 342.891 ft. Compute the adjusted elevations of BMs *B*, *C*, *D*, and *E*.

3.43. Reciprocal levels were used to carry a level line across a deep canyon from BM *X*, elev. 3262.514 ft, to BM *Y*, a new elevation point. From a setup near BM *X*, a backsight reading on BM *X* of 4.361 ft was followed by foresight readings on BM *Y* of 6.809 ft, 6.812 ft, 6.806 ft, and 6.801 ft, then a final backsight reading of 4.359 ft on BM *X*. The instrument was then set up near BM *Y*, a foresight reading of 5.666 ft on BM *Y* was followed by backsight readings on BM *X* of 3.257 ft, 3.250 ft, 3.255 ft, and 3.254 ft and a final foresight reading of 5.668 ft on BM *Y*. Prepare a set of field notes that includes computations to determine and check the elevation of BM *Y*.

3.44. Do the same as Prob. 3.43 except the known elevation of BM *X* is 2691.478 ft and the foresight readings to BM *Y* from the nearby instrument setup were 5.686 ft and 5.688 ft.

3.45. Profile levels were run along the centerline of a proposed roadway using an initial elevation of 426.15 ft at BM Start. Using the data listed below, prepare, compute, and check profile level notes. From the first setup, the BS to BM Start was 7.08 ft; IFSs were as follows: 6.9 ft to sta. 0+00, 8.4 ft to sta. 1+00, 4.1 ft to sta. 1+25, 7.8 ft to sta. 1+63, 9.6 ft to sta. 2+00; and FS to TP1 was 11.66 ft. From the second setup, the BS to TP1 was 5.99 ft; IFSs were as follows: 8.6 ft to sta. 2+33, 6.1 ft to sta. 2+61, 4.3 ft to sta. 3+00, 7.9 ft to sta. 3+15, 5.5 ft to sta. 4+00; and FS to BM Stop was 3.07 ft.

3.46. Do the same as Prob. 3.45 except BM Start elevation was 566.87 ft and the backsight from the second setup to TP 1 was 6.43 ft.

3.47. Explain why the elevation of an IFS should never be used as the beginning elevation for subsequent leveling observations.

CHAPTER 4

ANGLE MEASUREMENT

4.1 ANGLE TERMINOLOGY

This chapter treats both **horizontal angles** and **vertical angles**. Measurement of horizontal angles is discussed first; measurement of vertical angles is covered toward the end of the chapter. Boundary lines on property descriptions must be given as horizontal distances, so the horizontal angle at the intersection of two such boundary lines must obviously be measured on a horizontal plane. A horizontal angle thus measured is formed by the intersection of two vertical planes, the vertex of the angle lying on the vertical line of intersection of the two planes. In the surveying context, a horizontal angle is measured at the **vertex point** between a **backsight point** and a **foresight point,** as illustrated in Fig. 4.1, the three points, in all likelihood, lying at three different elevations. An angle thus measured will not only have magnitude, in degrees, minutes and seconds, it will also have a clockwise or counterclockwise *direction*. The reasons for stating the direction in which angles are taken will become quite apparent further along in the text. For the time being, it will simply be considered the proper thing to do.

Horizontal angles may be measured *directly,* by setting a transit or theodolite at the vertex point, then measuring between the backsight direction and the foresight direction. It is also possible to measure angles *indirectly,* using horizontal-distance measurements alone.

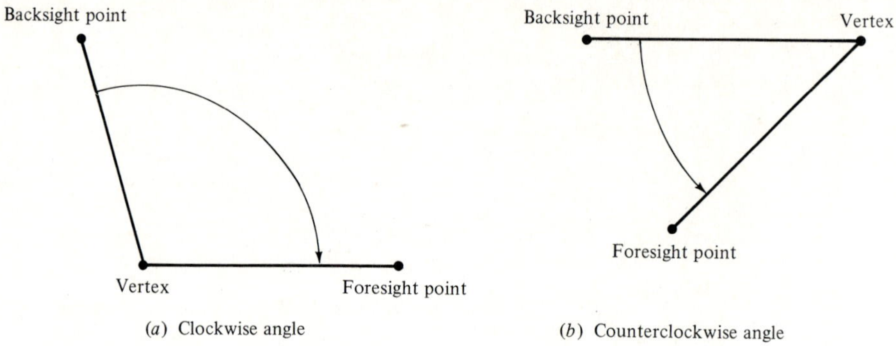

FIGURE 4.1
Plan view of horizontal angles.

4.2 HORIZONTAL ANGLES FROM HORIZONTAL DISTANCES

Any polygon can be surveyed by dividing it into triangles, then measuring the horizontal lengths of the sides and calculating the values of all angles. For example, the quadrilateral *ABCD*, shown in Fig. 4.2, is divided into triangles *ABC* and *ACD* by the diagonal *AC*. By convention, the sides opposite angles *A*, *B*, and *C* in triangle *ABC* are designated *a*, *b*, and *c*, respectively. Knowing lengths *a*, *b*, and *c*, angles *A*, *B*, and *C* may be computed using Eqs. 1.14*a*, *b*, and *c*, developed from the Law of Cosines in Section 1.11 and repeated here.

$$C = \cos^{-1} \frac{a^2 + b^2 - c^2}{2ab} \tag{1.14a}$$

$$B = \cos^{-1} \frac{a^2 + c^2 - b^2}{2ac} \tag{1.14b}$$

$$A = \cos^{-1} \frac{b^2 + c^2 - a^2}{2bc} \tag{1.14c}$$

Use of the above equations is demonstrated in Example 1.30 in Chap. 1.

Two methods of measuring a single angle with tape or EDM alone are shown in Fig. 4.3. In Fig. 4.3*a*, angle *A* lies at the intersection of lines *AB* and *AC*. Points *B′*

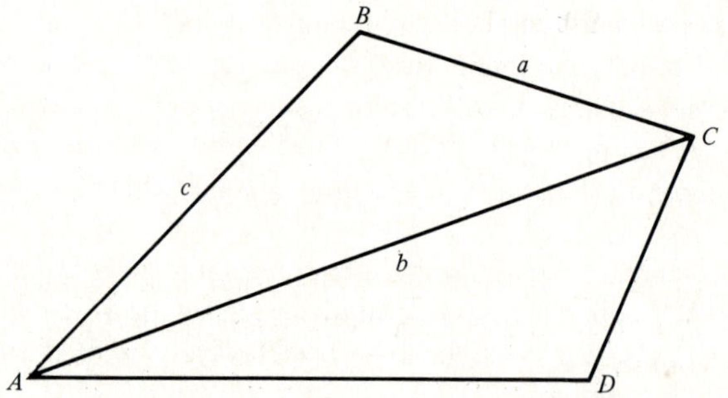

FIGURE 4.2
Quadrilateral ABCD divided into two triangles.

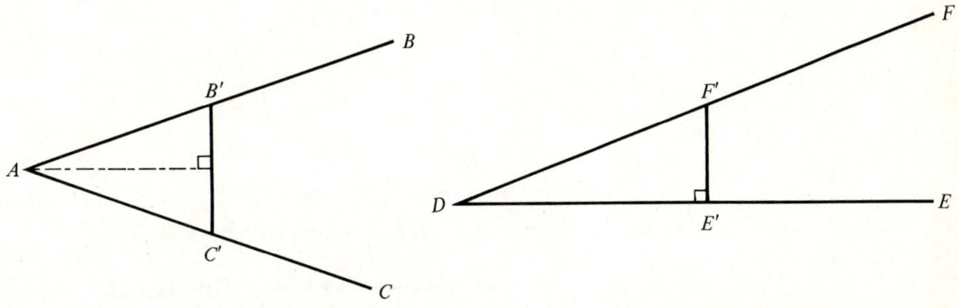

(a) Bisector method

(b) Perpendicular method

FIGURE 4.3
Single-angle determination by measuring horizontal distances.

and C' are laid off an equal horizontal distance from A, as shown, and the horizontal chord $B'C'$ is measured as well. Because $AB'C'$ is an isosceles triangle, the bisector of A must also bisect line $B'C'$. From the diagram, $\sin(A/2) = (B'C'/2)/AB' = B'C'/2AB'$. Therefore, $A/2 = \sin^{-1}(B'C'/2AB')$. Because angle A is twice the half-angle $A/2$:

$$\mathbf{A = 2\ \sin^{-1}\frac{B'C'}{2AB'}} \tag{4.1}$$

In Fig. 4.3*b*, point E' is laid out a convenient horizontal distance from D, a perpendicular is erected at E', and point F' is established at the intersection of the perpendicular with line DF. In the resulting triangle $DE'F'$, $\tan D = E'F'/DE'$, from which is obtained

$$\mathbf{D = \tan^{-1}\frac{E'F'}{DE'}} \tag{4.2}$$

Example 4.1. In Fig. 4.3*a*, points B' and C' are established 100.00 ft horizontally from A toward B and C, respectively. An 86.38-ft horizontal distance is then measured between B' and C'. Compute angle A to the nearest 0.1 minute.

Solution. Substituting into Eq. (4.1),

$$A = 2\ \sin^{-1}\frac{86.38}{2(100.00)} = 2(25.5881995°)$$

$$= 51.176399° = 51°10'35'' \Rightarrow\Rightarrow 51°10.6'$$

Example 4.2. In Fig. 4.3*b*, point E' is set 100.00 ft horizontally from D along line DE. The horizontally measured distance from E' to F' is 69.04 ft. To the nearest minute, compute angle D.

Solution. Substituting into Eq. (4.2),

$$D = \tan^{-1}\frac{69.04}{100.00} = 34.62120° \Rightarrow\Rightarrow 34°37'$$

Angles may also be *laid out* using an adaptation of the Fig. 4.3*b* method. Because $\tan D = E'F'/DE'$, it can also be stated that

$$\mathbf{E'F' = DE' \tan D} \tag{4.3}$$

If DE' is 100.00 ft, Eq. (4.3) becomes $E'F' = 100.00 \tan D$

Example 4.3. In Fig. 4.3*b*, point E' is established 100.00 ft horizontally from point D toward E and a perpendicular is erected at E' in the direction of intended line DF. If angle D is to be 38°28′, what horizontal distance $E'F'$ must be laid out to establish D?

Solution. Substituting into Eq. (4.3),

$$E'F' = 100.00 \tan 38°28' = 79.4486 \Rightarrow\Rightarrow 79.45 \text{ ft}$$

Although the above examples refer to acute angles, obtuse angles may be measured or laid out almost as easily. Simply lay out a backward extension of one of the lines forming the obtuse angle, as shown in Fig. 4.4. This creates an acute angle that is the supplement of the desired obtuse angle. The acute angle is then measured as before.

Example 4.4. In Fig. 4.4, point B' is established 100.00 ft horizontally from A along AB', the backward extension of AB, and point C' is established 100.00 ft horizontally from A along AC. The measured horizontal distance from B' to C' is 57.02 ft. What is the value of obtuse angle A, to the nearest 0.1 minute?

Solution. The acute angle at A is solved by substituting into Eq. (4.1):

$$A = 2 \sin^{-1}\frac{57.02}{2(100.00)} = 2(16.564826°)$$

$$= 33.12965° \Rightarrow\Rightarrow 33°07.8'$$

Obtuse angle A is the supplement of the acute angle, or

$$\text{Obtuse angle } A = 180°00.0' - 33°07.8' = 146°52.2'$$

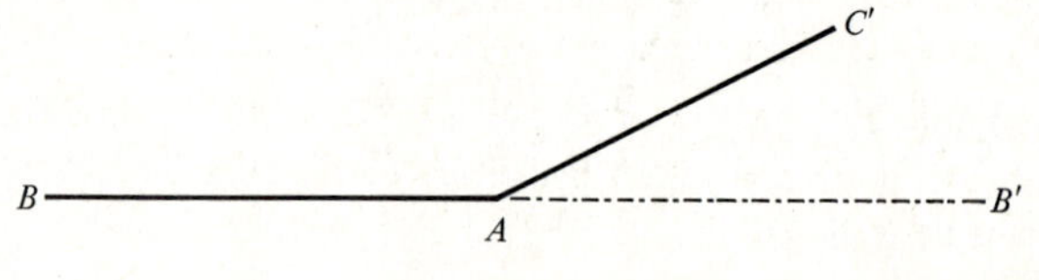

FIGURE 4.4
Establishing the acute supplement of an obtuse angle.

This is an appropriate point to bring up an alternative method of defining an angle. Consider Fig. 4.2. In the discussion of determining angles *A*, *B*, and *C*, it was assumed that "*A*" meant the angle at the vertex of triangle *ABC*. But *A* was originally the vertex of quadrilateral *ABCD*. To avoid confusion, it is better to give the vertex angle of triangle *ABC* the designation *BAC*, where the middle letter is the vertex point and the flanking letters are the backsight and foresight points. Thus, the desired obtuse angle in Example 4.4 is better given as *BAC*, while its acute supplement would be given as CAB'.

The above method also indicates whether an angle is taken clockwise or counterclockwise. By convention, the letter listed first is the backsight point and the last letter is the foresight point. Thus, the designation CAB', above, indicates that *AC* was the initial line and AB' was the line measured *to*. When measuring or laying out angles by the methods just discussed, this distinction is not too important; when measuring angles directly with transit or theodolite, it becomes quite important.

4.3 TRANSIT CONCEPT

A transit, stripped to its bare essentials, is a telescope that may be swung through both horizontal and vertical angles, those angles being directly measurable on circle scales that are an integral part of the transit. On an **engineer's transit,** horizontal and vertical angles are each read on their own main circle and on adjacent **vernier scales,** the latter providing a method of reading angles to a much greater precision than is possible from reading the divisions on the main circle alone. On more modern instruments, horizontal and vertical angles are read on nonvernier scales that are more precise and easier to use than verniers. According to present usage, the word "transit" refers to an open-design engineer's transit or to an enclosed-design transit that also uses verniers. Instruments having finely divided, optical-reading scales for refined angle readings are called **theodolites.**

Transits and theodolites have provisions for orienting and maintaining the zero-degree point of the **horizontal circle** in any desired direction. Usually, the backsight direction is indexed on zero degrees. Then, when the telescope is turned toward the foresight direction, the resultant reading on the horizontal circle is the horizontal angle between the backsight direction and the foresight direction.

Leveling a properly adjusted transit or theodolite automatically indexes the zero point so the vertical direction in which the telescope is pointed may be read on the **vertical circle** as a vertical angle. Vernier transits, and some theodolites, are designed so the vertical circle reading will be zero when the instrument is level and the telescope axis is horizontal. Thus, when the telescope is tilted upward, the vertical circle will read an **elevation angle** that increases positively from zero at the horizontal; when it is tilted downward, the circle will read a **depression angle** that increases negatively from zero at the horizontal. Other theodolites, however, have vertical circles that read zero when the telescope is pointed straight up and 90 degrees when the telescope is horizontal; vertical angles so defined are called **zenith angles** and are always positive.

4.4 VERNIER TRANSITS

CONSTRUCTION. The working mechanism of an engineer's transit is somewhat out in the open. Figures 4.5 and 4.6 show both sides of a typical engineer's transit. Above the tops of the leveling screws are two clamping rings, one over the other. Loosening the **lower clamp** allows the horizontal circle to be turned with respect to the transit base. Loosening the **upper clamp** allows the telescope to turn with respect to the horizontal circle. **Tangent screws** adjacent to each clamp provide fine adjustment in setting angle values or pointing the telescope. The horizontal circle, in the large, disc-shaped housing immediately above the upper clamp and tangent screw, is marked at ten-degree intervals that are further subdivided to a fraction of a degree. Refined angle readings are taken on either or both of two vernier scales that lie 180° apart on an inner circle concentric with the horizontal circle.

Two level bubbles, called **plate bubbles,** whose axes are 90° to one another, are mounted on top of the horizontal circle housing, as better seen in Fig. 4.5. One of the two vernier scales (the "A vernier") is viewed below the partially open flap to the right of the right-hand plate bubble in Fig. 4.5. The flap over the "B vernier" is seen fully open in Fig. 4.6 just left of the only visible plate bubble. The framework rising from the horizontal-circle housing holds the transit's horizontal axis, to which both

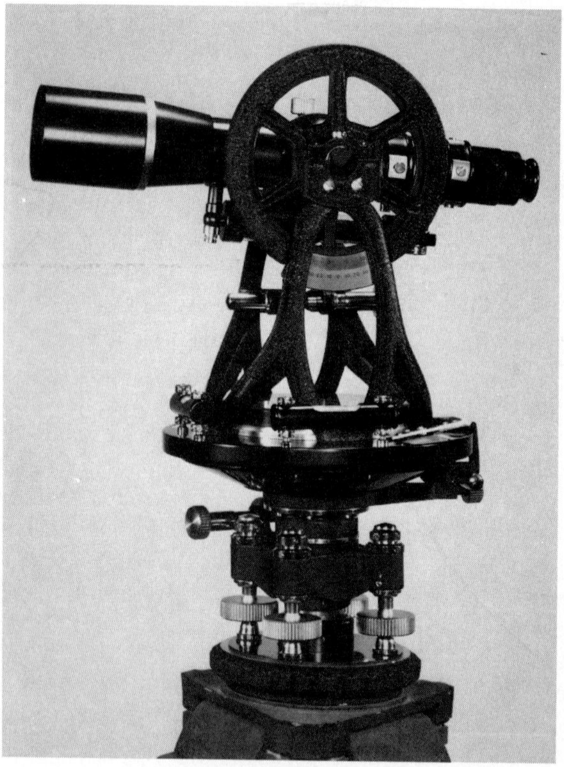

FIGURE 4.5
Engineer's transit, left side. (*Courtesy of Teledyne Gurley.*)

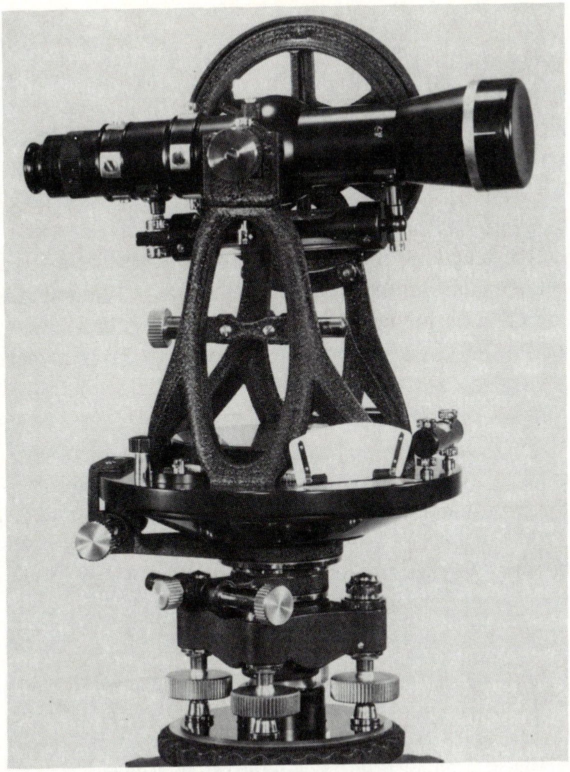

FIGURE 4.6
Engineer's transit, right side.
(*Courtesy of Teledyne Gurley.*)

the telescope and vertical circle are attached and on which both pivot. The markings on the vertical circle and the vertical-circle vernier are seen between the framework uprights in Fig. 4.5. The vertical circle is on the inside and its vernier is on the outside, unlike the horizontal-circle verniers, which are inside the horizontal circle. The vertical-circle clamp is seen at the top of the horizontal axis in Fig. 4.6 and the vertical tangent screw, which is positioned horizontally, is seen through the open telescope-support framework.

Except that it is designed to swing in a vertical plane, the telescope is quite similar to that of a dumpy level: The cross hairs include a pair of stadia hairs, the eyepiece and the main optics focus separately, and below the telescope barrel is a sensitive bubble whose axis lies parallel to the telescope optical axis.

VERNIERS. The horizontal circle is divided into whole degrees and further subdivided by 30-minute, 20-minute, or 15-minute marks, depending on the model of the transit. Refined readings to values much smaller than the smallest horizontal-circle division are made on verniers. Figure 4.7*a, b,* and *c* shows three different sets of horizontal circles (outside) and their accompanying vernier scales (inside). Note, in each case, that the vernier spaces are slightly closer together than the divisions on the horizontal circle. In Fig. 4.7*a,* for example, 30 vernier spaces cover the same arc as 29 spaces on the horizontal circle. The resultant *difference* in arc between the width of a vernier space and the width of a circle space is the vernier reading accuracy, also known as **least count,** which is defined as

$$\textbf{Least count} = \frac{s}{n} \qquad \textbf{(4.4)}$$

where s is the length of the smallest space on the circle scale in minutes of arc and n is the number of vernier spaces equal to $(n - 1)$ circle spaces. In Fig. 4.7*a,* the smallest division on the circle is half a degree, or 30 minutes, and 30 vernier spaces is equal to 29 circle spaces; therefore, s and n are both 30 and the least count is equal

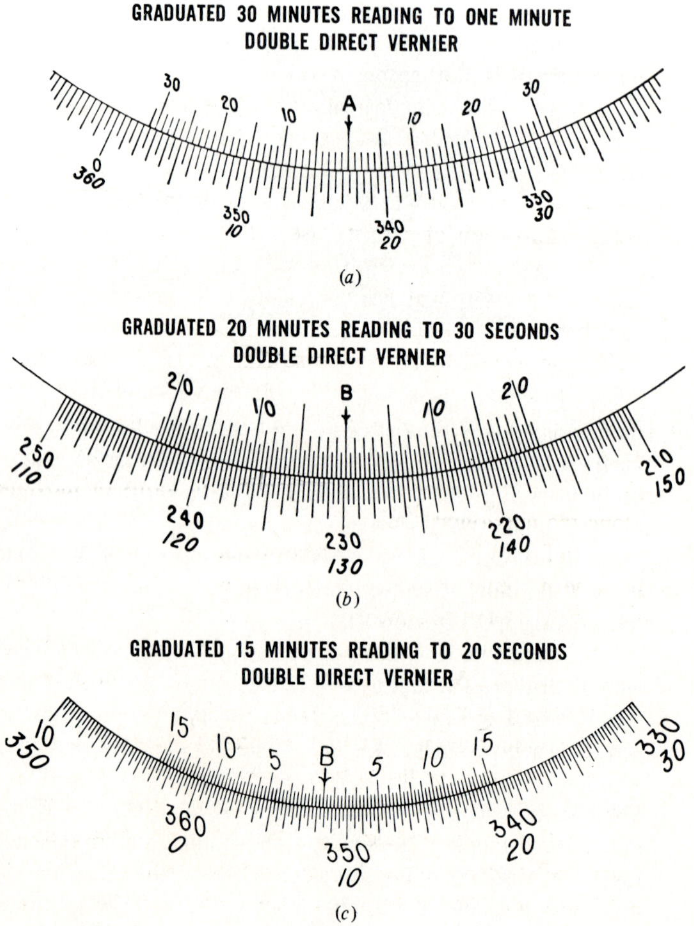

FIGURE 4.7
Examples of horizontal-circle verniers. (*Courtesy of Cubic Precision.*)

to 30′/30, or 1 minute. That means angles taken with this transit may be read to a 1-minute precision.

Example 4.5. What is the least count of the vernier in Fig. 4.7*b*?

Solution. The degree divisions on the circle scale are divided into three parts, each of which must be 20 minutes of arc; thus, $s = 20'$. The points of coincidence on the vernier are $10\frac{1}{2}$ minutes on the left side of the B arrow and $9\frac{1}{2}$ minutes on the right side. There are 40 vernier spaces between these two marks, corresponding to 39 spaces on the circle; thus, $n = 40$. Substituting into Eq. (4.4),

$$\text{Least count} = \frac{s}{n} = \frac{20'}{40} = \frac{1}{2}\ \text{minute} = 30\ \text{seconds}$$

However, least count is easier to determine than that. Looking at the vernier in Fig. 4.7*b*, it is seen that the smallest vernier division is half a minute; thus, by inspection, it is a 30-second vernier.

Before taking angle readings on the verniers in Fig. 4.7, it would be well to examine closely one of the vernier scales. In Fig. 4.7*a*, the arrow under "A" at the center of the vernier scale designates the zero point of the A vernier, but the zero does *not* coincide with an adjacent mark on the horizontal circle scale. However, five vernier marks left of zero it is seen that the vernier mark *is* in coincidence with its nearest neighbor on the horizontal circle. In the process of taking angles, the vernier scale moves left or right with respect to the stationary horizontal circle, depending on whether the angle is taken clockwise or counterclockwise. If the vernier in Fig. 4.7*a* were to be moved clockwise until the sixth mark clockwise from zero were in coincidence rather than the fifth mark, the transit telescope would have turned through a horizontal arc equal to an additional one-thirtieth of a vernier space. In the Fig. 4.7*a* case, this is one minute of arc, an amount of circular motion that is easy to detect on the vernier but essentially imperceptible in terms of movement of the A arrow line along the horizontal circle.

In Fig. 4.7*a*, of the two nearby lines on the circle scale, the A arrow is just past the line marking 342°30′ on the clockwise ring of numbers. Because the 5-minute line on the left (clockwise) side of the vernier scale is in exact alignment with a corresponding mark on the circle scale, the A arrow line must be exactly 5 minutes beyond the 342°30′ mark on the circle scale. Adding the two values, the clockwise angle reading must be $342°30' + 5' = 342°35'$.

Had the angle in Fig. 4.7*a* been taken counterclockwise instead, the right (counterclockwise) side of the vernier scale would have been used. In this case, it is seen that the 25-minute mark on the right side of the vernier is exactly aligned with a mark on the adjacent circle scale. Thus, the A arrow line must be 25 minutes in the counterclockwise direction beyond the appropriate circle scale value; that is, 25 minutes past 17°00′ on the counterclockwise (outer) ring of numbers. The counterclockwise reading must be $17°00' + 25' = 17°25'$. As a check, the counterclockwise value added to the clockwise value must equal 360°; in this case, $17°25' + 342°35' = 360°00'$.

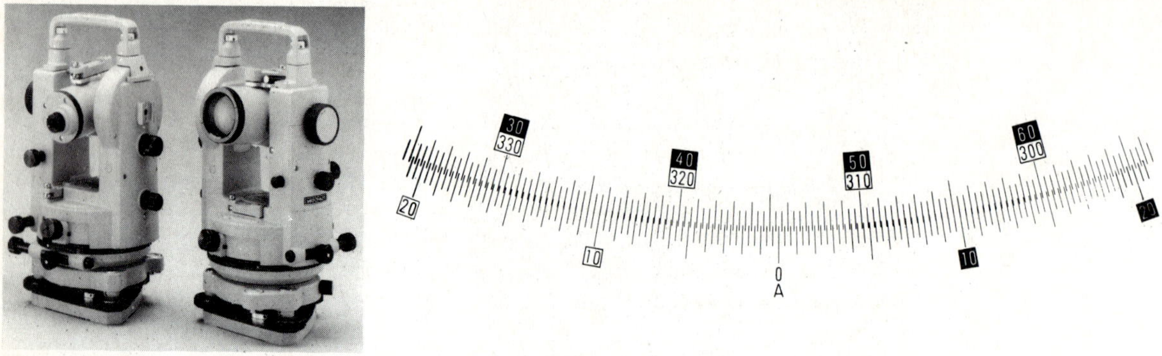

FIGURE 4.8
Topcon AG-20B transit and vernier scale. (*Courtesy of Topcon Instrument Corporation of America.*)

In Fig. 4.7*b*, the clockwise vernier reads 10′30″ and the B arrow is just past 229°40′ on the clockwise ring; therefore, the clockwise reading is 229°40′ + 10′30″ = 229°50′30″. In a similar manner, the counterclockwise reading is 130°00′ + 9′30′ = 130°09′30″. Checking, the clockwise and counterclockwise readings add to 360°00′00″.

In Fig. 4.7*c*, each minute on the vernier scale is divided into three parts; thus, the least count, and the reading precision, is 20 seconds. Coincidence occurs at 5′40″ on the clockwise side of the vernier and 9′20″ on the counterclockwise side. The B arrow falls between 351°30′ and 351°45′ on the clockwise ring, so the clockwise reading must be 351°30′ + 5′40″ = 351°35′40″. The counterclockwise reading is 8°15′ + 9′20″ = 8°24′20″. As a check, the two readings add to 360°00′00″.

Some newer vernier transits don't look much like engineer's transits. Figure 4.8 shows a Topcon AG-20B transit and its vernier. Note that the vernier scale is *outside* of, and slightly superimposed on the circle scale. The latter feature leads the eye more quickly to the line of coincidence of the two scales. Clockwise and counterclockwise readings are 314°30′20″ and 45°29′40″, respectively—the two readings add to 360°, as required.

Although reading angles on vernier scales may seem awkward and tedious at first, experienced surveyors are able to read angles on vernier transits almost as fast as most people read angles on scale-reading theodolites (discussed below), the latter being somewhat easier for beginners to use and read.

4.5 THEODOLITES

SCALE-READING THEODOLITES. The evolution from engineer's transits to today's highly sophisticated theodolites has taken place slowly. The first enclosed-body theodolites were first-order instruments using etched, optically viewed scales that allowed angles to be subdivided and read more precisely than was possible with vernier transits or with the open-micrometer, first-order instruments they replaced. Early-model theodolites intended for ordinary surveying incorporated about the same etched-scale technology as first-order theodolites but used simpler scales. Figure 4.9

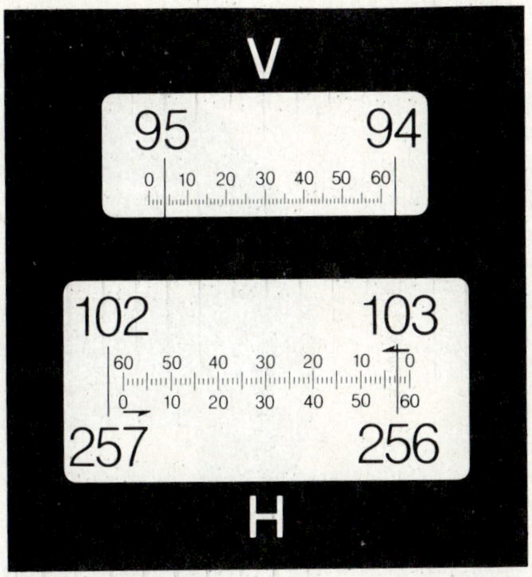

V: 95° 4′
H: ← 103° 2′40″
→ 256°57′20″

FIGURE 4.9
Reading scale from Sokkisha TS20A scale-reading theodolite. (*Courtesy of The Lietz Company.*)

shows the reading scale of one of these transitional theodolites. The vertical-angle window (which uses zenith angles) is on top and the horizontal-angle window is on the bottom. Within the bottom window, the two 0- to 60-minute scales are fixed within the field of view and the large numerals move as the instrument is turned clockwise or counterclockwise. The lower set of large numerals (256 and 257) denotes the clockwise angle and is paired with the lower of the two 0-to-60 scales (note the right-facing arrow at the zero point of the lower minute scale). Of the two vertical lines between the large numerals, only the line between 256 and 103 crosses the minute scale—it intersects the clockwise minute scale about four-tenths of the way between the 57-minute mark and the 58-minute mark; thus, the clockwise reading is $256° + 57.4' = 256°57.4'$. Reading counterclockwise, the upper minute-scale reading of 2.6 minutes is added to 103° to obtain 103°02.6′. As before, the clockwise and counterclockwise readings add to 360°00.0′.

A single-window, scale-reading theodolite such as the one whose scale is shown in Fig. 4.9 won't really afford 0.1-minute precision. The readings given above require more faith in interpolation than is merited—half-minute interpolation is probably a more reasonable expectation with this type of instrument.

DIGITAL MICROMETER THEODOLITES. Newer theodolites use separate scales: a main circle scale for degrees and a micrometer scale for minutes and seconds. Figures 4.10, 4.11, and 4.12 show three such instruments, along with views of their respective reading windows. With their telescopes in the direct position, the eyepiece for reading horizontal and vertical angles is just to the right of the main telescope eyepiece and the micrometer knob is on the outside of the right telescope support pillar—with telescopes reversed, the opposite is the case. The micrometers on all three instruments function in the same manner. Consider the horizontal-angle window in Fig. 4.10. When the

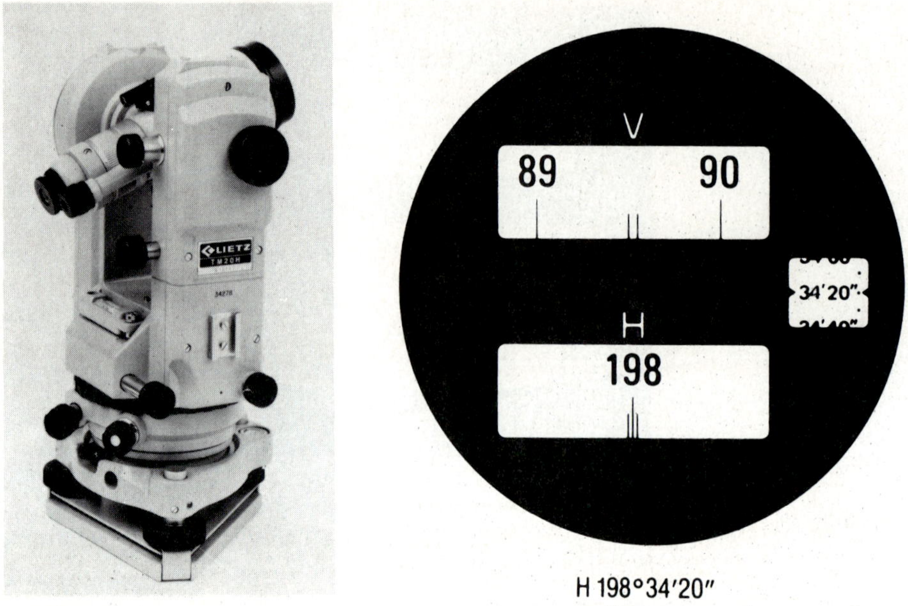

FIGURE 4.10
Lietz TM20H micrometer theodolite. (*Courtesy of The Lietz Company.*)

telescope was first centered on the right object (most digital micrometer theodolites provide clockwise angles only), the numeral 198 with its accompanying vertical line lay to one side of the short pair of index lines at the center of the main scale. Turning the micrometer knob until the line under 198 was centered between the index lines brought the correct horizontal minute-second reading into the micrometer-dial window at the right. The horizontal-angle reading is thus 198°34′20″. On this instrument, minute-second readings are numbered every 20 seconds with 10-second tick marks in between; thus, direct-reading precision for horizontal angles is 10 seconds. Vertical-angle precision on this model is also 10 seconds.

Figure 4.11 shows a digital-reading theodolite whose horizontal-angle reading (center) is 90°15′06″. The vertical-angle reading (at right) indicates a zenith angle of 86°05′36″. The reading precision of this instrument, both horizontally and vertically, is 6 seconds, with reasonable interpolation to 3 seconds.

The 86°05′36″ zenith-angle reading in Fig. 4.11 refers to the angle measured downward from a straight-up (zenith-zero) reference. The vertical angle referred to horizontal-zero is 90° minus the zenith angle. In the Fig. 4.11 case, $90° - 86°05'36'' = +3°54'24''$, the + sign indicating an elevation angle, that is, upward from horizontal-zero. Had the Fig. 4.11 zenith angle been, say, 96°05′36″, the angle referred to horizontal-zero would have been $90° - 96°05'36''$, or $-6°05'36''$, the − sign denoting a depression angle, that is, downward from horizontal-zero.

Unlike the engineer's transit, whose vertical-angle index is fastened to the telescope-support frame and is subject to misadjustment and errors in plate-bubble leveling, most theodolites have vertical circles with pendulum compensators that pro-

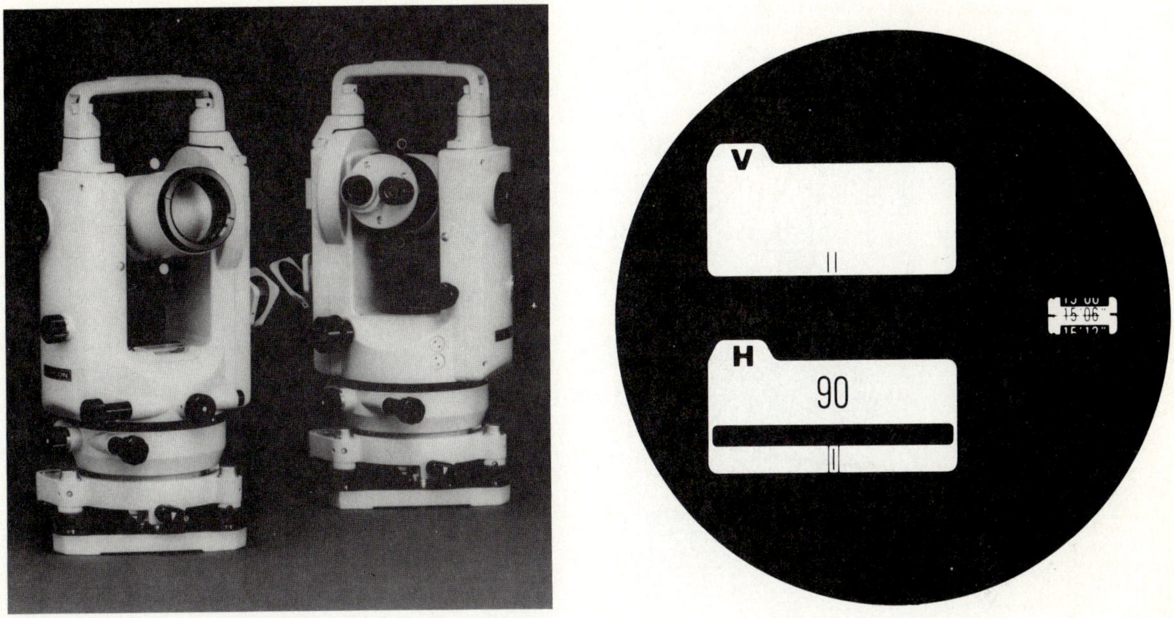

FIGURE 4.11
Topcon TL-6DE micrometer theodolite. (*Courtesy of Topcon Instrument Corporation of America.*)

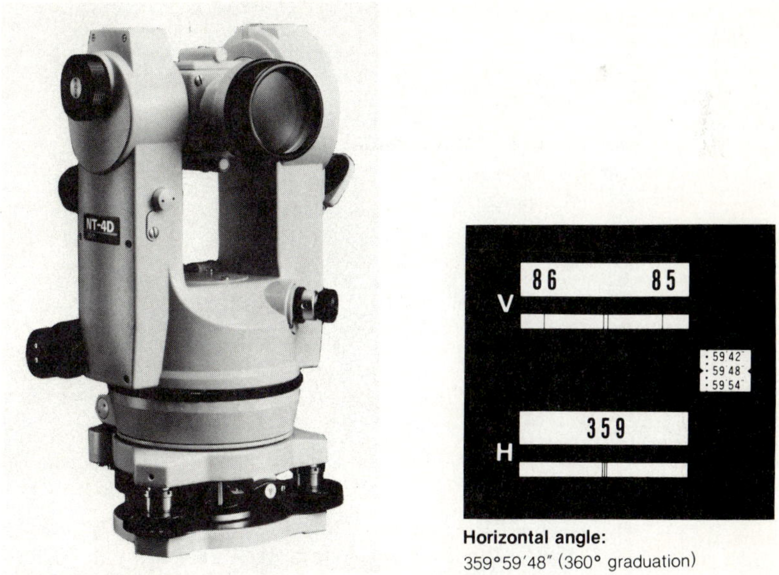

FIGURE 4.12
Nikon NT-4D micrometer theodolite. (*Courtesy of Nikon Incorporated, Surveying Instrument Division.*)

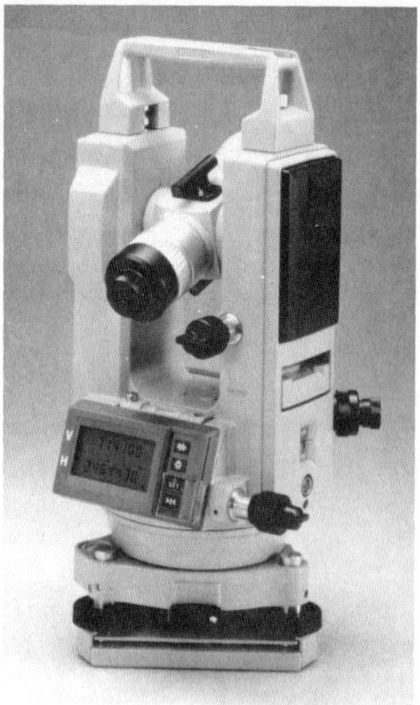

FIGURE 4.13
Lietz DT5 electronic digital theodolite. (*Courtesy of The Lietz Company.*)

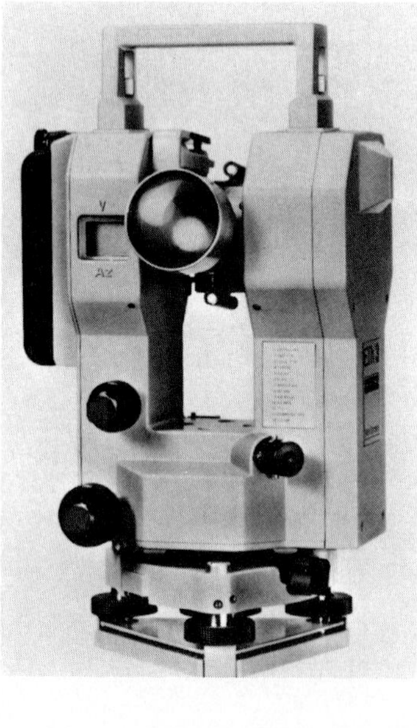

FIGURE 4.14
Zeiss ETh3 electronic digital theodolite. (*Courtesy of Carl Zeiss, Incorporated.*)

vide an index reference within 1 second of true vertical. The theodolites in Figs. 4.10 through 4.14 all have vertical compensators, as do the three total-station instruments pictured and described in Sec. 4.6.

Figure 4.12 shows another micrometer theodolite whose horizontal circle window shows the 359° mark centered in the horizontal index pair and a micrometer reading of 59′48″, for a horizontal clockwise angle reading of 359°59′48″. This instrument is direct reading to 3 seconds, on both horizontal and vertical angles, with potential for interpolation to values less than 3 seconds.

ELECTRONIC DIGITAL THEODOLITES. Digital micrometer theodolites require the observer to manually match a whole-degree index line to a fixed-index pair of lines on the main circle, producing a minutes-seconds reading on a secondary scale that is then added to the main-circle degree reading to produce the correct horizontal- or vertical-angle value. With an electronic digital theodolite, on the other hand, the observer need only zero the instrument on the backsight object, point the telescope at the foresight object, and read the horizontal and vertical angles directly, on a combined liquid-crystal display (LCD), to whatever precision the theodolite is capable. The angle readout derives from the instrument's automatic scan of precise, internal, digitally encoded circle scales.

The electronic theodolite shown in Fig. 4.13 automatically reads and displays either clockwise or counterclockwise horizontal angles as well as automatically indexed vertical angles, either zenith-zero or horizontal-zero. Angle display is in degrees-minutes-seconds to 5 seconds, gon to 1 mgon, or mils (6400 mils equal one circumference) to 0.02 mil, units being user-selectable. Replaceable AA batteries provide up to 15 hours continuous field operation. The instrument also has provision for angle readout to an optional electronic field book via RS232 hookup.

Figure 4.14 shows another electronic digital theodolite. It displays horizontal angles either clockwise or counterclockwise and vertical angles as zenith-zero, horizontal-zero, or percent slope. There are two compensators: one for vertical-angle indexing, the other to compensate for an out-of-level telescope horizontal axis. Readout units include degrees-minutes-seconds (to 2 seconds), decimal degrees, gon and mils, user-selectable. Replaceable NiCad batteries provide up to 18 hours field use. Interfacing to an external electronic recorder is available through an RS232 connector.

4.6 TOTAL-STATION INSTRUMENTS

Consolidating an advanced theodolite and an EDM into one instrument provides distinct advantages. For one thing, both distances and angles may be measured with a single instrument, thus saving setup and takedown time. In addition, if the theodolite telescope and the EDM optics are on the same axis (and the center of the prism array is used as the theodolite target) there is no need for an EDM slope-reduction computation of the sort discussed in Sec. 2.14. Finally, concurrent availability of slope distances along with the associated vertical and horizontal angles allows incorporation of microcircuitry to provide horizontal distances, vertical distances, horizontal and vertical coordinates, and other useful computations of importance to the plane surveyor.

Figure 4.15 shows a total-station instrument with micrometer-type, 6-second direct readout of both vertical angles and clockwise horizontal angles. Direct liquid-crystal-display readout is provided for single-prism slope distances to 1.8 km ± (5 mm + 3 ppm). A control panel provides for internal computation of horizontal and vertical distances along with horizontal and vertical coordinates. Optional data collection is available.

Figures 4.16 and 4.17 show total-station instruments that are based on electronic digital angle measurement rather than a micrometer knob. As is the case with electronic digital theodolites, they provide either clockwise or counterclockwise horizontal angles and zenith-zero or horizontal-zero vertical angles, with angular units selected by the user. Both provide internal computation of horizontal and vertical coordinates as well as horizontal and vertical distances, the latter being automatically corrected for curvature and refraction using the normal-atmospheric-condition equation. The total station in Fig. 4.16 has a 1-second or 0.2-mgon least count for horizontal and vertical angles and a single-reflector slope-distance range to 1.6 km ± (5 mm + 5 ppm). The instrument in Fig. 4.17 advertises a least count of 1 second or 1 mgon and a single-reflector slope-distance range to 2 km ± (3 mm + 2 ppm).

The total stations in Figs. 4.16 and 4.17 also provide digital data recording options, the former to a small external unit through an RS232 connector, the latter by way of a plug-in module that directly attaches to the instrument.

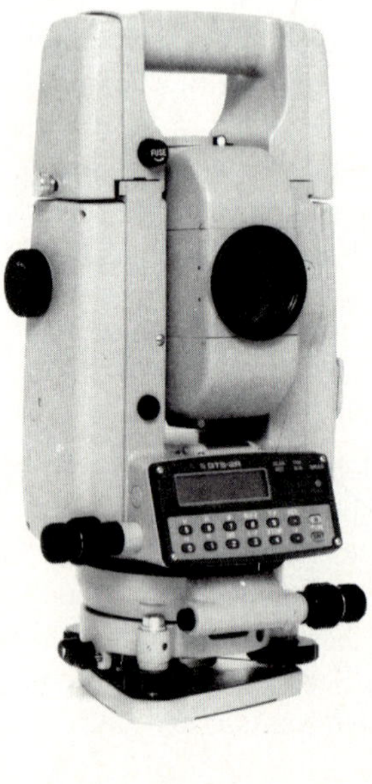

FIGURE 4.15
Topcon GTS-2R total station. (*Courtesy of Topcon Instrument Corporation of America.*)

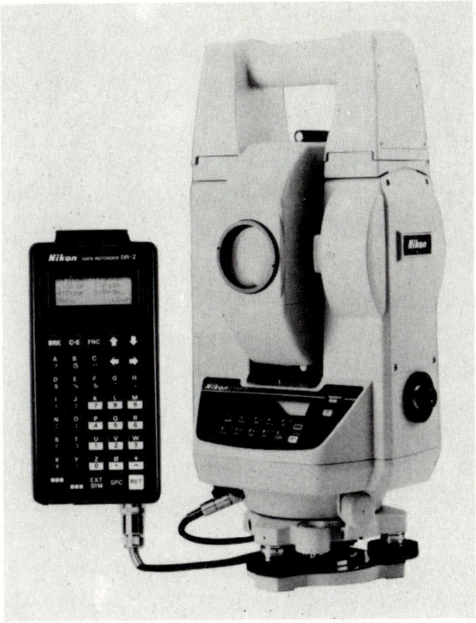

FIGURE 4.16
Nikon DTM-1 total station. (*Courtesy of Nikon Incorporated, Surveying Instrument Division.*)

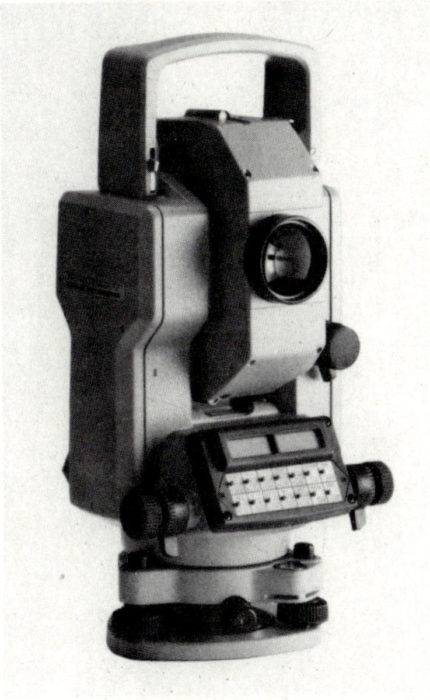

FIGURE 4.17
Wild TC1600 total station. (*Courtesy of Wild Leitz USA, Inc.*)

4.7 MEASURING HORIZONTAL ANGLES

The purpose of measuring horizontal angles, usually, is to establish the shape of a particular area and/or to establish the direction of a line. In either case, the instrument must be set up so its vertical center lies directly over a horizontal control point, a property corner, or a turning point in a traverse survey, such point being marked by a brass disc set in a permanent monument or by an aluminum surveyor's tack in a wooden hub. First, the problem of setting up over the point.

SETTING UP WITH A PLUMB BOB. Set the tripod approximately over the survey point with its top ring about level and attach the transit. Loosen the transit's leveling screws slightly, center the instrument within the shifting range of the baseplate, and tighten the leveling screws until just snug. Hang the plumb bob so its point is an inch or two higher than the elevation of the survey point. If the plumb bob is offset horizontally from the survey point by more than a couple of inches, move the tripod to place the plumb bob within an inch of the point. Alternately, until the plumb bob is nearly over the point, push the tripod leg that is on the opposite side of the survey point from the plumb bob deeper into the ground; this will move the plumb bob toward the point. Continue until the tripod is firmly set and the plumb bob is within a quarter-inch of the point. Loosen two adjacent leveling screws, shift the transit until the plumb bob is exactly over the point, then tighten the leveling screws until just snug.

With the instrument centered over the survey point, loosen either the upper or lower clamp and turn the transit until the plate bubble axes are parallel to opposing pairs of leveling screws, as illustrated in Fig. 4.18*a*. Center the plate bubbles by turning opposing leveling screws in equal and opposite directions (the bubble will move in the direction of motion of the left thumb). When both plate bubbles are

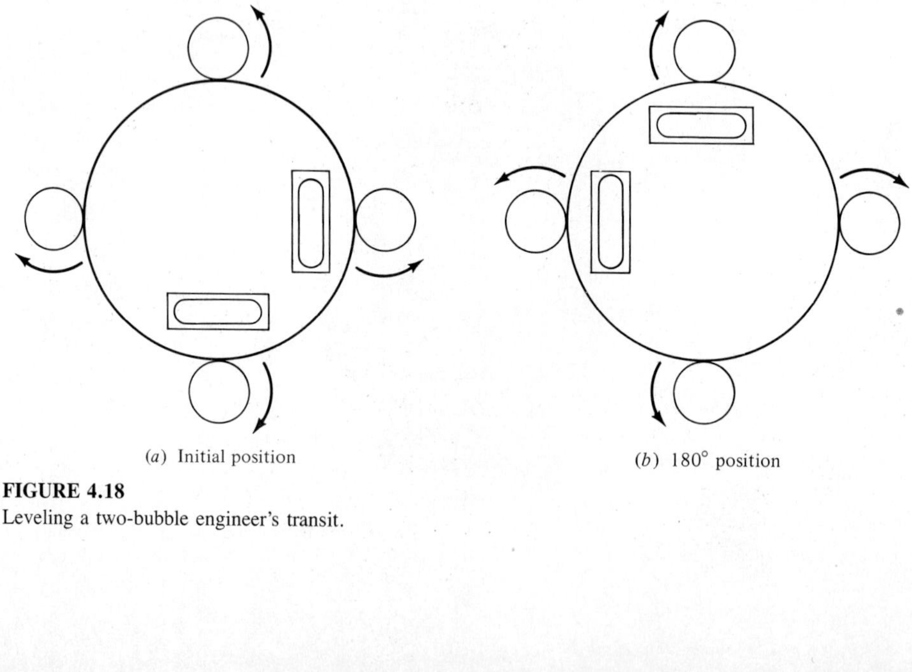

(*a*) Initial position (*b*) 180° position

FIGURE 4.18
Leveling a two-bubble engineer's transit.

precisely centered, turn the instrument 180° to the Fig. 4.18*b* orientation. If both plate bubbles remain centered, leveling is complete. If either or both are offset from the centered position, bring the bubble(s) halfway back to the centered position using opposing leveling screws as before.

SETTING UP WITH AN OPTICAL PLUMMET. An optical-plummet instrument may be set up in the manner just described but there is a much quicker method available. Fasten the instrument to a matching, adjustable-leg tripod with the instrument approximately centered in its shifting-plate range and the three leveling screws centered in their clockwise-counterclockwise range. Set the tripod firmly, with the instrument roughly over the point and the baseplate roughly level. Adjust the optical plummet's eyepiece focus and its in-and-out field focus until the ground surface and the reticle circle are both in focus and parallax has been eliminated. Using the leveling screws, bring the the image of the reticle circle onto the survey point.

At this stage, the bubble in the bull's-eye level will be off center in one direction or another. Starting with the tripod leg nearest in line with the offset direction of the bull's-eye bubble, extend or retract the adjustable leg until the bubble has moved closer to the bull's-eye circle and the bubble-to-circle axis is aligned with an adjacent tripod leg. Repeat the procedure until the bubble lies wholly within the circle. By centering the bull's-eye bubble in this manner, using tripod legs only, the instrument will have been brought almost directly over the survey point and nearly leveled in the process.

To level precisely, unclamp the upper instrument housing to allow rotation about the vertical axis, then turn the instrument so the plate bubble is parallel with two of the three leveling screws, as shown in Fig. 4.19*a*. Using those two screws only, center the plate bubble. Now turn the instrument 90° to the Fig. 4.19*b* position and bring the bubble to center again with the other leveling screw only. Look through the optical plummet, loosen the tripod clamp slightly, and move the instrument on its baseplate

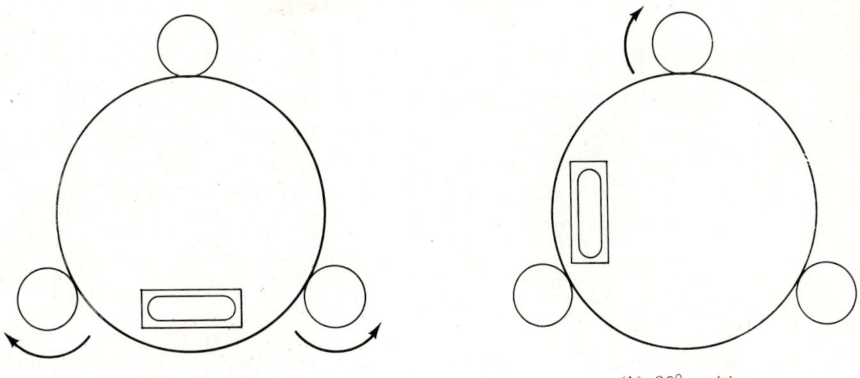

(*a*) Starting position (*b*) 90° position

FIGURE 4.19
Initial positions for leveling a three-screw instrument.

sufficient to bring the reticle circle image exactly on the survey point. The instrument is now directly over the survey point but it is not yet perfectly level; for one thing, moving the instrument with respect to the tripod baseplate has changed the level slightly, for another, refined leveling is still needed to compensate for a misadjusted plate level bubble.

Reclamp the tripod clamp and relevel the theodolite in the Fig. 4.19*a* and *b* positions as before. This time, after the plate bubble is centered in the initial position and in the 90° position, swing the instrument 180° away from the Fig. 4.19*a* position, check the plate bubble centering, and, if necessary, use the appropriate leveling screws to bring it halfway back to center. Repeat, in the orientation 180° away from the Fig. 4.19*b* position, again using the single leveling screw. When this procedure is completed, the instrument will be truly level and horizontal angles will be observed truly horizontally.

OBSERVING ANGLES. Consider Fig. 4.1, a plan view of a horizontal angle, with the instrument set up at the apex. To begin angle measurement, the instrument scale is usually set to read zero when the telescope cross hairs are exactly aligned with the center of the backsight point. Zeroing the scale and pointing the telescope require two separate operations.

1. Loosen the upper instrument clamp and rotate the instrument on its vertical axis until a near-zero reading is obtained on the vernier, scale, micrometer, or electronic readout, as appropriate. Tighten the upper instrument clamp and bring the angle reading exactly to zero using the upper tangent screw.
2. Loosen the lower instrument clamp. Turn the telescope until it points approximately to the center of the backsight point. Reclamp and bring the telescope vertical cross hair exactly on the center of the backsight point with the lower tangent screw. If both procedures were properly done, the angle reading will still be exactly zero—check to make sure.

To measure an angle once the instrument is zeroed as above:

3. Using upper clamp and tangent screw only, center the vertical cross hair exactly on the foresight point. Depending on the type of instrument being used, read the horizontal angle vernier as discussed in Sec. 4.4 or the theodolite horizontal-angle scale as covered in Sec. 4.5. Record the angle in the field book or electronic recorder, checking to verify agreement between the instrument reading and the recorded value.

4.8 IMPROVING ANGLE MEASUREMENT

Measuring a single angle can be a chancy business because there are a number of mistakes even experienced surveyors occasionally make. Mechanical errors can result from turning the lower tangent screw inadvertently, from clamping and unclamping in the wrong sequence, or from tightening a clamp insufficiently. Similarly, on instruments having both clockwise and counterclockwise circles, there is always the danger of reading numbers on the wrong circle or in the wrong direction. Reading errors also result when the surveyor drops a whole degree, misreads a vernier, or reads the wrong side of the vernier.

DOUBLE-CENTERING. To check against angle-measuring errors of this sort and to compensate for a number of possible manufacturing defects (see Sec. 4.11), it is usually best to measure an angle more than once as a check. The simplest method of doing this is known as **double-centering,** which means observing one angle "direct" (that is, with the telescope level vial below the telescope on an engineer's transit—or with the angle-reading optics to the right of the telescope in the case of a theodolite) and a second angle "reversed," or "plunged" (that is, with telescope inverted, so the level vial is on top of the telescope of the engineer's transit, or so the theodolite's angle-reading optics is to the left of the telescope optics). If direct and reversed angles are measured properly, their sum will be exactly twice the direct angle or very nearly so. Dividing the sum by two provides a significantly better value for the angle so observed. Continuing after step 3 above,

4. After taking the direct angle reading, "plunge" the telescope (that is, invert it by turning it on its horizontal axis). Using the lower clamp and tangent screw only, put the vertical cross hair exactly on the backsight object again as in step 2 (the direct-angle value will still be on the vernier or in the reading window). Now, using the upper clamp and tangent screw only, repeat step 3 to again center the vertical cross hair exactly on the foresight object. The reading is now the sum of the direct and reversed angles, giving a more accurate mean and compensating most of the small systematic errors resulting from certain small, unavoidable, manufacturing flaws. In addition, by comparing the final mean with the first reading, a mistake in reading either angle is readily detected.

STA	FROM	TO	INSTR	∠RDG	ANGLE
Hilltop	Pole		Direct	00°00′00″	
		Spire	Direct	46°14′20″	46°14′20″
		Spire	Reversed	92°29′00″	46°14′40″
				Mean	=46°14′30″ ✓

FIGURE 4.20
Recording a double-centered horizontal angle.

Figure 4.20 is a suggested six-column recording format for double-centered angles. The setup is over survey point Hilltop and a horizontal angle is measured (say, clockwise) from survey point Pole to survey point Spire. (It is usually the case that angles will be observed to a number of points from one setup; for the sake of simplicity only one angle is shown here.) The angle values under ∠RDG (short for "angle reading") are the transit-vernier or theodolite-scale readings from what is fairly obviously a 20-second instrument. The values under ANGLE are the differences between consecutive ∠RDG values; the 46°14′20″ ANGLE value is the difference between the 46°14′20″∠RDG value and the 00°00′00″∠RDG value while the 46°14′40″ ANGLE value is the difference between the 92°29′00″∠RDG value (the sum of the direct and reversed angles) and the 46°14′20″∠RDG value. An arithmetic check is derived from comparing the mean of the two ANGLE values with half the final ∠RDG value of 92°29′00″.

REPEATING ANGLES. For still greater accuracy with a given transit or theodolite, repeat the angles several times, observing an equal number of telescope-direct and telescope-reversed angles and dividing the resultant sum of the angles by the total number of angles observed. Thus, average angle = Σ/N, where N is the number of times the angle is observed and Σ is the sum of the angle values, which is the reading on the instrument after the final angle is observed, if the initial backsight is taken with zero on the instrument scale and if the sum of the angles didn't exceed 360°. Figure 4.21 illustrates a six-column field book record for that sort of situation, using a 20-second transit to observe six angles, three direct and three reversed. The column headed Rep designates number of "repetitions," or number of instrument pointings: 0 is the initial pointing on the backsight object, 1 is the first pointing on the foresight object, and 6 is the final foresight-object pointing (there is no need to record intermediate pointings).

Sta.	From/to	Rep.	Tel.	Circle rdg.	Mean vernier
Hilltop	Pole/	0	Dir	00°00′	00″
	/Spire	1	Dir	46°14′	20″
	/Spire	6	Rev	277°27′	20″

FIGURE 4.21
Recording repeating angles (sum less than 360°).

Example 4.6. Given the data in Fig. 4.21, compute the average angle and check it against the initial telescope-direct angle.

Solution. Compute by converting 227°27′20″ to a decimal value, dividing by number of repetitions, and converting back into degrees, minutes, and seconds. In this case, 277°27′20″ ÷ 6 = 46°14′33″.

The mean angle is very close to the value of the initial angle recorded in Fig. 4.21, a reasonable check on the repeating-angle procedure.

Observing angles by repetition does not require that the sum of the angles be less than 360°; neither is it necessary that the instrument reading be zero on the initial backsight-object pointing. The following all-purpose repeating-angle equation takes both factors into account:

$$\textbf{Average angle} = \frac{F + n(360°) - I}{N} \qquad \textbf{(4.5)}$$

Where F is the final angle reading, I is the horizontal-angle reading on the initial backsight pointing, and n is the number of times the 360° point has been passed in arriving at a sum equal to a total of N angle repetitions plus the initial backsight reading.

Figure 4.22 shows the recording format for a total of 12 angles, 6 direct and 6 reversed, the initial pointing being nonzero, and the 360° mark being passed at least once in adding up the 12 angle repetitions. Note also that the instrument used was a 20-second transit and that the observer read and recorded both the A and B vernier readings.

					Vernier	
Sta.	**From/to**	**Rep.**	**Tel.**	**Circle rdg.**	**A**	**B**
Pt *J*	Pt *Q*/	0	Dir	215°26′	20″	00″
	/Pt *W*	1	Dir	354°37′	40″	
	/Pt *W*	12	Rev	85°44′	20″	20″

FIGURE 4.22
Recording repeating angles (sum greater than 360°).

Example 4.7. Using the data in Fig. 4.22, compute the mean value of angle *QJW* and check against the initial value.

Solution. Determine the value for n and substitute into Eq. (4.5) along with F, I, and N. To find n, the number of times the 360° mark has been passed enroute to the mean F value of 85°44′20″, begin with the approximate value of *QJW*, multiply by 12, add the approximate value of I, and determine how many multiples of 360° may be subtracted from it and still leave a positive value less than 360°.

The approximate value of *QJW* is 354° − 215° = 139°. Multiplying by 12 and adding to the approximate value of I, 139°(12) + 215° = 1883°, which means that $n = 5$. That is, 1883 − 5(360) = 83, which, very roughly speaking, is the same as the final reading for F.

The numerator for Eq. (4.5) is found as follows:

$$\text{Mean value of } F + 5(360^\circ) = 85^\circ 44'20'' + 1800^\circ = 1885^\circ 44'20''$$
$$\text{Subtract the mean value of } I: \quad -\,215^\circ 26'10''$$
$$\text{Therefore, } F + n(360^\circ) - I = 1670^\circ 18'10''$$

Converting to a decimal value and dividing by 12,

$$1670^\circ 18'10'' \div 12 = 139^\circ 11'31''$$

To determine the value of the first angle, subtract the A-vernier reading of the initial backsight pointing from the A-vernier reading of the first foresight pointing, or

$$\text{First angle} = 354^\circ 37'40'' - 215^\circ 26'20'' = 139^\circ 11'20'' \checkmark$$

This is seen to closely check the mean-angle value.

4.9 MEAN-ANGLE ERROR

It may have occurred to the reader by now that the 1-second precision indicated in the answers to Examples 4.6 and 4.7 is probably not valid. The question then arises as to the sort of precision one might expect from observing N angles, direct and reversed, with an instrument whose precision in setting or reading angles is relatively poor. The answer is that it is not just the angle-reading or angle-setting precision that produces a significantly higher precision when measuring angles by repetition. Also involved is *pointing* precision, the precision with which the vertical cross hair may be set on the backsight or foresight object. When observing angles by repetition, the initial value is set into the reading scale just once and the final, sum-of-angles value is read from the reading scale just once (the first-angle value is only read for an approximate check)—the rest of the pointings are not read. Thus, the precision of the average angle is dependent on the precision of the initial setting, the precision of each pointing (twice as many pointings as angles, since each angle requires one pointing on the backsight object and one pointing on the foresight object), and the precision of the final reading. As will be explained in Chap. 5, the resultant error of the average angle may be expressed as

$$\boldsymbol{E} = \frac{\pm\sqrt{\boldsymbol{E}_i^2 + 2\boldsymbol{N}\boldsymbol{E}_p^2 + \boldsymbol{E}_r^2}}{\boldsymbol{N}} \tag{4.6}$$

where E = overall error, or precision

E_i = initial-setting precision

E_p = pointing precision

E_r = precision in reading the final value

N = total number of angle repetitions observed direct and reversed

("D&R" is the usual notation for indicating that angles are measured direct and reversed.)

Example 4.8. In Example 4.7, assume that the precision in both setting and reading a 20-second instrument is $\pm 10''$ and that the precision in pointing the telescope cross hair on an object is $\pm 2''$. Compute the precision of the mean angle for the 12 D&R angles.

Solution. E_i and E_r are $\pm 10''$, E_p is $\pm 2'$, and N is 12. Substituting into Eq. (4.6),

$$E = \frac{\pm\sqrt{(10)^2 + 2(12)(2)^2 + (10)^2}}{12} = \pm 1.43''$$

Because the computed value of E in the above case was close to 1 second, the 1-second result given in the solution to Example 4.7 is probably valid. Examination of Eq. (4.6) shows that, for a large N, the pointing precision is the dominant factor. In fact, as Example 4.8 shows, the average error for D&R repeating angles is less than the telescope-pointing precision if a large number of angles are observed. This is an important advantage of the repeating-angle procedure.

4.10 MEASURING VERTICAL ANGLES

The procedure for observing vertical angles with a transit or theodolite is not too unlike that for observing horizontal angles. For horizontal angles, the zero point is the backsight object; for vertical angles, the zero reference is either the telescope's horizontally held optical axis or it is the upper extension of the instrument's vertical axis, depending on whether the angle is horizontal-zero or zenith-zero. The surveyor measures direct and reversed horizontal angles as a self-check on angle measurement errors as well as to compensate for certain small irregularities in instrument construction. By the same token, the prudent surveyor will observe direct and reversed vertical angles for the purpose of verification as well as compensation.

INSTRUMENT PROCEDURE. Horizontal-zero and zenith-zero vertical angles are both observed in the same way but the numerical values are different; the following explanation involves horizontal-zero angles, using the convention that an upward (elevation) angle is positive while a downward (depression) angle is negative. To observe a vertical angle, direct and reversed:

1. Set up and level the instrument. With the telescope in the direct position, loosen either the upper or lower clamp and swing the telescope to align the vertical cross hair on the point to which the vertical angle will be measured.
2. Using the vertical clamp and tangent screw, bring the *center horizontal* cross hair exactly on the point. Read and record the direct vertical angle; for example, $+6°17'$.
3. "Plunge" the telescope (that is, invert it on its horizontal axis) and swing it 180° to the horizontal position of the point. Repeat the procedure in step 2 above, this time reading and recording the reversed vertical angle; for example, $(+)6°13'$ (assume the same algebraic sign as the direct reading, no matter what the indication on the vertical-circle scale).
4. Obtain the mean of the D&R readings; $+6°15'$, using the direct and reversed angle values in steps 2 and 3 above. This is the correct vertical angle.

INDEX ERROR. Because the values of the direct and reversed angles in the above explanation are relatively close to one another, there was probably not a significant error in measuring either angle. The fact that a measurable difference does exist, however, indicates there are one or more errors resulting from irregularities in manufacture and/or adjustment of the instrument. Lumped together, these errors comprise the **index error** of the transit or theodolite. The potential for significant index error is greater in an instrument with a vernier index on the vertical circle, as used in the transits in Figs. 4.5 and 4.8, than in the other instruments illustrated in this chapter, which are equipped with vertical-index compensators.

Index error is half the difference between the direct and reversed angle readings. Thus,

$$\textbf{Index error} = \frac{\textbf{direct angle} - \textbf{reversed angle}}{\textbf{2}} \tag{4.7}$$

Example 4.9. Compute index error from the direct and reversed vertical-angle readings given in the above explanation of taking vertical angles.

Solution. Substituting into Eq. (4.7),

$$\text{Index error} = \frac{+6°17' - (+)6°13'}{2} = +02'$$

INDEX CORRECTION. It is more useful to consider **index correction** than index error as such. Recall that a correction value is of the opposite algebraic sign from an error value, that is,

$$\textbf{Index correction} = -\textbf{index error}$$

$$= \frac{\textbf{reversed angle} - \textbf{direct angle}}{\textbf{2}} \qquad \textbf{(4.8)}$$

Example 4.10. Compute index correction for the Example 4.9 data.

Solution. Substituting into Eq. (4.8),

$$\text{Index correction} = \frac{6°13' - 6°17'}{2} = -02'$$

The index correction of a transit or theodolite is a convenient value to be aware of. Once determined, it may be applied to all subsequent vertical angles observed direct with that instrument—until there is reason to suspect that the index error has changed, usually because the instrument has been jarred out of adjustment. When applying corrections to horizontal-zero angles, exercise care in observing algebraic signs, remembering that an elevation angle is positive and a depression angle is negative.

Example 4.11. A depression angle of $-4°46'$ is observed with the Example 4.10 instrument, telescope direct. What is the corrected vertical angle?

Solution. Substitute into the basic correction equation, true value = observed value + correction, to produce

$$\text{Corrected vertical angle} = -4°46' + (-02') = -4°48'$$

It would be well to consider again the relationship between vertical angles observed horizontal-zero and those observed zenith-zero. For zenith angles, straight up is zero, so the instrument should read 90° with the telescope direct and its axis horizontal, 270° when the telescope is reversed and its axis horizontal. If a horizontal-zero vertical angle is designated **V**, direct zenith angles **Z**, and reversed zenith angles $\mathbf{Z}_r$, the three are related by the following equations:

$$\mathbf{Z = 360° - Z_r} \qquad \textbf{(4.9)}$$

$$\mathbf{V = 90° - Z} \qquad \textbf{(4.10)}$$

Example 4.12. A zenith angle, measured with telescope-reversed, reads 267°49.4′. What are the corresponding values of a direct-measured zenith angle and a horizontal-zero angle.

Solution. Substituting into Eq. (4.9),

$$\text{Direct-measured zenith angle} = \mathbf{Z} = 360° - 267°49.4' = 92°10.6'$$

Substituting Z into Eq. (4.10),

$$\text{Horizontal-zero vertical angle} = \mathbf{V} = 90° - 92°10.6' = -2°10.6'$$

4.11 ANGLE-MEASURING ERRORS

Angle-measuring errors fall into two categories: instrumental and personal. Instrumental errors are the inevitable result of certain inherent imperfections in manufacture or adjustment. They tend to be cumulative and they may or may not be compensated. Small personal errors are almost invariably random in nature and thus they tend to balance out. Large personal errors do not balance out; therefore, the observer must constantly check procedures and results to insure that no large personal errors creep in.

INSTRUMENT ERRORS—VERTICAL ANGLES

Instrument does not read zero (or 90°, in the case of zenith angles) when the telescope axis is horizontal. There are three possible causes: (1) the telescope optical axis is not parallel to the telescope bubble axis (on instruments that have a telescope bubble), (2) the angle-reading vernier is displaced along the arc of the vertical circle (on instruments having verniers), and/or (3) the vertical axis is inclined (or, for an instrument with a vertical compensator, the compensator is out of adjustment). As long as the resulting error is the same from one measurement to another, this may be compensated by observing vertical angles direct and reversed.

Telescope's horizontal (pivot) axis is not perpendicular to instrument's vertical axis. The telescope will not swing in a vertical plane even though the vertical axis of the instrument is truly vertical. The end result of this condition, also inherent in instrument manufacture, is vertical angles that are taken along a diagonal, rather than a vertical path. The tilt of the horizontal axis is usually quite small, so the error in vertical angle, which may be approximated as proportional to the cosine of the tilt angle, is ordinarily insignificant. This error cannot be compensated by observing angles direct and reversed. It is detected readily by observing the path taken by the cross-hair center in moving from a high object to a ground point with the telescope in the direct position, then repeating with telescope reversed to see how far apart the two ground points are. This can be adjusted by the observer (on some instruments) or by a competent technician at a repair facility.

INSTRUMENT ERRORS—HORIZONTAL ANGLES

Telescope optical axis is not perpendicular to telescope pivot axis. This irregularity in instrument construction will always be present to some degree and will be especially bothersome when extending a straight line by plunging the telescope, a problem that will be discussed at length in Sec. 6.1. The condition is compensated by observing horizontal angles direct and reversed.

Telescope pivot axis is not perpendicular to telescope optical axis. As indicated earlier, the impact on vertical angles is not particularly significant. It *is* significant for horizontal angles when the backsight and foresight objects are at significantly different relative elevations. In the case of horizontal angles, the error is reasonably compensated if direct and reversed angles are observed.

***A* and *B* verniers are not precisely 180° apart or verniers are not concentric with the horizontal circle.** In either case, there will be an error in a single-angle reading. This error, too, is compensated by measuring horizontal angles both direct and reversed and by taking the mean of the *A* and *B* vernier readings.

Horizontal circle is eccentric or graduations are not evenly spaced around the circle. The error may be compensated by taking the initial reading at several points around the circle rather than at zero only. The disadvantage to this procedure is that the initial backsight reading must be set into the instrument, and a foresight reading made, each time an angle is observed, thus losing the increased-precision advantage of the repeating-angle procedure discussed earlier.

PERSONAL ERRORS. The measured values resulting from any surveying operation will inevitably be different from the unknowable true values. Once instrument errors are reasonably compensated, residual differences between measured and true values are for the most part the result of an accumulation of personal errors by the instrument operator, some of which are avoidable and some not.

Not centering over the mark. This error has the greatest impact where distances are short or where the offset is at right angles to the line of sight.

Carelessness in pointing. Most instruments these days allow the observer to point on an object within about 3 seconds. A 1-minute error in pointing, which would be rather extreme, would result in a displacement of about 1 in. in 100 ft.

Parallax. This error, though potentially quite significant, is readily avoidable. It is eliminated by careful refocus of the cross hairs, then the field of view, until there is no longer cross-hair movement when the eye is moved laterally or vertically with respect to the eyepiece.

Misreading of angles. Double-centering provides an easy check.

Heat refraction. Surveying should not be attempted if the cross hairs waver on the object to such an extent that exact lineup is impossible. The cure is to work only during the cooler morning hours.

Sun heats up one side of the telescope barrel, distorting it, and deflecting the line of sight. The cure is to shade the instrument.

4.12 PROBLEMS

Refer to Fig. 4.2 for Probs. 4.1 and 4.2. The side dimensions, in feet, are $AB = 126.13$, $BC = 91.77$, $CD = 68.42$, $DA = 149.63$, $AC = 188.54$.

4.1. Using the data provided above, calculate the interior angles in triangle *ABC* to the nearest second and provide a check on the results.

4.2. Using the data provided above, calculate the interior angles in triangle *ACD* to the nearest second and provide a check on the results.

Refer to Fig. 4.3*a* for Probs. 4.3 and 4.4. Sides AB' and AC' are both 100.00 ft.

4.3. Using the data provided above and a $B'C'$ dimension of 47.81 ft, compute angle *A* to the nearest second.

4.4. Using the data provided above and a $B'C'$ dimension of 62.49 ft, compute angle *A* to the nearest 0.1 minute.

Refer to Fig. 4.3*b* for Probs. 4.5, 4.6, and 4.7. $DE' = 100.00$ ft.

4.5. Using the data provided above and an $E'F'$ dimension of 39.43 ft, compute angle *D* to the nearest second.

4.6. Using the data provided above and an $E'F'$ dimension of 56.72 ft, compute angle *D* to the nearest 0.1 minute.

4.7. What horizontal distance $E'F'$ must to laid out to establish an angle *D* of 59°12.6′?

4.8. In Fig. 4.4, AC' and AB' are laid out to be 100.00 ft and $B'C'$ is measured as 39.22 ft. What are the values of acute angle CAB' and its supplement, obtuse angle *BAC*, to the nearest second?

4.9. Inspection of a transit shows that the whole-degree spaces on the horizontal-circle scale are divided into thirds, that the longer markings on the vernier scale are numbered from 0 to 20, left and right of the index mark, and that the spaces between the longer markings are further subdivided into thirds. What is the least count of the instrument, in seconds?

4.10. Inspection of a transit shows that the whole-degree spaces on the horizontal-circle scale are divided into quarters, that the longer markings on the vernier scale are number from 0 to 15, left and right of the index mark, and that the spaces between the longer markings are further subdivided in half. What is the least count of the instrument, in seconds?

4.11. Why is the horizontal-circle vernier on a typical transit graduated to the left and right with zero in the center?

4.12. On a vernier transit with a least count of 20 seconds, the difference in arc length between the width of a vernier space and the width of a space on the horizontal circle is 0.000291 in. What is the diameter of the horizontal circle to the nearest 0.001 in.?

4.13. On a vernier transit with a least count of 30 seconds, the difference in arc length between the width of a vernier space and the width of a space on the horizontal circle is 0.000327 in. What is the diameter of the horizontal circle to the nearest 0.001 in.?

4.14. A horizontal angle was measured by double-centering with the initial reading set at 0°00′00″. After the first angle was taken, with telescope direct, the mean of the A and

B verniers was 69°21′40″. After the second angle, with telescope reversed, the mean of the A and B verniers was 138°43′40″. Compute the mean angle.

4.15. Do the same as Prob. 4.14 except the reading after the directly taken angle was 186°22′30″ and the reading after the reversed angle was 12°45′30″. Compute the mean angle.

4.16. A horizontal angle was measured by repetition with the initial reading set at 0°00′00″. After the first angle, with telescope direct, the mean of the A and B verniers was 72°22′15″. After taking one more direct angle and two reversed angles, the mean of the A and B verniers was 289°28′30″. Compute the mean angle.

4.17. A horizontal angle was measured by repetition with an initial reading (mean of the A and B verniers) of 45°13′40″. After the first angle was taken, with telescope direct, the mean of the A and B verniers was 97°46′00″. After taking one more direct angle and two reversed angles, the mean of the A and B verniers was 255°24′00″. Compute the mean angle.

4.18. A horizontal angle was measured six times, three direct and three reversed. The mean of the A and B verniers on the initial backsight was 38°25′15″. The A vernier reading after the first angle was taken was 199°14′00″. After the final repetition, the mean of the A and B verniers was 283°16′45″. Compute the mean angle.

4.19. A horizontal angle was measured eight times, four direct and four reversed. The mean of the A and B verniers after the initial backsight was 142°38′20″. The A vernier read 344°20′00″ after the first direct angle. After the final repetition, the mean of the A and B verniers was 316°12′20″. Compute the mean angle.

4.20. In measuring an angle by repetition, the initial backsight read 238°14.6′. After the first directly measured angle, the circle reading was 119°58.3′. After two more direct and three reversed angles, the circle scale read 248°38.6′. Compute the mean angle.

4.21. Assume the precision is setting and reading a scale-reading theodolite is ±6″ and the precision in placing the telescope cross hair on an object is ±2″. For an angle measured with six repetitions, what is the assumed error in the mean value?

4.22. Assume that the precision in setting and reading a 30-second transit is ±15″ and that the precision in placing the telescope cross hair on an object is ±3″. If two direct and two reversed angles are measured, what is the assumed error in the mean value?

4.23. A direct vertical angle of +4°15′ was measured to the top of a smokestack and a reverse reading of +4°11′ was observed to the same object. Compute the correct vertical angle and the index error.

4.24. A direct vertical angle of −2°41′ was measured to an object and a reverse reading of −2°43′ was observed to the same object. Compute the correct vertical angle and the index error.

4.25. If a direct-read vertical angle of −3°29′ was measured to an object with the instrument in Prob. 4.23, determine the correct vertical angle.

4.26. If a direct-read vertical angle of +6°47′ was measured to an object with the instrument in Prob. 4.24, determine the correct vertical angle.

4.27. Using the instrument in Prob. 4.23, what direct angle would have to be set on the vertical circle to lay off correct vertical angles of (*a*) + 3°42′ and (*b*) − 5°33′?

4.28. Using the instrument in Prob. 4.24, what direct angle would have to be set on the vertical circle to lay off correct vertical angles of (*a*) + 4°10′ and (*b*) − 3°52′?

4.29. A vertical angle was measured with a zenith-angle instrument. The direct reading was 86°14.3′ and the reverse reading was 273°46.1′. Compute the correct zenith angle, the correct horizontal-zero vertical angle, and the index error for zenith angles.

4.30. A vertical angle was measured with a zenith-angle instrument. The direct reading was 92°41.6′ and the reverse reading was 267°18.2′. Compute the correct zenith angle, the correct horizontal-zero vertical angle, and the index error for zenith angles.

4.31. Using the instrument in Prob. 4.29, what direct zenith angle would be set on the vertical scale to lay off a correct, horizontal-zero vertical angle of −3°21′42″?

4.32. Using the instrument in Prob. 4.30, what direct zenith angle would be set on the vertical scale to lay off a correct, horizontal-zero vertical angle of +4°12.8′?

4.33. Describe two observing procedures for taking horizontal angles that, if carried out together, will eliminate virtually all instrumental errors.

80° 12′ 13″
45° 15′ 17″
75° 27′ 30″

CHAPTER 5

SURVEYING ERRORS AND ADJUSTMENTS

5.1 RANDOM-ERROR CONCEPTS

RANDOM ERRORS. As discussed in Chaps. 2, 3, and 4, measurements of distances, elevations, and angles are subject to a variety of errors. One obvious example of measurement error involves simple procedural mistakes such as misreading a tape, EDM, rod, transit, or theodolite; but these errors are detected, and thus eliminated, by making whatever check measurements are appropriate. Systematic errors—those involving imperfections in the instrument itself—are compensated or adjusted as discussed earlier, the compensation or adjustment procedure being dependent on the instrument or device being used as well as the field conditions at the time of measurement. **Random errors**, on the other hand, can't be overcome in the same fashion as simple mistakes and systematic errors. Random errors are not only quite accidental and, thus, unpredictable, they are also inherent in the surveying process. That being the case, the surveyor must be aware of the potential for random errors and the fact that the *accumulation* of such unpredictable errors will be *mathematically* predictable.

PROBABILITY. Besides being unpredictable, random errors will be positive as often as negative. Given a large number of random errors, one may expect them to accumulate to some value. One possibility is for the errors to cancel, that is, for their sum to equal zero. In fact, of all of the possible combinations of cumulative errors arising from any series of measurements, the most probable combination is a zero sum. In other words, cancellation of random errors is more likely than any other single possibility, but given the large number of other possibilities, this is small comfort.

A convenient model for understanding random errors is the tossing of a coin. A balanced coin, randomly tossed, will land "heads" as often as "tails." For two coin tosses, there are four possible sequences (two heads, two tails, tails followed by heads and heads followed by tails), but only three two-coin combinations (two heads, two tails, and one of each). From this it is seen that there is one chance in four for two heads, one chance in four for two tails, but two chances in four for one of each to occur. Thus, in the normal course of events, there is a 50% probability of one head and one tail but only a 25% probability of either two heads or two tails.

If occurrence of a "head" or "tail" is considered equivalent to the occurrence of equal-sized random errors, a head may be defined as a random error of +1 and a tail may be defined as a random error of –1. The combination of two heads would have a numerical value of +2, the combination of one head and one tail would have a numerical value of zero, and the combination of two tails would have a numerical value of −2. Figure 5.1*a* shows the relative probabilities of each of the three 2-coin combinations; the *y* axis showing relative probability (number of occurrences) of each combination while numerical values of different combinations are ranged along the *x* axis.

It is easy to work out the number of two-toss combinations. With large numbers of coin tosses, however, use the equation

$$C_q = \frac{m!}{(m-q)!q!} \tag{5.1}$$

where C_q is the number of sequences in which q events occur in m tosses of the coin. If we were looking for the number of times exactly 9 heads occurred in 12 tosses, m would be 12 and q would be 9. The ! means **factorial**, or multiplying a number by one less than the number, then by two less than the number, then by three less, and so forth, until reaching 1; for example, $q! = q(q-1)(q-2)\cdots(1)$. The expression 0!, which occasionally occurs, is defined as $0! = 1$.

It is useful to consider the **probability** P_q of q events' occurring in the course of m coin tosses (or the probability of one specific sum of equal random errors). In this case,

$$P_q = \frac{C_q}{2^m} \tag{5.2}$$

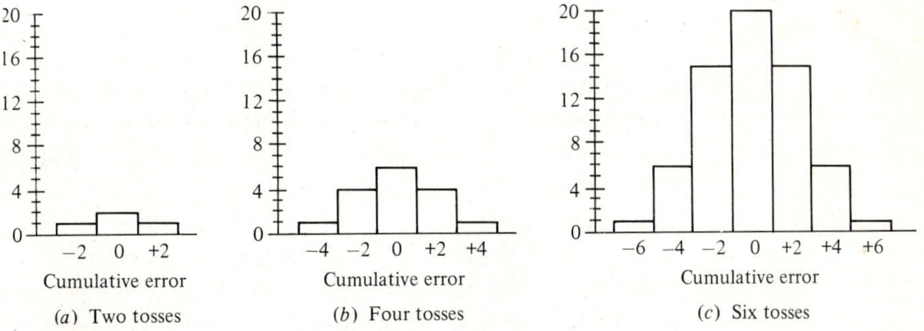

FIGURE 5.1
Plots of random error versus relative probability.

where C_q is obtained from Eq. (5.1) and 2^m is the total number of different sequences that can occur from m coin tosses.

Example 5.1. For four coin tosses, compute the number of sequences involving 4, 3, 2, 1, and 0 heads as well as the probability of equal numbers of heads and tails (note that this is the probability that four equal random errors will cancel).

Solution. For the four-heads combination, m is 4 and q is 4, so the expression $(m - q)$ is zero. Inserting these values into Eq. (5.1):

$$C_4 = \frac{4!}{0!(4!)} = 1 \qquad \text{(remembering that } 0! = 1\text{)}$$

As expected, there is only one combination of four heads in four tosses.

$$C_3 = \frac{4!}{1!(3!)} = \frac{4 \cdot 3 \cdot 2 \cdot 1}{(1)(3 \cdot 2 \cdot 1)} = 4$$

$$C_2 = \frac{4!}{2!(2!)} = \frac{4 \cdot 3 \cdot 2 \cdot 1}{(2 \cdot 1)(2 \cdot 1)} = 6$$

$$C_1 = \frac{4!}{3!(1!)} = 4 \qquad \text{(same values as } C_3\text{)}$$

and

$$C_0 = \frac{4!}{4!(0!)} = 1 \qquad \text{(same values as } C_4\text{)}$$

The probability of an equal number of heads and tails is obtained from Eq. (5.2) as follows:

$$P_q = \frac{C_q}{2^m} = \frac{6}{(2)^4} = 0.375 = 37.5\%$$

Example 5.2. For six coin tosses, compute the number of sequences involving 6, 5, 4, 3, 2, 1, and 0 heads as well as the probability of equal number of heads and tails.

Solution. Substituting into Eq. (5.1) as before, $C_6 = 1, C_5 = 6, C_4 = 15, C_3 = 20, C_2 = 15, C_1 = 6$ and $C_0 = 1$. Substituting into Eq. (5.2), $P_3 = 20/2^6 = 0.3125 = 31.25\%$.

The results of Examples 5.1 and 5.2 are shown graphically in Fig. 5.1*b* and *c*, with head and tail occurrences again assigned values of +1 and −1, respectively; that is, four heads out of four tosses is +4; one head (and five tails) out of six tosses is −4; and so forth.

Examination of Fig. 5.1, along with consideration of the results of Examples 5.1 and 5.2, show that for an even number of tosses, the combination of an equal number of heads and tails has the highest probability of occurrence, although that percentage probability diminishes as the number of coin tosses (random errors) increases. It is also seen that there is a diminishing probability of a given unequal number of heads

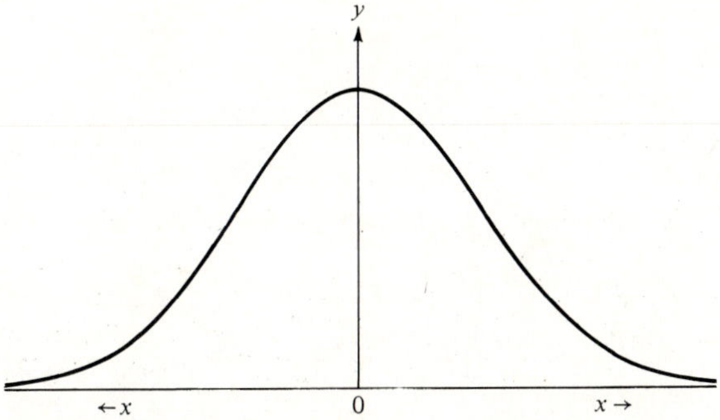

FIGURE 5.2
Probability curve.

and tails as the heads-tails imbalance increases, with the lowest probability occurring in the cases of all-heads or all-tails combinations. In addition, all of the graphs in Fig. 5.1 are symmetrical, two heads and four tails having the same probability as four heads and two tails, and so forth.

As coin tosses or equal-sized random errors approach an infinite number the graphs in Fig. 5.1 change from a blocky configuration to Fig. 5.2, the **probability curve**.

The equation of the probability curve is

$$\boldsymbol{y = ke^{-h^2x^2}} \tag{5.3}$$

where e is the natural logarithm base, y is probability of occurrence of a *cumulative* error x, and k and h are dimensional constants.

From Fig. 5.2 and Eq. (5.3), it is seen that y is a maximum when x is zero and that y becomes small when x is large, either positively or negatively. Because x is a cumulative error value, it can be said that (1) small cumulative errors are more likely to occur than large cumulative errors, (2) very large cumulative errors rarely occur, (3) a positive cumulative error is as likely as a negative cumulative error, and (4) the mean of all measurements, which occurs at $x = 0$, is the single most probable measurement value.

5.2 STANDARD ERROR

NORMAL DISTRIBUTION. The value of x in Eq. (5.3) cannot be determined because it represents the difference between a given measurement and an unknowable true value. To deal with the errors involved in a finite number of everyday measurements, it is simpler to replace x with v, known as the **residual**, the difference between a given measurement and the mean of all measurements taken. In equation form:

$$\boldsymbol{v_i = M_i - M} \tag{5.4}$$

where M_i is the ith of n measurements whose arithmetic mean has been determined to be M.

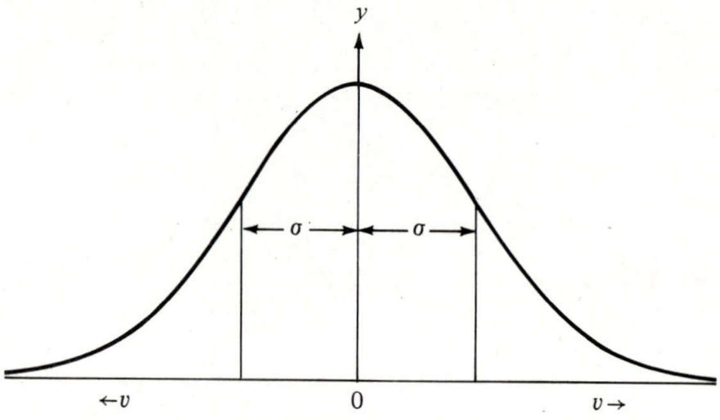

FIGURE 5.3
Normal distribution curve.

A **normal distribution curve** results if Eq. (5.3) is modified by substituting v for the indeterminate value x and by defining k and h. The equation of the new curve, shown in Fig. 5.3, is

$$y = \frac{e^{-v^2/2\sigma^2}}{\sigma\sqrt{2\pi}} \tag{5.5}$$

where k in Eq. (5.3) has been replaced by $1/\sigma\sqrt{2\pi}$ (the value of y when $v = 0$) and h^2 has been replaced by $1/2\sigma^2$. The term h, which is thus $1/(\sigma\sqrt{2})$, is called the **precision modulus**. The term σ, called **standard error** or **standard deviation**, is most important in the analysis of measurement errors. It is defined by

$$\sigma = \sqrt{\frac{\Sigma v_i^2}{n-1}} \tag{5.6}$$

where Σ indicates the sum of all v_i^2 from 1 to n, and n is the total number of measurements made.

The greater the precision of measurement, the smaller the average residual, v. The smaller the v, the smaller the standard error σ. The smaller the σ, the taller and thinner the normal distribution curve.

MEASURES OF PRECISION. Note the presence of the value σ on either side of the zero line in Fig. 5.3. The dimensions of σ are the same as the dimensions of the measurements to which σ refers and its value is such that 68.27% of the area under the normal distribution curve lies within $\pm 1\sigma$ of the mean. Thus, errors will fall within $\pm 1\sigma$ of the mean over two-thirds of the time. However, the figures 68.27% and $\pm 1\sigma$ are not as meaningful as one might like in the context of survey measurement—other **measures of precision** are handier to use. Consider the following definitions:

Fifty percent of the area under the curve lies within $\pm 0.6745\sigma$ of the mean. This multiple of σ is designated E_{50}, or **probable error**. Its significance is that, statistically, any given measurement will lie within $\pm 0.6745\sigma$ exactly half the time.

Ninety percent of the area under the curve lies within $\pm 1.6449\sigma$ of the mean. This σ multiple is designated $\mathbf{E}_{90}$.

Ninety-five percent of the area under the curve lies within $\pm 1.9599\sigma$ of the mean. This σ multiple is designated $\mathbf{E}_{95}$.

The mean of n measurements will deviate from a true mean of all possible measurements by a small error amount that is dependent on the precision of the measurement and the value of n. That error is called the **standard error of the mean**, or $\boldsymbol{\sigma}_M$, and

$$\boldsymbol{\sigma}_M = \frac{\boldsymbol{\sigma}}{\sqrt{n}} \tag{5.7}$$

Example 5.3. The distance from A to B is measured 10 times as follows: 728.93 ft, 728.89 ft, 728.92 ft, 728.91 ft, 728.87 ft, 728.90 ft, 728.95 ft, 728.86 ft, 728.92 ft, and 728.80 ft. From this, compute the most probable line length, the standard error of a single measurement, probable error, E_{90}, E_{95}, and the standard error of the mean.

Solution. The 10 measurements are listed in the first column and summed [1]; the mean measurement is found [2] and used to determine the 10 values for v (second column) and the 10 values for v^2 (third column), from whose sum σ is computed [3] and the remainder of the values determined [4].

Length of *AB*, ft	***v*,ft**	v^2
(Measured)	(Meas. − Mean)	
728.93	+0.035	0.001225
728.89	−0.005	0.000025
728.92	+0.025	0.000625
728.91	+0.015	0.000225
728.87	−0.025	0.000625
728.90	+0.005	0.000025
728.95	+0.055	0.003025
728.86	−0.035	0.001225
728.92	+0.025	0.000625
728.80	−0.095	0.009025
ΣAB = 7288.95 [1]	Σv = 0.000	Σv^2 = 0.016650

$$[2]\ M = \frac{\Sigma_{AB}}{n} = \frac{7288.95}{10} = 728.895 \text{ ft} \equiv \text{most probable line length}$$

$$[3]\ \sigma = \pm\sqrt{\frac{\Sigma v^2}{n-1}} = \pm\sqrt{\frac{0.01665}{9}} = \pm 0.043 \text{ ft} \equiv \text{standard error}$$

$$[4]\ E_{50} = 0.6745\sigma = \pm 0.029 \text{ ft} \equiv \text{probable error}$$

$$[4]\ E_{90} = 1.6449\sigma = \pm 0.071 \text{ ft}$$

$$[4]\ E_{95} = 1.9599\sigma = \pm 0.084 \text{ ft}$$

$$[4]\ \sigma_M = \frac{\sigma}{\sqrt{n}} = \pm 0.014 \text{ ft} \equiv \text{standard error of the mean}$$

Note: Σv should be zero (or close to zero in the case of round-off error). This checks the validity of the measurement sum, the computation of the mean, and the computation of the values for v.

In Example 5.3, the significant figures for M and σ are validly taken one beyond that of the individual measurements. By convention, all of the figures in the σ computation, valid or not, are subsequently used to compute the E_{50}, E_{90}, E_{95}, and σ_M values, which are then rounded down to the same final significant figures as M and σ.

5.3 REJECTION LIMIT

Before continuing further with standard errors, it would be well to reflect on the concept of a **rejection limit**, a test of the validity of individual measurements within a data set. In Example 5.3, all of the measured values lay within reasonable proximity of the mean; thus, none were questioned. Suppose, though, that the third value, instead of being 728.92 ft, had actually been listed as 729.82 ft. Clearly, that value could not have been accepted. It would appear that either there had been a transposition of the 8 and the 9 or the listed value was in error by exactly 1 ft, with no way of knowing which was the case.

In the above example, a measured value that differs from the mean by about a foot is an obvious stray that should be deleted from the set. More often, though, questionable measurements fall much closer to the mean, so a rule of thumb is needed concerning which values to accept and which to reject.

It has become common practice to reject measurements lying beyond 3σ of the mean of all measurements. Recall that the expression E_{95} signifies that 95% of all measurements may be expected to lie within $\pm 1.9599\ \sigma$ of the mean. Probability theory also tells us that 99.73% of all measurements may be expected to lie within $\pm 3\sigma$ of the mean. Put another way, there is only 1 chance in 370 for a given measurement to be more than 3σ above or below the mean. Those are pretty stiff odds, particularly when dealing with a relative handful of measurements. Thus, for any small set of measurements, a single measurement whose value lies beyond the 3σ rejection limit is probably wrong for reasons other than cumulative random error and, for that reason, should be rejected.

Example 5.4. Use the same data as in Example 5.3 except the first value is 728.63 ft rather than 728.93 ft. Compute the mean and standard deviation and determine which of the measured values, if any, should be rejected.

Solution. A quick comparison with the other values shows that the 728.63-ft value is questionable. For the few measurements involved in this problem, its use would significantly distort the mean, the v's, the sum of the v^2 values, and thus the value of σ

itself. For that reason, the 728.63-ft value is eliminated from a test computation of σ, as shown below.

Length of *AB*, ft	*v*,ft	v^2
(Measured)	(Meas. − Mean)	
(728.63)		
728.89	−0.001	0.000001
728.92	+0.029	0.000841
728.91	+0.019	0.000361
728.87	−0.021	0.000441
728.90	+0.009	0.000081
728.95	+0.059	0.003481
728.86	−0.031	0.000961
728.92	+0.029	0.000841
728.80	−0.091	0.008281
$\Sigma\, AB = 6560.02$	$\Sigma\, v = 0.001$	$\Sigma\, v^2 = 0.015289$

$$M = \frac{\Sigma AB}{n} = \frac{6560.02}{9} = 728.891 \text{ ft}$$

$$\sigma = \pm\sqrt{\frac{\Sigma v^2}{(n-1)}} = \pm\sqrt{\frac{0.015289}{8}} = \pm 0.044 \text{ ft}$$

From the above, $3\sigma = \pm 0.132$, so only those measured values between 728.759 ft and 729.023 ft lie within 3σ of the mean. This means the listed value of 728.63 ft should, indeed, be rejected.

5.4 STANDARD ERROR OF MULTIPLE MEASUREMENTS

The foregoing has examined the standard statistical errors that arise from repeated observations of the same measurement, say, the distance, difference in elevation, or horizontal angle between two points. Consider now the frequently encountered case of an overall value that is the sum or product of separate, repeated measurements.

STANDARD ERROR OF A SUM OF REPEATED MEASUREMENTS. A simple example involves determining the distance between two points when the measurement is made in separate segments, each segment being measured repeatedly and, thus, each segment having its own standard error. The solution equation is one of several that are derived in Appendix A from the basic error propagation equation:

$$\boldsymbol{\sigma}_s = \sqrt{\boldsymbol{\sigma}_1^2 + \boldsymbol{\sigma}_2^2 + \boldsymbol{\sigma}_3^2 + \cdots + \boldsymbol{\sigma}_n^2} \qquad \textbf{(5.8)}$$

where σ_s is the standard error of the sum and σ_n is the standard error of the *i*th measurement in the series.

Example 5.5. Line AE is measured in four segments with lengths and standard errors as follows: $AB = 487.62$ ft, $\sigma_{AB} = \pm 0.029$ ft; $BC = 637.34$ ft, $\sigma_{BC} = \pm 0.034$ ft; $CD = 314.08$ ft, $\sigma_{CD} = \pm 0.016$ ft; $DE = 763.91$ ft, $\sigma_{DE} = \pm 0.041$ ft. What is the length and standard error of line AE?

Solution. Line length is the sum of the segments: $487.62 + 637.34 + 314.08 + 763.91 = 2202.95$ ft.

Standard error is obtained by substituting into Eq. (5.8):

$$\sigma_{AE} = \sqrt{\sigma_{AB}^2 + \sigma_{BC}^2 + \sigma_{CD}^2 + \sigma_{DE}^2}$$

$$= \sqrt{(0.029)^2 + (0.034)^2 + (0.016)^2 + (0.041)^2} = \pm 0.063 \text{ ft}$$

MULTIPLE MEASUREMENTS WITH SAME STANDARD ERROR. In Eq. (5.8), if the standard error of each measurement is the same, so that $\sigma_1 = \sigma_2 = \sigma_3 = \cdots = \sigma_n \equiv \sigma$, for n measurements

$$\boldsymbol{\sigma_s = \sigma\sqrt{n}} \tag{5.9}$$

Example 5.6. A line of differential levels involved nine setups with each rod reading having a standard error σ of $\pm$ 0.005 ft. What is the standard error of the entire line?

Solution. There were two sights (one backsight and one foresight) at each setup, for 18 rod readings altogether. Substituting $n = 18$ into Eq. (5.9)

$$\sigma_s = \pm 0.005\sqrt{18} = \pm 0.021 \text{ ft}$$

ERROR OF THE MEAN OF A SET OF REPEATING ANGLES. There are three categories of standard errors that may be anticipated when observing repeating angles with a transit or theodolite: σ_i is the standard error in setting the initial backsight value, σ_p is the standard error in pointing the cross hair exactly on the backsight object or foresight object for each of n repeating angles, and σ_r is the standard error in reading the sum-of-angles value on the instrument scale. Because there are two pointings for each of the n angles (one backsight and one foresight pointing), the value σ_p is involved $2n$ times, whereas σ_i and σ_r are involved once. Substituting into Eq. (5.8) produces $\sigma_s = \sqrt{\sigma_i^2 + 2n\sigma_p^2 + \sigma_r^2}$, the overall error in measuring the multiple angles. Now, since the average angle is equal to the sum of the repeated angles divided by the number of repetitions, the standard error of the mean of the repeating angles is obtained by dividing σ_s by n to produce

$$\boldsymbol{\sigma_{MA} = \frac{\sqrt{\sigma_i^2 + 2n\sigma_p^2 + \sigma_r^2}}{n}} \tag{5.10}$$

This equation was introduced in a slightly different form in the previous chapter as Eq. (4.6). Its application is illustrated in Example 4.8.

STANDARD ERROR FOR AREAS. When dealing with separately measured sides of a rectangular area, X and Y, with separate standard errors σ_X and σ_Y, the equation for the standard error of the area, σ_A, derived in Appendix A, is

$$\sigma_A = \sqrt{Y^2\sigma_X^2 + X^2\sigma_Y^2} \tag{5.11}$$

Example 5.7. The two sides of a large rectangular plot were measured with an EDM and found to be 2673.24 ft and 4892.07 ft in length. The EDM's standard measurement error is assumed to be $\pm(0.025 \text{ ft} + 4 \text{ ppm})$. What is the standard error for the area?

Solution. For the shorter side, $\sigma_X = \pm[0.025 + 2673.24(4\times10^{-6})] = \pm0.036$ ft. For the longer side, $\sigma^Y = \pm[0.025 + 4892.07(4\times10^{-6})] = \pm0.045$ ft. Substituting into Eq. (5.11),

$$\sigma_A = \sqrt{(4892.07)^2(0.036)^2 + (2673.24)^2(0.045)^2} = 213.3 \text{ ft}^2$$

This may seem like a significant error, but considering the total area of the plot $[4892.07(2673.24) = 13,077,677 \text{ ft}^2]$, the error is less than two thousandths of 1%.

5.5 WEIGHTING

WEIGHTED MEASUREMENTS. It may have occurred to the reader that not all measurements are made with equal care or effort. If one measurement has been made with greater precision or has been repeated more times than another measurement, the more precise measurement is said to have greater **weight**. Relative weight is designated w. For a series of measurements M_i, each having relative weight w_i, a weighted mean M_w may be obtained from

$$M_w = \frac{w_1M_1 + w_2M_2 + \cdots + w_nM_n}{w_1 + w_2 + \cdots + w_n} = \frac{\Sigma w_iM_i}{\Sigma w_i} \tag{5.12}$$

Example 5.8. Distance AB was measured with a reinforced cloth tape, a steel tape, and an EDM, and found to be 566.93 ft, 566.72 ft, and 566.74 ft, respectively. If the cloth-tape measurement is assigned a relative weight of 1, the steel-tape measurement a relative weight of 6, and the EDM measurement a relative weight of 8, what is the weighted mean of the three measurements?

Solution. Substituting into Eq. (5.12),

$$M_w = \frac{1(566.93) + 6(566.72) + 8(566.74)}{1 + 6 + 8} = 566.745 \text{ ft}$$

WEIGHTING LINES OF LEVELS. The weights in Example 5.8 appear to be assigned arbitrarily, as is often the case. At times, weighting may be more precisely done. Mathematically, the weight of a measurement is proportional to the square of the precision modulus (Sec 5.2) and thus to the reciprocal of the square of the standard error; that is, $w \propto \frac{1}{\sigma^2}$. From this, note that the larger the value of σ, the smaller the weight, and vice versa.

Consider a line of levels. From Eq. (5.9), $\sigma_s = \sigma\sqrt{n}$. Because w is proportional to $1/\sigma_s^2$, it is also proportional to $1/n$. Since the number of sights is proportional to distance run, w is proportional to the reciprocal of the distance run, or $1/d$. This relationship may be used to determine the weighted mean of the differences in elevation determined by running along different paths.

Example 5.9. Levels were run from BM A to BM B by three separate routes of 1.15, 3.34, and 4.07 miles, with elevation differences of 71.791 ft, 71.770 ft, and 71.823 ft, respectively. Calculate the weighted mean of the elevation differences.

Solution. For convenience, each weight (distance reciprocal) is normalized by multiplying by the longest line length, giving that line a weight of 1 and the other lines proportionally larger weights; thus,

$$w_1 = \frac{4.07}{1.15} = 3.54 \qquad w_2 = \frac{4.07}{3.34} = 1.22 \qquad w_3 = \frac{4.07}{4.07} = 1.00$$

Substituting into Eq. (5.12),

$$M_w = \frac{3.54(71.791) + 1.22(71.770) + 1.00(71.823)}{3.54 + 1.22 + 1.00}$$

$$= 71.792 \text{ ft}$$

Given M_w, its standard error σ_w can be determined using a weighted version of Eq. (5.6):

$$\boldsymbol{\sigma_w = \sqrt{\frac{\Sigma(w_i v_i^2)}{\Sigma w_i(n-1)}}} \qquad \textbf{(5.13)}$$

Example 5.10. Using the data in Example 5.9, compute σ_w.

Solution. In this case, $v_i = M_i - M_w$, therefore

$$v_1 = 71.791 - 71.792 = -0.001 \qquad w_1 v_1^2 = 3.54 \cdot 0.001^2 = 0.00000354$$

$$v_2 = 71.770 - 71.792 = -0.022 \qquad w_2 v_2^2 = 1.22 \cdot 0.022^2 = 0.00059048$$

$$v_3 = 71.823 - 71.792 = +0.031 \qquad w_3 v_3^2 = 1.00 \cdot 0.031^2 = 0.00096100$$

$$\Sigma w_i v_i^2 = 0.00155502$$

Substituting into Eq. (5.13),

$$\sigma_w = \sqrt{\frac{\Sigma w_i v_i^2}{\Sigma w_i(n-1)}}$$

$$= \sqrt{\frac{0.00155502}{(3.54 + 1.22 + 1.00)(3-1)}} = \pm 0.0116 \text{ ft}$$

5.6 ADJUSTMENTS

ADJUSTMENT VERSUS STANDARD ERROR. Discussion of random errors has centered on the concept of standard error as a limiting condition of a given measurement. Thus, there is an expectation that random errors will accumulate in some predictable fashion and that the true value of a measurement will lie within a narrow, definable range of the measured value, that range being the standard error or a multiple thereof. Quite often, however, random-error accumulation is reasonably well determined by a check measurement of one kind or another. The assumption in that case is that the difference between two measurements (say, the outbound and return legs of a double-run line of levels) or between a set of measurements and a geometric check (say, the sum of the measured interior angles in a polygon) will equal the random-error accumulation. If this comfortable assumption is correct, individual measurements may be adjusted by distributing the measurement difference in some reasonable fashion.

ADJUSTMENT OF EQUAL-WEIGHT ANGLES. The sum of the interior angles in a plane polygon must be a multiple of 180°, specifically:

$$\Sigma = (n - 2)180° \tag{5.14}$$

where n is the number of angles or sides in the polygon. For example, the sum of the interior angles in a triangle ($n - 2 = 1$) is 180°, the sum of the interior angles in a four-sided figure ($n - 2 = 2$) is 360°, and so forth. The difference between the sum of the measured angles in a polygon and the geometric sum given by Eq. (5.14) is the amount to be adjusted among the measured angles. If the angles in a polygon are measured with equal care (equal-weight angles), it may be reasonably assumed that corrections will be equally apportioned.

Example 5.11. The measured angles in triangle ABC, listed below, have equal weight. Determine the values of the adjusted angles.

Angle	Measured value
A	49°51′30″
B	60°32′00″
C	69°37′00″

Solution. The basic error equation states that true value equals measured value plus correction; therefore, correction is true value minus measured value. In this case, the true sum of angles must be 180°. Total correction is found by subtracting the sum of the measured angles from 180°.

Angle	Measured value
A	49°51′30″
B	60°32′00″
C	69°37′00″
	Σ = 180°00′30″

Therefore,

$$\Sigma \text{ corrections} = 180° - 180°00'30'' = -30''$$

$$\text{Correction} \cdot \text{angle}^{-1} = \frac{\Sigma \text{ corrections}}{\text{no. of angles}}$$

$$= \frac{-30''}{3} = -10''\cdot\text{angle}^{-1}$$

The corrections (third column, below) are then applied to the measured angles to determine the adjusted angles (fourth column, below).

Angle	Measured value	Correction	Adjusted value
A	49°51′30″	−10″	49°51′20″
B	60°32′00″	−10″	60°31′50″
C	69°37′00″	−10″	69°36′50″
	Σ = 180°00′30″		Σ = 180°00′00″ ✓

Note that the adjusted angles above added up to 180°00′00″. This is an essential arithmetic check. If the angle sum differed from 180° by twice the intended total correction, it would have indicated that the wrong algebraic sign was used for the correction values.

Example 5.11 illustrates a problem that can arise in the adjustment process. The three angle values, 49°51′30″, 60°32′00″ and 69°37′00″, may be reasonably assumed to result from single-angle observations with a 30-second transit or from double-centered observations with a 1-minute transit. The reader can also assume a probable error of about ±15″ in any individual angle (half the least count in the case of single-angle observations or a quarter the least count in the case of a double-center observation). Yet, adjustment was performed to 10 seconds. The alternatives are (1) to split the 30-second adjustment between three angles as was done in Example 5.11 or (2) to throw all 30 seconds into a single angle. In the latter case, convention would require that the entire adjustment be fed into the largest angle, which would be 69°37′00″ in this case. Trouble is, there's no way of knowing which of the three angle measurements was most in error—to apply the correction to the wrong angle would distort the result even further. Thus, a reasonable case can be made that less potential harm is done by adjusting each of the angles, even though the end result may appear to have too high a significant-figure value.

ADJUSTMENT OF WEIGHTED ANGLES. Recall that the weight of a line of levels is inversely proportional to the number of setups (or to distance run). The weight of an angle, on the other hand, is usually considered directly proportional to the number of repetitions. But to adjust weighted angles, corrections are apportioned inversely with respect to relative weights of individual angles; thus, the largest correction is applied to the measured angle having the least weight.

Example 5.12. The angles in triangle *ABC* were later remeasured, with number of repetitions as shown. Compute the adjusted angles.

Angle	Repetitions	Mean angle
A	6	49°51′25″
B	4	60°32′00″
C	2	69°36′45″

Solution. See below. Total correction is determined as before and relative correction is determined for each angle in a three-part computation in the next column to the right of the mean-value column. First, the reciprocal of the number of repetitions is entered as a fraction, then each fraction is converted to a common-denominator form—in this case, twelfths—and the results added.

Angle	Rep.	Mean angle	Relative correction
A	6	49°51′25″	$\frac{1}{6} = \frac{2}{12}$
B	4	60°32′00″	$\frac{1}{4} = \frac{3}{12}$
C	2	69°36′45″	$\frac{1}{2} = \frac{6}{12}$
		Σ = 180°00′10″	$\Sigma = \frac{11}{12}$

Total correction = 180° − 180°00′10″ = −10″

The fraction denominators are then changed to the numerator of the sum fraction—in this case, 11—in order for the sum of the fractions to add to 1. *The new fractions are then multiplied by the total correction to obtain the weighted adjustment of each angle.* The adjustments are then applied to the measured angles to find the adjusted angles, which are then added as a check. (See below.)

Angle	Rep.	Mean angle	Relative correction	Adj.	Adjusted angle
A	6	49°51′25″	$\frac{1}{6} \Rightarrow \frac{2}{12} \Rightarrow \frac{2}{11}$	−1.8″	49°51′23.2″
B	4	60°32′00″	$\frac{1}{4} \Rightarrow \frac{3}{12} \Rightarrow \frac{3}{11}$	−2.7″	60°31′57.3″
C	2	69°36′45″	$\frac{1}{2} \Rightarrow \frac{6}{12} \Rightarrow \frac{6}{11}$	−5.5″	69°36′39.5″
		Σ = 180°00′10″	$\Sigma = \frac{11}{12} \Rightarrow \frac{11}{11}$		180°00′00.0″✓

5.7 PROBLEMS

5.1. A line was measured 10 times with a steel tape with the following results, in feet: 583.11, 583.13, 583.09, 583.02, 583.06, 582.98, 583.05, 583.15, 583.06, and 583.07. Compute the most probable line length, the standard error of a single measurement, the probable error, E_{90}, E_{95}, and the standard error of the mean.

5.2. Do the same as Prob. 5.1, except delete the first measurement.

5.3. Do the same as Prob. 5.1, except delete the first and second measurements.

5.4. An angle was measured by eight different observers with results as follows: 77°15′10″, 77°15′00″, 77°15′30″, 77°14′40″, 77°14′40″, 77°15′40″, 77°15′30″, and 77°14′50″. Similar equipment was used and each observer took one direct angle and one reversed. Compute the most probable value for the angle, the standard error of a single angle value, probable error, $E_{90}.E_{95}$, and the standard error of the mean.

5.5. Do the same as Prob. 5.4, except delete the eighth measurement.

5.6. Do the same as Prob. 5.4, except delete the seventh and eighth measurements.

5.7. A final exam in plane surveying was given to 640 students. The average grade (most probable value) was 73.49% with a standard error of ±9.24%. Assuming the grades fell into a normal distribution, how many students earned grades within $\pm 1\sigma$?

5.8. In Prob. 5.7, the bottom 5% of those taking the final exam failed it. Assuming a normal distribution of the exam results, what minimum grade, to the nearest 0.1%, was needed to keep from failing?

5.9. A line was taped once each by six different groups in a surveying class using tapes marked in feet. Their reported measurements were exactly as follows: 812.75, 812.74, 817.25, 812.76, 812.78, and 812.75. Analyze the data and determine the most probable value for the distance, the standard error of an individual measurement, and the standard error of the mean.

5.10. Five students used their known mean pace length to measure the same distance by pacing, with the results listed below. Analyze the data and determine the most probable value for the paced distance, the standard error of an individual measurement, and the standard error of the mean.

	Mean pace length, ft	Paces
Pacer 1	2.87	147.3
Pacer 2	2.43	174.9
Pacer 3	3.02	140.3
Pacer 4	2.77	123.1
Pacer 5	2.94	144.3

5.11. An angle was measured five times with a 20-second transit with the following results: 126°53′20″, 126°54′00″, 126°53′00″, 126°14′00″, and 126°53′20″. Analyze the data and determine the most probable value for the angle and the probable error of an individual measurement.

5.12. A line was measured in four sections. Most probable length and standard error for each section were as follows: 961.22 ± 0.044 ft, 433.12 ft ± 0.031 ft, 1545.90 ± 0.060 ft, and 355.40 ± 0.021 ft. What is the most probable length of the line and the standard error?

5.13. A line was measured in three sections. Most probable length and standard error for each section were as follows: 647.32 ± 0.051 ft, 829.68 ± 0.069 ft, and 461.31 ± 0.029 ft. What is the most probable length of the line and the standard error?

5.14. Section corners are spaced at 1-mile intervals along the boundary of a township (a township is a 6-mile square). If section corners are set with a standard error of ±1.80 ft, what is the standard error in the length of a township side?

5.15. A line of differential levels was run requiring 14 setups. The vernier target used on the rod gave a standard error of ±0.001 ft for each rod reading. What is the standard error of the entire line?

5.16. Differential levels meeting third-order standards were to be run from BM *A* to BM *B*, 3600 ft away, and closed back on BM *A*. If all sight distances were 180 ft, what is the maximum allowable standard error per rod reading?

5.17. The sides and standard errors of a rectangular parcel of land were measured as 426.39 ± 0.082 ft by 781.89 ± 0.144 ft. Compute the area of the parcel and the standard error of the area.

5.18. The sides and standard errors of a rectangular parcel of land were measured as 679.27 ± 0.066 ft by 231.86 ± 0.021 ft. Compute the area of the parcel and the standard error of the area.

5.19. The two sides of a rectangular tract of land were measured with an EDM and found to be 1492.68 ft and 3167.19 ft. According to the EDM's specifications, its standard measurement error was ±(0.016 ft + 5 ppm). What is the standard error for the area?

5.20. A 500.00-ft square tract is to be staked out to an accuracy of ±20 ft^2. Compute the maximum error that can be made in measuring each side to achieve the desired accuracy.

5.21. In Prob. 5.20, if a 100-ft steel tape was used to lay out the tract, what maximum standard error could be made in each tape measurement to achieve the desired accuracy?

5.22. A line was measured four times with a steel tape on a day when wind conditions varied from calm to strong gusts. Based on wind conditions at the time of measurement, weights were assigned to each measurement as follows: 426.47 ft, wt. 5; 426.43 ft, wt. 5; 426.64 ft, wt. 1; and 426.56 ft, wt. 2. What is the weighted mean of the four measurements?

5.23. Do the same as Prob. 5.22, except distances and weights were 781.99 ft, wt. 1; 781.62 ft, wt. 5; 781.80 ft, wt. 3; and 782.08, wt. 1.

5.24. Levels were run separately from BMs *A*, *B*, and *C*—all of which were established to second order, class I standards—to new BM *Z*. By way of BM *A* over a 2.61-mile route, the computed elevation of BM *Z* was 428.391 ft. By way of BM *B* over a 3.89-mile route, the computed elevation of BM *Z* was 428.408 ft. By way of BM *C* over a 1.77-mile route, the elevation of BM *Z* was 428.374 ft. Calculate the weighted mean elevation of BM Z.

5.25. Do the same as Prob. 5.24, except that distances *AZ*, *BZ*, and *CZ* were 3.06 miles, 4.59 miles, and 2.02 miles, respectively. Calculate the weighted mean elevation of BM *Z*.

5.26. Using the data from Prob. 5.24, calculate the standard error of the weighted mean, σ_w.

5.27. Using the data from Prob. 5.25, calculate the standard error of the weighted mean, σ_w.

5.28. The observed angles in a four-sided figure, double-centering with a 20-second instrument, were as follows: $A = 86°40'20''$, $B = 99°10'00''$, $C = 121°39'30''$, and $D = 52°28'50''$. Assuming equal weight, determine the values of the adjusted angles.

5.29. The observed angles in a five-sided figure, double-centering with a 30-second instrument, were as follows: $A = 68°22'15''$, $B = 115°47'45''$, $C = 92°51'00''$, $D = 148°33'30''$, and $E = 114°26'45''$. Assuming equal weight, determine the values of the adjusted angles.

5.30. The observed angles in Prob. 5.28, instead of being all double-centered, were measured as follows: *A* with four repetitions, *B* with two repetitions, *C* with six repetitions, and *D* with two repetitions. Compute the adjusted angles.

5.31. The observed angles in Prob. 5.29, instead of being all double-centered, were measured as follows: *A* with two repetitions, *B* and *C* with six repetitions, *D* and *E* with four repetitions. Compute the adjusted angles.

CHAPTER 6

TRANSIT FIELD OPERATIONS

6.1 HORIZONTAL-ANGLE CONVENTIONS AND PROCEDURES

The phrase "transit field operations" is of long standing; operations discussed here apply to theodolites as well as to transits. This chapter deals with conventions for observing horizontal angles and with several everyday horizontal-angle procedures necessary to the accomplishment of many types of surveys.

HORIZONTAL-ANGLE NOMENCLATURE. Up until now, the numerical values of horizontal angles have been given, in one context or another, without any specifics concerning the manner in which the angle was taken. The surveyor may measure horizontal angles in a number of ways, depending on the sort of survey that is underway and the type of instrument being used. Consider first the concept of **interior angles**, that is, the angles between adjacent sides in a closed figure. Figure 6.1*a* shows interior angles turned right (clockwise), while Fig. 6.1*b* shows the same angles turned left (counterclockwise), the numerical values of the angles being the same in both cases. For example, consider a theodolite set up at *A* for the purpose of measuring interior angle *A* in the quadrilateral *ABCD*. To measure *A* angle-right, initial (backsight) on *B*, turn the instrument telescope clockwise to *D*, and read the instrument's clockwise circle; to take *A* angle-left, initial on *D* instead of *B*, turn the telescope counterclockwise to *B*, then read the counterclockwise circle. For instruments without a counterclockwise circle, as is the case with many theodolites, interior angles would always be taken angle-right.

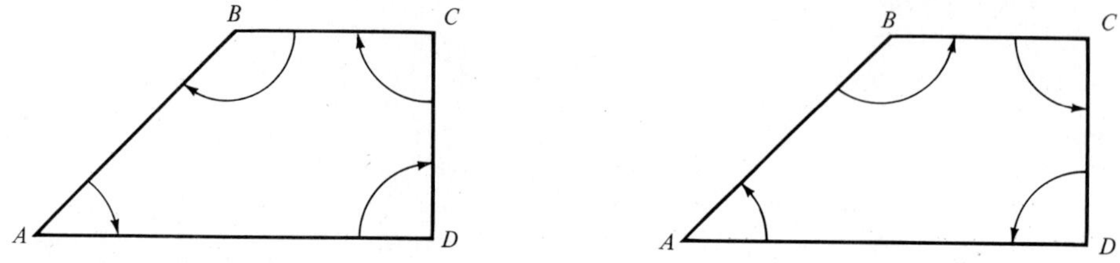

(*a*) Interior angles right (*b*) Interior angles left

FIGURE 6.1
Interior angles.

As will be discussed in detail in Chap. 8, a "traverse" starts from an initial point with a backsight in a known initial direction, then measures angles and distances to successive points along an existing course. Traverse surveys are often open-ended rather than closed as in Fig. 6.1. The angles in an open-ended survey may be turned **angle-right** or **angle-left**; but most often angle-right, as shown in Fig. 6.2, because so many theodolites have only clockwise circles.

Deflection angles begin on a backsight, the telescope is plunged—that is, flipped on its horizontal axis—and turned to the foresight point to measure the angle between the backsight extension and the foresight point. Figure 6.3 shows the Fig. 6.2 traverse observed with deflection angles. A deflection angle may be turned right (R), as at points *A* and *B*, or left (L), as at *C*. In either event, deflection angles *must* be labeled R or L, as appropriate, in the field record.

Recall the summation of angles in a closed figure. As discussed earlier, the sum of the interior angles in a closed, straight-sided figure must equal 180° multiplied by two less than the number of sides in the figure; that is, $\Sigma_{\text{interior angles}} = 180°(n - 2)$. Consider now that the angles in a closed figure may also be measured as deflection angles. Deflection angles right are defined as positive and deflection angles left are defined as negative. With these definitions in mind, the sum of the deflection angles in a closed, straight-sided figure must equal 360°; that is, $\Sigma_{\text{deflection angles}} = 360°$.

Example 6.1. Given the measured-angle values in Fig. 6.4, adjust as necessary, assuming equal-weight angles.

Solution. The sum of the deflection angles in *ABCDE* is found by adding the sum of right deflection angles *A, B, D*, and *E* and subtracting left deflection angle *C*. This value, subtracted from 360°, equals the total correction, which is then divided by 5 to find the correction to each angle.

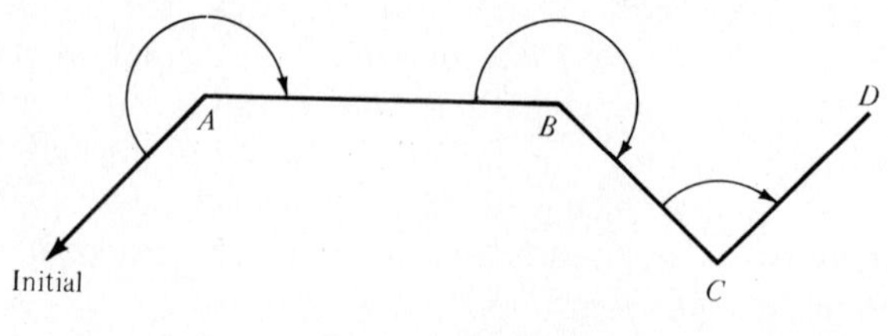

FIGURE 6.2
Angle-right traverse.

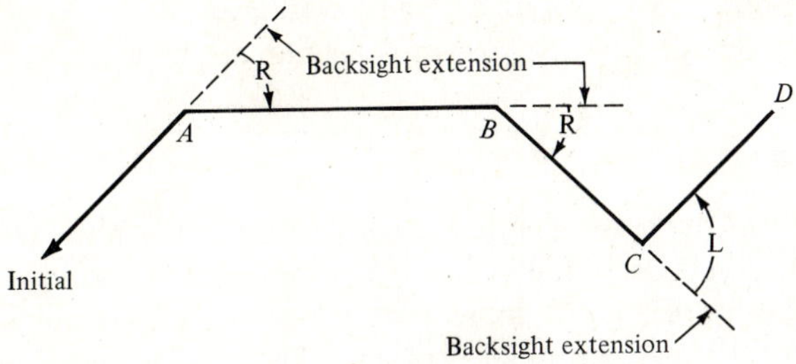

FIGURE 6.3
Deflection-angle traverse.

$$\text{Right-angle sum} = 114°28' + 83°26' + 119°32' + 116°43' = 434°09'$$

$$\text{Minus left-angle sum:} \quad \underline{-74°14'}$$

$$\text{Sum of deflection angles} = 359°55'$$

$$\text{Total correction} = 360°00' - 359°55' = +05'$$

Thus, correction = +05′/5 angles = +01′ per angle. Correcting:

$$A = +114°28' + 01' = +114°29'$$

$$B = +83°26' + 01' = +83°27'$$

$$C = -74°14' + 01' = -74°13'$$

$$D = +119°32' + 01' = +119°33'$$

$$E = +116°43' + 01' = +116°44'$$

$$\Sigma = +360°00' \checkmark$$

Deflection angles are double-centered in the same way as clockwise or counterclockwise angles. The first angle begins with telescope direct, sighted on the backsight object, and the telescope is then plunged; thus, the first deflection angle is read with the telescope in the reversed (inverted) position. Then the telescope, still reversed, is

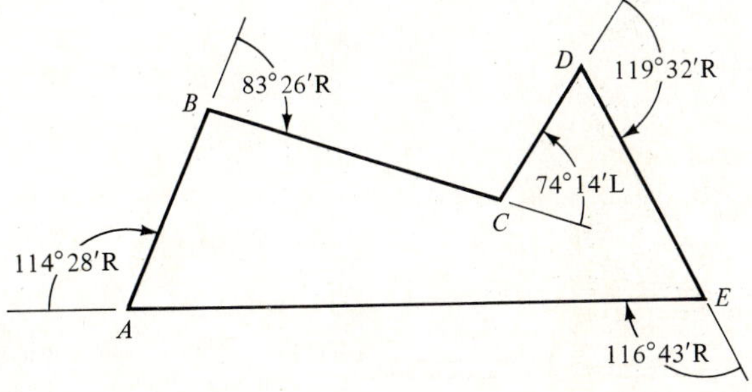

FIGURE 6.4
Measured deflection angles in a closed figure.

pointed on the backsight object and plunged to the direct position; thus, the second angle of a double-centered pair of deflection angles is read with telescope direct.

6.2 LAYING OUT ANGLES

The surveyor usually measures angles between already-established points. In construction surveys and subdivision layout, however, the surveyor must establish one or more points at desired locations with respect to existing points. In this case an angle is laid out between an initial backsight direction and the direction of the point that is to be established. However, a variety of errors tend to creep into the measurement process and the angle laid out will not be precisely the angle intended. To get around this, angles are laid out adapting the double-centering or repeating-angle procedure, preferably the latter.

Start with a backsight on the initial point, with a zero setting on the horizontal circle of the instrument, then bring the telescope right or left, as appropriate, to the proper angle reading. A preliminary point is then established along the telescope's line of sight at the proper horizontal distance from the setup point. That angle is the first repetition in a repeating-angle measurement as discussed in Sec. 4.8. Continue with the repeating-angle procedure, taking additional repetitions as necessary, direct and reversed. The mean of the repeating angles will establish the precise angle to the preliminary point. It is now easy to determine an offset distance from the preliminary point to establish precisely the location of the desired point. Figure 6.5 shows the relationship between the backsight direction, instrument setup, preliminary point, and the desired point.

Example 6.2. In figure 6.5, a clockwise angle *ABC* of 47°38′24″ is needed to establish point *C*, 208.37 ft from *B*. A 30-second transit was set up at *B*. With telescope direct, an initial angle *ABC* of 47°38′30″ was set on the instrument and a horizontal distance of 208.37 ft was measured from *B* to establish point *C*′. Continuing the repeating-angle method, the mean angle *ABC*′ was determined to be 47°38′33″. What is the right-angle offset distance from *C*′ needed to establish the correct position of *C* (and in which direction, clockwise or counterclockwise with respect to line *BC*′)?

Solution. The angular offset is the difference between the mean angle to point *C*′(47°38′33″) and the desired angle (47°38′24″), or 9″. Since the angle to *C*′ is greater

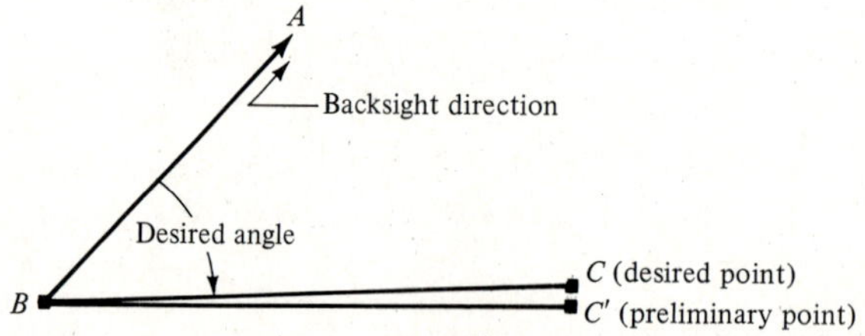

FIGURE 6.5
Laying out a desired angle.

than the desired angle to C, the offset direction must be counterclockwise. The offset distance in feet is equal to the offset angle in radians multiplied by distance BC in feet.

$$00°00'09'' = 0.000043633 \text{rad}$$

$$0.000043633 \text{ rad } (208.37 \text{ ft}) = 0.009 \text{ ft}$$

For very small angle values such as this, the same result is obtained by multiplying distance BC by the sine or tangent of angle α. That is,

$$\text{Offset} = BC \sin\alpha(\text{ or } BC \tan\alpha) = 208.37 \sin 00°00'09'' = 0.009 \text{ ft}$$

The results of Example 6.2 indicate that for most cases of laying out angles, the angular offset α will be so small that the preliminary point and the intended point will share a 2 in.-by-2 in. hub (the former being marked by a pencil cross and the latter by a surveyor's tack).

6.3 STRAIGHT-LINE EXTENSION

Straight-line extension (prolongation) is a very-much-simplified case of laying out an angle in which the desired angle is 180°. Consider a highway centerline that passes from point A to point B. It is now required that additional points, beginning with C, be established at desired intervals along the straight-line extension of AB. In a perfect world, with a perfectly adjusted instrument, a line of sight to C is established by setting up on B, sighting on A with telescope direct, then plunging the telescope (that is, turning it on its horizontal axis so it points in the opposite direction, toward C and beyond). Alas, due to instrument imperfection or misadjustment, precise extension of the centerline must be done by the method of double-centering.

Figure 6.6, a plan view of straight-line extension, illustrates the problem and the solution. The instrument is set up at B and sighted on A, telescope direct. Due to inevitable instrument misadjustment, the telescope's horizontal (plunge) axis will not be truly perpendicular to line of sight AB—the true angle between the telescope's horizontal axis and its optical axis is α. The geometry before and after plunging

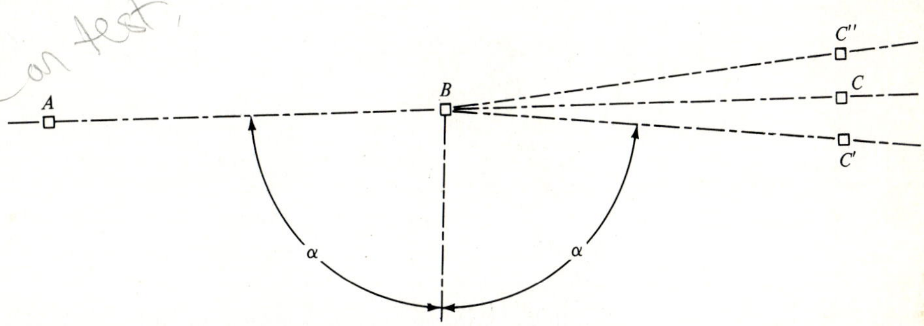

FIGURE 6.6
Straight-line extension by double-centering.

is shown in Fig. 6.6. Angle α is fixed and the telescope's plunge axis will remain stationary while the telescope is plunged; thus, the new pointing direction, BC', will differ from the backsight direction, BA, by 2α, counterclockwise from BA to BC'. Assume that point C' is set the desired distance, BC, from point B.

To compensate by double-centering, the telescope is turned, still reversed (that is, *without plunging*), until the line of sight is once again backsighted on point A. The telescope is plunged once again, to the direct position and C'' is set at the desired distance, BC, from point B. The angle between BA and the new direction BC'' is still 2α, but this time the angle is taken clockwise. The straight-line extension of BC *must* lie along the bisector of angle $C''BC'$, that is, it must lie halfway between points C'' and C'. For reasonably short distances between points B and C, all three points (C, C', and C'') should lie on a common hub. Additional centerline points on the straight-line extension of A-B-C are located in a similar manner.

Example 6.3. On a particular theodolite, the telescope optical axis makes an angle of 89°59′53″ with the telescope plunge axis. To the nearest thousandth of a foot, how far will point C' (see Fig. 6.6) lie from the true centerline if C' is set 100.00 ft from B?

Solution. $C'C = 100.00 \sin$ (angle $C''BC'/2$). Half of angle $C''BC'$ must equal 180°00′00″ − 2(89°59′53″) = 14″. Thus,

$$C'C = 100.00 \sin 00°00'14'' = 0.00679 \Rightarrow\Rightarrow 0.007\text{ft}$$

Example 6.4. Refer again to Fig. 6.6. In the process of extending a centerline 200.00 ft from B using the double-centering method, points C' and C'' were found to lie 0.019 ft apart on their common hub. Compute the angle α that the telescope optical axis makes with the telescope plunge axis.

Solution. From Example 6.3, half the angle $C''BC'$ is equal to 180° − 2α. Thus,

$$2\alpha = 180° - \frac{\text{angle } C''BC'}{2}$$

$$\alpha = 90° - \frac{\text{angle } C''BC'}{4}$$

$$\frac{\text{Angle } C''BC'}{2} = \sin^{-1}\frac{C'C''/2}{200.00} = \sin^{-1}\frac{0.019/2}{200.00}$$

$$= 9.8''$$

$$\frac{\text{Angle } C''BC'}{4} = \frac{9.8''}{2} = 4.9''$$

Thus, $\alpha = 90°00'00.0'' - 4.9'' = 89°59'55.1''$.

By now, the reader may believe that the difference between the true extension of the centerline and the centerline established by a single plunge of the telescope is so small that it may be ignored. This may be true for a single extension, but the effect is cumulative. The angular offset for a single plunging is twice the difference between angle a and a true perpendicular—in the Example 6.3 case, this is an angular deviation of 14 seconds each time the centerline is extended 100.00 ft. After just 10 extensions, which is less than a fifth of a mile, the cumulative angular deviation for 10 single plungings would amount to 140 seconds, or $2\frac{1}{3}$ minutes, and the offset distance would be about 10(0.007) ft, or 0.07 ft. Considering the fact that highway straightaways continue for miles, the cumulative offset from successive single plungings would be unacceptable.

6.4 OBSTACLE ON LINE

During the construction phase of a highway, pipeline, or transmission line, bulldozers are used to clear away obstacles along a construction right-of-way that has been staked out by the surveying crew. For a number of reasons, however, the surveyor doesn't have the luxury of a bulldozer to clear the line of sight during the stakeout process. A continuing problem for the surveyor, in most terrain, is a seeming multitude of obstacles that interfere with the instrument line of sight along the intended centerline. To overcome the problem, the surveyor requires methods for bypassing such obstacles while staking out survey points at desired intervals along the centerline. For highway stakeout, the survey points, called **stations**, are set 100 ft apart.

Figure 6.7 (next page) illustrates two methods of accomplishing this end. In Fig. 6.7a, points A, B, and C have been established at regular intervals (say, 100.00 ft) along a centerline but an obstacle lies between C and the next centerline point at D. To pass the obstacle, new points B' and C' are established at equal perpendicular distances from B and $C(BB' = CC')$; thus, line $B'C'$ is parallel to BC and the same length as BC. Point D' is established at the same fixed interval from $C'(C'D' = B'C')$ by double-centering, after which E' is set in the same manner and at the same interval. Points D and E are then established in the same manner as B' and C', at measured distances $D'D = E'E = BB' = CC'$; thus, DE is parallel to $B'C'D'E'$ and it is a true extension of centerline ABC.

The method of Fig. 6.7a usually involves measuring relatively short perpendicular distances, something better done with a tape than with an EDM. With a theodolite-EDM combination, it is sometimes more convenient to use the method of Fig. 6.7b, which depends on shallow deflection angles and longer distances to produce the squat isosceles triangle BIE'. Given the same situation as in Fig. 6.7a, with an obstacle lying just beyond C and the instrument still at B, a double-centered deflection angle α is taken to point I, set at a convenient location adjacent to the obstacle, and the EDM distance BI is measured. Setting up on I and backsighting on B, a double-centered deflection angle 2α is laid out, as described in Sec. 6.2, and point E' is set at EDM-measured distance $IE' = BI$ in order to complete the isosceles triangle BIE'. With the instrument set up at E', a double-centered deflection angle α is laid out to establish the continuation of centerline ABC to F and beyond. In order to establish points D and E, the distance between B and E' must first be computed.

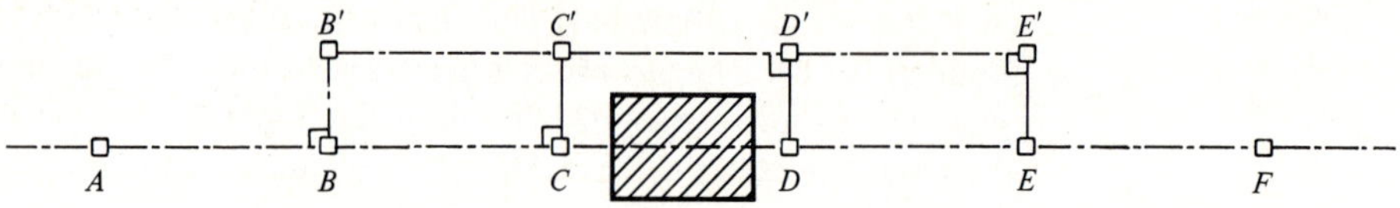

(*a*) Perpendicular-offset method

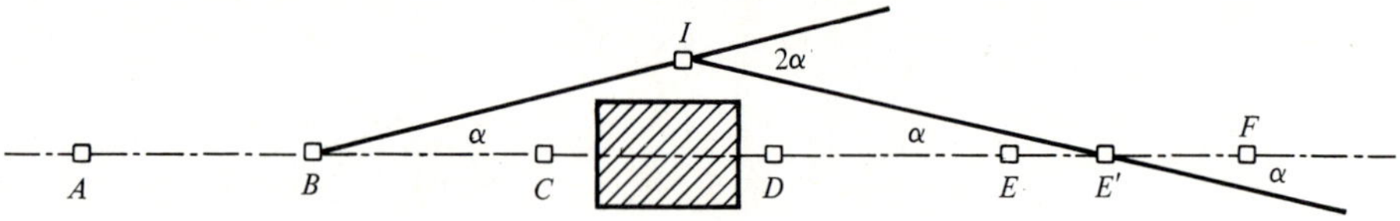

(*b*) Deflection-angle method

FIGURE 6.7
Passing an obstacle on line.

Since BIE' is an isosceles triangle, the baseline distance BE' must be twice the side length BI multiplied by the cosine of α; that is, $BE' = 2BI\cos\alpha$. With BE' known, the proper distances back along the centerline to E and D may be measured from E'.

Example 6.5. In Fig. 6.7*b*, deflection angle α is $4°18'57''$ and EDM distances BI and IE' are 158.94 ft. Compute distance BE' and the distances $E'E$ and $E'D$ to be measured back from E' to establish E and D, and the distance $E'F$ to be measured forward from E' to establish F, assuming 100.00-ft intervals between points along the centerline.

Solution.

$$BE' = 2BI\ \cos\alpha = 2(158.94)\cos 4°18'57'' = 316.98 \text{ ft}$$

Points C, D, E, and F lie 100.00, 200.00, 300.00, and 400.00 ft, respectively, from B. Therefore,

$$E'E = 316.98 - 300.00 = 16.98 \text{ ft back from } E'$$

$$E'D = 316.98 - 200.00 = 116.98 \text{ ft back from } E'$$

$$E'F = 400.00 - 316.98 = 83.02 \text{ ft forward from } E'$$

It is simpler, at times, to make the baseline of the Fig. 6.7*b* triangle an exact number of intervals, which means the distance BI must be precomputed.

Example 6.6. Given the same deflection angle α as in Example 6.5, compute the EDM distance BI that must be laid out if the triangle baseline is to be 300.00 ft.

Solution. In this case, the baseline is BE, and $BE = 2BI\cos\alpha$. Transposing, $BI = BE/2(\cos\ \alpha)$. Substituting,

$$BI = \frac{300.00}{2\cos 4°18'57''} = 150.43\text{ft}$$

In the event the surveyor wishes to establish a line parallel to a given centerline, whether an obstacle lies in the path of the original centerline or not, this can be accomplished the same way as line $B'C'D'E'$ was laid out in Fig. 6.7*a*. It can also be done as in Fig. 6.7*b*, by laying out a deflection angle at I equal to α rather than 2α.

6.5 BALANCING IN

It is frequently the case that a ridgetop point is to be located on-line between two points that are not intervisible. Figures 6.8*a* show a profile of the problem and Fig. 6.8*b* and *c* shows two plan views of the solution method. Referring to Fig. 6.8*a*, the surveyor's intention is to locate point C on top of a ridge that constitutes a visual obstacle between points A and B. Figures 6.8*b* and *c* show the location of point C directly on line between A and B. As an initial step, the transit or theodolite is set up on point C', the surveyor's best guess as to the true location of C. Using the straight-line-extension procedure from Sec. 6.3, the extension of AC' is found to miss point B, as expected. BB', the approximate perpendicular distance from B to the extension of AC', is then measured.

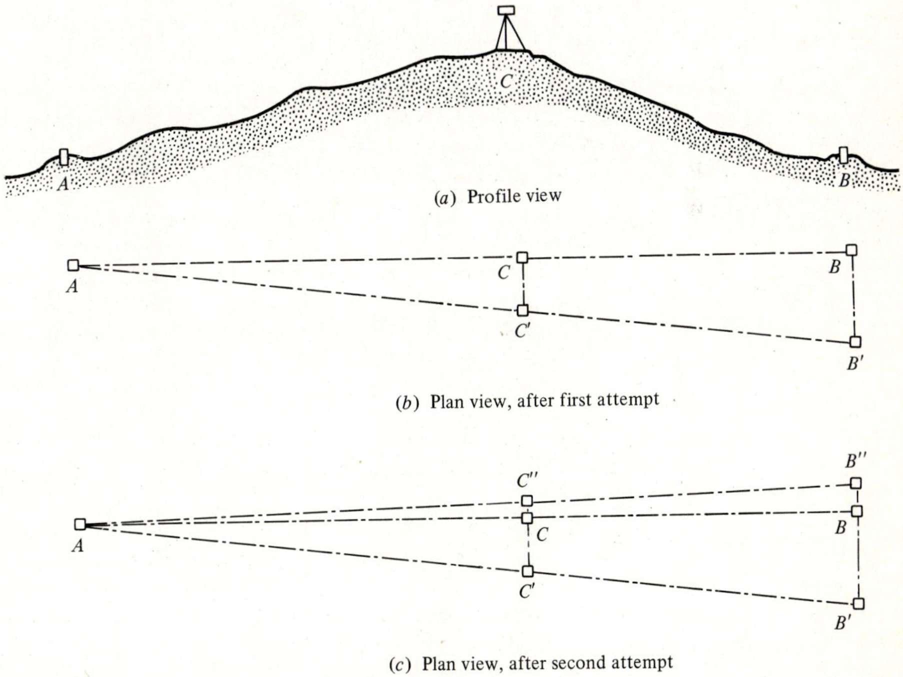

(*a*) Profile view

(*b*) Plan view, after first attempt

(*c*) Plan view, after second attempt

FIGURE 6.8
Balancing in on a line between nonintervisible points.

It is seen in Fig. 6.8*b* that $AC'C$ and $AB'B$ are similar triangles, giving the relationship: $CC'/BB' = AC/AB = AC'/AB'$. Solving for CC',

$$CC' = BB'\left(\frac{AC}{AB}\right) \tag{6.1}$$

If AC and AB are known, the solution is simple. Unfortunately, the ratio AC/AB is usually *not* known. The surveyor, from the instrument position at C', must then estimate distances AC and CB to arrive at the ratio AC/AB. Substituting this value of AC/AB into Eq. (6.1), CC' is computed. The surveyor then establishes point C'' a distance CC' from the initial setup point at C', in the same direction as from B' to B, and perpendicular to AB, as shown in Fig. 6.8*c*. From the new setup at C'', the surveyor then sights on A, plunges the telescope, and establishes B'' adjacent to B.

Figure 6.8*c* shows there are now three sets of similar triangles: (1) $AC'C$ and $AB'B$, (2) $AC''C$ and $AB''B$, and (3) $AC''C'$ and $AB''B'$. From these relationships it is seen that $C''C/C''C' = B''B/B''B'$. Transposing,

$$C''C = \frac{(C''C')(B''B)}{B''B'} \tag{6.2}$$

The instrument is then moved once again, this time a distance $C''C$ in the appropriate direction to establish the on-line point C. If the three distances $B'B, B''B$, and $C'C''[C'C$ from Eq. (6.1)] have all been measured perpendicular to line AB, the value for $C''C$ in Eq. (6.2) would be the precise distance from C'' to the true on-line point. Because of minor discrepancies in the three measurements, the final setup point will be offset from line AB by a very small distance; close enough that a double-centered check can be made (as in Sec. 6.3) and the instrument can be brought on-line by use of the shifting plate alone.

Example 6.7. Referring to Fig. 6.8, a surveyor determined an offset distance $B'B$ of 5.68 ft from her initial set up at C'. Estimating C to lie 60% of the distance from A to B, she used Eq. (6.1) to compute $C'C = (AC/AB)(B'B) = 0.60(5.68 \text{ ft}) = 3.41$ ft. Moving the setup point 3.41 ft in the appropriate direction, she then established point B'' a distance of 0.82 ft from B and on the opposite side of B from B'. How far should she now move her instrument setup position to arrive at the final setup over C?

Solution. In Eq. (6.2), the value $C''C'$ must be the approximated value $C'C$ from Eq. (6.1), or 3.41 ft, and the value $B''B'$ must be $B'B + B''B'$, or $5.68 + 0.82 = 6.50$ ft. Substituting into Eq. (6.2),

$$C''C = \frac{(3.41)(0.82)}{6.50} = 0.43 \text{ ft}$$

From similar triangles, the true ratio AC/AB must equal $C''C'/B''B'$, or 3.41/6.50—about $52\frac{1}{2}\%$. As a point of interest, had the surveyor's original estimate of AC/AB been 50% or 55%, point C'' would have been set somewhat closer to C.

6.6 RANDOM LINE

Occasionally, because of dense, intervening vegetation or irregular terrain, it becomes difficult to measure directly from one survey point to another. To get around this problem, a line may be run in a random direction, along a path that has a clear line of sight, to a point slightly offset from the intended point. In Fig. 6.9, dense brush obscures the line of sight between A and B, so random line AB' is run along a clear line of sight to offset point B', which "sees" both A and B. The instrument is then set up on B', and angle β is measured, along with distances $B'A$ and $B'B$. AB is computed from the law of cosines.

Example 6.8. Referring to Fig. 6.9, heavy brush obscures the line between points A and B. A random line of horizontal length 1824.65 ft is measured from A to point B'. With the instrument set up on B', the value of angle β is found to be 92°34′09″. A horizontal distance of 84.69 ft is then measured with a tape between B' and B. Compute the length of AB.

Solution. From the law of cosines, $AB = \sqrt{(AB')^2 + (B'B)^2 - 2(AB')(B'B)\cos\beta}$. Substituting,

$$AB = \sqrt{(1824.65)^2 + (84.69)^2 - 2(1824.65)(84.69)\cos 92°34'09''}$$

$$= 1830.40 \text{ ft}$$

The random-line method is also used to solve a somewhat different problem. In Fig. 6.10, point A on a property boundary has been established. With the instrument at A, angle α has been turned from existing point J to the direction of B, only to find that the line of sight to B is obscured. In order to establish point B, a random line is laid out along a clear line of sight to offset point B', from which $B'B$ may be easily measured—even though a short section of brush must be negotiated to clear a path. In this case, distance AB is already known. Distance AB' is measured, along with angle A (from existing point J to offset point B'), leaving $\alpha' = A - \alpha$. With angle α', distance AB and distance AB' all known, distance $B'B$ is computed from the law of cosines. With $B'B$ computed, angle β may be computed from the law of sines. With $B'B$ and angle β both computed, the surveyor may be set up on B', backsight on A, turn angle β to the direction of B, and lay out distance $B'B$ to establish point B.

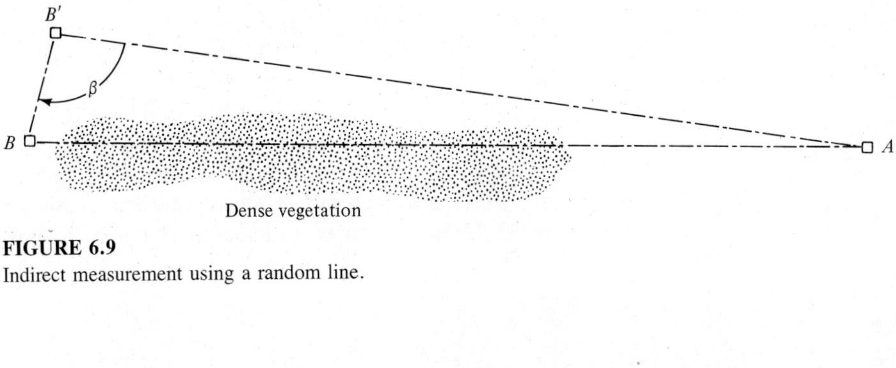

FIGURE 6.9
Indirect measurement using a random line.

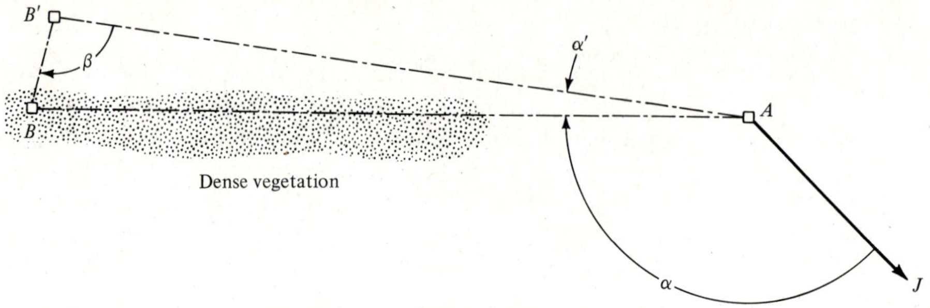

FIGURE 6.10
Establishing a survey point using a random line.

Example 6.9. In Fig. 6.10, a property boundary segment was to be laid out with angle $\alpha = 110°52'17''$ and side $AB = 855.63$ ft. The surveyor set up on A, turned angle α and found AB obscured by dense undergrowth. Point B' was then set 850.00 ft along a clear path from A. Measured angle JAB' is $116°01'33''$. Compute angle β and distance $B'B$ so that point B may be located from point B'.

Solution. Angle α' = angle JAB' − angle $\alpha = 116°01'33'' - 110°52'17'' = 5°09'16''$. Substituting into the law of cosines as before,

$$B'B = \sqrt{(AB)^2 + (AB')^2 - 2(AB)(AB')\cos\alpha}$$

$$= \sqrt{(855.63)^2 + (850.00)^2 - 2(855.63)(850.00)\cos 5°09'16''}$$

$$= 76.901\ldots \Rightarrow\Rightarrow 76.90 \text{ ft}$$

From the law of sines: $B'B/\sin\alpha' = AB/\sin\beta$. Rearranging to solve for β, $\sin\beta = \sin\alpha'(AB/B'B)$, $\beta = \sin^{-1}[\sin\alpha'(AB/B'B)]$. Substituting,

$$\beta = \sin^{-1}\left[\sin 5°09'16''\left(\frac{855.63}{76.901\ldots}\right)\right] = \sin^{-1} 0.999601781$$

Remember that $\sin\beta = \sin(180° - \beta)$. Using the calculator's "$\sin^{-1}$" or "arcsin" key to determine the angle whose sine is 0.999601781, the calculator gives the answer as $88°22'59''$, the smaller of the two possibilities, the other being $180°00'00'' - 88°22'59''$, or $91°37'01''$. But distances AB and AB' are 855.63 ft and 850.00 ft, respectively, an indication that angle β is obtuse, rather than acute. Thus,

$$\beta = 180° - 88°22'59'' = 91°37'01''$$

To locate B, set up on B', backsight on A, turn a clockwise angle of $91°37'01''$ to establish the B'-to-B line, then measure a horizontal distance of 76.90 ft along the line of sight to establish point B.

The same method is used to "recover" an existing survey point whose location has been obscured. In this case, points J and A have been found, B lies in dense undergrowth. A random line is laid out in the same manner to offset point B' so that clearing the final approach to B is much easier than it would have been from A.

6.7 PROBLEMS

6.1. Deflection angles were measured at the corners of a six-sided closed figure as follows: A = 84°13′R, B = 101°46′R, C = 37°29′L, D = 123°41′R, E = 95°37′R, F = 7°42′L. Assuming equal-weight angles, adjust as necessary to provide proper closure.

6.2. Do the same as Prob. 6.1, except the angles are A = 37°28′L, B = 137°31′R, C = 161°47′R, D = 88°09′L, E = 65°55′R, F = 120°36′R.

6.3. Convert the adjusted deflection angles obtained in Prob. 6.1 to interior angles and provide a check.

6.4. Convert the adjusted deflection angles obtained in Prob. 6.2 to interior angles and provide a check.

6.5. What is the maximum value a deflection angle can have?

6.6. You wish to set a point 392.16 ft from a property corner at a clockwise angle of 76°14′33″ from an adjacent property line. Using a 20-second transit, you measure a clockwise angle of 76°14′40″ from an initial setting of zero and set a tack in a hub at the above distance along the preliminary direction. An additional direct angle was measured to the tack, plus two reversed angles, and the final circle reading was 304°58′20″. By what amount and in what direction (clockwise or counterclockwise) must a right-angle offset be measured from the tack to establish the desired point?

6.7. A surveyor is required to set a point 450.89 ft from a control point at an angle of 122°38′17″, turned clockwise from the adjacent base line. With a 30-second transit, she turns an angle of 122°38′30″ and sets a tack at the proper distance from her setup. She measured one additional direct angle and two reversed, arriving at a final circle reading of 130°35′00″. In what direction (clockwise or counterclockwise) and at what distance must she measure a right-angle offset from the tack to establish the desired point?

6.8. A novice surveyor extended existing line AB to point C and thence to point D by setting up on B, sighting on point A with the telescope direct, then plunging the telescope and setting point C. He moved the setup to C, sighted on point B with telescope direct, then plunged the telescope and set point D. If the telescope optical axis makes an angle of 89°59′47″ with the telescope's transverse (plunge) axis, what is the angular error between line AB and line CD?

6.9. After establishing point C in Prob. 6.8, the novice recognized the potential error in his method. After carefully setting up on point C, he reversed the previous procedure; that is, he sighted on point B with telescope reversed, plunged, and set point D with telescope direct. Assuming distances BC and CD were equal, did this bring point D back on the extension of line AB? If not, how far was D from the extension of line AB if distances BC and CD were each 200.00 ft?

6.10. In Prob. 6.8, if distances, AB, BC, CD, etc., were each 200.00 ft and the described observing procedure were continued repeatedly, what radius circle (nearest mile) would hypothetically pass through all points and eventually return and pass again through A?

6.11. Using the perpendicular-offset method of Fig. 6.7*a*, distances $B'C'$, $C'D'$, and $D'E'$ were all 100.00 ft; however, distance BB' was mistakenly laid out 0.10 ft shorter than CC'. What angular error in the extension of ABC would result?

6.12. Referring to Fig. 6.7*b*, if deflection angle α is 5°29′43″, compute distances BI and IE' that must be laid out if the base of triangle BIE' is to be precisely 400.00 ft.

6.13. In Prob. 6.12, assuming there is a clear path, what distance must be taped from I, perpendicular to line BE', to set a point on the line midway between B and E'?

6.14. A builder wished to place a point on a side property line 60.00 ft from the front property marker. He found that the front and rear property corners, though not intervisible, could be seen from the 60.00-ft distance where the point was to be placed. He set up his transit 60.00 ft from the front corner and at his best estimate for being on line. By double-centering, he extended the front corner-to-transit-position line and found that it missed the rear property corner by a perpendicular distance of 4.02 ft. If the side property line was recorded as 225.62 ft in length, what perpendicular distance must the transit be moved to place it on-line?

6.15. In Fig. 6.8, a surveyor set preliminary point C' 65 paces from point A and 88 paces from B. After extending line AC' she found the BB' offset was 7.64 ft. *Calculating* the appropriate offset from C' and setting point C'', she again extended the line but found that BB'' measured 0.71 ft in the same direction from B to B'. What distance must the instrument be moved from C'' to place it exactly on AB?

6.16. Points A and B in Fig. 6.9 were not intervisible, so a random line was run 2637.39 ft along a clear path to B', a point from which B is visible. Measured distance BB' was 28.81 ft and angle $AB'B$ was measured as 81°47′39″. Compute the length of line AB.

6.17. Referring to Fig. 6.10, a surveyor was to establish property corner B 733.80 ft from A, but found line AB to be obstructed. He then turned an angle 2°41′00″ further clockwise from direction AB to a clear path on which point B' was set 748.00 ft from A. Compute angle $AB'B(\beta)$, and distance $B'B$ to be measured to establish desired point B.

6.18. In Prob. 6.17, it is now necessary to divide property line AB into thirds by measuring right-angle offsets from AB' to AB. At what two distances along AB' are the two offset-measurement points to be established and what are the two perpendicular-offset distances?

CHAPTER 7

DIRECTIONS

7.1 MERIDIANS, AZIMUTHS, AND BEARINGS

In earlier discussions, measurements have been defined in terms of a distance or angle *from* a starting point *to* an end point, the starting point being the initial reference from which a measurement is taken. Consider the concept of direction, which also requires a beginning reference, but of a slightly different sort. "Northwest," for example, is west of the north direction; specifically, 45° counterclockwise from north. The idea of "north" must be further defined in terms of **true north** or **magnetic north**, depending on whether a given direction is a **true direction** or a **magnetic direction**. The discussion that follows will focus on true north and true direction. Magnetic direction, with its underlying concept of magnetic north as a function of the magnetic field of the earth, is covered in some depth in Sec. 7.3.

MERIDIANS. The north-south line on which the observer is located is the observer's **true meridian**, a true meridian being any great circle passing through the earth's north and south geographic poles. True north is the direction of the geographic north pole along the true meridian.

True direction, from the surveyor's standpoint, is always referenced to a true meridian. True direction, in turn, may be defined as a **true azimuth** or a **true bearing**. In this text, true azimuths and true bearings will be referred to simply as "azimuths" and "bearings," while magnetic azimuths and magnetic bearings, discussed further along, will be referred to by those names.

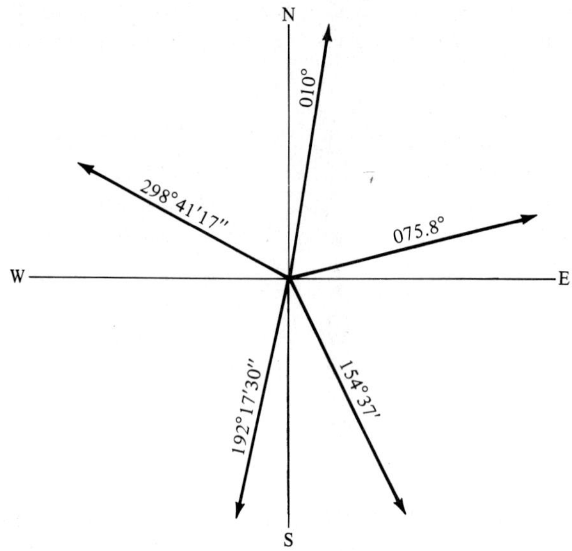

FIGURE 7.1
Azimuths.

AZIMUTHS. For most surveyors, **azimuth** is the direction of a line measured clockwise from the true north meridian that passes through the origin of the line. (The geodetic surveyor, until 1989, defined azimuth as the clockwise direction of a line with respect to true south; now, however, the geodetic reference is also true north.) Figure 7.1 displays five azimuths that originate from north. For clarity, an east-west axis is included with the north-south axis.

Consider these five examples one at a time, clockwise from north. The first azimuth is 010°, indicating that it points in a direction lying 10° clockwise from the true-north reference meridian. The second, also in the north-to-east quadrant, is 075.8°; thus, it points 75.8° clockwise from true north. The third, 154°37′, the fourth, 192°17′30″, and the fifth, 298°41′17″ are similarly defined as angular values measured clockwise from true north.

BACK AZIMUTH. A **back azimuth** is the direction of a line looking back on its origin from its terminus; that is, 180° from the initial azimuth direction. The second example from Fig. 7.1 is used to illustrate the azimuth–back azimuth principle. In

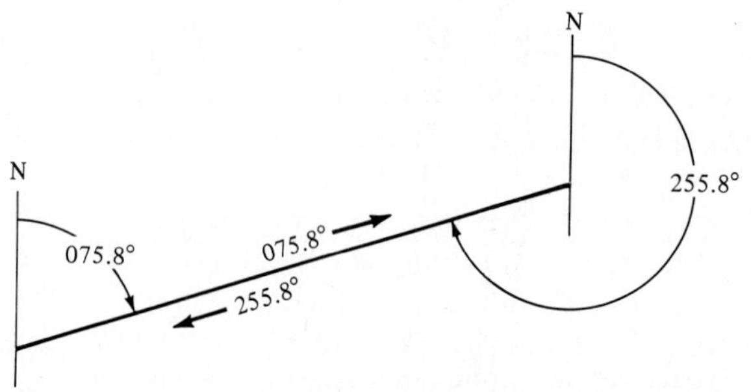

FIGURE 7.2
Back azimuth.

Fig. 7.2, the initial azimuth, shown at left, is 075.8°; thus, the back azimuth must be 255.8°, which is 075.8° + 180.0°. Conversely, the back azimuth of 255.8° is 075.8°; that is, 255.8° − 180.0°.

CONVERGENCE OF THE MERIDIANS. It is not strictly true that a line's back azimuth is precisely 180° from its azimuth. All true meridians are great circles passing through the poles, so they must come together, or "converge" at the poles. This convergence is increasingly apparent as the east-west spacing between meridians decreases as the meridians approach the poles. For ordinary surveys, in small areas, in nonpolar latitudes, and in which all lines have relatively short east-west components, the effects of meridional convergence may be neglected. Put another way: For such small-area surveys, all meridians are assumed to be parallel.

BEARINGS. A bearing is the direction of a line with respect to the north-south true meridian passing through the line's origin. It is *always* measured 90° *or less* east or west of true north or true south. In labeling bearings, "N" or "S" is listed first ("N" if the initial reference direction is north, "S" if the initial reference direction is south), followed by the angle value east or west of the north or south direction, and finally by "E" or "W" to denote whether the angle is taken east or west of the meridian. Figure 7.3 shows the Fig. 7.1 directions labeled as bearings.

Again taking directions in order, clockwise from north, the first bearing is N10°E, indicating a direction of 10° east of north. The second example, also in the first (northeast) quadrant, is N75.8°E. The third direction, azimuth 154°37′, carries a bearing designation of S25°23′E; that is, the direction is 25°23′ east of true

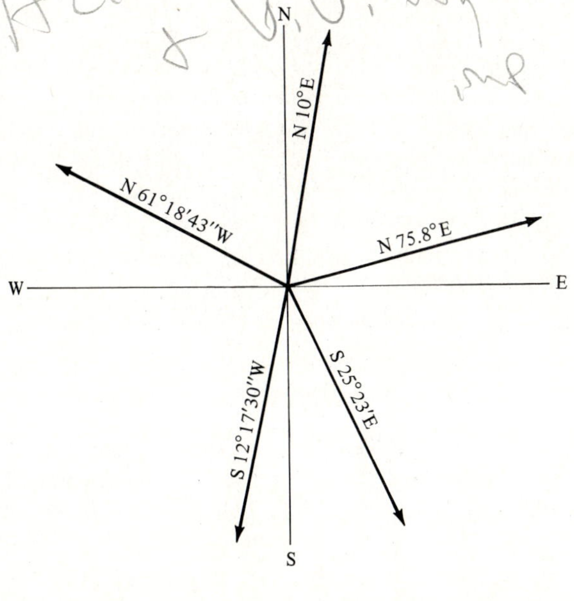

FIGURE 7.3
Bearings.

south (note here that the azimuth and the bearing angle add to 180°). In the fourth example, an azimuth of 192°17′30″, the bearing is S12°17′30″W, or 12°17′30″ west of south (note that the bearing angle here is 180° less than the azimuth). The final bearing, for azimuth 298°41′17″, is N61°18′43″W, or 61°18′43″ west of north (note that the azimuth and the bearing angle add to 360°).

The exact north direction may be referred to as **due north**, or as either N00°00′E or N00°00′W. Exact east is **due east**, N90°00′E or S90°00′E. Exact south and west are designated in a similar fashion.

BACK BEARINGS. A back bearing, analogous to a back azimuth, is the reciprocal of the bearing. As such, a back bearing carries the bearing designation in reverse; that is, the cardinal-point designations are reversed and the numerical value of the angle is left alone. Thus, the back bearing of N75.8°E must be S75.8°W, a bearing of S25°23′E has a back bearing of N25°23′W, and so forth.

AZIMUTH-TO-BEARING CONVERSION. Rules for converting azimuths to bearings vary as a function of the quadrant in which the azimuth lies. In the northeast quadrant, the azimuth value is taken as-is, accompanied by appropriately placed N and E designators. An azimuth of 063°22′ thus takes a bearing designation of N63°22′E.

In the southeast quadrant, the azimuth value is subtracted from 180° to obtain the bearing angle; thus, an azimuth of 172°14.4′ becomes S7°45.6′E (180°00.0′ − 172°14.4′ = 7°45.6′).

In the southwest quadrant, 180° is subtracted from the value of the azimuth. An azimuth of 249°34′03″, labeled as a bearing, would be S69°34′03″W (249°34′03″ − 180°00′00″ = 69°34′03″).

In the northwest quadrant, the azimuth value is subtracted from 360°. Thus, the azimuth 305°07′51.7″ converts to a bearing whose value is N54°52′08.3″W (360°00′00.0″ − 305°07′51.7″ = 54°52′08.3″).

BEARING-TO-AZIMUTH CONVERSION. Conversion of bearings to azimuths is carried out in the opposite way than azimuth-to-bearing conversion is done. In the northeast quadrant, the bearing angle is the azimuth: N15°E ⇒⇒ 015°. In the southeast quadrant, the bearing angle is subtracted from 180°: S80°30′ E ⇒⇒ 099°30′ (180°00′ − 80°30′ = 99°30′). In the southwest quadrant, the bearing angle is added to 180°: S24°16.5′W ⇒⇒ 204°16.5′ (180°00.0′ + 24°16.5′ = 204°16.5′). In the northwest quadrant, the bearing angle is subtracted from 360°: N36°45′W ⇒⇒ 323°15′ (360°00′ − 36°45′ = 323°15′).

7.2 DIRECTION COMPUTATION AND CHECKING

DIRECTIONS FROM INTERIOR ANGLES. Figure 7.4 shows four-sided polygon *ABCD* with known, adjusted interior angles and a known bearing of side *AB*. The procedure for computing and checking the bearings of the other three sides shall now be discussed.

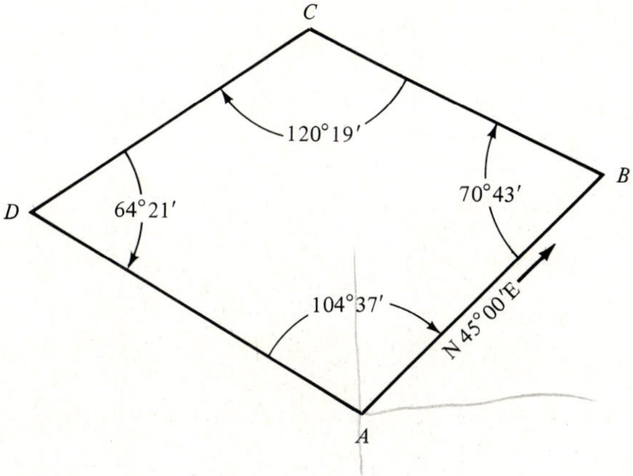

FIGURE 7.4
Quadrilateral *ABCD* with direction of side *AB* known.

First, consider corner *A* alone, as shown in Fig. 7.5, with the north direction designated by an arrow along a segment of the meridian that passes through *A*. The bearing of *AB* (that is, the bearing *from A to B*) is N45°00′E; thus, the azimuth of *AB* is 045°00′. The azimuth angle for *AB* is shown in Fig. 7.5 by a clockwise arrow from the meridian to line *AB*.

Corner *B* is shown in Fig. 7.6. Line *AB*, labeled with the values of both its azimuth (045°00′) and back azimuth (225°00′), terminates at the meridian that passes through corner *B*. The value of back azimuth *AB*, which may also be designated as azimuth *BA*, is shown by a clockwise arrow that starts at the meridian through *B*.

If the clockwise arrow to azimuth *BA* were to be continued to the direction of *BC*, it would turn through an angle equal to the numerical value of azimuth *BC*. From Fig. 7.6, the azimuth of *BC* must equal the azimuth of *BA* plus angle *B*; thus, azimuth *BC* is 225°00′ + 70°43′ = 295°43′. Bearing *BC*, because it lies in the fourth quadrant, must be N64°17′W (360°00′ − 295°43′ = 64°17′).

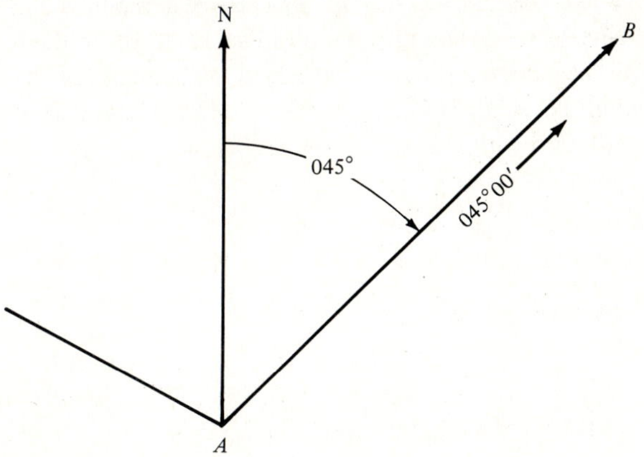

FIGURE 7.5
Direction of line *AB* at corner *A* of quadrilateral *ABCD*.

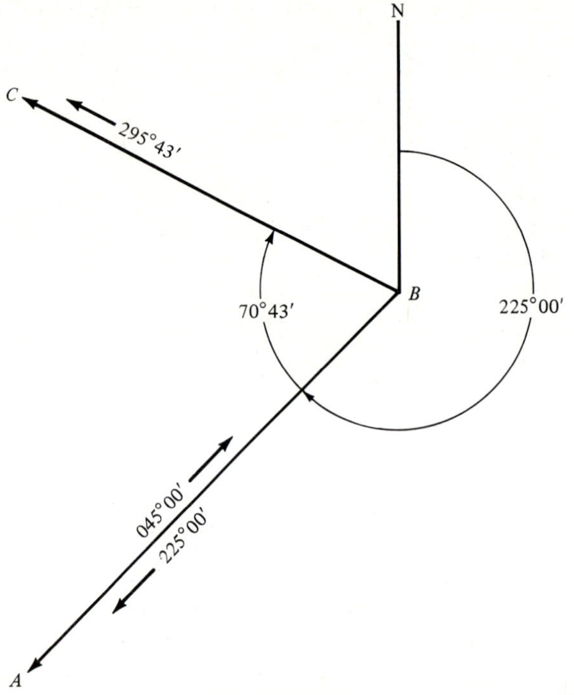

FIGURE 7.6
Directions of lines *AB* and *BC* at corner *B* of quadrilateral *ABCD*.

In computing bearings from angles and an initial known bearing, it is almost always simpler to compute in azimuth format and convert the azimuths to bearings. At a polygon corner, to compute a departing azimuth from a known entering azimuth and a known angle: Convert the entering azimuth to a back azimuth, then add the value of the angle to obtain the forward-going azimuth.

For quadrilateral *ABCD*, Fig. 7.4, the following format simplifies azimuth computation. The known bearing of *AB* (N45°00′E) is entered on the first line, along with its converted azimuth value (045°00′). Back azimuth *BA* (225°00′) is entered on the second line in the Azimuth column. The value of angle *B* (+70°43′) is entered on the third line under Azimuths and added to the value of back azimuth *BA* to produce the azimuth of *BC* (225°00′ + 70°43′ = 295°43′), which is entered on the fourth line along with the bearing of *BC* (N64°17′W).

Side	Azimuths	Bearings
AB	045°00′	N45°00′E
BA	225°00′	
∠*B*	+70°43′	
BC	295°43′	N64°17′W

Completing the computation around quadrilateral *ABCD*:

Side	Azimuths	Bearings
AB	045°00′	N45°00′E
BA	225°00′	
∠*B*	+70°43′	
BC	295°43′	N64°17′W
CB	115°43′	
∠*C*	+120°19′	
CD	236°02′	S56°02W
DC	056°02′	
∠*D*	+64°21′	
DA	120°23′	S59°37′E
AD	300°23′	
∠*A*	+104°37′	
AB	405°00′	
AB	045°00′	

Note, in the next-to-last line in the above computation, that the summation of back azimuth *AD* and angle *A* had produced an azimuth value greater than 360°, a natural consequence when corner angles continue to be added into the computation; this was easily rectified by subtracting 360° from the initially computed azimuth of line *AB* to arrive at its correct azimuth of 045°. Note also that the last-line value for azimuth *AB* checks the known initial value for *AB*. Recall that for *n*-sided polygons, interior angles $= (n - 2)180°$, which is the case here. The check indicates that *either* the arithmetic is correct *or* the computation contains precisely compensating errors—of the two possibilities, it is likelier that the arithmetic is correct.

DIRECTIONS IN A CLOCKWISE-LABELED FIGURE. In Fig. 7.4, with corners labeled progressively in a counterclockwise direction around the polygon, it is convenient to consider the corner angles to be clockwise, or positive, in determining

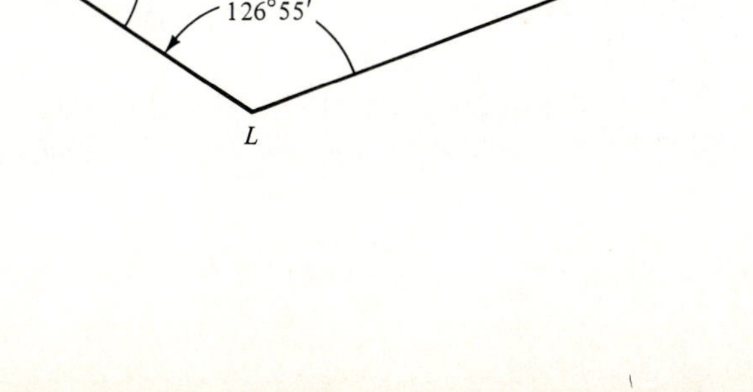

FIGURE 7.7
Quadrilateral *JKLM*, Example 7.1.

successive directions. In the opposite case, in which corners are labeled clockwise progressively around a closed figure, successive corner angles may be assumed to be counterclockwise, and thus negative. The computation format and procedure for corner angles that are assumed to be counterclockwise is identical to the foregoing except that angle values are subtracted.

Example 7.1. In quadrilateral JKLM (Fig. 7.7), the corner angles and the bearing *JK* are known. Compute the other three bearings.

Solution. The known bearing and azimuth of *JK* are entered on the first line, back azimuth *KJ* on the second line. Angle *K*, entered on the third line, is *subtracted* from back azimuth *KJ* to produce azimuth *KL*, which is then converted to bearing *KL*. The rest of the computation continues below in the same manner.

Side	Azimuths	Bearings
JK	114°48′	S65°12′E
KJ	294°48′	
∠*K*	−44°19′	
KL	250°29′	S70°29′W
LK	070°29′	
LK	430°29′	
∠*L*	−126°55′	
LM	303°34′	N56°26′W
ML	123°34′	
∠*M*	−91°06′	
MJ	032°28′	N32°28′E
JM	212°28′	
∠*J*	−97°40′	
JK	114°48′ ✓	

Note the arithmetic check and note also that 360° had to be added to azimuth *LK* to produce a positive value when angle *L* was subtracted.

DIRECTIONS FROM DEFLECTION ANGLES. Directions are computed from deflection angles somewhat more simply than is the case for angles-left or angles-right. In Fig. 7.8, a traverse begins at point *Q* with a backsight on point *P* (azimuth *QP* is known), then continues with deflection angles to points *R*, *S*, and *T*.

Considering initial point *Q* first, the backward azimuth to *P* is a known value, 245°17′00″. The forward azimuth, needed for the computation of azimuth *QR*, is thus 065°17′00″ (245°17′00″ − 180° = 065°17′00″). From Fig. 7.8, the azimuth of line *QR* must equal the azimuth of the forward extension of *PQ* (dashed line) plus the value of angle *Q*, a deflection angle *right*. Had *Q* been a left deflection angle, its value would have been subtracted from the value of the forward azimuth *PQ*.

At a traverse corner, a deflection angle right is added to the value of the forward-going azimuth entering that corner to produce the azimuth departing from that corner; a deflection angle left is subtracted from the value of the forward-going azimuth entering that corner to produce the value of the azimuth departing from that corner.

The computation format for deflection angles is shown below for the situation in Fig. 7.8.

Line	Azimuth	
QP	245°17′00″	(known backward azimuth)
PQ	065°17′00″	(*QP*, converted to required azimuth *PQ*)
∠*Q*	+45°06′30″R	(note addition of deflection angle right)
QR	110°23′30″	(sum of azimuth *PQ* and angle *Q*)
∠*R*	+55°49′00″R	
RS	166°12′30″	
∠*S*	−59°27′30″L	(note subtraction of deflection angle left)
ST	106°45′00″	(no check for this open-ended traverse)

It is obviously quicker and simpler to compute directions from deflection angles rather than angles left or angles right. It is not necessary to convert forward azimuths to back azimuths; all that is needed is a forward-going azimuth at the starting point.

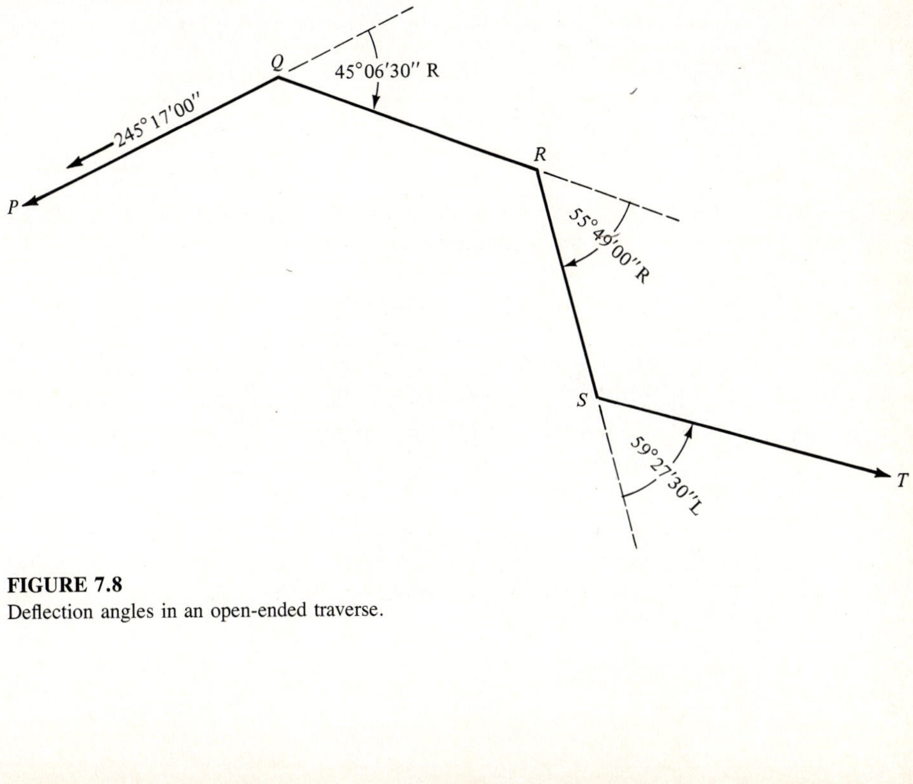

FIGURE 7.8
Deflection angles in an open-ended traverse.

7.3 MAGNETIC FIELD OF THE EARTH

IDEALIZED MAGNETIC FIELD. Many of us recall seeing a bar magnet placed on a sheet of paper on which iron filings had been scattered. Iron filings lying within the bar magnet's external **field** become magnetized and they will align with the **lines of magnetic flux** the magnet generates. A bar magnet has a north and south pole and its field will be fairly symmetrical about the axis connecting the poles. The pattern of alignment of the magnetized filings will outline the roughly elliptical shape of the magnetic field lying in the plane of the sheet of paper. If the bar magnet could somehow be made shorter and shorter, while maintaining a constant magnet intensity, the external pattern of its field would become more circular.

The earth's external magnetic field behaves as though its magnetism derives from a very short, very powerful bar magnet located close to the earth's center. The solid line in Fig. 7.9 is the earth's true north-south axis and the dashed line passing through the axis of the hypothetical bar magnet is the magnetic north-south axis about which the earth's magnetic field is centered. In Fig. 7.9, the field is represented as truly symmetrical, so the dashed line perpendicular to the magnetic axis at the center of the bar magnet may be said to pass through the earth's "magnetic equator." Note that the earth's field is not horizontal except at the magnetic equator; at all other points, it intersects the surface at an angle, the angle of intersection being known as **dip**.

DECLINATION. A magnetized needle suspended at its balance point will align itself with the earth's magnetic field at the point of suspension. To convert the suspended needle to a compass needle, a counterbalance is needed at the high end of the needle, positioned to make it horizontal. A counterbalanced compass needle will align itself with the horizontal component of earth's magnetic field; that is, it will point toward

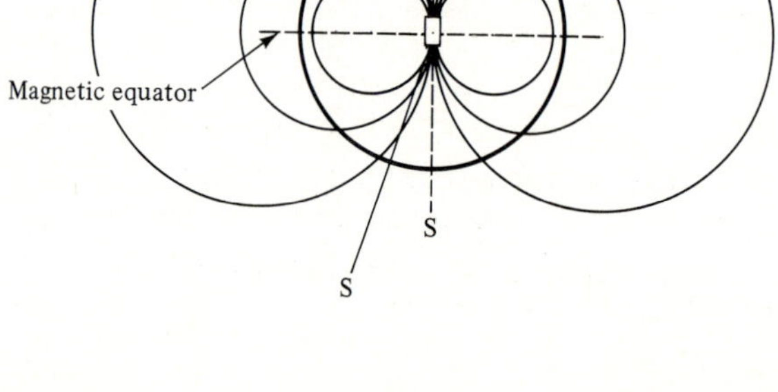

FIGURE 7.9
Simplified two-dimensional diagram of earth's magnetic field.

magnetic north at the point of observation. The angle between the direction of *local* magnetic north and true north is the **declination** for that locality. Declination is designated by direction as well as by its angular value. *If the compass needle points west of true north, declination is west; if the compass needle points east of true north, declination is east.*

The earth's magnetic field is far less symmetrical than indicated by Fig. 7.9. Both dip and declination vary quite irregularly over the earth's surface, due to the fact that magnetic material within the earth is not homogeneously dispersed.

DECLINATION IN THE UNITED STATES. Because directions on older property descriptions are often given as magnetic bearings, the value of declination in such cases is important to the surveyor. Declination varies significantly from one point to another. Figures 7.10 and 7.11 show maps of declination values for the conterminous United States, Alaska, and Hawaii as of January 1, 1985. Solid lines, called **isogonic lines**, are lines of constant declination. The heaviest of these is the **agonic line**, or line of zero declination. Declination changes in both space and time; therefore, isogonic lines will drift. Some indication of rate of drift is seen in the dashed lines in Fig. 7.10 and 7.11, which are lines of constant annual change.

In Fig. 7.10, the isogonic line labeled 15°W passes through Rhode Island. Assuming the line passes through the center of Providence, R. I., we may say that the declination there was 15°00′ west at the beginning of 1985. Interpolating, the annual change for Providence is approximately 3.5 minutes west per year. Assuming no change in the rate of change, it can be assumed that a total change of 14 minutes westerly occurred by the beginning of 1989. Thus, the start-of-1989 declination for downtown Providence would have been 15°14′ west (15°00′W plus 14′ W = 15°14′W).

The surveyor should never rely on declination maps as generalized as those in Figs. 7.10 and 7.11. In reality, isogonic lines run nowhere near that smoothly. More precise and detailed information on local values of declination, both present and past (in some localities, as far back as the 1740s), as well as annual change, is available from the National Geophysical Data Center, NOAA, Code E/GC4/FLI, 325 Broadway, Boulder, CO 80303; telephone (303) 497-6124.

7.4 DECLINATION COMPUTATIONS

BACKGROUND. Not too many years ago, surveyors in the United States routinely used transit compasses to establish a reference direction for field surveys. Now, with the establishment of horizontal-control networks throughout the United States during this century, true-north reference directions of one kind or another are available within, or close by, any project area on which surveys are to be run. Reflecting this change in surveying procedure, most modern angle-taking instruments no longer contain compasses. Nevertheless, many older property records are based on a magnetic-north reference; thus, modern surveyors who work with property boundaries so described are occasionally compelled to convert from magnetic direction to true direction—or from true to magnetic, as necessary.

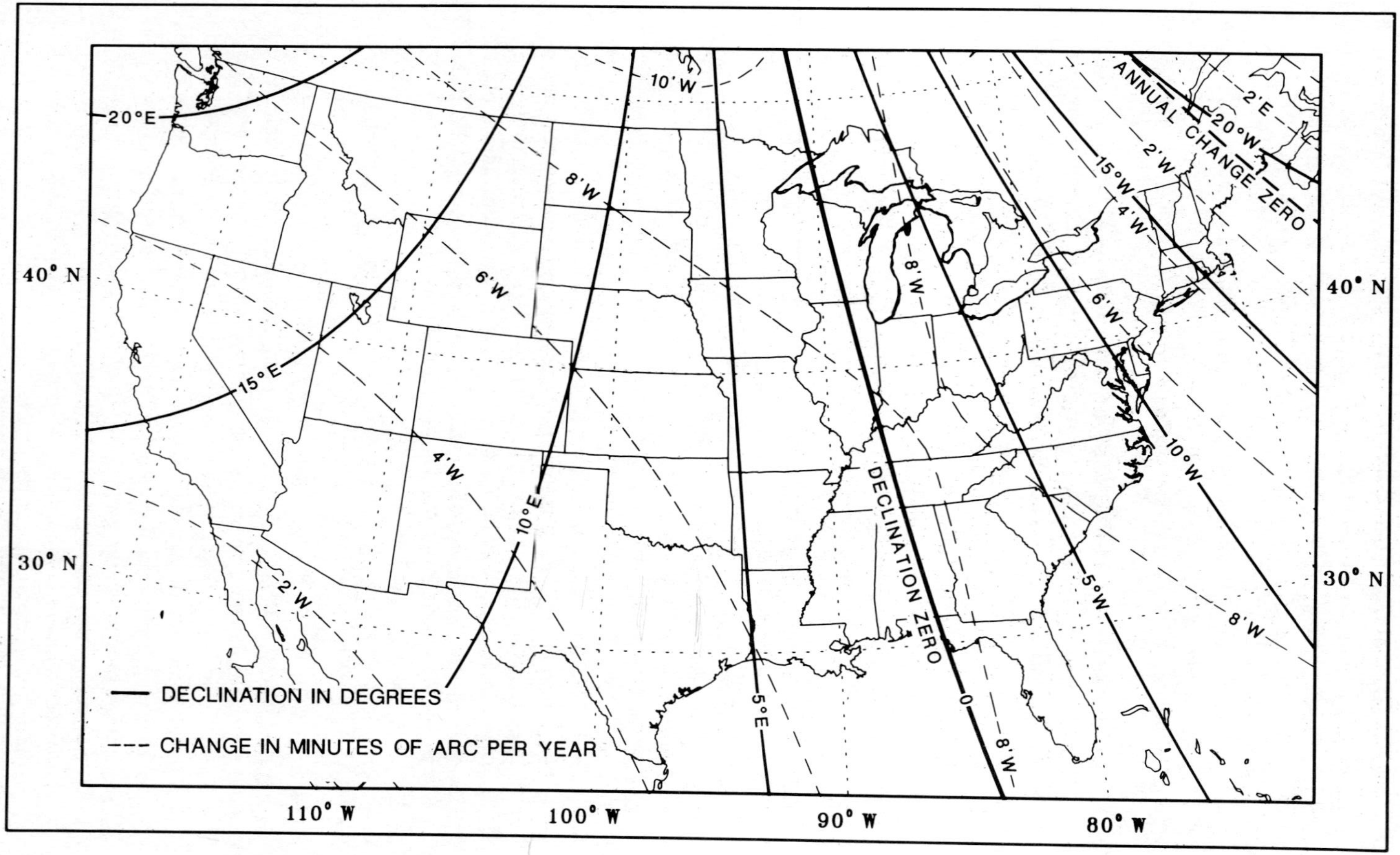

FIGURE 7.10
1985 declination in the conterminous United States. *(Courtesy of U.S. Geological Survey, Department of Interior.)*

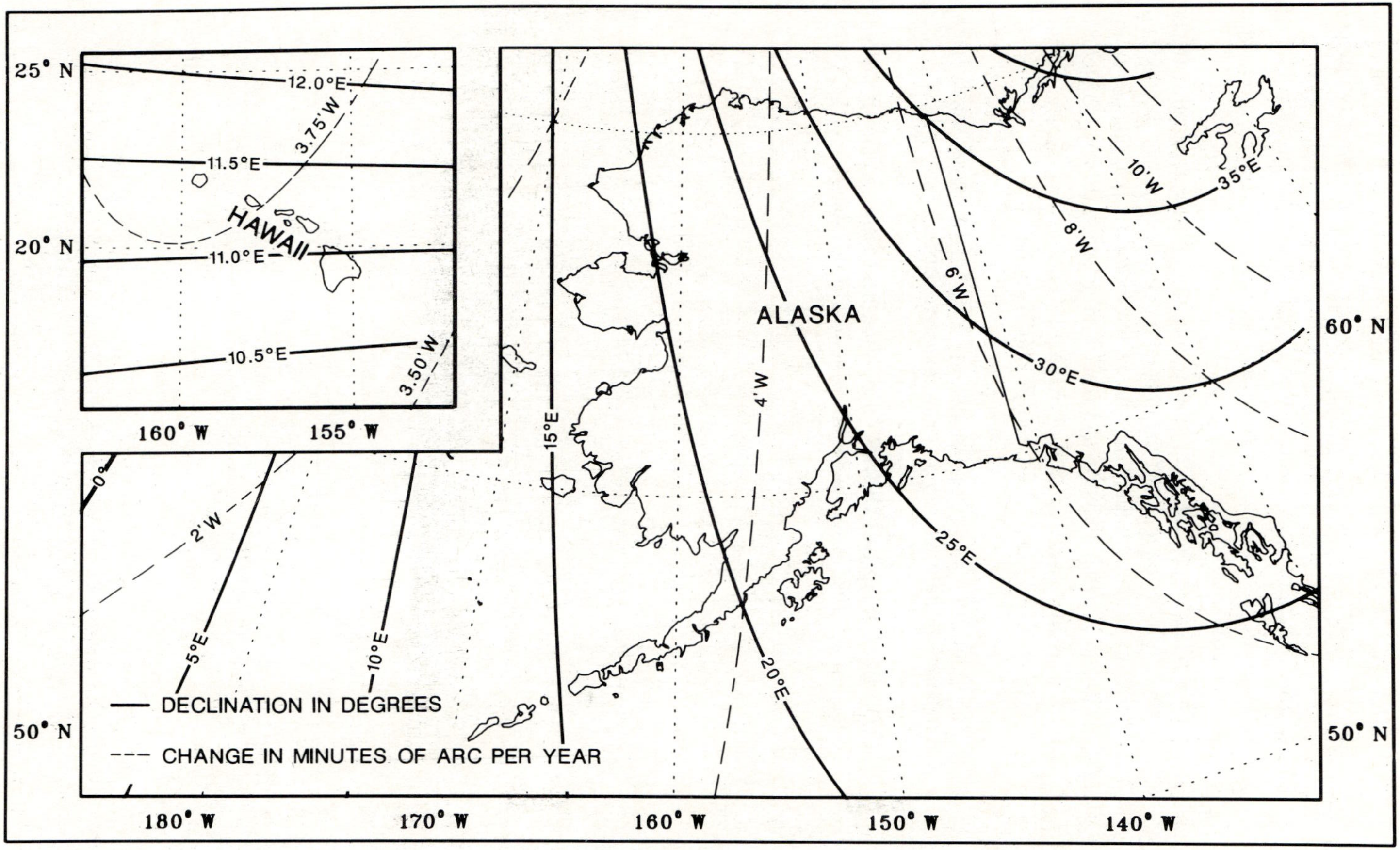

FIGURE 7.11
1985 declination in Alaska and Hawaii. *(Courtesy of U.S. Geological Survey, Department of Interior.)*

CONVERSION GUIDELINES. Conversion back and forth between magnetic and true directions can be confusing. A few rules are in order to keep the process straight:

1. True directions, referenced to true north, do not change over time.
2. Magnetic directions are referenced to local magnetic north, a direction that changes over time.
3. When converting magnetic direction at one time to magnetic direction at another time, it is the direction of the magnetic reference that has changed, not the position of the line along the ground. As an intermediate step, convert from old magnetic direction to true direction and from true direction to new magnetic direction.
4. It is usually better to change magnetic bearings to magnetic azimuths before inserting declination values into the computation.

A diagram is a most useful tool for conversion between magnetic and true directions, or between old and new magnetic directions.

CONVERSION PROCEDURES. It is often necessary to convert an old property-line direction from its listed magnetic value to a true value.

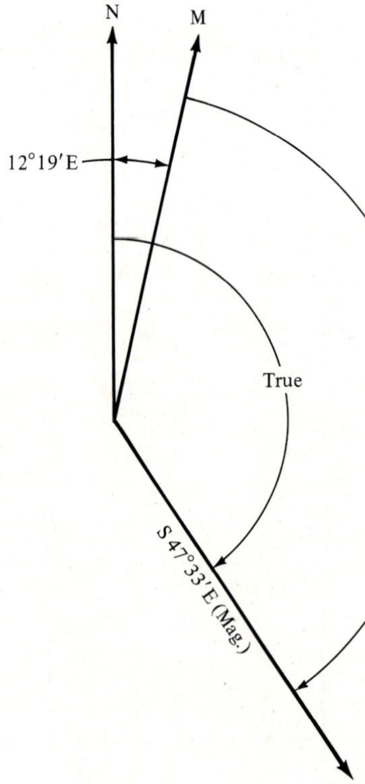

FIGURE 7.12
Diagram for Example 7.2.

Example 7.2. A 1914 Denver, Colo., courthouse record shows a property-line bearing to be S47°33′E, magnetic. From the National Geophysical Data Center (NGDC), it is learned that the magnetic declination on the date of the 1914 survey was 12°19′E. Find the true azimuth of the property line.

Solution. In Fig. 7.12, true north is designated by the N arrow and magnetic north for 1914 is designated by the M arrow. The 12°19′ angle between them is the east declination (east, because magnetic north is east of true north). The S47°33′E magnetic bearing is converted to a magnetic azimuth of 132°27′, as indicated by the clockwise arrow from the M line to the property line. The true azimuth to the line is therefore designated by the clockwise arrow from the N line to the property line, whose angular length is the sum of the magnetic azimuth and 1914 east declination values. Thus:

$$\text{True azimuth} = \text{magnetic azimuth} + \text{east declination}$$

$$= 132°27' + 12°19' = 144°46'$$

Consider a similar situation in which the declination is west instead of east.

Example 7.3. A property boundary is listed on an old record at Caribou, Maine, as N88°04′W, magnetic. It was learned that the declination at the time of the property survey was 20°38′W. What is the true bearing of the property line?

Solution. The N88°04′W magnetic bearing of the property line is changed to a magnetic azimuth of 271°56′, designated in Fig. 7.13 by the clockwise arrow from the M line to the property-line direction. The value of the clockwise arrow showing true azimuth in Fig. 7.13 is seen to be the difference between the magnetic azimuth value and the west declination value. Thus:

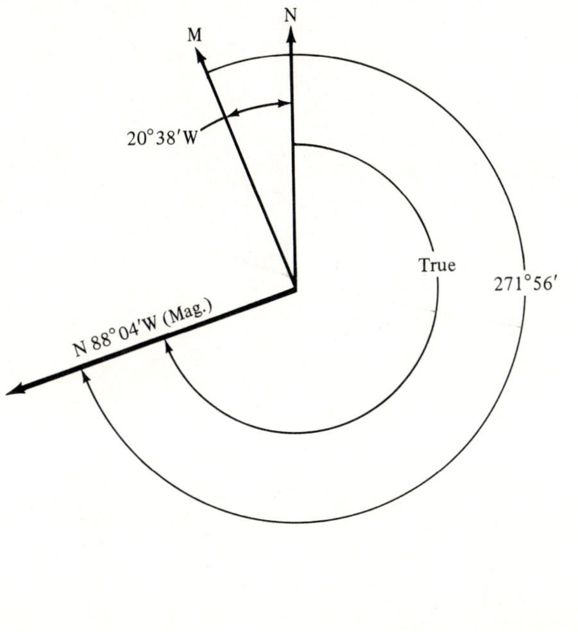

FIGURE 7.13
Diagram for Example 7.3.

$$\text{True azimuth} = \text{magnetic azimuth} - \text{west declination}$$

$$= 271°56' - 20°38' = 251°18'$$

$$\Rightarrow\Rightarrow \quad \text{True bearing} = \text{S}71°18'\text{W}$$

It is sometimes necessary to determine present magnetic direction as well as true direction from an old magnetic direction.

Example 7.4 The courthouse record for a 1902 property survey at Ketchikan, Alaska, lists the magnetic direction of a property boundary as S82°44′W. If the 1902 declination was 37°08′E and the present declination is 26°17′E, find the true bearing and the present magnetic bearing of the property line.

Solution. Consider that the diagram in Fig. 7.14 was drawn by first laying out the true north reference line N, then, from the two different declination values, the reference lines for 1902 magnetic north (M_{1902}) and present magnetic north (M_{NOW}). The property-line direction was then established by measuring a 262°44′ clockwise angle (equivalent to the bearing of S82°44′W) from 1902 magnetic north. From this, it is seen that the true azimuth is the clockwise angle from true north to the property line and that the present magnetic azimuth is the clockwise angle from present magnetic north to the property line. Thus:

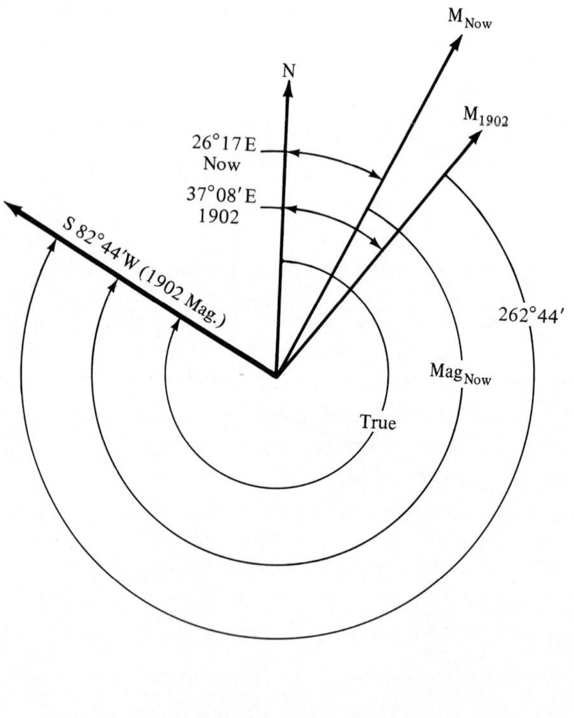

FIGURE 7.14
Diagram for Example 7.4.

$$\text{True azimuth} = 1902 \text{ magnetic azimuth} + 1902 \text{ east declination}$$
$$= 262°44' + 37°08'\text{E} = 299°52'$$
$$\Rightarrow\Rightarrow \quad \text{True bearing} = \text{N}60°08'\text{W}$$
$$\text{Present magnetic azimuth} = \text{true azimuth} - \text{present east declination}$$
$$= 299°52' - 26°17'\text{E} = 273°35'$$
$$\Rightarrow\Rightarrow \quad \text{Present magnetic bearing} = \text{N}86°25'\text{W}$$

At times a survey can be speeded up by using a surveyor's compass.

Example 7.5. A surveyor at a remote location on Niihau Island, Hawaii, has recovered a point on the northwest end of a property line whose listed 1932 magnetic bearing is S64°01′E. Having learned that the 1932 declination was 15°47′E and that the present declination is 11°23′E, what bearing should be laid out, using a surveyor's compass set up on the recovered point, to find the old survey point marking the southeast end of the line?

Solution. As before, the diagram (Fig. 7.15) is drawn by laying out the true-north direction and the old and new magnetic-north directions. From old magnetic north, the angle of the old magnetic azimuth, 115°59′, is used to draw the property line. From the diagram

$$\text{True azimuth} = \text{old magnetic azimuth} + 1932 \text{ east declination}$$

and

$$\text{Present magnetic azimuth} = \text{true azimuth} - \text{present east declination}$$

Substituting true azimuth from the first equation for the value of true azimuth in the second equation:

$$\text{Present magnetic azimuth} = \text{old magnetic azimuth} + 1932 \text{ east declination}$$
$$- \text{ present east declination}$$
$$= 115°59' + 15°47'\text{E} - 11°23'\text{E} = 120°23'$$
$$\Rightarrow\Rightarrow \text{Present magnetic bearing} = \text{S}59°37'\text{E}$$

Even for an imprecise operation similar to Example 7.5, care should be taken in making a magnetic observation. The compass needle is attracted not only to local magnetic north, but to whatever magnetic material may be lying about; oftener than not, on the person of the surveyor or perhaps a vehicle parked too close. Pocket items may be tested for suspected magnetism by holding them next to the compass needle: If there is detectable deflection of the needle, the offending item(s) should be set out of the way during the observation. Needles can become sluggish as they demagnetize over time or as friction increases between the pivot and the needle bearing, usually because of failure to lift the needle from its pivot when the instrument is being transported. Minor sluggishness is often overcome by gently tapping the glass cover with an eraser on a nonmagnetic pencil.

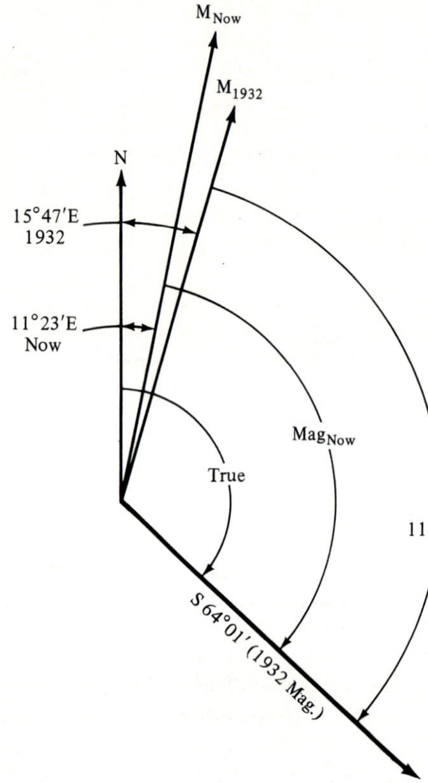

FIGURE 7.15
Diagram for Example 7.5.

7.5 MAGNETIC DISTURBANCES

Local magnetic directions can be substantially in error as a result of fixed or variable disturbances in the local magnetic field. An example of a variable disturbance is a strong but fluctuating direct-current electric field that would deflect a compass needle in the same way as an electromagnet. Fortunately, direct-current transmission is not used on a scale large enough to be of concern to the surveyor.

Fixed local disturbances, often called magnetic "anomalies," are another matter. These result from deposits of magnetic ore or from such man-made causes as concentrations of magnetic material (for example, discarded auto bodies) in dumps or sanitary landfills.

The site of a fixed disturbance can be isolated by laying out points along a zigzag line through the suspected area, as shown in Fig. 7.16. Enough points are established to insure that adjacent points on each end of the line are out of the disturbed area; that is, so bearings between such points are in agreement within a very few minutes.

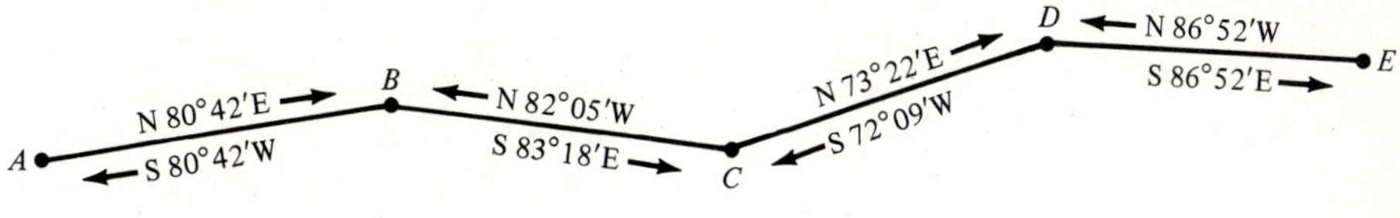

FIGURE 7.16
Direction observations near a magnetic anomaly.

Example 7.6. Inconsistencies in observed magnetic bearings in a survey area indicated the presence of a local magnetic disturbance. Points A through E were established and magnetic bearings were measured between them, with results as shown in Fig. 7.16. By analyzing the magnetic bearings, determine which station is magnetically disturbed, by what amount, and in what direction, east or west.

Solution. Assume (1) magnetic directions observed from a magnetically disturbed station are incorrect, (2) magnetic directions observed from a magnetically undisturbed station will be correct, and (3) in both cases, the horizontal angles calculated from magnetic directions will be correct. With this in mind, note that the back bearing from B to A agrees with the forward bearing from A to B so neither A nor B is disturbed. Since station B is not disturbed, the forward bearing to C (S83°18′E) is also correct. Note that the back bearing from C to B (N82°05′W) disagrees with forward bearing BC by 1°13′, indicating that point C is magnetically disturbed. Because the counterclockwise angle between magnetic north at C and line CB is 82°05′ instead of 83°18′, magnetic north at C must be 1°13′ west of magnetic north at B; thus, the local magnetic anomaly at C is 1°13′W.

Similarly, forward magnetic bearing DE is in agreement with back magnetic bearing ED, indicating that the magnetic disturbance is limited to station C. Further, magnetic bearing CD is *greater* than magnetic bearing DC by 1°13′. For this to be so, magnetic north at C must lie *west* of magnetic north at D, confirming that the magnetic anomaly at C is indeed 1°13′W.

Significant magnetic disturbances of solar origin make observations with a magnetic compass unreliable or sometimes impossible. Solar activity of this sort is continuously monitored and predicted by the Space Environment Services Center of the National Oceanic and Atmospheric Administration, Boulder, Colo. Information on the level of activity in the earth's magnetic field for the last 24 hours and the forecast for the coming 24 hours can be obtained on a round-the-clock basis by calling (303) 497-3235 and listening to the recorded message. If a short-wave radio is available, the same information is broadcast in a 40-second message exactly 18 minutes after each hour on short-wave station WWV, which may be tuned in at 2.5, 5, 10, 15, and 20 megahertz (MHz). The condition of the geomagnetic field will be described as "quiet," "unsettled," "active," "minor storm," "major storm," or "severe storm." The magnetic compass should be used for determining direction only when conditions are quiet or, at worst, unsettled.

7.6 PROBLEMS

7.1. Explain the difference between an angle of 30°00′00″ and a direction of 030°00′00″.

7.2. Calculate the back azimuth of each of the following azimuths:
(*a*) 121°47′16″ (*b*) 304°22′ (*c*) 63°12′32″ (*d*) 215°41′

7.3. Calculate the forward azimuth of each of the following back azimuths:
(*a*) 268.3° (*b*) 77°32′19″ (*c*) 167°46′ (*d*) 342°56′.

7.4. The forward and back azimuths of a very long line are 123°47′18″ and 303°47′21″, respectively. Explain the difference in the two values.

7.5. Given the following bearings, compute back bearings:
(*a*) S63°22′W (*b*) S21°42′29″E (*c*) N38°29.2′E (*d*) N77°56′W

7.6. Given the following back bearings, compute forward bearings:
(*a*) N63°48′E (*b*) S22°11′49″W (*c*) N58°12′W (*d*) S10°33.8′E

7.7. Convert the azimuths in Prob. 7.2 to bearings.

7.8. Convert the back azimuths in Prob. 7.3 to back bearings.

7.9. Convert the bearings in Prob. 7.5 to azimuths.

7.10. Convert the back bearings in Prob. 7.6 to back azimuths.

7.11. Proceeding in a counterclockwise direction around a five-sided polygon, the adjusted interior angles are as follows: $\angle A$ = 123°12′, $\angle B$ = 92°46′, $\angle C$ = 185°37′, $\angle D$ = 33°56′, $\angle E$ = 104°29′. If the bearing of side *AB* is S47°44′W, what are the azimuths of all five sides?

7.12. In Prob. 7.11, what are the bearings of all the sides?

7.13. Do the same as Prob. 7.11 except side *AB* bears N63°21′W.

7.14. In Prob. 7.13, what are the bearings of all the sides?

7.15. Proceeding in a clockwise direction around a six-sided polygon, adjusted interior angles are as follows: $\angle A$ = 70°19′, $\angle B$ = 238°38′, $\angle C$ = 50°51′, $\angle D$ = 69°23′, $\angle E$ = 215°07′, $\angle F$ = 75°42′. If the bearing of side *AB* is due east, what are the azimuths of all six sides?

7.16. In Prob. 7.15, what are the bearings of all the sides?

7.17. Do the same as Prob. 7.15 except side *AB* bears S49°39′W.

7.18. In Prob. 7.17, what are the bearings of all sides?

7.19. Convert the interior angles in Prob. 7.11 to deflection angles, then determine the azimuths of the sides.

7.20. Convert the interior angles in Prob. 7.15 to deflection angles, then determine the azimuths of the sides.

7.21. The bearings of the sides of a closed polygon, listed in clockwise order around the figure, are *AB* = S47°15′E, *BC* = S39°46′W, *CD* = N03°17′W, *DE* = N82°58′W, and *EA* = N50°34′E. Calculate the interior angles in the figure.

7.22. The bearings of the sides of a closed polygon, listed in counterclockwise order around the figure, are *AB* = S02°27′E, *BC* = N51°13′E, *CD* = S58°41′E, *DE* = N13°44′W, and *EA* = S77°33′W. Calculate the interior angles in the figure.

7.23. From the declination chart in Fig. 7.10, estimate the maximum angular difference in declination within the 48 states.

7.24. From Fig. 7.10, if the declination at Panama City, Fla., was 0° on January 1, 1985, what was the declination on July 1, 1989?

7.25. If the declination at Kansas City, Mo., was 5°05′ east on January 1, 1985, and 4°37′ east on January 1, 1989, what was the amount and direction of the average annual change between those two dates?

7.26. A survey was done in Tucson, Ariz., in 1915 when the declination was 13°40′E. The magnetic bearings of the boundaries were measured as N42°31′W, S28°56′W, S53°20′E, and N26°55′E. Calculate the true bearings of the boundaries.

7.27. In Prob. 7.26, if the declination at Tucson is now 12°17′E, what magnetic bearings must be used to retrace the boundaries?

7.28. A survey was done in Sitka, Alaska, in 1909 when the declination was 30°12′E. The magnetic bearings of the boundaries were measured as S19°22′W, N57°41′W, N32°16′E, and S49°14′E. Calculate the true bearings of the boundaries.

7.29. In Prob. 7.28, if the declination in Sitka is now 27°14′E, what magnetic bearings must be used to retrace the boundaries?

7.30. A survey was done in Baltimore, Md., in 1911 when the declination was 5°36′W. The magnetic bearings of the boundaries were measured as N55°10′E, S07°27′E, N44°06′W, and due west. Calculate the true bearings of the boundaries.

7.31. In Prob. 7.30, if the declination in Baltimore is now 9°57′W, what magnetic bearings must be used to retrace the boundaries?

7.32. The true bearing of a line is N42°12′W. The 1985 declination of your location was 8°15′E. If the annual change has been 7′W, what is the approximate magnetic bearing today?

7.33. The magnetic bearing of a boundary line was measured several times over a period of a few minutes with significant differences in results. What conditions could have caused the differences?

7.34. When forward and back magnetic bearings measured from both ends of the same line disagreed by a significant amount, additional points were established on both sides of the original points and magnetic forward and back bearings were read. The readings that were obtained are shown in the sketch below. Analyze the data and determine (*a*) which stations were disturbed, (*b*) in which direction (east or west) the north end of the needle was deflected from its proper position, and (*c*) the angular amount of the disturbance.

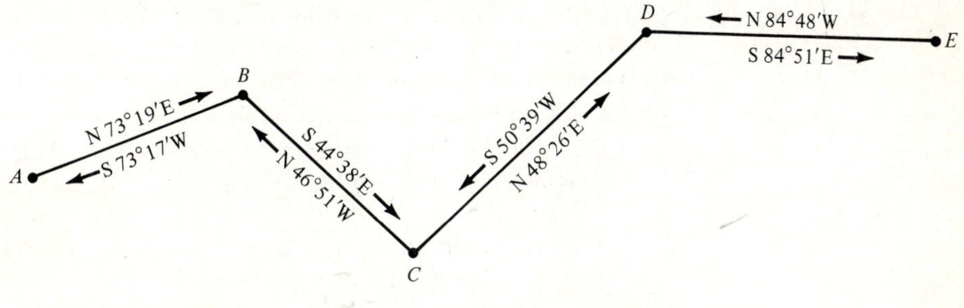

7.35. Do the same as Prob. 7.34 except readings are as shown below.

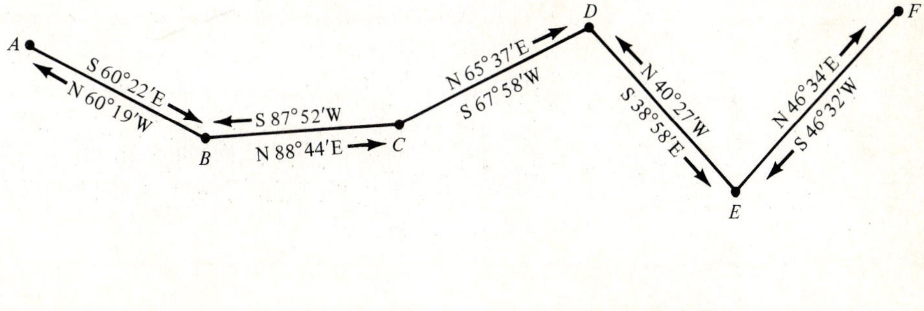

CHAPTER 8

TRAVERSE

8.1 INTRODUCTION

Distance measurement was introduced in Chap. 2 and methods for converting angles to directions were demonstrated in Chap. 7. Now, the two concepts are combined to produce a **traverse**. A traverse advances from points of known position to points whose positions were previously unknown. It does this by successive determination of both the distance and the direction from a known point to an unknown point. A line of known distance and direction is a traverse **leg**.

Figure 8.1 illustrates the concept. Traverse leg *AB* advances from point *A*, whose *xy* coordinates are known, to point *B*, whose *xy* coordinates may be determined from known distance and direction *AB*. The procedure for determining the *x* and *y* differences between *A* and *B* is discussed in Sec. 8.3.

Traverse is perhaps the most useful single application in surveying. Traverse and traverse computations are used for

- Laying out the locations of property boundary points from property descriptions
- Recovering existing property boundary points
- Determining areas within property boundaries
- Establishing survey control points within an area to be mapped
- Establishing control within a construction area to determine the volume of earthwork to be brought in or taken away
- Establishing center lines and right-of-way boundaries for transportation projects

Traverses and traverse computations, though fairly straightforward, involve a number of steps subsequent to measurement of angles and distances in the field. These operations, discussed in this chapter, include:

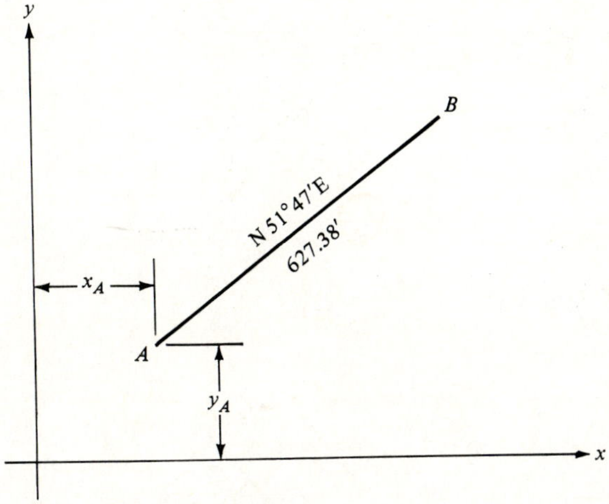

FIGURE 8.1
Single-leg traverse from A to B.

1. Angle closure and balancing, if feasible, then determination of directions
2. Computation of latitudes (north-south components of leg lengths) and departures (east-west components of leg lengths) from directions and distances
3. Determination of closure and precision
4. Adjustment of latitudes and departures to compensate for closure error
5. Determination of adjusted coordinate positions of traverse points
6. Computation of area, if applicable

Traverses are classified as open or closed. An **open traverse** begins at a known position point and ends at a point whose position was not previously known. A **closed traverse** either closes on its origin (**loop traverse**) or starts at one known point and ends at another known point (**connecting traverse**). The computed position of the end point of a closed traverse rarely agrees with the known end-point position. The distance between the actual end point and the computed end point is the traverse error and its magnitude, compared to total traverse length, determines traverse precision.

8.2 ANGLE CLOSURE AND BALANCING

ANGLE ADJUSTMENT. A closed traverse permits a check on the final direction, allowing adjustment of the traverse angles and, thus, of the traverse leg directions. For a figure with N interior angles,

$$\text{Total angular correction} = (N - 2) \cdot 180^\circ - \Sigma \text{ interior angles}$$

If total angular correction exceeds the instrument's reading precision (least count) multiplied by the square root of N, *all the angles in the figure should be remeasured.*

Assuming equal-weight adjustments, total correction is divided by N to obtain correction per angle, which is distributed equally within the figure. In closed-loop

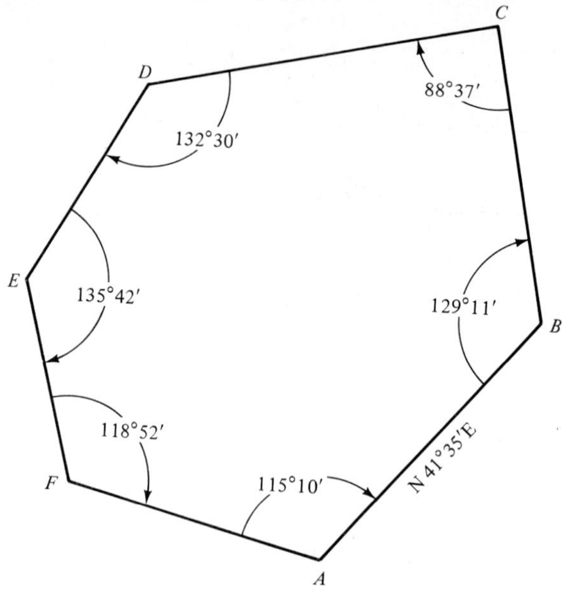

FIGURE 8.2
Closed-loop traverse using interior angles.

traverse *ABCDEF* in Fig. 8.2, observed field angles are as shown and the bearing of leg *AB* is known. The adjustment is tabulated in Figure 8.3.

In Fig. 8.3, corner angles in the second column are added and their 720°02′ sum found to exceed the correct interior-angle sum for a six-sided figure by 02′; thus, correction is 720°00′ − 720°02′ = −02′, or 120″. Total correction is divided by 6 to obtain corrections of −20″ per angle, which are then entered in the third column and applied to the observed-angle values to produce the corrected angles in the fourth column. The sum of the corrected angles is 720°00′00″—it is important that this check *always* be performed.

Station	Measured angle	Correction	Adjusted angle
A	115°10′	−20″	115°09′40″
B	129°11′	−20″	129°10′40″
C	88°37′	−20″	88°36′40″
D	132°30′	−20″	132°29′40″
E	135°42′	−20″	135°41′40″
F	118°52′	−20″	118°51′40″
	Σ = 720°02′		720°00′00″ ✓

Correction = 720°00′ − 720°02′ = −02′ = −120″

$$\text{Correction per angle} = \frac{-120''}{6 \text{ angles}} = -20'' \cdot \text{angle}^{-1}$$

FIGURE 8.3
Adjustment of interior angles in closed-loop traverse.

Line	Adjusted azimuth	Adjusted bearing
AB	041°35′00″	N41°35′00″E
BA	221°35′00″	
+∠B	+129°10′40″	
BC	350°45′40″	N09°14′20″W
CB	170°45′40″	
+∠C	+ 88°36′40″	
CD	259°22′20″	S79°22′20″W
DC	079°22′20″	
+∠D	132°29′40″	
DE	211°52′00″	S31°52′00″W
ED	031°52′00″	
+∠E	135°41′40″	
EF	167°33′40″	S12°26′20″E
FE	347°33′40″	
+∠F	118°51′40″	
FA	466°25′20″	
⇒⇒FA	106°25′20″	S73° 34′40″E
AF	286°25′20″	
+∠A	+115°09′40″	
AB	401°35′00″	
⇒⇒AB	041°35′00″ ✓	

FIGURE 8.4
Computing directions from adjusted interior angles in a closed-loop traverse.

Using the adjusted interior angles, azimuths and bearings are computed for all six sides, starting with known side *AB*, as shown in Fig. 8.4. the procedure, demonstrated earlier in Sec. 7.2, begins at corner *B*, azimuth *BA* being the back azimuth of the azimuth obtained from known bearing *AB*. Computation of azimuths continues around polygon *ABCDEF*, the final computed value of azimuth *AB* being used as an arithmetic check (same as initial azimuth *AB*).

For deflection-angle traverses, it is easier to compute azimuths from unadjusted angles, then adjust azimuths. In the Fig. 8.5 traverse, initial azimuth *FG* and final azimuth *EM* are known but deflection angles *F, A, B, C, D* and *E* are from unadjusted field observations.

Azimuths are computed from deflection angles as demonstrated in Sec. 7.2; that is, computation begins with conversion of known back azimuth *FG* to forward azimuth *GF*, to which the value of *F*, a deflection angle *right*, is *added* to produce azimuth *FA*. Successive computations of azimuths *AB, BC, CD, DE,* and *EM* continue in the same manner, as shown in the first two columns of Fig. 8.6, with right deflection angles always added to the forward-going azimuth and left deflection angles always subtracted. The known value of azimuth *EM* (306°43′00″) is subtracted from the 306°46′00″ computed value of *EM* to obtain the total cumulative error in computed azimuth *EM* (+3′00″, or +180″). The total *correction*, which must be of the opposite algebraic sign of the total *error*, is therefore −180″. Assuming equal-weight angles, correction per angle must be total correction divided by number of deflection angles; that is −180″/6 angles = −30″· angle^{-1}. Because successive azimuth values are

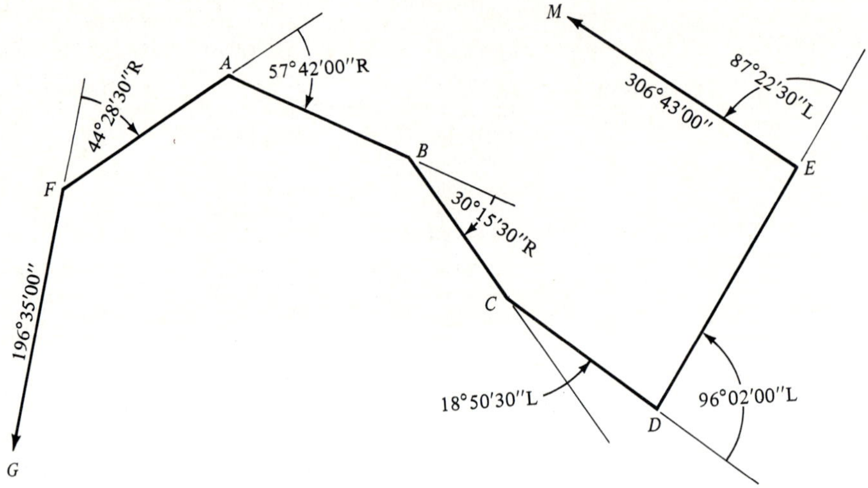

FIGURE 8.5
Closed, connecting traverse using deflection angles.

Line	Azimuth	Correction	Adjusted azimuth
FG	196°35′00″	(fixed)	
GF	016°35′00″		
+∠F	+44°28′30″R		
FA	061°03′30″	−30″	061°03′00″
+∠A	+57°42′00″R		
AB	118°45′30″	−1′00″	118°44′30″
+∠B	+30°15′30″R		
BC	149°01′00″	−1′30″	148°59′30″
−∠C	−18°50′30″L		
CD	130°10′30″	−2′00″	130°08′30″
−∠D	−96°02′00″L		
DE	034°08′30″	−2′30″	034°06′00″
⇒⇒DE	394°08′30″		
−∠E	−87°22′30″L		
EM	306°46′00″	−3′00″	306°43′00″ ✓
EM_{known}	306°43′00″		

Total Error = +3′00″ ⇒⇒ Total Correction = −3′00″ = −180″

$$\Rightarrow\Rightarrow \text{Correction per angle} = \frac{-180^{\circ}}{6} \text{ angles} = -30' \cdot \text{angle}^{-1}$$

FIGURE 8.6
Adjustment of azimuths from deflection angles.

dependent on the accumulation of successive deflection angles, the corrections applied to azimuths must be cumulative as well. Thus, the corrections to azimuths shown in the third column of Fig. 8.6 are cumulative multiples of the per-angle correction. The correction to the final computed azimuth *EM* is therefore the total correction determined earlier from comparison of the computed and known values of azimuth *EM*, providing a partial arithmetic check.

8.3 LATITUDES AND DEPARTURES

A traverse leg has both length and direction and it may be described in terms of its north-south and east-west components. **Latitude** is the north or south distance component of a traverse leg. If the leg's origin-to-terminus direction is *northerly*, its latitude is *positive*; if *southerly*, its latitude is *negative*. **Departure** is the east or west distance component of a traverse leg. If the leg's origin-to-terminus direction is *easterly*, departure is *positive*; if *westerly*, departure is *negative*. Figure 8.7 shows line *OA*, whose direction is N29°17′30″E and whose length is 637.49 ft. The line's north-south component is indicated by the vertical dashed line in Fig. 8.7 and its east-west component is shown by the horizontal dashed line. The two letter designators for this particular bearing are N and E, indicating that line *OA* proceeds in northerly and easterly directions; thus, its latitude and departure are both positive.

The numerical value of the latitude is the line length multiplied by the cosine of the bearing angle. The numerical value of the departure is the line length multiplied by the sine of the bearing angle. That is,

$$\textbf{Latitude} = \textbf{leg length}\ \ \textbf{cos}\,\textbf{(bearing angle)} \tag{8.1a}$$

$$\textbf{Departure} = \textbf{leg length}\ \ \textbf{sin}\,\textbf{(bearing angle)} \tag{8.1b}$$

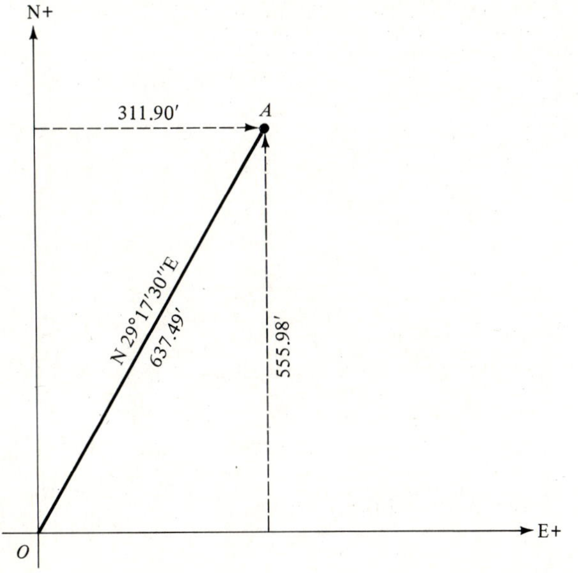

FIGURE 8.7
Latitude and departure from traverse leg using bearing.

Thus, for line *OA*,

$$\text{Latitude } \boldsymbol{OA} = 637.49\,\mathbf{cos}\,29°17'30'' = +555.98 \text{ ft}$$

$$\text{Departure } \boldsymbol{OA} = 637.49\,\mathbf{sin}\,29°17'30'' = +311.90 \text{ ft}$$

With bearing nomenclature, algebraic signs of latitude and departure are given by the north, south, east and west designators, north and east being positive while south and west are negative. If the direction of line *OA* had been S29°17′30″W rather than N29°17′30″E, the numerical values for latitude and departure would have been the same but both signs would have been negative rather than positive.

When using azimuths, algebraic signs of latitude and departure are assigned automatically when using the equations:

$$\textbf{Latitude} = \textbf{leg length cos (azimuth)} \tag{8.2a}$$

$$\textbf{Departure} = \textbf{leg length sin (azimuth)} \tag{8.2b}$$

Figure 8.8 shows a line originating at *J* and ending at *K*. Its direction is 218°17′22″ and its length is 219.47 ft. Because the direction is in the third quadrant, latitude and departure are both negative. Applying Eqs. (8.2a) and (8.2b) to line *JK*:

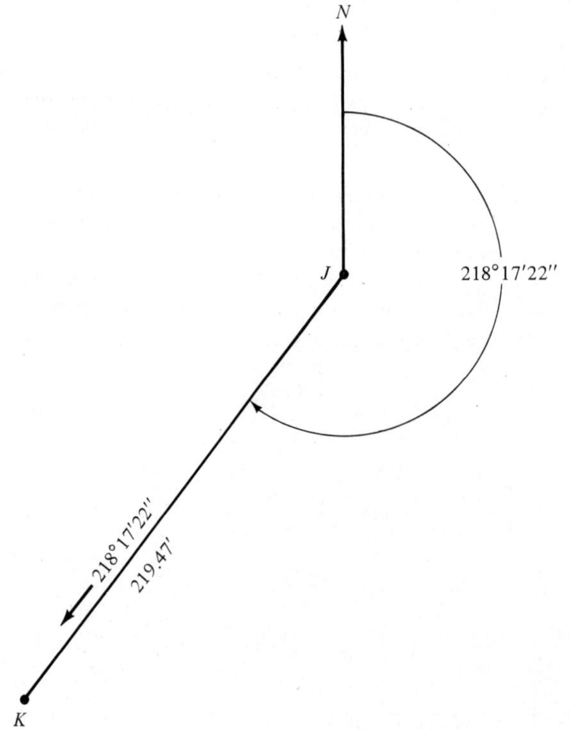

FIGURE 8.8
Latitude and departure from traverse leg using azimuth.

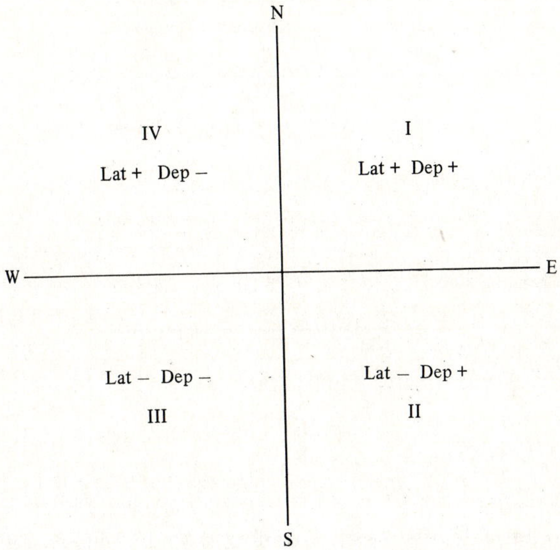

FIGURE 8.9
Algebraic signs for latitude and departure—by quadrant.

$$\text{Latitude} = 219.47 \cos 218°17'22'' = -172.26 \text{ ft}$$

$$\text{Departure} = 219.47 \sin 218°17'22'' = -135.99 \text{ ft}$$

Algebraic signs for the sines and cosines of angles in the four quadrants were discussed in Sec. 1.9. The diagram in Fig. 8.9 shows the sign relationships, by quadrant. The act of keying the azimuth value into the calculator provides the correct signs for both sine and cosine and, thus, for both latitude and departure.

Example 8.1. A connecting traverse is run from known station Hill to known station Dale via unknown stations Alfa, Beta, and Gama. From the data in Fig. 8.10, compute latitudes and departures for all legs.

Solution. The format in Fig. 8.11 lists station names on alternate lines, with leg lengths and adjusted azimuths on intervening lines. Latitudes and departures are computed from Eqs. (8.2*a*) and (8.2*b*).

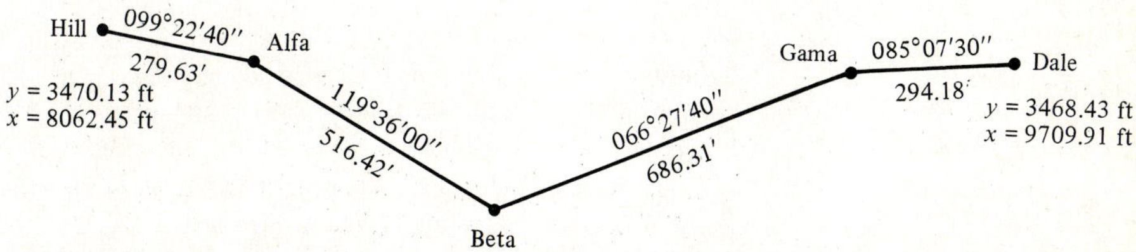

FIGURE 8.10
Closed, connecting traverse from Hill-to-Dale.

Sta.	Length, ft	Adjusted azimuth	Latitude, ft	Departure, ft
Hill				
	279.63	99°22′40″	−45.56	+275.89
Alfa				
	516.42	119°36′00″	−255.08	+449.02
Beta				
	686.31	066°27′40″	+274.09	+629.20
Gama				
	294.18	085°07′30″	+ 25.00	+293.12
Dale				

FIGURE 8.11
Latitude and departure calculations, Hill-to-Dale.

8.4 CLOSURE AND PRECISION

The sum of the latitudes between the starting and ending stations of a connecting traverse is the computed y distance between the two stations; the sum of the departures is the computed x distance. If the correct x and y coordinates of the starting and ending stations are known at the outset, the correct y distance and x distance must be the correct y- and x-coordinate differences, respectively, from the starting to the ending station. Thus, the latitude error must be the computed latitude sum minus the fixed y-coordinate difference between the two stations; the departure error must be the computed departure sum minus the fixed x-coordinate difference. Knowing the latitude and departure errors, total **closure error** is computed from

$$\textbf{Closure error} = \sqrt{\textbf{latitude error}^2 + \textbf{departure error}^2} \tag{8.3}$$

Traverse precision is the closure-to-length ratio, defined by

$$\textbf{Precision} = \frac{\textbf{closure error}}{\textbf{total traverse length}} \tag{8.4}$$

where total traverse length is the sum of the leg lengths. The value of precision is usually expressed as 1:n, where n is the reciprocal of the decimal value obtained from Eq. (8.4).

Example 8.2. Compute closure and precision for the Hill-to-Dale traverse in Example 8.1.

Solution. First, total traverse length and the sums of the latitudes and departures are obtained by adding up the appropriate column values in Fig. 8.11, as shown in Fig. 8.12.

Next, the fixed y- and x-coordinate differences from Hill-to-Dale are obtained from the known coordinate values given in Fig. 8.10.

$$y_{\text{Dale}} - y_{\text{Hill}} = 3468.43 - 3470.13 = -1.70\text{ft}$$

$$x_{\text{Dale}} - x_{\text{Hill}} = 9709.91 - 8062.45 = +1647.46\text{ft}$$

Sta.	Length, ft	Adjusted, Azimuth	Latitude, ft	Departure, ft
Hill				
	279.63	99°22′40″	−45.56	+275.89
Alfa				
	516.42	119°36′00″	−255.08	+449.02
Beta				
	686.31	066°27′40″	+274.09	+629.20
Gama				
	294.18	085°07′30″	+25.00	+293.12
Dale	Σ 1776.54		−1.55	+1647.23

FIGURE 8.12
Latitude and departure sums, Hill-to-Dale traverse.

The latitude error is the computed Hill-to-Dale latitude sum minus the correct Hill-to-Dale y difference:

$$\text{Latitude error}_{\text{Hill-to-Dale}} = -1.55 - (-1.70) = +0.15\text{ft}$$

The departure error is the computed Hill-to-Dale departure sum minus the correct Hill-to-Dale x difference:

$$\text{Departure error}_{\text{Hill-to-Dale}} = +1647.23 - 1647.46 = -0.23\text{ft}$$

Substituting into Eq. (8.3):

$$\text{Closure error} = \sqrt{0.15^2 + 0.23^2} = 0.27459 \Rightarrow\Rightarrow 0.27\text{ft}$$

Substituting into Eq. (8.4):

$$\text{Precision} = \frac{0.27459\text{ft}}{1776.54\text{ft}} = 0.00015456$$

Converting the 0.00015456 decimal value to the 1:n ratio:

$$\text{Precision} = 1 : (1/0.00015456) \cong 1 : 6500$$

The calculator result for the above equation was 1:6469.78; however, because the total error is known to no better than hundredths of a foot, or two significant figures, the precision, in this case, could only be expressed to the nearest hundred. That is the reason for the symbol "≅," which means "approximately equal to."

8.5 COMPASS RULE ADJUSTMENT

Closure errors are almost certain to occur in any traverse, making it necessary to distribute corrections back through the traverse in a consistent, logical manner that takes into account the sorts of errors that could arise. Small closure errors result from an accumulation of small random errors in observation of angles and distances. For the instruments in use today, the effects of distance errors and angle errors are considered to be about equal, so an adjustment method is required that assumes a condition of equality. Another requirement—until recently, at least—is for the adjustment method to be simple enough to apply using just a hand-calculator capability.

The almost universally accepted adjustment procedure is called the **compass rule**. Proposed by the navigator Nathaniel Bowditch in the early 1800s, it assumes that errors occur in direct proportion to the distance run and requires that latitude and departure adjustments to a given leg be proportional to leg length. Expressed mathematically:

$$\frac{\text{Latitude correction for a given leg}}{\text{Leg length}} = \frac{\text{total latitude correction}}{\text{total length of traverse}}$$

$$\frac{\text{Departure correction for a given leg}}{\text{Leg length}} = \frac{\text{total departure correction}}{\text{total length of traverse}}$$

Total latitude and departure corrections have the opposite algebraic sign from total latitude and departure errors. Solving for latitude and departure correction using equations in the above form is somewhat tedious. The following modified form permits entry and reuse of constant multipliers:

$$\textbf{Latitude correction for a given leg} = \left[\frac{\textbf{total latitude correction}}{\textbf{traverse length}}\right] \textbf{leg length} \quad \textbf{(8.5)}$$

$$\textbf{Departure correction for a given leg} = \left[\frac{\textbf{total departure correction}}{\textbf{traverse length}}\right] \textbf{leg length} \quad \textbf{(8.6)}$$

The bracketed values are keyed into the calculator as constants by which successive leg lengths are multiplied to obtain successive values of latitude and departure corrections.

Example 8.3. Compute Hill-to-Dale compass-rule corrections.

Solution. Total latitude and departure corrections, which are of the opposite algebraic sign from the total latitude and departure errors given in Fig. 8.12, are −0.15 ft and +0.23 ft, respectively. Constant multipliers for latitude and departure corrections are −0.15/1776.54 and +0.23/1776.54. Using Eqs. (8.5) and (8.6), these values are multiplied by successive leg lengths to produce the correction values shown in the fifth and sixth columns in Fig. 8.13, the second column being leg length while the third and fourth columns are latitudes and departures computed earlier.

The sums of latitude and departure corrections in Fig. 8.13 are *not* the *correct* values for total latitude and departure corrections, that is, −0.15 and +0.23 ft, respectively. This is because of error induced by rounding the results in Eqs. (8.5) and (8.6) to the nearest hundredth of a foot. This can be avoided by examining each result to, say, four decimal places and deciding which of the values may be adjusted the needed hundredth of a foot, up or down.

Consecutive to-the-fourth-decimal-place values for latitude correction are −0.0236, 0.0436, −0.0231, and −0.0248. From this, it is seen that the fourth value, the latitude correction for the Gama-to-Dale leg, is on the verge of being rounded to −0.03. This is shown in Fig. 8.14, bracketed, as an adjusted latitude correction value. Similarly, the Hill-to-Alfa departure correction should be adjusted from +0.04 ft to +0.03

Sta.	Length, ft	Latitude, ft	Departure, ft	Corrections Lat., ft	Corrections Dep., ft
Hill					
	279.63	−45.56	+275.89	−0.02	+0.04
Alfa					
	516.42	−255.08	+449.02	−0.04	+0.07
Beta					
	686.31	+274.09	+629.20	−0.06	+0.09
Gama					
	294.18	+ 25.00	+293.12	−0.02	+0.04
Dale					
	Σ 1776.54			−0.14	+0.24

FIGURE 8.13
Trial latitude and departure corrections, Hill-to-Dale.

ft, also shown bracketed in Fig. 8.14. Adjusted latitudes and departures in the last two columns were obtained by adding latitude and departure corrections to original latitude and departure values.

Figure 8.14 shows the completed compass-rule adjustment, with the sums of the latitude and departure corrections in agreement with total latitude and departure errors determined earlier. The next step is to establish coordinate values for Alfa, Beta, and Gama, the three intermediate traverse stations. Before doing this, however, it would be well to consider that there are other adjustment methods that may produce results superior to the compass rule but that would be very tedious for the surveyor to attempt with a pocket or desk calculator.

Sta.	Length, ft	Computed Latitude, ft	Computed Departure, ft	Corrections Lat., ft	Corrections Dep., ft	Adjusted Latitude, ft	Adjusted Departure, ft
Hill							
	279.63	−45.56	+275.89	−0.02	[+0.03]	−45.58	+275.92
Alfa							
	516.42	−255.08	+449.02	−0.04	+0.07	−255.12	+449.09
Beta							
	686.31	+274.09	+629.20	−0.06	+0.09	+274.03	+629.29
Gama							
	294.18	+25.00	+293.12	[−0.03]	+0.04	+24.97	+293.16
Dale							
	Σ 1776.54			−0.15	+0.23		

FIGURE 8.14
Final latitude and departure corrections, Hill-to-Dale.

8.6 OTHER METHODS OF ADJUSTMENT

Any traverse adjustment procedure must satisfy a minimum of three conditions:

1. The sum of the latitudes must equal the y-coordinate difference between the starting and end points of the traverse.
2. The sum of the departures must equal the x-coordinate difference between the starting and end points of the traverse.
3. The cumulative set of observed angles in the traverse must be in agreement with the known starting and finishing directions. For a traverse beginning at one point and ending at another, the sum of the deflection angles must equal the difference in direction between the known forward azimuths at the initial and terminal points. For an n-sided traverse closing on its origin, the sum of the n interior angles must equal 180° multiplied by $(n - 2)$.

These conditions may be satisfied by adjusting the measured angles, by adjusting the measured distances, or by a procedure that results in the adjustment of both angles and distances. The compass rule, which assumes that angles and distances are measured with about equal precision, utilizes an adjustment procedure that permits equivalent adjustment of both distance and direction.

The **transit rule**, an older method which won't be gone into here, presupposes relatively precise angle measurement and relatively sloppy distance measurement, an improbable situation given the capabilities of modern instruments.

The **Crandall method**, proposed at the turn of the present century, is a relatively complex procedure based on two related assumptions: (1) that angles are measured precisely and distances are not and (2) that distance error varies with the square root of the distance. Although both may be true in the case of taped distances, EDM distances are as precisely determined as are theodolite angles. Further, in the case of EDMs whose reflector constant has been compensated, distance error varies linearly with distance. At any rate, based on those no-longer-valid assumptions, the Crandall method distributes angle misclosure equally, then holds all directions fixed, closing the figure by weighted least-squares adjustment of measured distances.

The **least-squares (minimum variance) method** is a maximum-likelihood solution based on probability theory. It is considered the most statistically rational of the lot. The least-squares method adjusts transverse points so that the *weighted* sum of the squares of the adjustments to measured angles and measured distances is a minimum. The use of weights, as discussed in Sec. 5.5, takes cognizance of possible differences in the precision of angle and distance measurements. The procedure is too involved for most surveyors using ordinary calculators.

In light of recent advances in distance-measurement technology, many theoreticians have examined the premises involved in the various adjustment methods available. Notable among these is Sprinsky, who suggested in a 1987 paper on the subject that an adaptation of the **Tuttle method**, proposed at about the same time as Crandall's, is a reasonable compromise between the less precise compass rule and the more rigorous and complex method of least squares.

The Tuttle adaptation, not reproduced here, is unsuitable for the hand-held calculator but could well be considered by surveyors using microcomputers or advanced programmable calculators. For those interested, the Sprinsky paper is cited in the Bibliography.

At this writing, there is a good deal of microcomputer software on the market that claims to provide for least-squares adjustment of traverse observations. Unfortunately, some of it uses the simpler Crandall method, also a least-squares procedure, but one that is based on a currently questionable set of assumptions. Anyone considering purchase of least-squares adjustment software should make sure it uses the minimum-variance algorithm and has good, understandable documentation.

8.7 COORDINATE DETERMINATION

Adjusted latitudes and departures are added, sequentially, beginning with the known coordinates of the initial station, to obtain the y- and x-coordinate values of all subsequent traverse stations.

Example 8.4. Compute the coordinates of all stations in the Fig. 8.10 traverse beginning with the known coordinates of station Hill and using the adjusted latitudes and departures from Fig. 8.14.

Solution. As tabulated in Figure 8.15, the y coordinate of station Alfa is the y coordinate of station Hill plus the adjusted latitude of the Hill-to-Alfa leg, the adjusted latitude for a leg being the computed latitude of the leg plus the latitude correction for the leg. The y coordinate of Beta must equal the y coordinate of Alfa plus the adjusted latitude of the Alfa-to-Beta leg, and so forth, arriving at terminal station Dale, where the computed y coordinate of Dale must equal its known y coordinate. x coordinates are computed in a similar manner and afford a similar arithmetic check at the end. y and x coordinates are shown in the final two columns of Fig. 8.15, the coordinate values alternating with successively added values of adjusted latitudes and departures.

				Corrections		Coordinates	
Sta.	Length, ft	Latitude, ft	Departure, ft	Lat., ft	Dep., ft	Y, ft	X, ft
Hill						3470.13	8062.45
	279.63	−45.56	+275.89	−0.02	+0.03	−45.58	+275.92
Alfa						3424.55	8338.37
	516.42	−255.08	+449.02	−0.04	+0.07	−255.12	+449.09
Beta						3169.43	8787.46
	686.31	+274.09	+629.20	−0.06	+0.09	+274.03	+629.29
Gama						3443.46	9416.75
	294.18	+25.00	+293.12	−0.03	+0.04	+24.97	+293.16
Dale						3468.43	9709.91
	Σ 1776.54			−0.15	+0.23	✓	✓

FIGURE 8.15
Computation of coordinates, Hill-to-Dale.

Check marks under the final two columns in Fig. 8.15 indicate the computed coordinates for Dale are the same as the known coordinates of that station. Two checks are thus provided: first, that all of the arithmetic was carried out correctly and, second, that the correction values were given the right algebraic sign. If the correction signs for either latitude or departure had been reversed, final-station computed coordinates would have differed from final-station known coordinates by *exactly* double the closure values determined earlier.

8.8 SUMMARY OF TRAVERSE COMPUTATIONS

OPEN TRAVERSE. For an open traverse (terminating at a point whose position is unknown) the following computing sequence is needed:

1. Compute leg directions from initial direction and measured angles, then tabulate directions along with the measured lengths of legs.
2. Compute latitude and departure for each leg.
3. Compute the coordinates of successive stations, starting with the known, or arbitrarily assigned coordinates for the initial station.

Example 8.5. A new municipal water well is to be located on a local coordinate system. An open traverse is run to the well from known station Glen via points *A* and *B*, with the coordinates of Glen, the starting azimuth, measured deflection angles, and measured horizontal distances as shown in Fig. 8.16. Compute the local coordinates of new station Well.

Solution. Computations are shown in tabular form in Fig. 8.17. Azimuths are computed from forward azimuths as demonstrated in Secs. 7.2 and 8.2 and tabulated in the second column of Fig. 8.17. Leg lengths, listed in the third column, are used with azimuths to compute the latitudes and departures listed in the fourth and fifth columns. Successive *y* and *x* coordinates, beginning with the known coordinates of station Glen, are tabulated in the final two columns using successive computed latitude and departure values.

No computation of closure or precision is possible in Example 8.5. For an open traverse, the surveyor simply assumes the measurements are all correct. This is ordinarily a risky proposition, one that could be avoided by running the traverse in both directions so it closes on the initial station and provides a check on the observations.

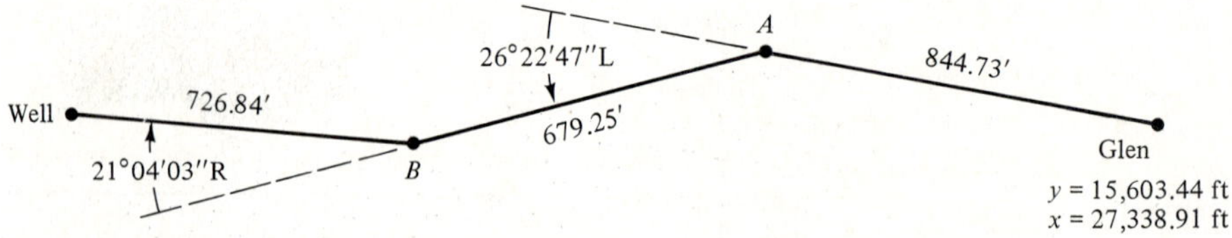

FIGURE 8.16
Open traverse, Glen to Well.

					Coordinates	
Sta.	Azimuth	Distance, ft	Latitude, ft	Departure, ft	Y, ft	X, ft
Glen					15,603.44	27,338.91
	280°36′18″	844.73	+155.46	−830.30		
A	−26°22′47″L				15,758.90	26,508.61
	254°13′31″	679.25	−184.66	−653.67		
B	+21°04′03″R				15,574.24	25,854.94
	275°17′34″	726.84	+67.05	−723.74		
Well					15,641.29	25,131.20

FIGURE 8.17
Computation of open traverse Glen to Well.

CLOSED CONNECTING TRAVERSE. Computation of a closed, connecting traverse was demonstrated earlier in the Hill-to-Dale examples. The summarized procedure is as follows:

1. Compute leg directions from initial direction and measured angles.
2. Adjust leg directions as necessary to bring the computed terminal direction into agreement with the known terminal direction.
3. Compute latitudes and departures from adjusted leg directions and measured leg lengths, using Eqs. (8.1*a*) and (8.1*b*) or Eqs. (8.2*a*) and (8.2*b*), as appropriate.
4. Determine total latitude and departure errors by comparing sums of latitudes and departures with the known y- and x-coordinate differences between the initial and terminal traverse stations.
5. Compute total error and precision using Eqs. (8.3) and (8.4).
6. Reverse the signs of the total latitude and departure errors to obtain the total latitude and departure corrections. Compute compass-rule corrections using Eqs. (8.5) and (8.6).
7. Add compass-rule corrections to previously computed latitudes and departures to obtain adjusted latitude and departures. Apply adjusted latitudes and departures to successive y and x coordinates, starting with the known coordinates of the initial traverse station. Check the computed coordinate values of the terminal traverse station against the known coordinate values for that station.

CLOSED-LOOP TRAVERSE. The computing procedure for a traverse that closes on its origin is easier than for a closed connecting traverse because both the angle-balancing and latitude-departure adjustment concepts are simpler. Proceed as follows:

1. Balance angles using the equation Σ interior angles $= (n-2)180°$ or Σ deflection angles $= 360°$, as appropriate.
2. Compute adjusted leg directions from initial direction and adjusted angles.
3. Compute latitudes and departures from adjusted leg directions and measured leg lengths, using Eqs. (8.1*a*) and (8.1*b*) or Eqs. (8.2*a*) and (8.2*b*), as appropriate.

4. Because the traverse closes on itself, the sums of the latitudes and departures are the latitude and departure errors.
5. Compute total error and precision using Eqs. (8.3) and (8.4).
6. Reverse the signs of the total latitude and departure errors to obtain the total latitude and departure corrections. Compute compass-rule corrections using Eqs. (8.5) and (8.6).
7. Add compass-rule corrections to previously computed latitudes and departures to obtain adjusted latitude and departures. Apply adjusted latitudes and departures to successive y and x coordinates, starting with the known coordinates of the initial traverse station. The end-of-loop computed coordinate values for the initial station are used as a check against the known starting coordinates of the initial station.

Example 8.6. A closed-loop traverse is run from corner A to corners B, C, D, E, and back to A, with measured deflection angles, measured distances, and initial azimuth AB as shown in Fig. 8.18. Balance the angles and compute the traverse, including total error and precision. Given the arbitrary starting coordinates for corner A as shown in Fig. 8.18, compute and check the adjusted coordinates of the remaining corners.

Solution. Balancing of angles and determination of leg directions are shown in Fig. 8.19.

Direction computations in this example could have been handled as shown at the end of Sec. 7.2, by computing azimuths first, then adjusting azimuths. Instead, Fig. 8.19 demonstrates the angles-first adjustment method. Measured deflection angles are listed in the second column, angles right taking a positive sign, angles left being negative. Their

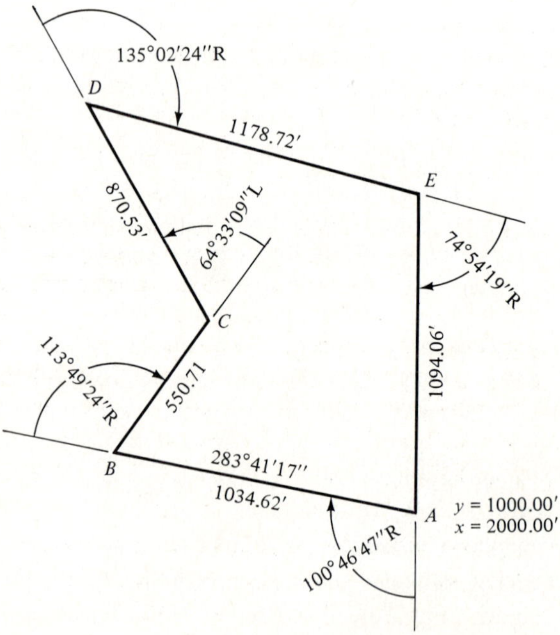

FIGURE 8.18
Closed-loop, deflection-angle traverse, Example 8.6.

Sta.	Measured angle	Correction	Adjusted angle	Adjusted azimuth
A	+100°46′47″R	+03″	+100°46′50″R	
				283°41′17″
B	+113°49′24″R	+03″	+113°49′27″R	
				[397°30′44″]
C	−64°33′09″L	+03″	−64°33′06″L	
				332°57′38″
D	+135°02′24″R	+03″	+135°02′27″R	
				[468°00′05″]
E	+74°54′19″R	+03″	+ 74°54′22″R	
				182°54′27″
A			+100°46′50″R	
ΣR =	+424°32′54″	+15″	360°00′00″	283°41′17″
ΣL =	− 64°33′09″			

359°59′45″⇒⇒ Error = −15″

$$\Rightarrow\Rightarrow \text{Correction} = \frac{+15''}{5 \text{ angles}} = +03'' \cdot \text{angle}^{-1}$$

FIGURE 8.19
Directions from deflection angles, Example 8.6.

sum being 15 seconds less than 360°00′00″, the total *error* is −15 seconds and total *correction* is +15″. Resultant per-angle corrections of +03 seconds are listed opposite each angle in the third column, with adjusted deflection angles in the fourth column. Known forward azimuth *AB*(283°41′17″) is fixed; azimuths of sides *BC, CD, DE,* and *EA* are determined by sequentially adding adjusted deflection angles *B, C, D,* and *E* (positive, right; negative, left) and listing the results in the last column. Bracketed azimuth values indicate over-360° results; thus, *BC* is actually 037°30′44″ rather than 397°30′44″, while side *DE* is really 108°00′05″ instead of 468°00′05″ (indicated values minus 360°). Note that the adjusted value for angle *A* is listed a second time opposite the lower "*A*" line of Fig. 8.19 so it may be added to the adjusted azimuth of side *EA* and thus produce the check value for side *AB*.

The two available arithmetic checks are shown as the last entries of the two final columns: The first verifies that the sum of the adjusted deflection angles is indeed 360°00′00″; the second verifies that the around-the-loop computed value for azimuth *AB* is the same as the known initial azimuth value for *AB*.

The rest of the computations are shown in an all-on-one-page format in Fig. 8.20, including latitudes, departures, total error, precision, compass-rule adjustment, and final coordinate determination. The procedure is identical to the previously demonstrated Hill-to-Dale computation except that (1) the sums of latitudes and departures in closed loops *are* the total latitude and departure errors and (2) y and x coordinates at the initial and end points are identical because they describe the same station.

Note, also, that cumulative rounding error for latitude and departure corrections in the Fig. 8.20 example were both zero (that is, sums of the rounded, computed latitude and departure corrections were equal to, and of opposite sign to, the total latitude and departure errors). Thus, unlike the Hill-to-Dale example, no further adjustment of latitude and/or departure corrections was needed.

					Corrections		Coordinates	
Sta.	Length, (ft)	Azimuth	Latitude, (ft)	Departure, (ft)	Lat. (ft)	Dep (ft)	y (ft)	x (ft)
A							1000.00	2000.00
	1034.62	283°41′17″	+244.83	−1005.24	−0.03	−0.03	+244.80	−1005.21
B							1244.80	994.79
	550.71	037°30′44″	+436.84	+335.34	−0.02	+0.01	+436.82	+335.35
C							1681.62	1330.14
	870.53	332°57′38″	+775.38	−395.75	−0.02	+0.02	+775.36	−395.73
D							2456.98	934.41
	1178.72	108°00′05″	−364.27	+1121.02	−0.03	+0.03	−364.30	+1121.05
E							2092.68	2055.46
	1094.06	182°54′27″	−1092.65	−55.49	−0.03	+0.03	−1092.68	−55.46
A							1000.00	2000.00
Σ	4728.64		+0.13	−0.12	−0.13	+0.12	✓	✓
					✓	✓		

$$\text{Closure} = \sqrt{0.13^2 + 0.12^2} = 0.17692 \Rightarrow\Rightarrow 0.18\text{ft}$$

$$\text{Precision} = \frac{0.17692}{4728.64} \approx 1 : 26{,}700$$

FIGURE 8.20
Computation of closed-loop traverse, Example 8.6.

HORIZONTAL CLOSURE SPECIFICATIONS FOR TRAVERSE. The National Geodetic Survey has published accuracy standards for traverse surveys similar to those for level surveys. Minimum closure specifications are listed below in a much-abbreviated version.

First order	1:100,000
Second order, Class I	1:50,000
Second order, Class II	1:20,000
Third order, Class I	1:10,000
Third order, Class II	1:5,000

As with level-survey standards, the recognized standards of traverse accuracy listed above may be achieved only with precise instruments and procedures. All distances must be measured with an EDM of appropriate precision and proper corrections must be applied; angles must be taken with a first-order theodolite to achieve first-order results, or with a second-order theodolite to achieve second- and third-order results. If an indicated closure of 1:100,000 or better were somehow achieved by taking single angles with a 1-minute transit and measuring single-run distances with a steel tape, the survey could not be classified as first order because of the inherent imprecision in both equipment and methods. The Hill-to-Dale precision of 1:6500, computed in Example 8.2, does *not* signify third order, class II results because angles were taken with an instrument that read only to ±20″ rather than ±1″. Similarly, the 1:26,700 precision in Example 8.6 could be classified as second order, class II only if the angles had been taken with a second order instrument, with eight or more repetitions, and using a 5-second rejection limit on individual angles. Traverses requiring high precision must always be run using precise instruments and exact methods. Ordinary traverses, where high precision is not required, may be run using more common instruments and methods.

8.9 AREA BY COORDINATES

Consider the *xy*-coordinate system of Example 8.6, whose origin lies south and west of all the corners of the polygon *ABCDE*, as shown in Fig. 8.21. For the purpose of illustration, the numerals along the *y* axis mark the location of the *y*-coordinate values for the five corners. The *x*-coordinate values are shown as dimension lines originating at the *y* axis. Note that there are five trapezoids thus formed whose bases lie along the *y* axis. In order, starting with the trapezoid formed by *y*-axis points 1 and 2 and traverse corners *B* and *A*, they are 12*BA*, 23*CB*, 35*DC*, 45*DE*, and 14*EA*. Figure 8.22 shows the Fig. 8.21 polygon with the 14*EA* and 45*DE* trapezoids shaded in. Similarly, Fig. 8.23 shows the Fig. 8.21 polygon with the 12*BA*, 23*CB*, and 35*DC* trapezoids shaded in. Now, if the shaded area in Fig. 8.23 is subtracted from the shaded area in Fig. 8.22, only polygon *ABCDE* remains. Thus, from the standpoint of area,

$$ABCDE = 14EA + 45DE - 12BA - 23CB - 35DC$$

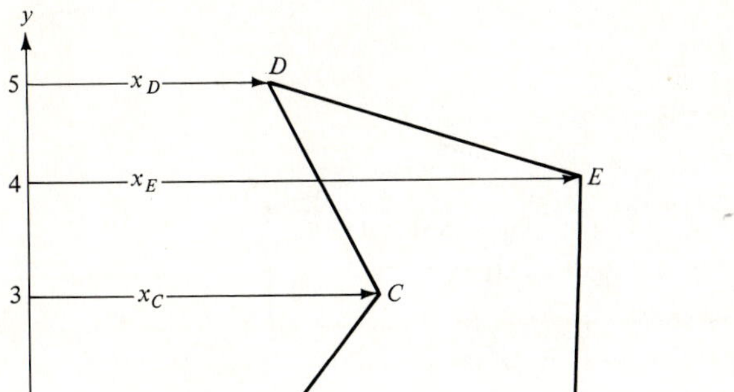

FIGURE 8.21
Polygon *ABCDE* (Example 8.6) in xy =coordinate system.

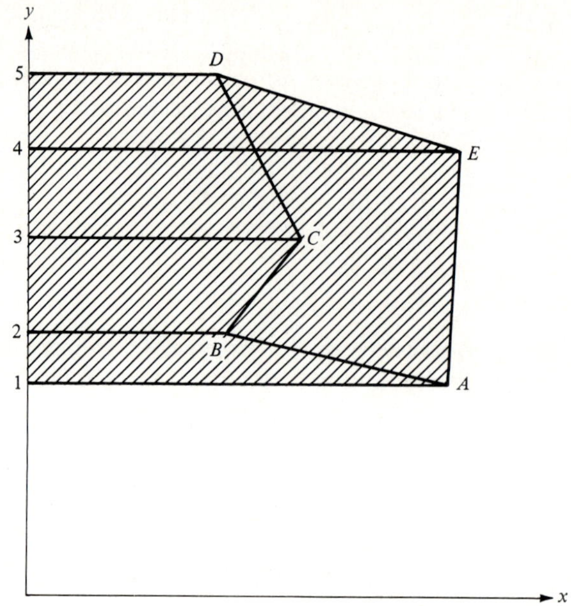

FIGURE 8.22
Polygon *ABCDE* with trapezoids 14*EA* and 45*DE* shaded.

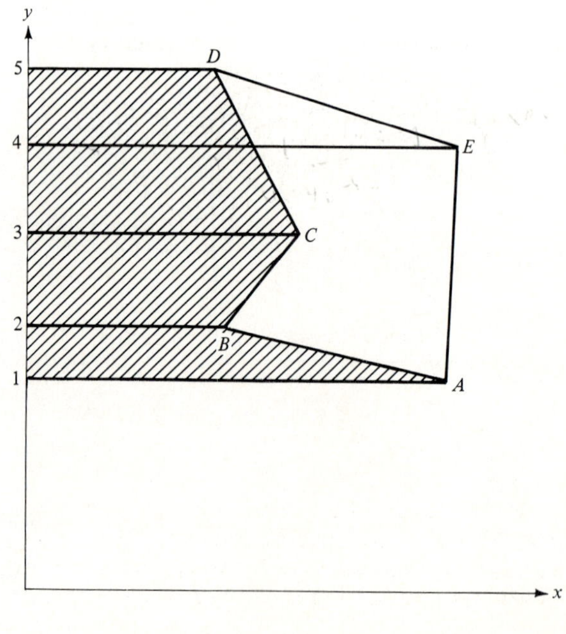

FIGURE 8.23
Polygon *ABCDE* with trapezoids 12*BA*, 23*CB*, and 35*DC* shaded.

Recall that the area of a trapezoid is the width of the base multiplied by the average height of the two parallel sides. The area of trapezoid 14*EA*, expressed in terms of coordinated values, is, therefore,

$$\text{Area of } 14EA = \frac{(y_E - y_A)(x_E + x_A)}{2}$$

$$= \frac{y_E x_E + y_E x_A - y_A x_E - y_A x_A}{2}$$

Similarly

$$\text{Area of } 45DE = \frac{y_D x_D + y_D x_E - y_E x_D - y_E x_E}{2}$$

$$\text{Area of } 12BA = \frac{y_B x_B + y_B x_A - y_A x_B - y_A x_A}{2}$$

$$\text{Area of } 23CB = \frac{y_C x_C + y_C x_B - y_B x_C - y_B x_B}{2}$$

$$\text{Area of } 35DC = \frac{y_D x_D + y_D x_C - y_C x_D - y_C x_C}{2}$$

Substituting into the equation $ABCDE = 14EA + 45DE - 12BA - 23CB - 35DC$, canceling as appropriate, and collecting terms:

Area of *ABCDE* =

$$\frac{y_A(x_B - x_E) + y_B(x_C - x_A) + y_C(x_D - x_B) + y_D(x_E - x_C) + y_E(x_A - x_D)}{2} \qquad \textbf{(8.7)}$$

Equation (8.7) may be expressed as a set of instructions:

1. Letter or number corners consecutively around the traverse.
2. Multiply each *y*-coordinate value by the difference between the following *x* coordinate and the preceding *x* coordinate, always subtracting preceding from following.
3. The sum of the products of the *y* coordinates and the *x*-coordinate differences is the "double area." The correct area of the traverse polygon is half the double-area sum.

Example 8.7. Compute the area of polygon *ABCDE* in Example 8.6. Give the answer in both square feet and acres.

Solution. *y*- and *x*-coordinate values are taken from Fig. 8.20, the solution to Example 8.6. Coordinate and coordinate-difference values are substituted into Eq. (8.7) using the format shown in Fig. 8.24.

The Δx value on line *A* is the *x* coordinate of point *B* minus the *x* coordinate of point *E*; that is, the preceding *x* coordinate (x_E in this case) subtracted from the following *x* coordinate (x_B in this case). All other Δx values were obtained in the same manner. In arriving at the five Δx values, all five *x* coordinates are used *twice*, once as a positive

Sta.	y, ft	x, ft	Δx	Double area, ft^2
A	1000.00	2000.00	−1060.67	−1,060,670.000
B	1244.80	994.79	−669.86	−833,841.7280
C	1681.62	1330.14	−60.38	−101,536.2156
D	2456.98	934.41	+725.32	+1,782,096.734
E	2092.68	2055.46	+1065.59	+2,229,938.881
			Sums: 0.00	2,015,987.671

Dividing double area by 2 ⇒ 1,007,993.836 ⇒ 1, 007, 994 ft^2
Dividing ft^2 by 43,560 ft^2 · acre^{-1} ⇒⇒ 23.1404 acres

FIGURE 8.24
Format for computation of area by coordinates.

value and once as a negative value. Thus, the sum of the Δx values *must* be zero, an important arithmetic check.

The final-column sum has no algebraic sign; it could be negative as often as positive, depending on irrelevant factors such as whether the stations were lettered clockwise or counterclockwise around the traverse. Thus, an algebraic sign that may appear for the double-sum result is essentially meaningless.

Note that the numerical double-sum values in the final column of Fig. 8.24 exceed the number of digits on a pocket calculator, with resulting one-digit round-off loss in several cases. When using very large coordinate values, such as those found in state plane coordinate systems (Chap. 9), important significant digits are lost due to calculator round-off when multiplying the y coordinate by Δx. Desktop computers, with their 16-significant-figure availability, are not bothered by round-off error of this sort. For hand calculators, though, it is better to reduce each y-coordinate value by a constant amount before doing the $y\ \Delta x$ computations that produce double-area values.

For example, had the original y-coordinate values in Fig. 8.20 been exactly 1,000,000 ft greater than those listed—the sort of thing that happens all the time with state plane coordinate values—it would be prudent to reduce each y coordinate by exactly 1,000,000 ft for the purpose of doing the area-by-coordinates computation with a hand calculator. The result would have been the same as in Fig. 8.24, but without having a numerical overflow problem in the calculator and a resulting rounding error.

8.10 PROBLEMS

8.1. Discuss procedures that can be followed to reduce the possibility of measurement error on an open traverse.

8.2. State which type of traverse—a closed, connecting traverse or a closed-loop traverse—provides the more rigorous check on closure and explain why.

8.3. Corner angles J, K, L, M, and N, lettered clockwise, consecutively, around a closed-loop traverse, were measured with an instrument whose least count is 1 minute. The measured values of the interior angles were as follows: $J = 76°14'$, $K = 118°37'$, $L = 137°40'$, $M = 95°58'$, $N = 111°27'$. Determine whether or not these angles need to be measured again. If they do *not* need remeasuring, adjust the angles and determine the bearings of the other four sides if the known bearing of side *JK* is N64°27′00″E.

8.4. Corner angles *R*, *S*, *T*, *U*, *V*, and *W*, lettered counterclockwise, consecutively, around a closed-loop traverse, were measured with an instrument whose least count is 6 seconds. The measured values of the deflection angles were as follows: *R* = 67°15′24″ L, *S* = 92°30′12″ L, *T* = 114°06′30″ L, *U* = 53°47′48″ R, *V* = 76°33′06″ L, *W* = 63°23′18″ L. Determine whether or not these angles need to be measured again. If they do *not* need remeasuring, adjust the angles and determine the azimuths of the other five sides if the known azimuth of side *RS* is 248°08′54″.

8.5. Interior angles *A* through *E*, designated clockwise consecutively around a closed-loop traverse, were measured by repetition with an instrument whose least count is 1 minute. Measured values are as follows: *A* = 54°20′30″, *B* = 105°43′00″, *C* = 68°15′15″, *D* = 89°37′45″, *E* = 222°02′30″. Determine whether or not these angles need to be measured again. If they do *not* need remeasuring, adjust the angles and determine the bearings of the remaining sides if the known bearing of side *AB* is S38°22′08″W.

8.6. Do the same as Prob. 8.5 except the bearing of *AB* is N42°38′42″W.

8.7. Interior angles *A* through *E*, designated counterclockwise consecutively around a closed-loop traverse, were measured by repetition with an instrument whose least count is 20 seconds. Measured values are as follows: *A* = 106°46′15″, *B* = 88°15′40″, *C* = 39°02′10″, *D* = 264°31′25″, *E* = 41°25′05″. Determine whether or not these angles need to be measured again. If they do *not* need remeasuring, adjust the angles and determine the bearings of the remaining sides if the known bearing of side *AB* is N55°11′18″W.

8.8. Do the same as Prob. 8.7 except the bearing of *AB* is S22°39′37″E.

8.9. For the closed, connecting traverse shown below, known initial bearing *LK* is N85°33′17″W and known final bearing *PQ* is N52°15′23″E. Assuming a 20-second least count, determine whether or not the traverse angles need to be measured again. If they do *not* need remeasuring, adjust the bearings of connecting lines *LA*, *AB*, and *BP*.

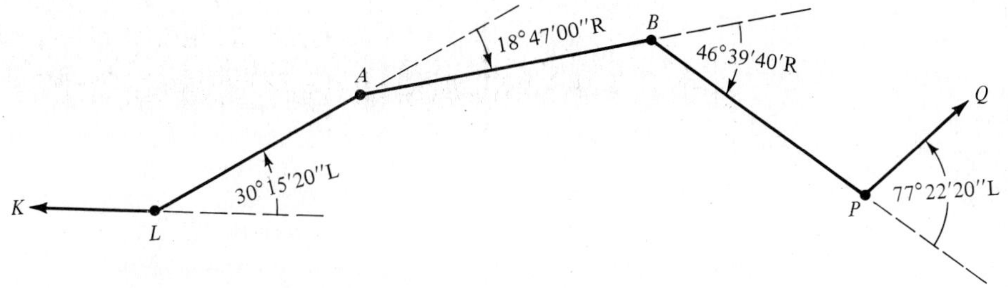

8.10. Do the same as Prob. 8.9 except known bearing *LK* is S29°42′25″W and known bearing *PQ* is N12°27′59″W.

8.11. For the closed, connecting traverse shown on the next page, known initial azimuth *FG* is 222°33′49″ and known final azimuth *MN* is 309°51′11″. Assuming a 30-second least count, determine whether or not the traverse angles need to be measured again. If they do *not* need remeasuring, adjust the azimuths of connecting lines *GA*, *AB*, and *BM*.

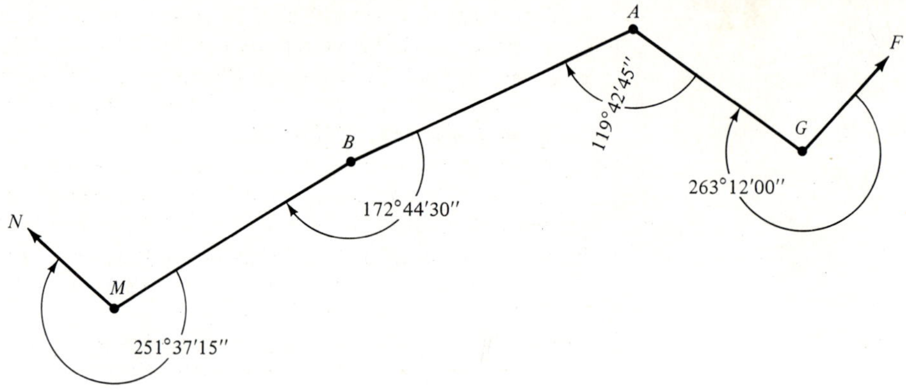

8.12. Do the same as Prob. 8.11 except known azimuth FG is 147°18′ 37″ and known azimuth MN is 234°34′43″.

8.13. Referring to Prob. 8.5, the field measurements of the sides were $AB = 229.32$ ft, $BC = 330.09$ ft, $CD = 249.36$ ft, $DE = 168.05$ ft, $EA = 153.49$ ft. Calculate the latitudes and departures of the sides, the closure error, and the survey precision.

8.14. Referring to Prob. 8.6, the field measurements of the sides were $AB = 288.94$ ft, $BC = 415.91$ ft, $CD = 314.19$ ft, $DE = 211.65$ ft, $EA = 193.40$ ft. Calculate the latitudes and departures of the sides, the closure error, and the survey precision.

8.15. Referring to Prob. 8.7, the field measurements of the sides were $AB = 255.02$ ft, $BC = 386.43$ ft, $CD = 233.66$ ft, $DE = 234.78$ ft, $EA = 338.51$ ft. Calculate the latitudes and departures of the sides, the closure error, and the survey precision.

8.16. Referring to Prob. 8.8, the field measurements of the sides were $AB = 400.38$ ft, $BC = 606.70$ ft, $CD = 366.85$ ft, $DE = 368.68$ ft, $EA = 531.46$ ft. Calculate the latitudes and departures of the sides, the closure error, and the survey precision.

8.17. Referring to Prob. 8.9, the field measurements of the sides were $LA = 372.46$ ft, $AB = 455.78$ ft, $BP = 419.19$ ft. The known (fixed) coordinates of point L are $y = 9421.63$ ft and $x = 15,879.48$ ft. The known (fixed) coordinates of point P are $y = 9372.04$ ft and $x = 16,989.84$ ft. Calculate the latitudes and departures of the sides, the closure error, and the survey precision.

8.18. Referring to Prob. 8.10, the field measurements of the sides were $LA = 299.76$ ft, $AB = 506.43$ ft, $BP = 476.81$ ft. The known (fixed) coordinates of point L are $y = 6802.86$ ft and $x = 4389.22$ ft. The known (fixed) coordinates of point P are $y = 7785.90$ ft and $x = 4976.62$ ft. Calculate the latitudes and departures of the sides, the closure error, and the survey precision.

8.19. Referring to Prob. 8.11, the field measurements of the sides were $GA = 422.17$ ft, $AB = 648.73$ ft, $BM = 587.41$ ft. The known (fixed) coordinates of point G are $y = 15,672.72$ ft and $x = 8288.11$ ft. The known (fixed) coordinates of point M are $y = 15,341.17$ ft and $x = 6855.75$ ft. Calculate the latitudes and departures of the sides, the closure error, and the survey precision.

8.20. Referring to Prob. 8.12, the field measurements of the sides were $GA = 651.07$ ft, $AB = 717.63$ ft, $BM = 679.10$ ft. The known (fixed) coordinates of point G are $y =$ 5481.73 ft and $x = 11,809.33$ ft. The known (fixed) coordinates of point M are

$y = 3711.00$ ft and $x = 11,627.95$ ft. Calculate the latitudes and departures of the sides, the closure error, and the survey precision.

8.21. Referring to Prob. 8.13, adjust the latitudes and departures using the compass rule and calculate the coordinates of points *B*, *C*, *D*, and *E*. The coordinates of point *A* are $y = 429.63$ ft and $x = 858.74$ ft.

8.22. In Prob. 8.14, adjust the latitudes and departures using the compass rule and calculate the coordinates of points *B*, *C*, *D*, and *E* if the coordinates of *A* are $y = 360,472.19$ ft. and $x = 3,294,476.35$ ft.

8.23. In Prob. 8.15, adjust the latitudes and departures using the compass rule and calculate the coordinates of points *B*, *C*, *D*, and *E* if the coordinates of *A* are $y = 5658.97$ ft and $x = 3288.44$ ft.

8.24. In Prob. 8.16, adjust the latitudes and departures using the compass rule and calculate the coordinates of points *B*, *C*, *D*, and *E* if the coordinates of *A* are $y = 296,821.39$ ft and $x = 2,433,807.69$ ft.

8.25. In Prob. 8.17, adjust the latitudes and departures using the compass rule and calculate adjusted coordinates for points *A* and *B*.

8.26. In Prob. 8.18, adjust the latitudes and departures using the compass rule and calculate adjusted coordinates for points *A* and *B*.

8.27. In Prob. 8.19, adjust the latitudes and departures using the compass rule and calculate adjusted coordinates for points *A* and *B*.

8.28. In Prob. 8.20, adjust the latitudes and departures using the compass rule and calculate adjusted coordinates for points *A* and *B*.

8.29. Using the National Geodetic Survey standards, what order of accuracy was obtained in the Prob. 8.13 traverse?

8.30. Using the National Geodetic Survey standards, what order of accuracy was obtained in the Prob. 8.14 traverse?

8.31. Using the National Geodetic Survey standards, what order of accuracy was obtained in the Prob. 8.15 traverse?

8.32. Using the National Geodetic Survey standards, what order of accuracy was obtained in the Prob. 8.16 traverse?

8.33. Using the National Geodetic Survey standards, what order of accuracy was obtained in the Prob. 8.17 traverse?

8.34. Using the National Geodetic Survey standards, what order of accuracy was obtained in the Prob. 8.18 traverse?

8.35. Using the National Geodetic Survey standards, what order of accuracy was obtained in the Prob. 8.19 traverse?

8.36. Using the National Geodetic Survey standards, what order of accuracy was obtained in the Prob. 8.20 traverse?

8.37. In Prob. 8.21, calculate, by coordinates, the area of the figure to integer square feet and to thousandths of an acre.

8.38. In Prob. 8.22, calculate, by coordinates, the area of the figure to integer square feet and to thousandths of an acre.

8.39. In Prob. 8.23, calculate, by coordinates, the area of the figure to integer square feet and to thousandths of an acre.

8.40. In Prob. 8.24, calculate, by coordinates, the area of the figure to integer square feet and to thousandths of an acre.

CHAPTER 9

WORKING WITH COORDINATES

9.1 INTRODUCTION

Older surveying texts treated rectangular coordinates in a rather offhand manner, providing very little information about plane coordinate systems or their beneficial application by surveyors. With the development and widespread use of computer software that deals readily with two- and three-dimensional coordinate geometry, it has become imperative for surveyors to have a reasonable understanding of coordinates and their applications.

Section 8.9 illustrated how the area of a closed-loop traverse may be computed from the coordinates of its corner points—coordinates that were obviously based on a very localized *xy* system set up to deal only with that single traverse. Consider, however, the advantages of a *common* coordinate system that provides, within a local or regional context, coordinate addresses of corner points on all of the property surveys, as well as the locations of all rights-of-way and utility lines. Already established common coordinate systems, as well as recent technological advances for providing and managing common-coordinate data, are discussed further along in this chapter. For the moment, it would be well to consider a number of relatively simple surveying applications of coordinate geometry.

9.2 COORDINATE APPLICATIONS

INVERSES. Recall, from Chap. 8, that successive coordinate positions are determined by application of adjusted latitudes and departures, the latitudes and departures having been determined from distances and directions between successive points. Consider now that if successive coordinate values can be found from distances and directions, then the reverse is also true; that is, that distances and directions between successive

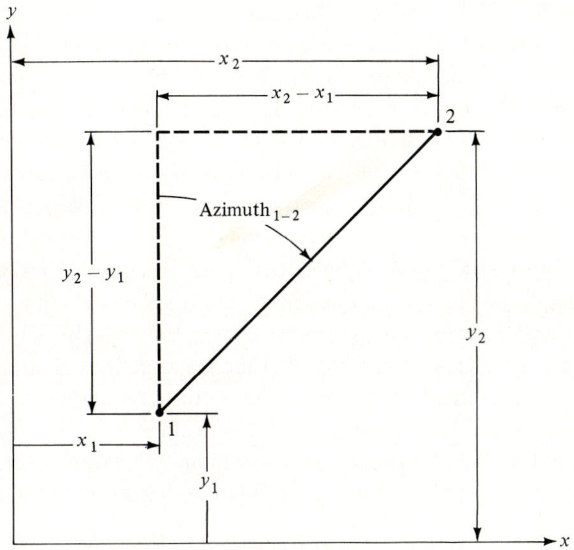

FIGURE 9.1
The inverse problem.

points may be determined from the values of those coordinate points. The procedure for doing this is called an **inverse** computation. Used to determine the distance and direction between any two points on a common coordinate system, it is perhaps the most useful single application of coordinates.

Consider the *xy* coordinate system in Fig. 9.1, with points 1 and 2 defined by the coordinate values x_1, y_1 and x_2, y_2, respectively. The direction and length of the line *from* point 1 *to* point 2 are obtained from the companion equations:

$$\textbf{Azimuth}_{1-2} = \tan^{-1} \frac{x_2 - x_1}{y_2 - y_1} \quad \textbf{(9.1)}$$

$$\textbf{Length}_{1-2} = \sqrt{(x_2 - x_1)^2 + (y_2 - y_1)^2} \quad \textbf{(9.2)}$$

Example 9.1. From Fig. 8.20 coordinate data, compute the azimuth and length of line *BE*.

Solution. From Fig. 8.20, the coordindates are $x_B = 994.79$ ft, $y_B = 1244.80$ ft; $x_E = 2055.46$ ft, $y_E = 2092.68$ ft. Substituting into Eq. (9.1),

$$\text{Azimuth}_{BE} = \tan^{-1} \frac{2055.46 - 994.79}{2092.68 - 1244.80}$$

$$= \tan^{-1} \frac{1060.67}{847.88} = 051°21'42.5''$$

Substituting into Eq. (9.2),

$$\text{Length}_{BE} = \sqrt{1060.67^2 + 847.88^2} = 1357.911 \text{ ft}$$

Note that the above results are taken one significant figure beyond that of the distance and azimuth input data used to solve the Fig. 8.20 example. This seeming inconsistency is necessary for two reasons. First, the adjusted coordinate values are assumed to be precisely known. Second, an inverse computation is usually the first step in a chain of computations leading to solution of a more complex problem; thus, by convention, an artificially high order of significant figures is carried along until the end of the problem and the rounding down of the final result.

TWO-POINT LINE EQUATION. Computer solution of problems involving lines, either straight or curved, requires that such lines be described in equation form. When dealing with coordindates, the most convenient straight-line equation is the **two-point form**, briefly discussed in Chap. 1. Figure 9.2 defines coordinate points and directions used in development of the two-point form. The known coordinates of points 1 and 2 are designated, respectively, by x_1, y_1 and by x_2, y_2. The azimuth of line 1-2 is designated ϕ. The values x and y—without subscripts—are defined as the unknown coordinates of a point lying on the line passing through points 1 and 2. Now

$$\tan\phi = \frac{x_2 - x_1}{y_2 - y_1} \qquad \text{and} \qquad \tan\phi = \frac{x - x_1}{y - y_1}$$

Equating the two expressions for tan ϕ:

$$\frac{x_2 - x_1}{y_2 - y_1} = \frac{x - x_1}{y - y_1}$$

which may also be written

$$(x_2 - x_1)(y - y_1) = (y_2 - y_1)(x - x_1)$$

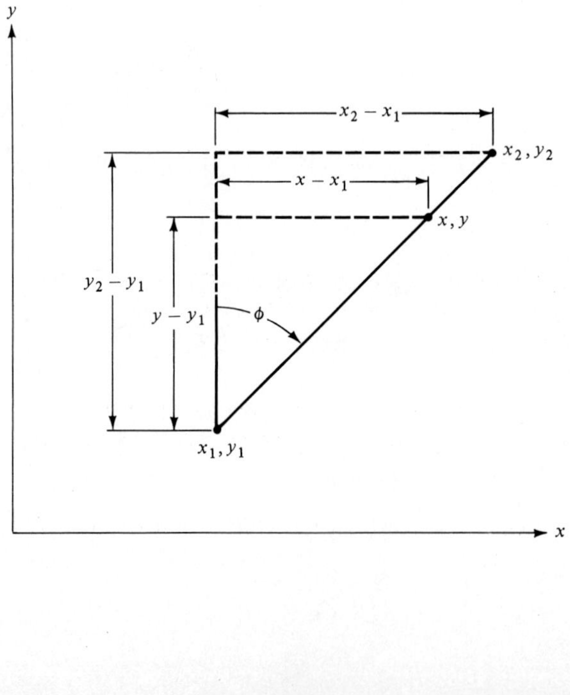

FIGURE 9.2
Two-point form labeling.

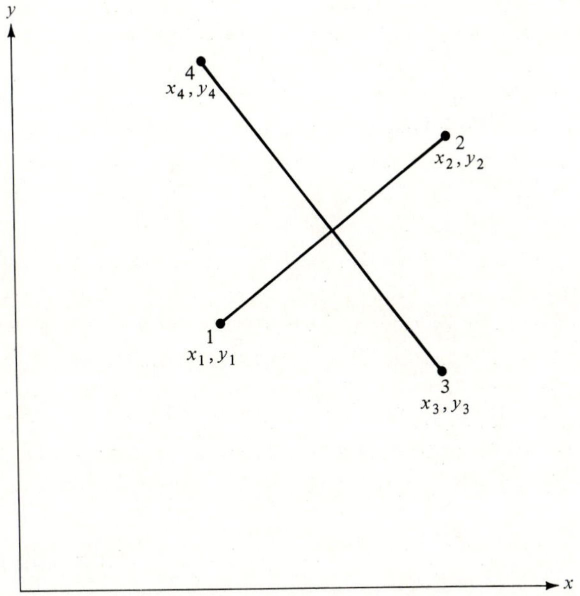

FIGURE 9.3
Two lines on a common-coordinate system.

and rearranged as follows in terms of unknowns x and y:

$$(y_2 - y_1)x + (x_1 - x_2)y + x_2y_1 - x_1y_2 = 0 \tag{9.3}$$

Figure 9.3 shows a second line, defined by known points 3 and 4 in the same manner that known points 1 and 2 define the first line. The second line may be expressed in the following, similar form:

$$(y_4 - y_3)x + (x_3 - x_4)y + x_4y_3 - x_3y_4 = 0 \tag{9.4}$$

Note that Eqs. (9.3) and (9.4) are simply forms of the line equation $ax + by + c = 0$. In Eq. (9.3), for example, $a = y_2 - y_1$, $b = x_1 - x_2$, and $c = x_2y_1 - x_1y_2$.

LINE INTERSECTION

Simultaneous Equation Solution. As discussed in Chap. 1, the x and y value for the point of intersection of the two lines of the form $ax + by + c = 0$ may be solved by the method of simultaneous equations.

Example 9.2. Figure 9.4 shows points A, B, D, and E from Fig. 8.18 and their coordinate values from Fig. 8.20. Compute the coordinates of point I, the intersection of lines BE and AD.

Solution. Substituting into Eq. (9.3) for line BE,

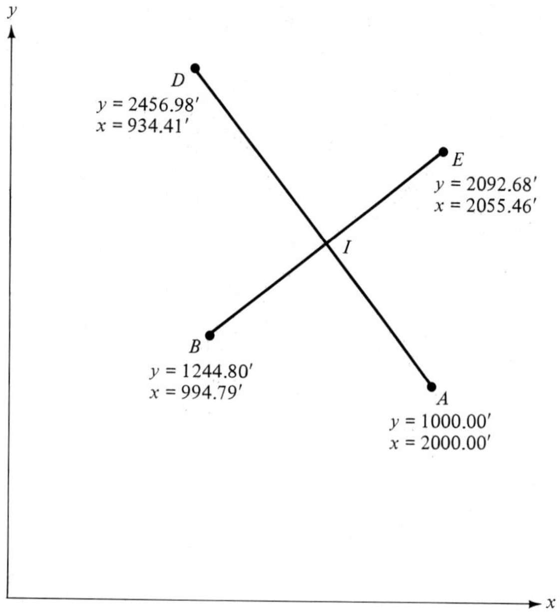

FIGURE 9.4
Intersection of lines *BE* and *AD*.

$$(y_E - y_B)x + (x_B - x_E)y + x_E y_B - x_B y_E = 0$$

$$(2092.68 - 1244.80)x_1 + (994.79 - 2055.46)y_1$$

$$+2055.46(1244.80) - 994.79(2092.68) = 0$$

$$\Rightarrow\Rightarrow \quad 847.88x_1 - 1060.67y_1 + 476,859.4708 = 0$$

Substituting into Eq. (9.4) for line *AD*,

$$(y_D - y_A)x + (x_A - x_D)y + x_D y_A - x_A y_D = 0$$

$$(2456.98 - 1000.00)x_1 + (2000.00 - 934.41)y_1$$

$$+934.41(1000.00) - 2000.00(2456.98) = 0$$

$$\Rightarrow\Rightarrow \quad 1456.98x_1 + 1065.59y_1 - 3,979,550.0000 = 0$$

Placing the two equations in simultaneous equation form,

$$847.88x_1 - 1060.67y_1 = -476,859.4708$$

$$\underline{1456.98x_1 + 1065.59y_1 = +3,979,550.0000}$$

Eliminating the x_1 values by multiplying the top equation by 1456.98, the bottom equation by 847.88, then subtracting, y_1 is found to be

$$y_1 = 1661.5663 \Rightarrow\Rightarrow 1661.57\text{ft}$$

Substituting the *unrounded* value of y_1 into the *upper* simultaneous equation and solving for x_1:

$$x_1 = 1516.1509 \Rightarrow\Rightarrow 1516.15\text{ft}$$

Substituting the *unrounded* values of both x_1 and y_1 into the left side of the *lower* simultaneous equation produces

$$1456.98(1516.1509) + 1065.59(1661.5663) = 3,979,549.972 \checkmark$$

This checks the right-side value of the lower equation to the eighth significant figure; thus, the arithmetic is valid.

The simultaneous equation method words reasonably well for small coordinate values, though it is somewhat tedious—more so than the solved values in Example 9.2 indicate. State plane coordinate values (Sec. 9.3) are likely to run in the millions of feet, which, when taken to a hundredth of a foot, results in nine or more significant figures. Substitution of nine-significant-figure values into Eqs. (9.3) and (9.4), then trying to solve with ordinary hand calculators, will result in substantial rounding overflow during intermediate steps, with potential loss of significant figures in the final result. For this reason, simultaneous equations involving full state plane coordinate values should be solved on computers capable of significant-figure handling well beyond the abilities of most hand calculators. An alternative approach, discussed in Chap. 8, is to reduce the x- and y-coordinate values to more reasonable numbers by subtracting an appropriate x constant from all x values and an appropriate y constant from all y values before performing the computation, then adding the constants back into the solved values of x and y.

Law of Sines Solution. Computers with **COGO** (**CO**ordinate **GeO**metry) software will quickly and easily work the above problem. Where no computer is available, it may be done by using a triangle consisting of the intersection point and one point on each of the two lines, solving with the law of sines, then computing intersection coordinates from latitudes and departures.

Example 9.3. Solve Example 9.2 using the law of sines.

Solution. Figure 9.5 shows triangle *ABI* and the coordinates of points *A*, *B*, *D*, and *E*. The solution procedure is as follows; (1) Solve for the azimuth and length of side *AB*. (2) Solve for azimuths *AD* and *BE*, from which angles α, β and γ are determined using azimuth differences. (3) With all three angles and the length of side *AB* known, the lengths of *AI* and *BI* are determined using the law of sines. (4) The coordinates of *I* are solved by adding the latitude and departure of line *AI* to the y and x coordinates of *A*, respectively. (5) Adding latitude *BI* to the y coordinate of *B* and departure *BI* to the x coordinate of *B* provides an arithmetic check.

1. Substituting into Eq. (9.1),

$$\text{Azimuth}_{AB} = \tan^{-1}\frac{994.79 - 2000.00}{1244.80 - 1000.00}$$

$$= \tan^{-1}\frac{-1005.21}{244.80}$$

$$= -76°18'47.2'' \Rightarrow\Rightarrow 283°41'12.8''$$

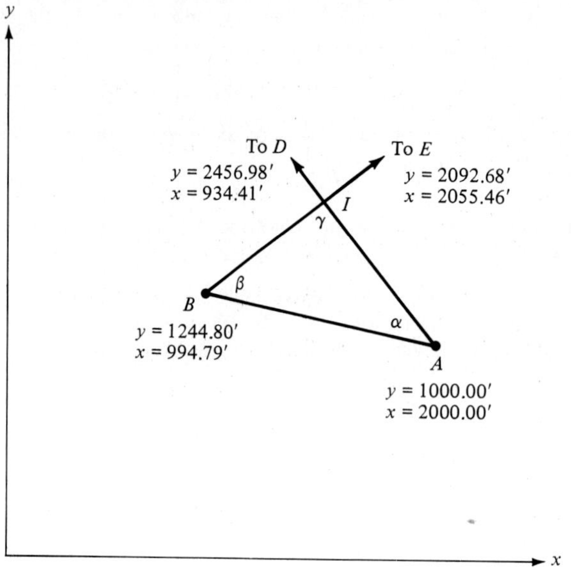

FIGURE 9.5
Intersection triangle *ABI*.

and

$$\text{Azimuth}_{BA} = 283°41'12.8'' - 180°00'00.0'' = 103°41'12.8''$$

Substituting into Eq. (9.2),

$$\text{Length}_{AB} = \sqrt{244.80^2 + 1005.21^2} = 1034.589\text{ft}$$

2. Substituting into Eq. (9.1),

$$\text{Azimuth}_{AD} = \tan^{-1}\frac{934.41 - 2000.00}{2456.98 - 1000.00}$$

$$= \tan^{-1}\frac{-1065.59}{1456.98}$$

$$= -36°10'50.1'' \Rightarrow\Rightarrow 323°49'09.9''$$

and

$$\text{Azimuth}_{DA} = 323°49'09.9'' - 180°00'00.0'' = 143°49'09.9''$$

$$\angle\alpha = \text{azimuth}_{AD} - \text{azimuth}_{AB}$$

$$= 323°49'09.9'' - 283°41'12.8'' = 40°07'57.1''$$

From the solution to Example 9.1:

$$\text{Azimuth}_{BE} = 051°21'42.5''$$

and

$$\text{Azimuth}_{EB} = 051°21'42.5'' + 180°00'00.0'' = 231°21'42.5''$$

$$\angle\beta = \text{azimuth}_{BA} - \text{azimuth}_{BE}$$

$$= 103°41'12.8'' - 051°21'42.5'' = 52°19'30.3''$$

$$\angle\gamma = \text{azimuth}_{EB} - \text{azimuth}_{DA}$$

$$= 231°21'42.5'' - 143°49'09.9'' = 87°32'32.6''$$

Adding α, β, and γ to provide an arithmetic check,

$$40°07'57.1'' + 52°19'30.3'' + 89°32'32.6'' = 180°00'00.0'' \quad \checkmark$$

3. Using the law of sines,

$$\frac{AI}{\sin\beta} = \frac{AB}{\sin\gamma}$$

$$AI = \sin\beta\left(\frac{AB}{\sin\gamma}\right)$$

$$= \sin 52°19'30.3''\left(\frac{1034.589}{\sin 87°32'32.6''}\right)$$

$$= 819.622 \text{ ft}$$

Similarly

$$\frac{BI}{\sin\alpha} = \frac{AB}{\sin\gamma}$$

$$BI = \sin\alpha\left(\frac{AB}{\sin\gamma}\right)$$

$$= \sin 40°07'57.1''\left(\frac{1034.589}{\sin 87°32'32.6''}\right)$$

$$= 667.466 \text{ ft}$$

4. Determining y_I by adding latitude$_A$ to y_A,

$$y_I = AI\cos(\text{azimuth}_{AD}) + y_A = 819.622\cos 323°49'09.9'' + 1000.00$$

$$= 1661.566 \Rightarrow\Rightarrow 1661.57 \text{ ft}$$

Determining x_I, by adding departure $_{AI}$ to x_A,

$$x_I = AI\sin(\text{azimuth}_{AD}) + x_A = 819.622\sin 323°49'09.9'' + 2000.00$$

$$= 1516.151 \Rightarrow\Rightarrow 1516.15 \text{ ft}$$

5. As an arithmetic check, y_I and x_I are determined separately from the coordinates of B and the latitude and departure of line BI.

$$y_I = 667.466 \cos 051°21'42.5'' + 1244.80 = 1661.566 \text{ ft}$$

$$x_I = 667.466 \sin 051°21'42.5'' + 994.79 + 1516.151 \text{ft}$$

Note that the azimuth and length of line *AB*, as determined in Example 9.3 from an inverse taken on the coordinates of points *A* and *B*, does *not* agree with the original, entering azimuth and length of line *AB* used in the Fig. 8.20 solution. This is because the *A* and *B* coordinates come from *adjusted* latitudes and departures while the entering azimuth and length correspond to the *unadjusted* latitude and departure for line *AB*. These adjusted directions and lengths are used in the legal description of property and are sometimes referred to as "deed" length and bearing (or azimuth).

NORMAL LINE EQUATION. Consider the general form of the line equation, $ax + by + c = 0$. Again, for a line defined by points 1 and 2 with coordinates x_1, y_1 and $x_2 y_2$, constant a is $y_2 - y_1$, constant b is $x_1 - x_2$, and constant c is $x_2 y_1 - x_1 y_2$. Dividing the general form of the line equation by $\pm\sqrt{a^2 + b^2}$ produces the **normal form** of the line equation:

$$\left(\frac{a}{\pm\sqrt{a^2 + b^2}}\right) x + \left(\frac{b}{\pm\sqrt{a^2 + b^2}}\right) y + \frac{c}{\pm\sqrt{a^2 + b^2}} = 0 \qquad \textbf{(9.5)}$$

This awkward-seeming form has at least one special property: The expression $c/\pm\sqrt{a^2 + b^2}$ is the "normal" distance, or perpendicular distance from the origin (at $x = 0, y = 0$) to the line. This relationship is shown in Fig. 9.6, where the value n is $c/\pm\sqrt{a^2 + b^2}$.

POINT-TO-LINE DISTANCE

Normal Solution. Point *P* in Fig. 9.6 lies a distance p perpendicular to the line. Distance p is determined from a slightly modified version of Eq. (9.5) as follows:

$$\left(\frac{a}{\pm\sqrt{a^2 + b^2}}\right) x_P + \left(\frac{b}{\pm\sqrt{a^2 + b^2}}\right) y_P + \frac{c}{\pm\sqrt{a^2 + b^2}} = p \qquad \textbf{(9.6)}$$

where x_P and y_P are the coordinates of point *P* and p is the normal distance to the line. The $\pm$ preceding $\sqrt{a^2 + b^2}$ takes the sign of b. If b is positive, use $+\sqrt{a^2 + b^2}$; if negative, use $-\sqrt{a^2 + b^2}$.

The normal *direction* from a point to a line is the direction of the *perpendicular* to the line. The azimuth of the line is computed from Eq. (9.1). The azimuth of its perpendicular is the azimuth of the line $\pm 90°$. The azimuth of the perpendicular is obtained directly from the following modification of Eq. (9.1), based on the fact that the tangent of a line is the negative reciprocal of its perpendicular:

$$\textbf{Azimuth}_p = \tan^{-1} \frac{y_2 - y_1}{x_1 - x_2} \qquad \textbf{(9.7)}$$

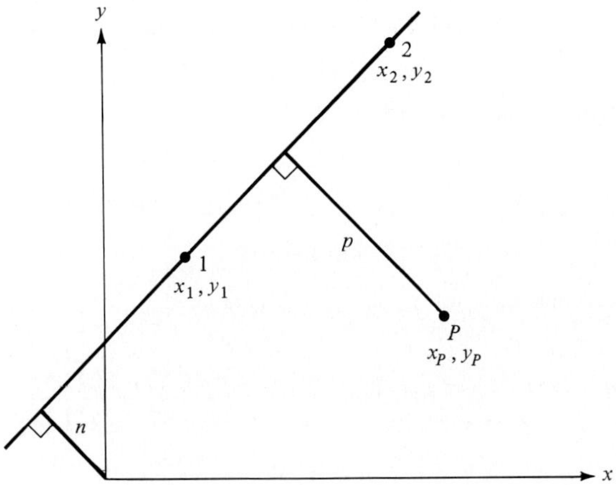

FIGURE 9.6
Normal (perpendicular) distance to a line.

Example 9.4. Figure 9.7 shows points B, C, and E from fig. 8.18 and their coordinates from Fig. 8.20. Compute the perpendicular distance and direction of the point B from line EC.

Solution. Constants a, b, and c in Eq. (9.6) are $y_C - y_E$, $x_E - x_C$, and $x_C y_E - x_E y_C$, respectively. Substituting known coordinate values,

$$a = 1681.61 - 2092.68 = -411.06$$

$$b = 2055.46 - 1330.14 = +725.32$$

$$\sqrt{a^2 + b^2} = +833.70224 \qquad \text{(positive because } b \text{ is positive)}$$

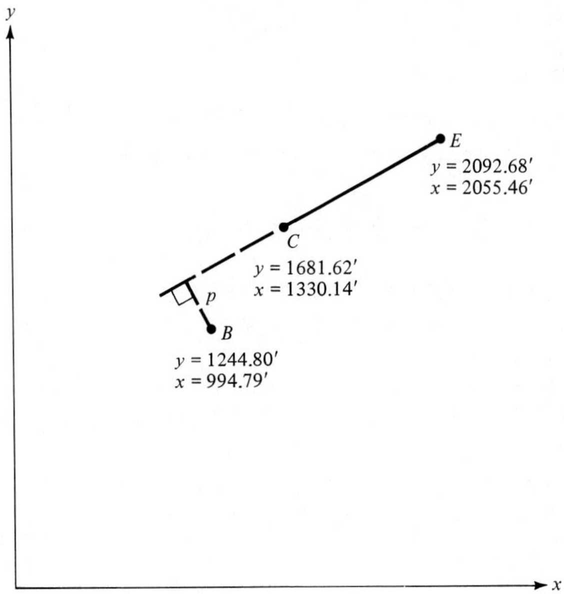

FIGURE 9.7
Normal distance from B to line CE.

$$c = 1330.14(2092.68) - 2055.46(1681.62) = -672,945.27$$

$$x_p = x_B = 994.79 \text{ ft} \qquad \text{and} \qquad y_p = y_B = 1244.80 \text{ ft}$$

Substituting into Eq. (9.6),

$$p = \left(\frac{-411.06}{+833.70224}\right) 994.79 + \left(\frac{725.32}{+833.70224}\right) 1244.80 + \left(\frac{-672,945.27}{+833.70224}\right)$$

$$= -214.687 \Rightarrow\Rightarrow 214.69 \text{ ft}$$

(The minus sign in the above result is ignored.)

Substituting the coordinates of points E and C into Eq. (9.7),

$$\text{Azimuth}_p = \tan^{-1} \frac{1681.62 - 2092.68}{2055.46 - 1330.14}$$

$$= \tan^{-1} \frac{-411.06}{725.32}$$

$$= -29°32'29.4'' \Rightarrow\Rightarrow 330°27'30.6''$$

But the azimuth *from* the line *to* the point is required, while Eq. (9.7) provides the azimuth from the point to the line. Thus, the azimuth to point B from line EC is the back azimuth of B to line EC, or 150°27'30.6''.

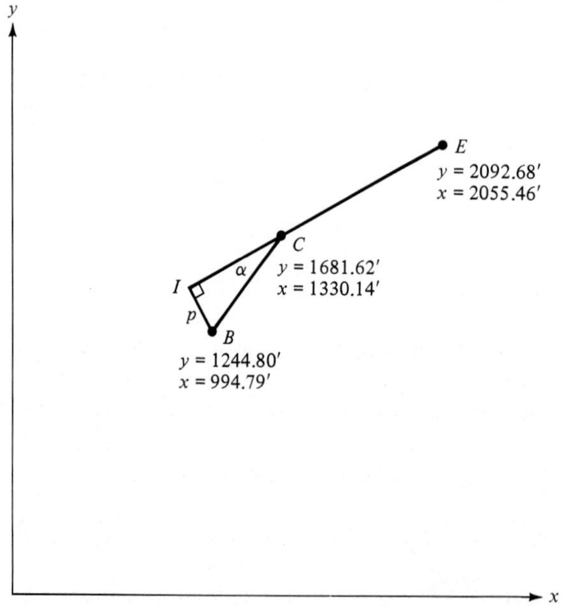

FIGURE 9.8
Right triangle *CBI* on line *EC* extension.

Right-Triangle Solution. The above solution is easily handled by a computer programmed with COGO software. With a hand-held calculator, it is easier to approach the problem with a simple right triangle as in Fig. 9.8.

Example 9.5. Solve Example 9.4 using right-triangle trigonometry.

Solution. In Fig. 9.8, the perpendicular from B intersects the extension of line EC at point I. Angle α is the difference between azimuth EC and azimuth CB. The length of line IB is $CB \sin \alpha$; its azimuth is 90° less than the azimuth of EC.

Substituting the coordinates of points C and B into Eqs. (9.1) and (9.2) for line CB,

$$\text{Azimuth}_{CB} = \tan^{-1} \frac{994.79 - 1330.14}{1244.80 - 1681.62}$$

$$= \tan^{-1} \frac{-335.35}{-436.82} = 217°30'49.4''$$

$$\text{Length}_{CB} = \sqrt{335.35^2 + 436.82^2} = 550.701 \text{ ft}$$

Substituting the coordinates of points E and C into Eq. (9.1) for line EC,

$$\text{Az}_{EC} = \text{Az}_{CI} = \tan^{-1} \frac{1330.14 - 2055.46}{1681.62 - 2092.68}$$

$$= \tan^{-1} \frac{-725.32}{-411.06} = 240°27'30.6''$$

$$\text{Angle } \alpha = \text{azimuth}_{CI} - \text{azimuth}_{CB} = 240°27'30.6'' - 217°30'49.4'' = 22°56'41.2''$$

$$\text{Perpendicular}_{IB} = CB \sin \alpha$$

$$= 550.701 \sin 22°56'41.2'' = 214.687 \Rightarrow\Rightarrow 214.69 \text{ ft}$$

$$\text{Azimuth}_{IB} = \text{azimuth}_{CI} - 90° = 240°27'30.6'' - 90° = 150°27'30.6''$$

Note that this very straightforward solution provides the same results as the more esoteric solution involving Eqs. (9.6) and (9.7).

9.3 MISSING DATA

At times, survey parties are unable to make all required angle and distance observations in the course of a traverse survey. It also happens that part of the record of an earlier survey is missing. If it isn't feasible to observe omitted values or to reobserve lost values, reconstruction of the missing portion is possible as long as no more than two elements are missing. The missing elements can be directions, distances, or one of each.

Always, in missing-data computations, assume that the values of known data are known precisely; for example, that a listed azimuth of 246°47′ is exactly 246°47′00.0″ or that a listed distance of 609.38 ft is exactly 609.3800 ft.

There are three variations of the two-missing-element situation:

Case 1. Two elements missing on a common side
Case 2. One missing element on each of two adjacent sides
Case 3. One missing element on each of two nonadjacent sides

CASE 1—BOTH ELEMENTS MISSING ON A COMMON SIDE. The only possible combination is one distance and one direction. The problem is solved readily by determining the coordinates of the two ends of the missing side, then computing an inverse to find the missing length and the missing direction.

Example 9.6. In Fig. 9.9, the coordinates of Hill and Dale are known, as well as the lengths and azimuths of Hill-Alfa, Alfa-Beta and Beta-Gama, but no data exist for Gama-Dale. Compute the length and azimuth of the missing Gama-to-Dale leg.

Solution. The coordinates of Gama are determined by computing successive latitudes and departures along the Hill-Alfa-Beta-Gama traverse and adding their values successively to the coordinates of Hill, Alfa, and Beta, as shown in the last four columns of Fig. 9.10. The x- and y-coordinate differences from Gama to Dale are then used to compute the Gama-to-Dale inverse (bottom of Fig. 9.10).

Although the data for Example 9.6 were taken from Fig. 8.10, the Gama-to-Dale coordinate differences obtained above do not match those computed in Chap. 8 from the Fig. 8.10 data. This is because the Gama-to-Dale coordinate differences in the Chap.8 computation came from *adjusted* latitudes and departures while the missing-data computation of Example 9.6 defines "known" distances and directions—and therefore computed latitudes and departures—as exact. The result of the assumption that all known data are exact is that cumulative random errors involved in the observation of the "known" elements must all end up in the solved-for missing elements.

CASE 2A—LENGTHS OF ADJACENT SIDES BOTH MISSING. This fairly common situation involves angles taken from two known stations to a third station over which it is not possible to set up an instrument. A triangle is formed in which both elements of one side are known, but only the directions are known on the other two sides.

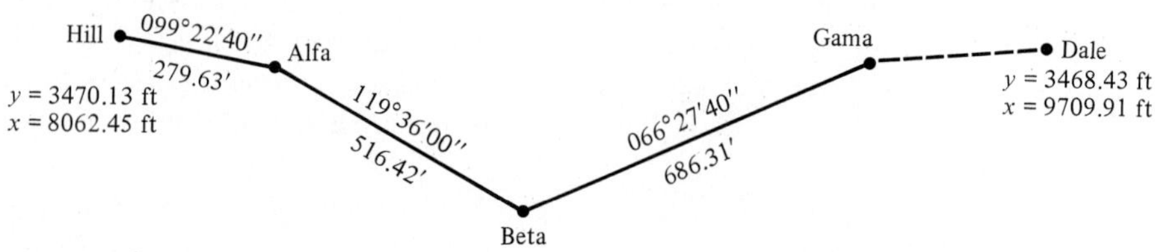

FIGURE 9.9
Distance and direction missing on one traverse leg.

					Coordinates	
Sta.	Length, ft	Azimuth	Latitude, ft	Departure, ft	y, ft	x, ft
Hill					3470.130	8062.450
	279.63	099°22'40"	− 45.564	+275.893		
Alfa					3424.566	8338.343
	516.42	119°36'00"	−255.081	+449.025		
Beta					3169.485	8787.368
	686.31	066°27'40"	+274.093	+629.202		
Gama					3443.578	9416.570
	———	———				
Dale					3468.430	9709.910
			Coordinate differences, Gama to Dale:		+24.852	+293.340

Substituting y- and x-differences into Equations 9.1 and 9.2:

$$\text{Azimuth}_{\text{Gama to Dale}} = \tan^{-1}\frac{+293.340}{+24.852} = 085°09'26.7''$$

$$\text{Length}_{\text{Gama to Dale}} = \sqrt{293.340^2 + 24.852^2} = 294.391 \text{ ft}$$

FIGURE 9.10
Computing omitted length and direction, common side.

Example 9.7. In Fig. 9.11, known stations Park and Dock are intervisible and the top of a church steeple is visible from both. From the data in Fig. 9.11, compute the coordinates of Steeple.

Solution. The direction and length of Park-Dock are computed from an inverse. The angle at Steeple is obtained by subtracting the sum of angles at Park and Dock from 180°. Knowing all three angles and the length of Park-Dock, the lengths of the other two sides are computed from the law of sines. Azimuths Park-Steeple and Dock-Steeple are found by applying the angles at Park and Dock to the Park-Dock and Dock-Park azimuths. The coordinates of Steeple are computed, and checked, by applying the appropriate latitudes and departures to the coordinates of Park and Dock. For convenience, Park, Dock, and Steeple are designated *P*, *D*, and *S*, respectively, in the following computation.

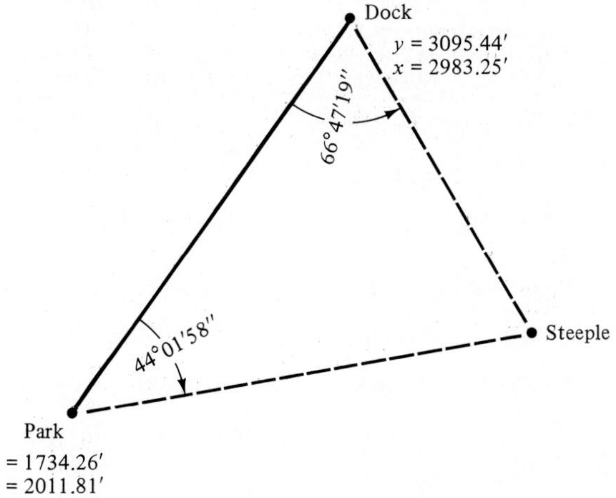

FIGURE 9.11
Missing distances on adjacent sides.

Substituting the coordinates of Park and Dock into Eqs. (9.1) and (9.2),

$$\text{Azimuth}_{PD} = \tan^{-1}\frac{2983.25 - 2011.81}{3095.44 - 1734.26}$$

$$= \tan^{-1}\frac{+971.44}{+1361.18} = 035°30'52.2''$$

$$\text{Azimuth}_{DP} = \text{Azimuth}_{PD} + 180° = 035°30'52.2'' + 180° = 215°30'52.2''$$

$$\text{Length}_{PD} = \sqrt{971.44^2 + 1361.18^2} = 1672.276 \text{ ft}$$

$$\angle S = 180° - (\angle P + \angle D) = 180° - (44°01'58'' + 66°47'19'') = 69°10'43''$$

From the law of sines:

$$\frac{DS}{\sin P} = \frac{PD}{\sin S}$$

$$\Rightarrow\Rightarrow \quad DS = \sin P\left(\frac{PD}{\sin S}\right)$$

Substituting,

$$DS = \sin 44°01'58''\left(\frac{1672.276}{\sin 69°10'43''}\right) = 1243.562 \text{ ft}$$

From the law of sines:

$$\frac{PS}{\sin D} = \frac{PD}{\sin S}$$

$$\Rightarrow\Rightarrow \quad PS = \sin D\left(\frac{PD}{\sin S}\right)$$

Substituting,

$$PS = \sin 66°47'19''\left(\frac{1672.276}{\sin 69°10'43''}\right) = 1644.301 \text{ ft}$$

$$\text{Azimuth}_{PS} = \text{azimuth}_{PD} + \angle P = 035°30'52.2'' + 44°01'58'' = 079°32'50.2''$$

Coordinates of steeple:

$$\begin{aligned} y_S = y_P + \text{latitude}_{PS} &= y_P + PS\cos(\text{azimuth}_{PS}) \\ &= 1734.26 + 1644.301\cos 079°32'50.2'' \\ &= 1734.26 + (+298.316) = 2032.58 \text{ ft} \\ x_S = x_P + \text{departure}_{PS} &= x_P + PS\sin(\text{azimuth}_{PS}) \\ &= 2011.81 + 1644.301\sin 079°32'50.2'' \\ &= 2011.81 + (+1617.014) = 3628.82 \text{ ft} \end{aligned}$$

Checking the coordinate computation for Steeple,

$$\text{Azimuth}_{DS} = \text{azimuth}_{DP} - \angle D = 215°30'52.2'' - 66°47'19'' = 148°43'33.2''$$

$$\begin{aligned} y_S = y_D + \text{latitude}_{DS} &= y_D + DS\cos(\text{azimuth}_{DS}) \\ &= 3095.44 + 1243.562\cos 148°43'33.2'' \\ &= 3095.44 + (-1062.864) = 2032.58 \text{ ft} \ \checkmark \end{aligned}$$

$$x_S = x_D + \text{departure}_{DS} = x_D + DS \sin(\text{azimuth}_{DS})$$
$$= 2983.25 + 1243.562 \sin 148°43'33.2''$$
$$= 2983.25 + (+645.574) = 3628.82 \text{ ft}$$

The above missing-data problems were solved by use of coordinates; however, missing data on adjacent or nonadjacent sides may also be obtained from latitudes and departures without resorting to the use of coordinates, as demonstrated in Example 9.8.

CASE 2B—DIRECTIONS OF ADJACENT SIDES BOTH MISSING. Figure 9.12 illustrates a five-sided loop traverse with the lengths of all legs known but with the bearings of two adjacent legs missing.

Example 9.8. From the data in Fig. 9.12, solve for missing bearings *DE* and *CD*.

Solution. Latitudes and departures from *E* to *A*, *A* to *B*, and *B* to *C* are added to obtain *y*- and *x*-coordinate differences from *E* to *C*, which are used for the inverse computation of the azimuth and length of *EC*. The problem is thus reduced to triangle *ECD* with all sides known and the azimuth from *E* to *C* also known. The angles in the final triangle are solved using the law of cosines and the missing directions are then easily determined from the known azimuth and angles.

Finding $y_C - y_E$ by computing and adding latitudes from *E* to *C*,

$$y_C - y_E = \text{latitude}_{EA} + \text{latitude}_{AB} + \text{latitude}_{BC}$$
$$= EA \cos(\text{bearing}_{EA}) + AB \cos(\text{bearing}_{AB}) + BC \cos(\text{bearing}_{BC})$$
$$= -287.04 \cos 73°44' - 236.85 \cos 04°17' - 344.69 \cos 57°22'$$
$$= -80.402 - 236.188 - 185.878 = -502.468 \text{ ft}$$

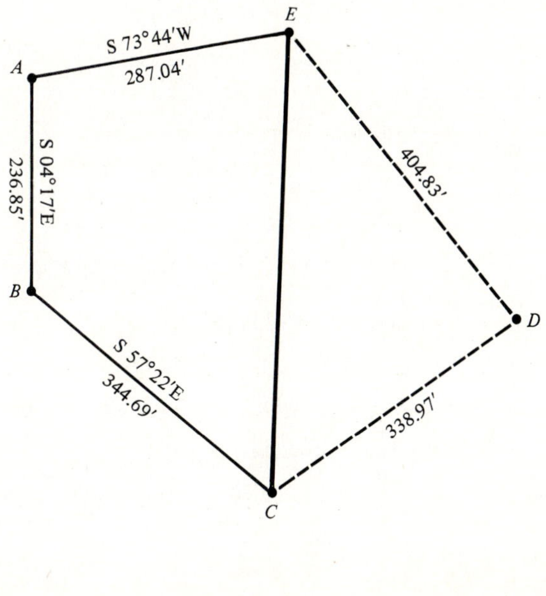

FIGURE 9.12
Missing directions on adjacent sides.

Finding $x_C - x_E$ by computing and adding departures from E to C,

$$
\begin{aligned}
x_C - x_E &= \text{departure}_{EA} + \text{departure}_{AB} + \text{departure}_{BC} \\
&= EA \sin(\text{bearing}_{EA}) + AB \sin(\text{bearing}_{AB}) + BC \sin(\text{bearing}_{BC}) \\
&= -287.04 \sin 73°44' + 236.85 \sin 04°17' + 344.69 \sin 57°22' \\
&= -275.549 + 17.690 + 290.277 = +32.418 \text{ ft}
\end{aligned}
$$

Substituting $x_C - x_E$ and $y_C - y_E$ into Eq. (9.1),

$$\text{Azimuth}_{EC} = \tan^{-1} \frac{+32.418}{-502.468} = -003°41'29.3''$$

With the x-coordinate difference positive and the y-coordinate difference negative, the azimuth is in the second quadrant; thus:

$$\text{Azimuth}_{EC} = 180° - 003°41'29.3'' \Rightarrow\Rightarrow 176°18'30.7''$$

$$\text{Azimuth}_{CE} = \text{azimuth}_{EC} + 180° = 356°18'30.7''$$

Substituting $x_C - x_E$ and $y_C - y_E$ into Eq. (9.2),

$$\text{Length}_{EC} = \sqrt{32.418^2 + 502.468^2} = 503.513 \text{ ft}$$

Considering *only* the final solution triangle ECD, whose sides are now known, the angles at E, C, and D are computed from the law of cosines demonstrated in Sec. 4.2:

$$
\begin{aligned}
\angle E &= \cos^{-1} \frac{ED^2 + EC^2 - CD^2}{2(ED)(EC)} \\
&= \cos^{-1} \frac{404.83^2 + 503.513^2 - 338.97^2}{2(404.83)(503.513)} \\
&= 42°05'39.2''
\end{aligned}
$$

$$
\begin{aligned}
\angle C &= \cos^{-1} \frac{EC^2 + CD^2 - ED^2}{2(EC)(CD)} \\
&= \cos^{-1} \frac{503.513^2 + 338.97^2 - 404.83^2}{2(503.513)(338.97)} \\
&= 53°11'14.0''
\end{aligned}
$$

$$
\begin{aligned}
\angle D &= \cos^{-1} \frac{CD^2 + ED^2 - EC^2}{2(CD)(ED)} \\
&= \cos^{-1} \frac{338.97^2 + 404.83^2 - 503.513^2}{2(338.97)(404.83)} \\
&= 84°43'06.8''
\end{aligned}
$$

Checking:

$$\angle E + \angle C + \angle D = 42°05'39.2'' + 53°11'14.0'' + 84°43'06.8'' = 180°00'00.0''$$

$$
\begin{aligned}
\text{Azimuth}_{CD} &= \text{azimuth}_{CE} + \angle C \\
&= 356°18'30.7'' + 53°11'14.0'' = 409°29'44.7'' \\
&= 049°29'44.7''
\end{aligned}
$$

$$\Rightarrow\Rightarrow \quad \text{Bearing}_{CD} = \text{N}49°29'44.7''\text{E}$$

$$\text{Azimuth}_{DE} = \text{azimuth}_{DC} + \angle D = (\text{azimuth}_{CD} + 180°) + \angle D$$

$$= 229°29'44.7'' + 84°43'06.8'' = 314°12'51.5''$$

$$\Rightarrow\Rightarrow \quad \text{Bearing}_{DE} = \text{N}45°47'08.5''\text{W}$$

Checking:

$$\text{Azimuth}_{EC} = \text{azimuth}_{ED} + \angle E = (\text{azimuth}_{DE} - 180°) + \angle E$$

$$= 134°12'51.5'' + 42°05'39.2'' = 176°18'30.7'' \checkmark$$

A final check on computed missing data in a closed-loop traverse is to insert the solved values back into the traverse computation, compute the missing latitudes and departures, then verify that the latitude and departure sums equal zero. Recall that $y_C - y_E$ was found earlier to be +32.418 ft and $x_C - x_E$ was found to be −502.468 ft. Computing sums of latitudes and departures from C to D and from D to E,

$$\text{Latitude}_{CD} + \text{latitude}_{DE} = CD\cos(\text{azimuth}_{CD}) + DE\cos(\text{azimuth}_{DE})$$

$$= 338.97\cos 049°29'44.7'' + 404.83\cos 314°12'51.5''$$

$$= +502.468 \text{ ft} \checkmark$$

$$\text{Departure}_{CD} + \text{departure}_{DE} = CD\sin(\text{azimuth}_{CD}) + DE\sin(\text{azimuth}_{DE})$$

$$= 338.97\sin 049°29'44.7'' + 404.83\sin 314°12'51.5''$$

$$= -32.418 \text{ ft} \checkmark$$

CASE 2C—ONE DIRECTION AND ONE LENGTH MISSING FROM ADJACENT SIDES. This situation, perhaps the most interesting of the three adjacent-side combinations, is seen more often in surveying texts than in the real world. It is solved by reducing the problem to a single triangle, as before, then solving that triangle using the law of sines. There is one pitfall: When the law of sines is used to solve for a missing *angle*, the solution is ambiguous. In polygon *ABCDE*, (Fig. 9.13), the direction of *DE* is known, but not its length, and the length of *AE* is known, but not its direction. Thus, *DE* may be considered as a line that originates at *D* and continues indefinitely in the *D*-to-*E* direction. Similarly, line *AE* may be represented as an arc of radius *AE* with its center at *A*. As such, radius *AE* intersects line *DE* at *two* points: *E* and *E'* in Fig. 9.13. The value appearing in the calculator window as the law of sines solution will be acute angle *E* rather than obtuse angle *E'* (the supplement of angle *E*), because that is what the calculator is designed to do. The surveyor, in solving the problem, must realize that the resultant angle can be either *E* or (180° − *E*). Distinguishing between the two requires additional information.

Example 9.9. Known data for polygon *ABCDE* are as shown in Fig. 9.14 and in tabular form below. It is also known that an on-site surveyor estimated that the length of side *DE* was a bit under 500 ft. From the measured and estimated data, solve for the exact distance *DE* and the azimuth of *EA*.

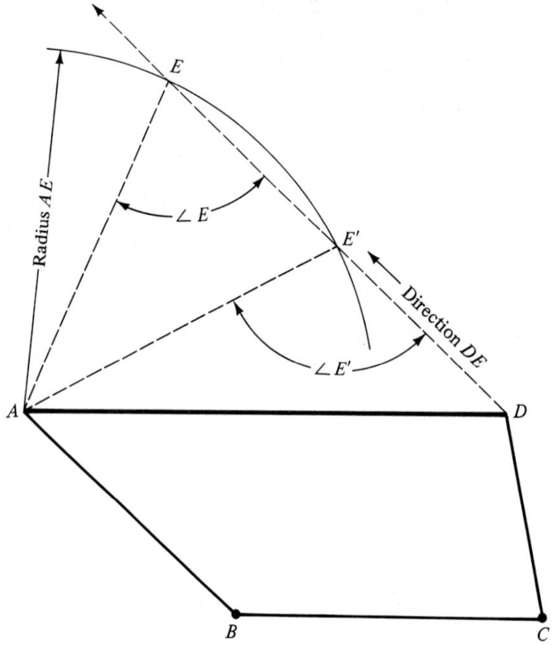

FIGURE 9.13
Law-of-sines ambiguity.

Side	Length, ft	Azimuth
AB	617.84	133°09′
BC	630.44	091°07′
CD	422.89	351°56′
DE	—	314°42′
EA	753.38	—

Solution. Compute the azimuth and distance of line *AD* by inverse computation, taking $x_D - x_A$ and $y_D - y_A$ as the sums of the longitudes and latitudes, respectively, from *A* to *D*. These values are computed in tabular format below.

Side	Length, ft	Azimuth	Latitude, ft	Departure, ft
AB	617.84	133°09′	−422.547	+450.755
BC	630.44	091°07′	−12.286	+630.320
CD	422.89	351°56′	+418.706	−59.342
DE	—	314°42′		
EA	753.38	—		
			Σ = −16.127	+1021.733
			$(y_D - y_A)$	$(x_D - x_A)$

Substituting into Eq. (9.1),

$$\text{Azimuth}_{AD} = \tan^{-1}\frac{+1021.733}{-16.127} = -89°05'44.6''$$

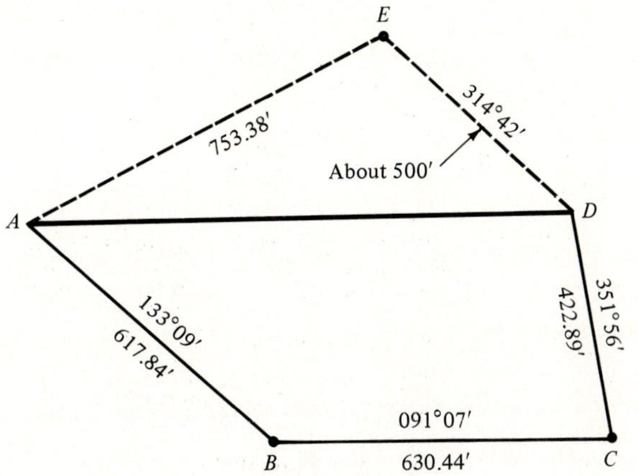

FIGURE 9.14
One missing length and one missing direction on adjacent sides.

Because $x_D - x_A$ is positive and $y_D - y_A$ is negative, azimuth$_{AD}$ is in the second quadrant; thus azimuth$_{AD}$ = 180° − 89°05′44.6″ = 090°54′15.4″ and azimuth$_{DA}$ = azimuth$_{AD}$ + 180° = 270°54′15.4″. Substituting into Eq. (9.2),

$$\text{Length}_{AD} = \sqrt{1021.733^2 + 16.127^2} = 1021.860 \text{ ft}$$

Going to the final solution triangle DAE,

$$\angle D = \text{azimuth}_{DE} - \text{azimuth}_{DA}$$

$$= 314°42'00.0'' - 270°54'15.4'' = 43°47'44.6''$$

From the law of sines:

$$\frac{\sin E}{AD} = \frac{\sin D}{AE}$$

$$\sin E = \left(\frac{AD}{AE}\right) \sin D$$

$$E = \sin^{-1}\left[\left(\frac{AD}{AE}\right) \sin D\right]$$

Substituting,

$$E = \sin^{-1}\left[\left(\frac{1021.860}{753.38}\right) \sin 43°47'44.6''\right] = 69°50'20.1''$$

From the above data the value of E could be *either* 69°50′20.1″ *or* its supplement, which is 180° − 69°50′20.1″ = 110°09′39.9″. Since side DE was estimated to be something under 500 ft in length, it is apparent from the side dimensions in Fig. 9.14 that the obtuse angle is the proper angle; that is, E = 110°09′39.9″, from which now comes

$$\text{Azimuth}_{EA} = \text{azimuth}_{ED} + \angle E$$

$$= 134°42'00.0'' + 110°09'39.9'' = 244°51'39.9''$$

Now, $\angle A$ must be $180° - \angle E - \angle D$. Substituting,

$$\angle A = 180°00'00.0'' - 110°09'39.9'' - 43°47'44.6'' = 26°02'35.5''$$

From the law of sines:

$$\frac{DE}{\sin A} = \frac{AE}{\sin D}$$

$$DE = \left(\frac{AE}{\sin D}\right) \sin A$$

Substituting,

$$DE = \left(\frac{753.38}{\sin 43°47'44.6''}\right) \sin 26°02'35.5'' = 477.930 \text{ ft}$$

A simple check of the computed values for the length of *DE* and the direction of *AE* is to substitute back into the tabular calculation of traverse *ABCDE* as follows:

Side	Length, ft	Azimuth	Latitude, ft	Departure, ft
AB	617.84	133°09′	−422.547	+450.755
BC	630.44	091°07′	−12.286	+630.320
CD	422.89	351°56′	+418.706	−59.342
DE	(477.930)	314°42′	(+336.173)	(−339.712)
EA	753.38	(244°51′39.9″)	(−320.047)	(−682.020)
		Σ =	−0.001	+0.001

The minor discrepancies in the sums of the latitudes and departures are the result of cumulative rounding error.

CASE 3—ONE MISSING ELEMENT ON EACH OF TWO NONADJACENT SIDES. Once a graphic manipulation is performed on the original polygon, Case 3 turns out to be nearly identical to Case 2, as shown in Fig. 9.15. Figure 9.15*a* diagrams the basic Case 3 situation, the two dashed lines indicating that nonadjacent sides *CD* and *EA* are each missing one element.

In Fig. 9.15*b*, parallelogram *CDEE′* has been added, side *CE′* being parallel to fully known side *DE* and side *E′E* parallel to partially known side *CD*. From geometry: (1) side *CE′* must have the same length and direction as fully known side *DE* and (2) side *E′E* must have the same length and direction as partially known side *CD*.

In Fig. 9.15*c*, sides *CD* and *DE* have been eliminated and line *AE′* has been inserted to produce the final solution triangle *AE′E*, in which side *E′E* has the length and direction of partially known side *CD* and side *EA* is the original, partially known side *EA*. The solution triangle is shown by itself in Fig. 9.15*d*.

Given the final solution triangle, the Case 3 problem is solved in the same manner as the Case 2 solutions illustrated above, depending on whether the missing elements are both lengths, both directions, or one of each.

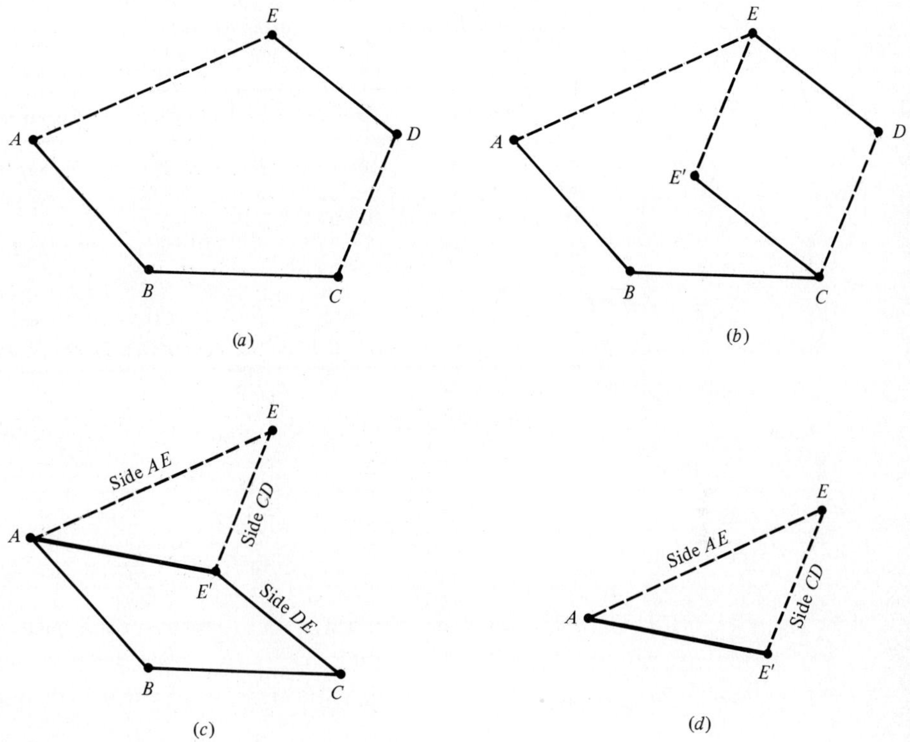

FIGURE 9.15
Converting a nonadjacent-side missing-data case to an adjacent-side case.

Example 9.10. Figure 9.16 shows the Fig. 9.15*a* situation with known elements labeled. Compute the lengths of sides *CD* and *EA*.

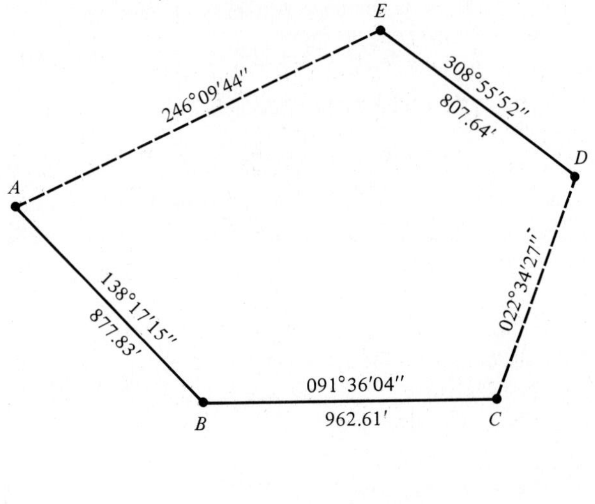

FIGURE 9.16
Data for Example 9.10.

Solution. The original pattern is reduced to final-triangle form by computing E'-to-A latitude and departure differences as follows:

Side	Length, ft	Azimuth	Latitude, ft	Departure, ft
AB	877.83	138°17′15″	−655.294	+584.102
BC	962.61	091°36′04″	−26.896	+962.234
CD	—	022°34′27″		
DE	807.64	308°55′52″	+507.509	−628.265
EA	—	246°09′44″		
			Σ = −174.681	+918.071
			$(y_{E'} - y_A)$	$(x_{E'} - x_A)$

Substituting into Eq. (9.1),

$$\text{Azimuth}_{AE'} = \tan^{-1}\frac{x_{E'} - x_A}{y_{E'} - y_A}$$

$$= \tan^{-1}\frac{+918.071}{-174.681} = -79°13'37.7''$$

With $x_{E'} - x_A$ positive and $y_{E'} - y_A$ negative, the azimuth is in the second quadrant; thus:

$$\text{Azimuth}_{AE'} = 180° - 79°13'37.7'' = 100°46'22.3''$$

$$\text{Azimuth}_{E'A} = 180° + \text{azimuth}_{AE'} = 280°46'22.3''$$

Substituting into Eq. (9.2),

$$\text{Length}_{AE'} = \sqrt{918.071^2 + 174.681^2} = 934.5415 \text{ ft}$$

$$\angle A = \text{azimuth}_{AE'} - \text{azimuth}_{AE}$$

$$= 100°46'22.3'' - 066°09'44'' = 34°36'38.3''$$

$$\angle E' = (\text{azimuth}_{CD} + 360°) - \text{azimuth}_{E'A}$$

$$= 382°34'27'' - 280°46'22.3'' = 101°48'04.7''$$

$$\angle E = \text{azimuth}_{EA} - \text{azimuth}_{CD}$$

$$= 246°09'44'' - 202°34'27'' = 43°35'17''$$

$$\text{Sum of} \angle \text{ s} = 34°36'38.3'' + 101°48'04.7'' + 43°35'17'' = 180°00'00.0'' \checkmark$$

From the law of sines:

$$\frac{EA}{\sin E'} = \frac{AE'}{\sin E}$$

$$EA = \left(\frac{AE'}{\sin E}\right) \sin E'$$

$$= \left(\frac{934.5415}{\sin 43°35'17''}\right) \sin 101°48'04.7'' = 1326.801 \text{ ft}$$

and

$$\frac{CD}{\sin A} = \frac{AE'}{\sin E}$$

$$CD = \left(\frac{AE'}{\sin E}\right) \sin A$$

$$= \left(\frac{934.5415}{\sin 43°35'17''}\right) \sin 34°36'38.3'' = 769.892 \text{ ft}$$

Again, the appropriate check is to insert the two computed lengths into the tabular format and compute latitude and departure totals:

Side	Length, ft	Azimuth	Latitude, ft	Departure, ft
AB	877.83	138°17'15"	−655.294	−584.102
BC	962.61	091°36'04"	− 26.896	+962.234
CD	(769.892)	022°34'27"	(+710.905)	(+295.545)
DE	807.64	308°55'52"	+507.509	−628.265
EA	(1326.801)	246°09'44"	(−536.225)	(−1213.616)
		Σ =	−0.001	0.000

The other two variations of Case 3 are solved in the same manner as just illustrated: by reducing the original figure to the final solution triangle, then concluding the computation as illustrated in the comparable Case 2 solutions.

9.4 RECENT DEVELOPMENTS IN COORDINATE USE

As indicated in the introduction to this chapter, it is advantageous to local governments if natural and man-made features are given *xyz*-coordinate "addresses" (the *z* coordinate referring to elevation above a chosen datum). If all such features have their addresses on the same coordinate system, and if those addresses are stored in a computer's files of coordinate data, it becomes a simple matter to determine, on command, the distance and direction between any two or more such features that are of interest. Coordinate-position records are particularly important for determining the positional relationships between structures—both existing and proposed—and what has come to be called the "infrastructure"; that is, existing and proposed rights-of-way of transportation routes and utilities, with much of the utility network lying underground.

Bringing this about requires (1) an address system that all parties agree on, (2) a convenient way of converting an existing address mode to the agreed-upon system and for storing and manipulating converted address data, and (3) a fast and relatively inexpensive procedure for determining coordinate locations with good precision. These three necessary adjuncts are now available. The first is the **state plane coordinate** system, which was developed in the 1930s. The second is the **geographic information system (GIS)**, a relatively new computer-based procedure that catalogs natural and man-made features in terms of their positions on a common frame of reference and stores the positional data. The third is the **global positioning system (GPS)**, which provides direct output of distances, azimuths, and positions with accuracies rivaling that achievable with the best currently available instruments and the most exacting methods. At this writing, GPS equipment costs somewhat more than a good total-station instrument but its use lets surveyors perform a number of field-positioning and distance-measuring operations at less cost than possible with conventional instruments and methods. In time, further development should bring down significantly the cost of GPS and, in so doing, broaden its application.

The remaining sections of this chapter provide an overview intended to give the reader a general understanding of what the three systems consist of and what may be expected of them. Detailed treatment is reserved for advanced surveying texts.

9.5 STATE PLANE COORDINATES

Primary horizontal control in the United States is established by the National Geodetic Survey of the National Oceanic and Atmospheric Administration. Horizontal control points are designated by their **geographic coordinates**—often called **geodetic coordinates**—which are based on **latitude** and **longitude**. "Latitude," in this context, means angular distance north or south of the equator, the equator being 0° latitude and the north and south poles being 90°N and 90°S, respectively. Longitude is angular distance east or west of the Greenwich meridian, a true north-south meridian passing through the observatory at Greenwich, England. The longitude at Greenwich is thus defined as 0°, with longitude values increasing both east and west of Greenwich until the 180° meridian is reached on the opposite side of earth from the Greenwich meridian. The conterminous United States lies roughly between 67°W and 124°W longitude and between 26°N and 49°N latitude.

Latitude and longitude are ideal for describing widely spaced points on our generally spherical earth (the earth actually approximates a nearly spherical ellipsoid). Plane surveyors, who work in small areas where the earth's shape may be ignored, find it simpler to use plane rectangular coordinates of the sort discussed at length in previous sections. In the 1930s, the U.S. Coast and Geodetic Survey (now the National Geodetic Survey) proposed separate state plane coordinate systems for each state that would express geodetic positions in the convenient plane coordinate terminology that local surveyors prefer to work with.

PROJECTIONS. The survey points we deal with lie on the earth's surface, which is essentially spherical. Representation of those points on a flat surface requires use of a **map projection**, which is a plane, cylindrical or conical surface onto which points on the earth's surface are mathematically "projected" from the center of the earth. Figure 9.17 is a cross section of a plane that intersects a spherical surface. Points A through F on the sphere have their projected counterparts A' through F' on the plane. The

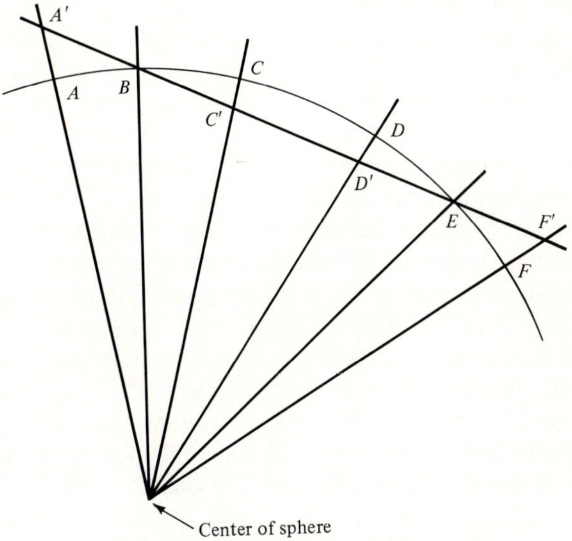

FIGURE 9.17
Projection, onto an intersecting plane, of points on a spherical surface.

two surfaces intersect at B and E, where points on the sphere coincide with projected points. Distance $A'B$ on the projection is obviously greater than true distance AB on the spherical surface; thus, the scale of line $A'B$ is larger than true scale. Distance $C'D'$ on the projection is obviously shorter than true distance CD, so the scale of line $C'D'$ is smaller than true scale.

Scale variation is the price one pays for having a spherical surface portrayed on a flat piece of paper or on a mathematical plane. Any projection will distort its representation of a spherical surface in a manner and to a degree that is dependent on the type of projection used and the area of the spherical surface involved. For mapping use, it is generally better to use a **conformal** projection; that is, one on which angles between short intersecting lines are shown correctly, which in turn makes small areas appear in their correct shape.

The National Geodetic Survey chose two conformal projections for the state plane coordinate systems: the **Lambert** and the **Transverse Mercator**. The former is a conical conformal projection whose cone axis coincides with earth's north-south axis; the latter is a cylindrical conformal projection whose cylinder axis lies in the equatorial plane.

Lambert Conformal Projection. A conical projection surface that is tangent to earth is called a **tangent projection**. Its scale is true only along the circle of tangency, becoming increasingly larger—and thus increasingly distorted—with greater distance from the tangent circle. To minimize scale distortion, a **secant projection** is used whose projection surface intersects the spherical surface along two circles of latitude, called **standard parallels**. In Fig. 9.17, areas on the projection lying between the standard parallels have scales smaller than true; areas beyond the standard parallels have projection scales larger than true. Thus, the overall area within a Lambert secant projection contains both too-small and too-large scales—too small between the parallels and too large beyond the parallels—but the maximum numerical scale distortion is kept to a minimum.

Figure 9.18 shows the National Geodetic Survey design for the Lambert secant projection. To hold scale distortion no worse than one part in 10,000, the north-south extent of a projection **zone** is held to a maximum of 158 miles (that distance is much exaggerated in Fig. 9.18). In actual state systems, usually comprising multiple zones, zone maxima are often less than 158 miles—sometimes much less—with corresponding reductions in maximum scale distortion.

Figure 9.19 illustrates a developed portion of the conic projection of Fig. 9.18, "developed" meaning that the conical surface has been unrolled and laid flat. Figure 9.19 diagrams the geometric basis for the rectangular coordinates on a Lambert projection. The coordinate origin is at lower left, so all x and y values within the boundaries of a particular state plane coordinate zone will be positive. True earth meridians are shown on the projection as straight lines that converge on point B, the apex of the Lambert conical surface. The **central meridian**, which lies a distance C from the y axis, is the only true meridian to parallel the y axis; all others deviate from the y axis by an angle θ. Parallels of latitude on earth are shown on the projection as concentric circles whose center lies at B. The distance from B to the latitude of point P is designated by radius R_p and the distance from B to the x axis is designated R_b.

The equations for computing Lambert state plane coordinates from geographic (latitude and longitude) coordinates are

$$x_p = C + R_p \sin\theta \tag{9.8}$$

$$y_p = R_b - R_p \cos\theta \tag{9.9}$$

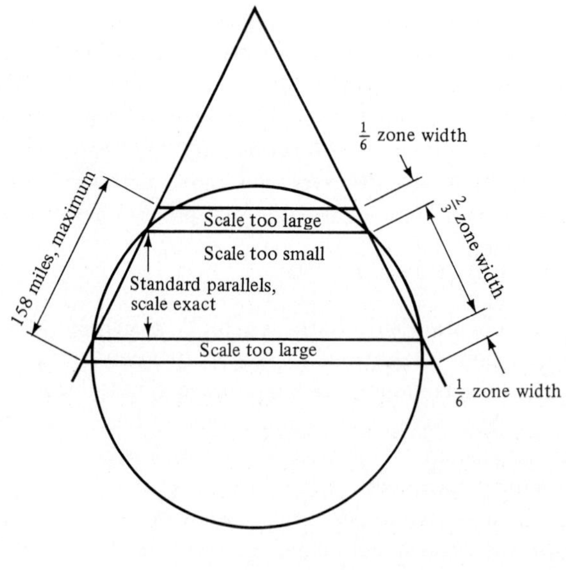

FIGURE 9.18
Lambert conformal projection.

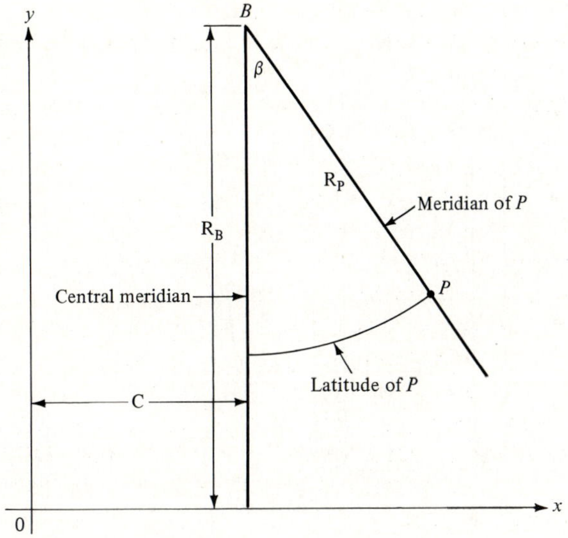

FIGURE 9.19
Geometry of the Lambert projection surface.

It was once necessary to undertake a tedious extrapolation of tables to find R and C values corresponding to latitudes and longitudes, then work through the computations with the above equations. The reader will be pleased to learn that programs are now available—for both desktop computers and hand-held computers such as the HP-41—that perform the entire task with a few keystrokes. This software, along with its additional capabilities, is discussed further along.

Transverse Mercator conformal projection. The National Geodetic Survey's transverse Mercator secant projection is in some ways quite similar to its Lambert secant projection. Figure 9.20 illustrates the secant form of the transverse Mercator projection, represented by a cylinder whose diameter is slightly smaller than the earth's and whose axis lies along the plane of the equator. Projection scale is true only along the two intersection circles. Scale is smaller between the circles and larger beyond the circles. For the same reason as for the Lambert projection, the National Geodetic Survey designated transverse Mercator projection zones whose east-west extent is limited to 158 miles.

Figure 9.21 shows a northern-hemisphere segment of the transverse Mercator projection in Fig. 9.20, with the coordinate origin at lower left to ensure that all x and y values are positive within the boundaries of the state plane coordinate zone. In Fig. 9.21, the central meridian is a north-south line but all other earth meridians are curves that converge at the north pole. Parallels of earth latitude are parabolas whose curvature increases as the pole is approached.

The x distance to the central meridian is x_0, usually a multiple of 100,000 ft. $\Delta\lambda''$ is difference in longitude, in seconds, between the central meridian (λ_m) and the longitude of point $P(\lambda_p)$, defined as $\Delta\lambda'' = \lambda_m - \lambda_p$. x'_p is the x distance

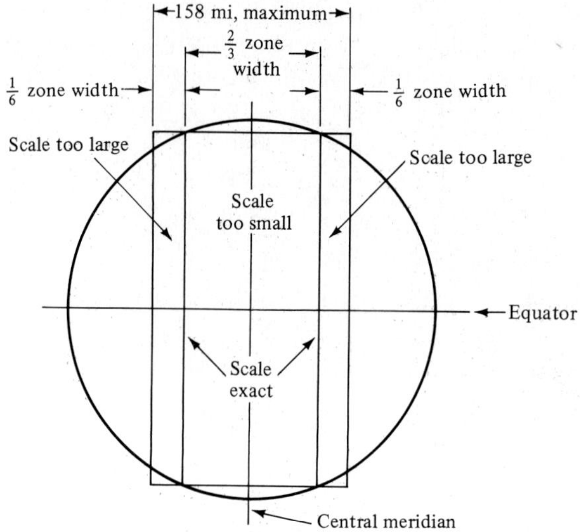

FIGURE 9.20
Transverse Mercator conformal projection.

from the central meridian to point *P*; positive if east of the central meridian, negative if west. $\Delta\alpha$ is the angle between grid north and the true meridian at a given point. The equations for computing transverse Mercator state plane coordinates from geographic coordinates are

$$x'_p = H\Delta\lambda'' \pm ab \tag{9.10}$$

$$x_p = x_0 + x'_p \tag{9.11}$$

$$y_p = y_0 + V\left(\frac{\Delta\lambda}{100}\right)^2 + c \tag{9.12}$$

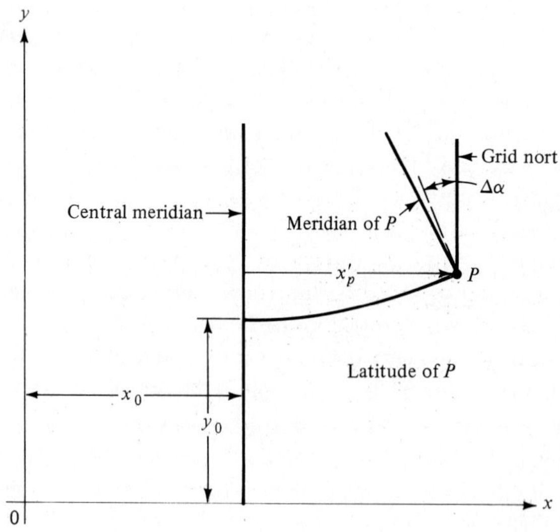

FIGURE 9.21
Geometry of the transverse Mercator projection surface.

where the values of H, V, y_0, and a are based on longitude while b and c are based on latitude. As is the case with Lambert state plane coordinate values, transverse Mercator state plane coordinate values no longer require tedious interpolation of tables; they may be quickly determined with desktop computers or with advanced-design hand-held computers, using software that is readily available.

GRID AZIMUTHS. As seen in Figs. 9.19 and 9.21, true meridians converge on the poles so they must cross the north-south grid lines on either state plane coordinate projection. On the Lambert projection, the crossing angle is θ (approximately); on the transverse Mercator, it is $\Delta\alpha$. For the Lambert projection:

$$\textbf{Grid azimuth} = \textbf{true azimuth} - \boldsymbol{\theta} + \textbf{second term} \tag{9.13}$$

where θ is positive if taken east of the central meridian, negative if west of the central meridian. The second term, available in tables or in conversion software, may be neglected for lines shorter than 5 miles.

For the transverse Mercator projection:

$$\textbf{Grid azimuth} = \textbf{true azimuth} - \boldsymbol{\Delta\alpha''} - \textbf{second term} \tag{9.14}$$

$$\boldsymbol{\Delta\alpha''} = \boldsymbol{\Delta\lambda''} + \boldsymbol{g} \tag{9.15}$$

where g is a very small longitude-dependent value. The second term in Eq. (9.14), whose value is included in conversion software along with that of g, may be neglected for lines under 5 miles long.

GROUND DISTANCE VERSUS GRID DISTANCE. As indicated previously, there are almost invariably differences, often significant, between distances measured on the earth's surface and corresponding distances on projection surfaces. To translate from one to the other requires an intermediate conversion from ground distance to **geoidal distance** (the **geoid** being the earth's mean-sea-level reference surface), then from geoidal distance to **grid distance** on the projection surface. However, the Lambert and transverse Mercator projections are both referred to the **ellipsoid**, which is an ellipse of revolution closely approximating the geoid; thus, grid distance derives from, first, the conversion from ground distance to the **ellipsoidal distance**, then from ellipsoidal distance to grid distance. For ordinary surveying purposes, the newest ellipsoid, called the **geodetic reference system of 1980**, or **GRS 80**, matches the geoid closely enough that geoidal (sea-level) distances may be assumed the same as ellipsoidal distances and, for grid-distance conversion purposes, elevations of points above sea level may be assumed the same as elevations above the GRS 80 ellipsoid. Figure 9.22 shows the relationship between ground distance, ellipsoidal/sea-level distance, and grid distance.

Figure 9.22 is a cross section of the earth showing ground points A and B, corresponding points A' and B' on the geoidal/GRS 80 surface, and the grid positions of A and B at A'' and B'' on the projection surface. If the mean elevation of line

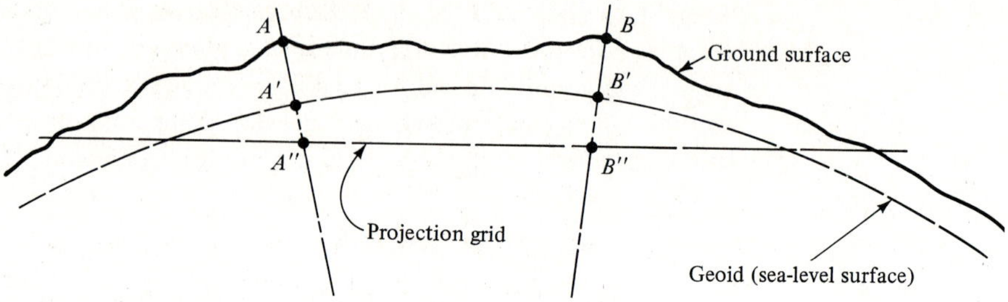

FIGURE 9.22
Ground length, sea-level length, and grid length.

AB above sea level is h, and its horizontal measured distance is H_m, the sea-level distance H_s is obtained from

$$H_s = H_m \left(\frac{R_s}{R_s + h} \right) \tag{9.16}$$

where R_s is the mean radius of the earth and the expression in parentheses is called the **sea-level factor**. Grid distance D_g is then determined from

$$D_g = H_s \times \textbf{scale factor} \tag{9.17}$$

where **scale factor** is the linear distortion imposed on sea-level distance in the process of projecting it on the surface of the state plane. As discussed earlier, scale factor varies throughout a state plane coordinate zone, its value at any point being a function of the location of the point, and its value between any two points being a function of the locations of those points.

Sea-level factor and scale factor may be multiplied together to produce a single **grid factor** (grid factor = sea-level factor × scale factor) so that

$$D_g = H_m \times \textbf{grid factor} \tag{9.18}$$

Until recently, scale factor was interpolated from tables; now, it comes with the same software package that is used to determine grid coordinates from geographic coordinates (and vice versa).

STATE PLANE COORDINATE CONVERSION SOFTWARE. The National Geodetic Survey has recently developed the state plane coordinate system of 1983 (SPCS 83), based on the GRS 80 ellipsoidal surface and a new sea-level datum, the North American Datum of 1983 (NAD 83). States are now enacting legislation to switch over to SPCS 83. As of this writing, 26 states have implemented SPCS 83, 10 more have drafted legislation to do so, and 7 others have begun prelegislation studies.

Computer software for converting ground positions, directions, and distances to SPCS 83 counterparts (and vice versa) is available from the NGS as follows:

Item SPCS83, for desktop computers, provides all conversions, for either the Lambert or the transverse Mercator projection system, for any state zone, but for SPCS 83 only. State plane coordinates are entered or computed to 1-mm precision; latitudes and longitudes are entered or computed to 0.3-mm precision.

Item 42–41, for HP-41CV computers, provides the same SPCS 83 conversions, but just for those states using the Lambert projection.

Item 54–41 also works with the HP-41CV, but for just those states using the transverse Mercator projection.

Items 43–41 and *49–41*, also for the HP-41CV, cover Alaska's three-projection system.

There is also HP-41CV software available for use with the older state plane coordinates, based on the North American Datum of 1927 (NAD 27). Item *3/4-41* is for states using transverse Mercator, item *5/6-41* is for states using Lambert, while items *7-41, 8-41*, and *9/10-41* are for the Alaska system.

The above software may be obtained by writing or calling:

National Geodetic Information Branch
N/CG174, Rockwall Building, Room 24
National Geodetic Survey, NOAA
Rockville, Maryland 20852
Telephone: 1-301-443-8631

SYSTEMS USED BY INDIVIDUAL STATES. Figure 9.23 lists states by projection type. Two states use both the Lambert and the transverse Mercator. Alaska uses three projections, including the oblique Mercator◇. Some states that have enacted SPCS 83 legislation have at the same time enacted changes in their zone boundaries, number of zones, or zone constants. Such states are denoted in Fig. 9.23 with an asterisk(*). Numbers in the projection columns indicate the number of projection zones in a state's system as of this writing.

State	Projection	State	Projection	State	Projection
Alabama	T.M. 2	Kansas	Lam. 2	North Carolina	Lam. 2
Alaska	O.M. 1◇	Kentucky	Lam. 2	North Dakota	Lam. 2
"	T.M. 8	Louisana	Lam. 3	Ohio	Lam. 2
"	Lam. 1	Maine	T.M. 2	Oklahoma	Lam. 2
Arizona	T.M. 3	Maryland	Lam. 1	Oregon	Lam. 2
Arkansas	Lam.2	Massachusetts	Lam. 2	Pennsylvania	Lam. 2
California*	Lam. 6	Michigan	Lam. 3	Rhode Island	T.M. 1
Colorado	Lam. 3	Minnesota	Lam. 3	South Carolina*	Lam. 1
Connecticut	Lam. 1	Mississippi	T.M. 2	South Dakota	Lam 2
Delaware	T.M. 1	Missouri	T.M. 3	Tennessee	Lam. 1
Florida	T.M. 2	Montana*	Lam. 1	Texas	Lam. 5
"	Lam. 1	Nebraska*	Lam. 1	Utah	Lam. 3
Georgia	T.M. 2	Nevada	T.M. 3	Vermont	T.M. 1
Hawaii	T.M. 5	New Hampshire	T.M. 1	Virginia	Lam. 2
Idaho	T.M. 3	New Jersy	T.M. 1	Washington	Lam. 2
Illinois	T.M. 2	New Mexico	T.M. 3	West Virginia	Lam. 2
Indiana	T.M. 2	New York	T.M. 3	Wisconsin	Lam. 3
Iowa	Lam. 2	" "	Lam. 1	Wyoming	T.M. 4

FIGURE 9.23
State plane coordinate systems in the United States.

Except in the case of Alaska, individual zones within state systems are aggregations of counties, grouped east-west for a Lambert zone and grouped north-south for a transverse Mercator zone. Thus, zone boundaries are irregular to the same extent as are state boundaries and the boundaries of peripheral counties within state zones.

UNIVERSAL TRANSVERSE MERCATOR PROJECTION. The **universal transverse Mercator** (UTM) system is intended as a common, worldwide coordinate system. It covers the earth's 360 degrees of longitude with 60 adjacent 6-degree-wide transverse Mercator zones. The geometry of the zones is the same as that of the U.S. transverse Mercator system but coordinate values are expressed in meters. Its chief advantage is that it permits a commonality between countries, though this is diminished somewhat in that different reference ellipsoids are used for different countries or groups of countries. Another problem is the broad spacing of its intersection circles, resulting in scale distortions of 1:2500 at the central meridian and as much as 1:1250 at the zone boundaries, as opposed to the worst-case 1:10,000 scale distortion in the U.S. state plane coordinate system.

At the time of conversion to the GRS 80 ellipsoid and the NAD 83 datum, the National Geodetic Survey considered switching the state plane coordinate system to the UTM system to achieve the advantage of a universal system with universal equations. After considerable study, the NGS decided to maintain the existing system because its scale factors change less rapidly than do those of the UTM system and because UTM zone boundaries do not coincide with state boundaries or with county-based zone boundaries.

9.6 GEOGRAPHIC INFORMATION SYSTEMS

CONCEPT. Ours might be characterized as the "information age," with an enormous availability of acquired data relating to our natural and man-made environments. Much of this data is "spatial," which means it is referenced to positions on the surface of the earth. For example, there is an abundance of spatial data on topography, agriculture, forests, wetlands, river systems, rangeland, parkland, watersheds, shorelines, transportation networks, urban boundaries, and the like. People involved in regional land-use planning or land management generally refer to these spatial information categories as **thematic data** and to their portrayal as **thematic mapping**. Thematic data must be considered concurrently to support rational decisions or to resolve conflicts between local and regional interests. With the introduction of the Landsat series of environmental satellites—whose multispectral scanners afforded a monumental capability for acquiring a variety of spatially oriented thematic data—it became possible to obtain, store, and portray a number of up-to-date thematic "layers," each keyed to a common geographic reference. Figure 9.24 is a schematic illustration of the concept of thematic layering, many of the layers being derivable from Landsat digital data.

Spatial thematic data, from whatever source, may be stored in computers and called up later in thematic-map form, either as a video display or as a color map printed on an automated plotter. Systems that store, recall, manipulate, and display spatial thematic data are called **geographic information systems (GIS)**. GIS data

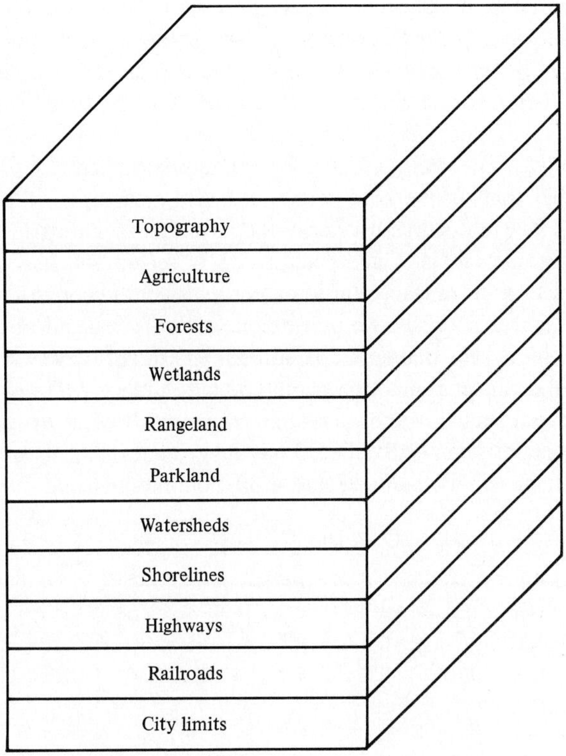

FIGURE 9.24
Thematic layering on a typical regional GIS.

are not really "layered" within the computer, as indicated in Fig. 9.24. Instead, each element of thematic data has a coordinate address, and for any given address there is a corresponding thematic data cell for each of the thematic categories.

LOW-PRECISION GIS. Most of the earlier systems, intended for large-area analysis, involved data acquired at small scale over large areas; thus, positional accuracy was, and still is, relatively low. The ground size of a Landsat data unit, called a "pixel" (for "picture element"), was originally the size of a football field—though current Landsat technology has reduced pixel size to about that of the average residential building lot. Most non-Landsat thematic data used in regional GISs is taken from maps whose scale is on the order of 1:24,000, with a locational precision of around 20 ft at best.

HIGHER-PRECISION GIS. Most of us have seen what happens when machinery digging a trench for a new utility line severs a telephone cable or another utility line. There is a certain Laurel and Hardy quality, too, about TV cable installers accidentally trenching through a phone line, then the phone company repair crew cutting the TV cable when they come to repair the phone line, and so forth, until someone finally takes a bit more care. Its entertainment value aside, this scenario can and should be prevented. What is needed is a GIS of another sort: one that layers buildings, property lines, utilities, streets, zoning and flood-plain boundaries, pipelines, and all the other interrelated appurtenances of civilization with which someone must deal, in one way or another, and fairly frequently at that. It has recently come to pass that the GIS concept is being taken up by local interests, usually municipal governments, who need to keep track of the absolute and relative locations of the underground infrastructure, topography, buildings, streets, property lines, geodetic control, and similar relevant data. The need is for thematic data whose addresses are much more precisely determined than is the case with the broad-area regional GIS. Figure 9.25 shows schematically the thematic layering appropriate for an urban GIS, with its rather different thematic labels. Once such a GIS is set up and its coordinate-addressed data are in place, it becomes a simple matter to select the corner coordinates of any project area that is to be worked on or studied, then to recall all of that area's thematic data to determine where potential spatial conflicts might occur.

The earlier regional GISs could be put in place with only the most casual regard for positional accuracy and no real need to consult with surveyors. On the other hand, high-precision urban GISs should have the input of local surveyors, both public and private. Increasingly, as more and more communities and metropolitan areas develop

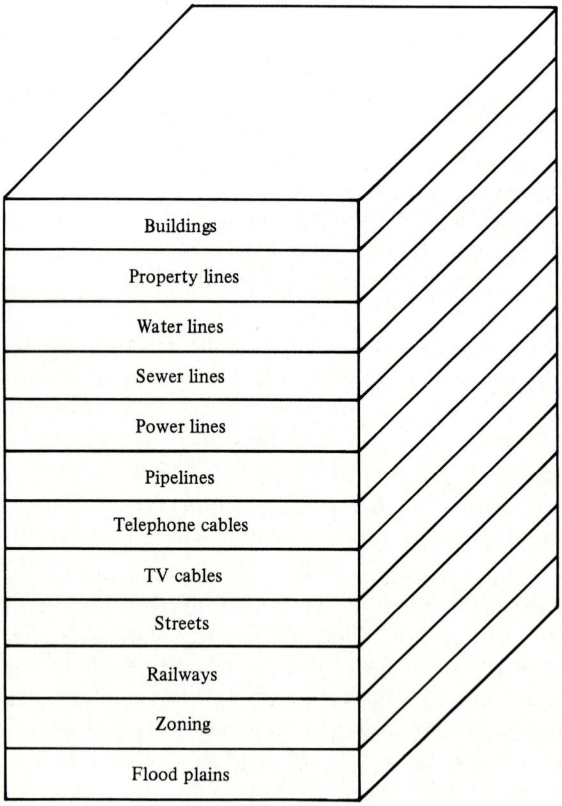

FIGURE 9.25
Thematic layering on a typical urban GIS.

their own GIS, surveyors must learn to provide coordinate addresses for all positions they determine within a given project area. Of necessity, every category of thematic data used in an urban GIS must use the same coordinate address system. The state plane coordinate system is most appropriate for use in an urban GIS: The system is in place, it is understood by most surveyors, and, when incorporating a thematic layer of geodetic control points, it readily translates between rectangular coordindates and latitude-longitude coordinates.

9.7 GLOBAL POSITIONING SYSTEM

CONCEPT. The global positioning system (GPS), mentioned earlier, is a military satellite positioning system with a rapidly developing set of civilian applications. Its initial operational plan called for 18 GPS satellites in six separate orbital planes, with three satellites evenly spaced about each orbit. At this writing it appears likely that there will be four satellites in each orbit, for a total of 24. The orbital planes are inclined 60° to the equatorial plane and the satellites orbit at an altitude of 20,200 km, which provides a 12-hour orbital period.

Each satellite broadcasts a coded signal on two carrier frequencies: L1 at 1575.42 MHz, L2 at 1227.60 MHz. The L1 carrier frequency is modulated at 293 m and 29.3 m; the L2 carrier at 29.3 m only. The 293-m modulation is called the C/A code or standard positioning service (SPS) and is available to all users; the 29.3-m modulation is called the P code or precise positioning service (PPS) and, as might be expected, is intended for Department of Defense users only. The codes permit ground receivers to lock onto the broadcast signal and to receive the exact transmission time of each modulation-frequency pulse. The received signals may then be used to determine either the absolute position of the receiver or its position relative to another receiver situated on a point of known position.

ABSOLUTE POSITION. With transmission time and reception time known, the instrument's computer can determine "pseudoranges" to each satellite on which the receiver is locked. Pseudorange r is satellite distance determined from

$$r = V_r \Delta t \tag{9.19}$$

where V_r is velocity of propagation of the satellite signal and Δt is time difference between transmission and reception. Pseudorange data from one satellite provide one equation with four unknowns: the x, y, and z coordinates of the receiver position, and the time bias, which is the lead or lag between the time kept on the satellite and the time kept in the receiver. Taking pseudoranges to four satellites provides four simultaneous equations that are solved to establish receiver position. If the ephemeris of each satellite is known (that is, the satellite path with respect to time), the position of the receiver may be computed with the same precision. The ephemeris is known to within about 20 m; thus position accuracy from C/A-code pseudoranges is about ±20 m.

The P code allows more precise phase matching from its 29.3-m modulation wavelength and it permits computation and reduction of ionospheric errors through use of two carrier frequencies. This in turn produces a more precise pseudorange measurement than the C/A code and a higher absolute position accuracy.

"Codeless" techniques have been developed, however, that unblock the P code, providing nonmilitary users with access to both the 29.3-m modulation wavelength and the L2 frequency. One of these techniques even provides phase measurement using the 20-cm carrier-frequency wavelength, with resulting subcentimeter measurement precision.

RELATIVE POSITION. Two receivers in reasonable proximity to one another, each locked on the same set of satellites at the same time, will obtain positions whose ephemeris errors are in the same amount and in the same direction. In this case, their relative positions will be quite precise; that is, the azimuth and distance from one to the other will be precisely known. If one receiver is set up over a known geodetic control point, the position of the other receiver may be determined with about the same accuracy as that of the known geodetic location. This method, called **translocation**, can provide distance measurements and positions to about 1 cm or better when codeless techniques are used.

Distance measurement with GPS does not require a clear line of sight between end points, only between the receiver antenna and the satellites. Similarly, GPS positioning by translocation does not require that the point of known position be visible from the point whose position is being determined. Since no time and effort is needed for clearing lines or for elevating the GPS instruments above obstructions, there is a significant cost saving that can override the relatively high cost of GPS equipment. Its use in many cases can result in lower costs per survey point compared to conventional means, particularly on surveys with moderate-to-long line lengths.

GPS technology is still fairly new. Its full constellation of 18 (or perhaps 24) satellites is not yet in orbit at this writing, due in part to the 1986 Challenger disaster and the subsequent delay in scheduled GPS satellite launches by the space shuttle program. Even with a partial suite of satellites, however, refinement of GPS techniques and instruments is an ongoing process. Some newer GPS equipment allows maintenance of a continuous lock on several satellites from a moving receiver. This technique, called **kinematic surveying**, permits distance measurement and position determination—with less-than-centimeter accuracy—more quickly and economically than with static methods. At present, with the existing satellite constellation, the static method requires about an hour to attain lock and to acquire a set of four-satellite position data. With the kinematic method, once the initial fix is obtained at the first station and lock is maintained as the receiver is moved, positions at other points may be acquired in a minute or less over station. With a full constellation it should be possible to routinely obtain static fixes in as little as 20 minutes and kinematic measurements in well under a minute. As is the case with static methods, kinematic surveying affords accuracies of a few millimeters, making its use economically feasible on relatively small survey areas.

FIGURE 9.26
Wild-Magnavox WM102 receiver and antenna in field use (*Courtesy of Surveying Division of Wild Leitz USA, Inc.*)

INSTRUMENTATION. Several manufacturers are currently offering GPS equipment. Small, lightweight instruments now available provide codeless, two-channel operation and a capability for locking onto five or more satellites simultaneously. Figure 9.26 shows the Wild-Magnavox WM102 system set up for use in the field, the receiver on the ground in front of the operator and the antenna mounted on a tripod over the survey point. The WM102 provides P-code access, including both the L1 and L2 carriers, along with multiple channels for simultaneous lockon with up to 12 satellites.

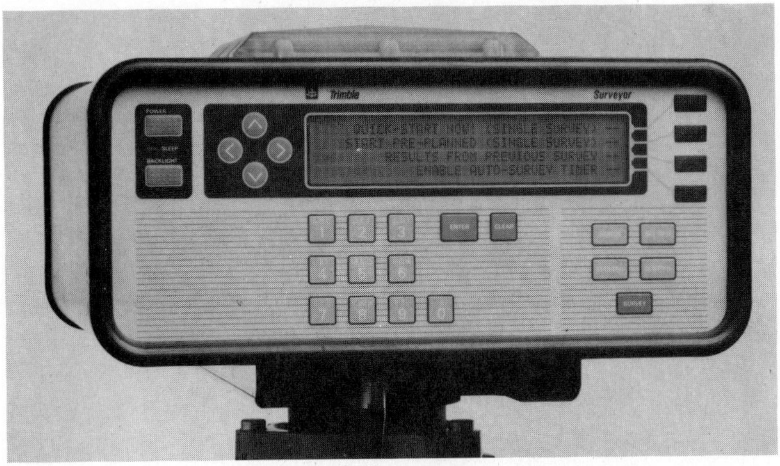

FIGURE 9.27
Trimble 4000ST receiver. (*Courtesy of Trimble Navigation, Ltd.*)

FIGURE 9.28
Ashtech XII receiver. (*Courtesy of Ashtech, Inc.*)

The Trimble 4000ST receiver (Fig. 9.27) can continuously track eight satellites on the L1 frequency with options for 12 channels of both L1 and L2 operation. An antenna is built into the top of the receiver/data-logger case, with plug-in for an external antenna for kinematic measurement. Data logging is internal. Optional software provides output in state plane coordinates.

The Ashtech XII receiver and antenna (Figs. 9.28 and 9.29) provide for simul-

FIGURE 9.29
Ashtech XII receiver and antenna in kinematic mode. (*Courtesy of Ashtech, Inc.*)

taneous satellite lockon with 12 channels on the L1 carrier frequency and 12 channels on L2. The instrument is available in two models: one with a self-contained computer with internal solid-state memory for data storage and one that operates in conjunction with an external laptop computer and records on the computer's diskette memory. Figure 9.29 shows single-operator kinematic operation, the receiver in a shoulder bag and the antenna carried on a pole extension to facilitate maintaining a clear line of sight to the satellites.

The GPS horizontal-position output for current instruments is normally displayed in units of latitude and longitude but software is available that converts from geographic coordinates to state plane coordinates. As already indicated, GPS survey techniques can be used for quickly and inexpensively extending geodetic control throughout an area of interest; this means GPS can be used to strengthen horizontal and vertical control for high-precision geographic information systems.

9.8 PROBLEMS

9.1. The coordinates of point A are $x = 1127.66$ ft and $y = 1547.63$ ft. The coordinates of point B are $x = 1215.08$ ft and $y = 1114.82$ ft. Compute the length, azimuth, and bearing of line AB.

9.2. The coordinates of point C are $x = 1609.77$ ft and $y = 3244.36$ ft. The coordinates of point D are $x = 1223.05$ ft and $y = 3513.44$ ft. Compute the length, azimuth, and bearing of line CD.

9.3. The coordinates of point F are $x = 34,419.22$ ft, $y = 11,624.51$ ft. The coordinates of point G are $x = 34,462.08$ ft, $y = 12,161.79$ ft. Compute the length, azimuth, and bearing of line FG.

9.4. The coordinates of point M are $x = 419,604.83$ ft, $y = 1,660,236.15$ ft. The coordinates of point N are $x = 419,182.12$ ft, $y = 1,660,066.29$ ft. Compute the length, azimuth, and bearing of line MN.

9.5. The coordinates of a TV tower are $x = 7249.47$ ft, $y = 4899.12$ ft. The coordinates of a smokestack are $x = 15,833.71$ ft, $y = 2659.46$ ft. Compute the distance, azimuth, and bearing from the TV tower to the stack.

9.6. The coordinates of transmission tower T_1 on one side of a river are $x = 5440.29$ ft, $y = 14,862.72$ ft. The coordinates of transmission tower T_2 on the other side of the river are $x = 5681.95$ ft, $y = 14,733.07$ ft. Compute the distance, azimuth, and bearing from tower T_1 to tower T_2.

9.7. The coordinates of point A are $x = 841.76$ ft, $y = 582.59$ ft. The coordinates of point Q are $x = 324.68$ ft, $y = 911.07$ ft. Compute the distance, azimuth, and bearing of line AQ.

9.8. The coordinates of point P are $x = 1127.89$ ft, $y = 708.92$ ft. The coordinates of point T are $x = 257.36$ ft, $y = 616.15$ ft. Compute the distance, azimuth, and bearing of line PT.

9.9. Referring to Probs. 9.7 and 9.8, compute the coordinates of the intersection of lines AQ and PT using the two-point line equation.

9.10. Solve Prob. 9.9 using the law of sines solution.

9.11. The xy coordinates (in that order) of four points are as follows: point A (12,532.08′, 8109.67′), point R (16,168.17′, 3922.44′), point S (3847.11′, 4229.82′), and point P

(5268.46′, 9707.48′). Using the two-point line equation, compute the coordinates of the intersection of line *AS* with line *RP*.

9.12. Solve Prob. 9.11 using the law of sines solution.

9.13. From the coordinates computed in Prob. 8.21, compute the deed lengths and bearings of the property lines for figure *ABCDE*.

9.14. From the coordinates computed in Prob. 8.22, compute the deed lengths and bearings of the property lines for figure *ABCDE*.

9.15. From the coordinates computed in Prob. 8.23, compute the deed lengths and bearings of the property lines for figure *ABCDE*.

9.16. From the coordinates computed in Prob. 8.24, compute the deed lengths and bearings of the property lines for figure *ABCDE*.

9.17. The coordinates of point *A* are $x = 2739.88$ ft and $y = 216.49$ ft. The coordinates of point *B* are $x = 1127.39$ ft and $y = 2481.70$ ft. The coordinates of point *P* are $x = 2011.69$ ft and $y = 882.41$ ft. Using the normal line equation, compute the perpendicular distance and direction from point *P* to a line passing through points *A* and *B*.

9.18. Solve Prob. 9.17 using the right-triangle solution.

9.19. Using the solution from either Prob. 9.17 or 9.18, determine the coordinates of the point of intersection of the line passing through points *A* and *B* with the perpendicular from point *P*.

9.20. The coordinates of point *P* are $x = 994.08$ ft and $y = 736.53$ ft. The coordinates of point *S* are $x = 1375.76$ ft and $y = 319.40$ ft. The coordinates of point *M* are $x = 943.11$ ft and $y = 867.32$ ft. Using the normal line equation, compute the perpendicular distance and direction from point *M* to a line passing through points *P* and *S*.

9.21. Solve Prob. 9.20 using the right-triangle solution.

9.22. Using the solution from either Prob. 9.20 or 9.21, determine the coordinates of the point of intersection of the line passing through points *P* and *S* with the perpendicular from point *M*.

9.23. Using the solution from Prob. 9.19, determine the distance from the point of intersection to point *B*.

9.24. Using the solution from Prob. 9.22, determine the distance from the point of intersection to point *P*.

9.25. A line bearing N62°19′43″W passes through point *K*, whose coordinates are $x = 2819.43$ ft and $y = 1666.97$ ft. Compute the perpendicular distance and direction to this line from point *L*, whose coordinates are $x = 1122.63$ ft and $y = 2315.70$ ft.

9.26. In Prob. 9.25, determine the coordinates of the point of intersection of the perpendicular from point *L* with the line passing through point *K* and determine the distance from the point of intersection to point *K*.

Compute the missing data in the closed polygons represented in Probs. 9.27 through 9.36.

9.27. Available measurements for clockwise figure *ABCD* are as follows:

Side	Length, ft	Bearing
AB	87.25	N49°58′45″E
BC	237.03	S66°32′17″E
CD	—	—
DA	158.52	N33°31′58″W

9.28. Available measurements for counterclockwise figure *ABCDE* are as follows:

Side	Length, ft	Bearing
AB	202.58	N73°35′35″W
BC	205.03	S67°56′41″W
CD	320.46	S20°32′54″E
DE	—	—
EA	211.05	N75°27′31″E

9.29. Available measurements for clockwise figure *ABCDE* are as follows:

Side	Length, ft	Bearing
AB	—	N56°21′51″E
BC	147.55	S07°15′55″E
CD	215.01	S65°59′23″E
DE	265.19	S72°30′29″W
EA	—	N39°04′11″W

9.30. Available measurements for counterclockwise figure *ABCD* are as follows:

Side	Length, ft	Bearing
AB	213.35	S65°32′01″W
BC	324.04	S37°38′17″E
CD	—	N15°24′56″E
DA	—	N17°22′22″W

9.31. In counterclockwise figure *ABCDE*, point *C* lies southeast of point *B* and the available measurements are as follows:

Side	Length, ft	Azimuth
AB	203.66	092°41′17″
BC	198.19	—
CD	360.93	—
DE	278.77	268°50′50″
EA	225.80	183°46′08″

9.32. In clockwise figure *ABCDE*, point *E* lies northwest of line *AD* and available measurements are as follows:

Side	Length, ft	Bearing
AB	205.74	S19°41′33″E
BC	174.86	S22°56′46″W
CD	267.37	N74°46′56″W
DE	200.22	—
EA	201.94	—

9.33. In counterclockwise figure *ABCDE*, bearing *BC* is known to lie southeasterly and available measurements are as follows:

Side	Length, ft	Bearing
AB	149.75	N81°31′02″E
BC	145.73	—
CD	—	N22°47′00″W
DE	204.98	S77°40′35″W
EA	166.03	S07°24′07″E

9.34. In clockwise figure *ABCDE*, the paced distance of side *EA* was found to be somewhat less than 500 ft and available measurements were as follows:

Side	Length, ft	Azimuth
AB	431.01	071°24′04″
BC	278.87	187°46′18″
CD	406.37	129°02′50″
DE	501.21	—
EA	—	335°58′02″

9.35. Available measurements for counterclockwise figure *ABCD* are as follows:

Side	Length, ft	Azimuth
AB	—	263°44′17″
BC	580.03	160°33′59″
CD	—	033°37′12″
DA	314.84	000°49′54″

9.36. Available measurements for counterclockwise figure *ABCDE* are as follows:

Side	Length, ft	Azimuth
AB	311.97	291°46′36″
BC	315.75	—
CD	493.51	164°49′17″
DE	423.93	—
EA	325.02	080°49′42″

CHAPTER 10

RAPID SURVEYING METHODS

10.1 INTRODUCTION

Preceding chapters emphasized the need for precise measurements that lead to distances, coordinate positions, and elevations stated to the nearest hundredth of a foot. A drawback to this level of precision is the time involved in both measurement and computation. Luckily, a number of surveying applications don't require hundredth-of-a-foot precision, particularly in the areas of construction and mapping. For these activities, other methods provide acceptable precision and a more rapid rate of progress in making field measurements.

The traditional lower-precision, higher-speed method involves **stadia**, discussed briefly in Chap. 3 in the context of measuring backsight and foresight distances in three-wire leveling. Stadia work combines a level rod with a transit or theodolite to approximately measure both horizontal distances and elevation differences. A newer method uses a highly portable, rod-mounted EDM reflector called a **prism pole**. When used with a theodolite-EDM combination, a prism pole affords still-faster measurement of distances and elevation differences, and somewhat more accurately than possible with stadia. When used with a typical total-station instrument, the prism pole makes it possible to obtain three-dimensional coordinates about as quickly as distances and elevation differences alone.

10.2 BASIC STADIA COMPUTATIONS

CONCEPT. Stadia methods allow fast measurements for low-precision surveys (say, on the order of 1:500). Figure 10.1 is a very idealized optical diagram of an internal-focusing instrument telescope. The distance between stadia hairs on the *telescope reticle* is i, the difference between the upper and lower stadia-hair positions on the *rod face* is I, the distance from the observer's eye to the stadia reticle is the instrument focal length f and the distance from the observer's eye to the rod face is D. There are two similar isosceles triangles: one with base i and height f, the other with base I and height D. From geometry, the ratio of f to i is the ratio of D to I; thus,

$$\frac{D}{I} = \frac{f}{i}$$

$$\Rightarrow\Rightarrow \quad D = \left(\frac{f}{i}\right) I$$

The value $\frac{f}{i}$ is called the **stadia interval factor**, designated K. For most instruments, $K = 100.0$. The value of K is normally marked on the instrument case. For a modern, internal-focusing instrument:

$$\boldsymbol{D = KI} \tag{10.1}$$

and for the majority of instruments—those whose stadia interval factor is 100.0—distance D is the convenient value $100.0I$. In this text, K is assumed to be 100.0 unless stated otherwise.

A few old, external-focusing instruments are still in use. External focus places the f dimension *beyond* the cross-hair reticle and changes the optical geometry from that shown in Fig. 10.1. The end result, for these instruments, would be a modification of Eq. (10.1), in which $D = KI + C$, where C is about 1 ft. For most stadia work with an external-focusing instrument, the quantity C may be neglected entirely.

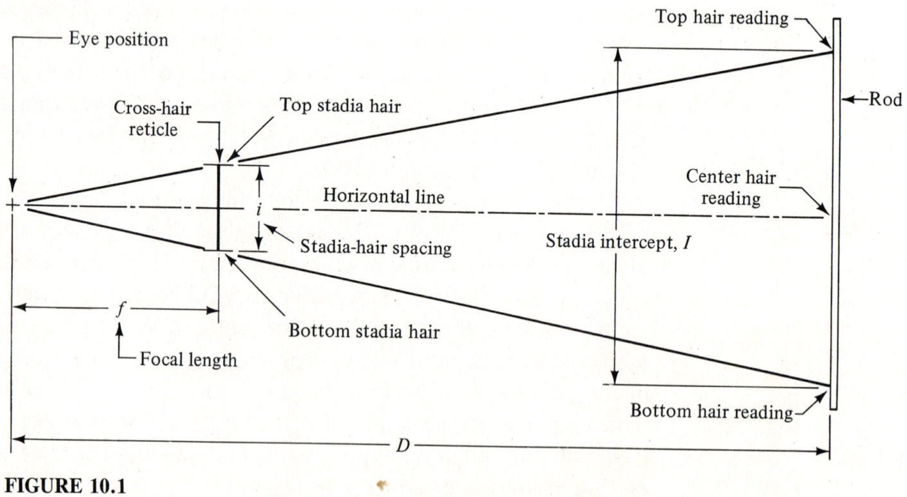

FIGURE 10.1
Idealized optical diagram of the stadia principle, internal-focusing telescope.

Example 10.1. In Fig. 10.1, the rod reading at the top stadia hair is 6.75 ft and the rod reading for the bottom stadia hair is 3.27 ft. What is the distance to the rod?

Solution. The intercept is 6.75 − 3.27, or 3.48 ft. Since the value of K isn't given, assume it to be 100.0. Substituting into Eq. (10.1),

$$D = 100.0(3.48) = 348 \text{ ft}$$

With rod readings in the above example taken to the nearest 0.01 ft., I is a three-significant-figure value. Thus, distance D is valid only to the nearest foot, a factor-of-100 degradation in precision compared to distance measurements discussed in earlier chapters.

INCLINED STADIA SIGHTS. The line of sight illustrated in Fig. 10.1 is horizontal. With an inclined line of sight, the smaller triangle is unchanged but the larger triangle is no longer isosceles and distance D from the instrument center to the position of the center hair image on the face of the rod is no longer a horizontal distance. Figure 10.2 shows the resultant geometry when the instrument telescope is tilted an angle α from the horizontal.

In Fig. 10.2, I is the stadia intercept actually seen and read, but I' is what the intercept reading would be if it were feasible for the rod to be held perpendicular to the line of sight; that is, with rod tilt equal to slope angle α. For the usual slope-angle values encountered in the field, the value I' is quite reasonably approximated by $I \cos\alpha$. Because distance D must be KI', it may be stated

$$D = KI' = KI \cos\alpha$$

From Fig. 10.2, however, $H = D\cos\alpha = (KI \cos\alpha)\cos\alpha$, from which is obtained

$$\mathbf{H = KI \cos^2 \alpha} \qquad \textbf{(10.2a)}$$

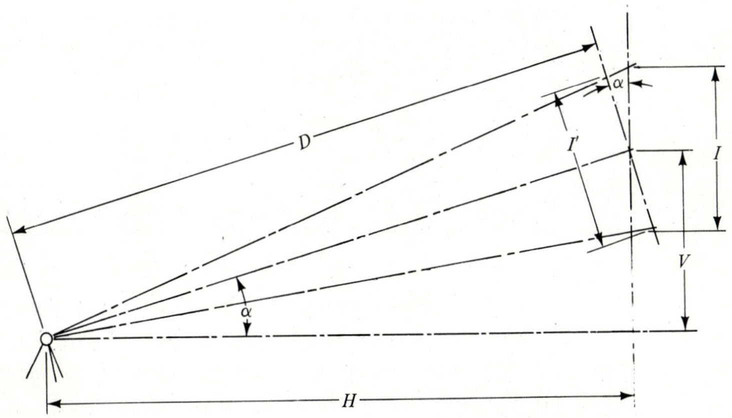

FIGURE 10.2
Inclined stadia sight.

For instruments that measure zenith angle (Z) rather than vertical angle (α), angle Z may be used directly for determination of horizontal distance H by modifying Eq. (10.2a) to read

$$H = KI \sin^2 Z \tag{10.2b}$$

Similarly, from Fig. 10.2, the vertical distance V between the instrument center and the center hair position on the rod face must be $D \sin\alpha$. Substituting for D, $V = (KI \cos\alpha)\sin\alpha = KI \sin\alpha \cos\alpha$. The trigonometric identity $\sin 2\alpha = 2 \sin\alpha \cos\alpha$ can be divided by 2 to produce $\sin\alpha \cos\alpha = (\sin 2\alpha)/2$. Substituting this value of $\sin\alpha \cos\alpha$ into the expression for V,

$$V = \frac{KI \sin 2\alpha}{2} \tag{10.3a}$$

This equation may also be used with zenith angles. In this case

$$V = \frac{KI \sin 2Z}{2} \tag{10.3b}$$

Example 10.2. A 30-second transit with a stadia interval factor of 98.4 was used to take a stadia sight on a vertically held rod. The top hair reading was 7.04 ft and the bottom hair reading was 2.68 ft. If the vertical angle was $+4°39'30''$, what were the horizontal and vertical distances from the transit center to the center hair reading point on the rod?

Solution. The stadia interval is $7.04 - 2.68 = 4.36$ ft. Substituting into Eq. (10.2) and (10.3),

$$H = 98.4(4.36)\cos^2(+4°39'30'') = 426 \text{ ft}$$

$$V = \frac{98.4(4.36)\sin[2(+4°39'30'')]}{2} = 34.7 \text{ ft}$$

For instruments whose stadia intercept factor is 100.0, Eqs. (10.2) and (10.3) become $D = 100.0I \cos^2\alpha$ and $V = 50.0I \sin 2\alpha$.

Example 10.3. Another instrument was used to make the following observations for a single stadia sight on a vertically held rod: Top hair, 6.19 ft; bottom hair, 3.33 ft; $\alpha = -3°42'12''$. Compute H and V.

Solution. Since K isn't given, it is assumed to be 100.0. $I = 6.19 - 3.33 = 2.86$ ft. Substituting,

$$H = 100.0(2.86)\cos^2(-3°42'12'') = 285 \text{ ft}$$

$$V = 50.0(2.86)\sin[2(-3°42'12'')] = -18.4 \text{ ft}$$

Had the solution to Example 10.3 been, say -8.4 ft rather than -18.4, the result would still have been valid to just a tenth of a foot rather than to three significant figures. In stadia work, the rod is usually set on soil rather than on solid, repeatable high points. As is the case with profile levels (Sec. 3.10), it is improper to assume a vertical precision greater than ± 0.1 ft for ordinary ground points.

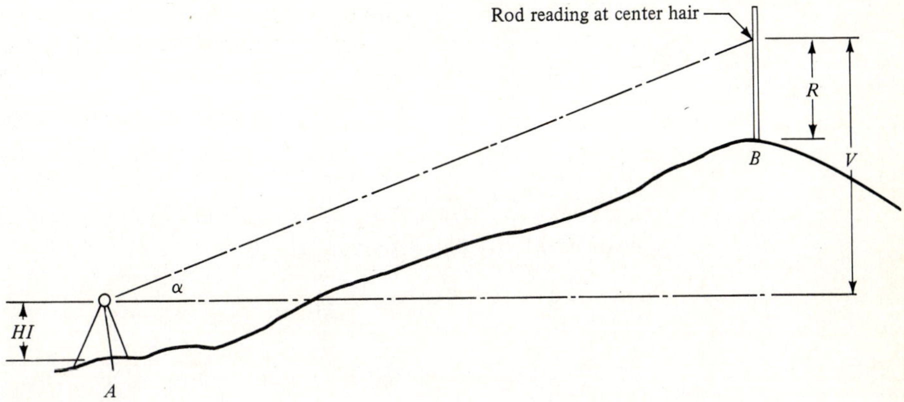

FIGURE 10.3
Elevation difference between two points, using stadia.

DIFFERENCE IN ELEVATION BETWEEN TWO POINTS. Examples 10.2 and 10.3 both dealt with V, the vertical distance between the instrument center and the point on the rod on which the telescope's center cross hair fell. What is ordinarily wanted, however, is the difference in elevation between the instrument setup point and the point on which the rod is placed. Figure 10.3 illustrates the usual situation, with the height-of-instrument (HI) and the center-hair reading (R) also being measured, along with slope angle α and stadia intercept I (not shown in the figure). Using Fig. 10.3, the difference in elevation between instrument point A and rod point B is the HI measurement plus V (the algebraic sign of V is determined by the sign of α) minus the center hair rod reading R. Thus,

$$\Delta\text{Elevation}_{A\text{-to-}B} = HI + \frac{KI \sin 2\alpha}{2} - R \qquad (10.4)$$

Example 10.4. The following stadia observations were taken to a rod set over point B from a transit whose HI was 5.34 ft over point A: $HI = 5.34$ ft, $\alpha = -3°08'$, top hair reading = 6.14 ft, bottom hair reading = 3.72 ft. Compute the difference in elevation from A to B and the elevation at B if the elevation at A is 622.9 ft above MSL.

Solution. K (not stated) must be 100.0. $I = 6.14 - 3.72 = 2.42$ ft. Since the top and bottom stadia hairs are equally spaced above and below the center hair on the cross-hair reticle, the center hair reading on the rod should fall halfway between the top hair reading and the bottom hair reading; that is, R is the *average* of the top and bottom hair readings. Thus

$$R = \frac{\text{top hair reading} + \text{bottom hair reading}}{2}$$

$$= \frac{6.14 + 3.72}{2} = \frac{9.86}{2} = 4.93 \text{ ft}$$

Substituting into Eq. (10.4),

$$\Delta \text{Elev}_{A\text{-to-}B} = 5.34 + 50.0(2.42)\sin[2(-3°08')] - 4.93$$

$$= 5.34 - 13.21 - 4.93 = -12.80 \Rightarrow\Rightarrow -12.8 \text{ ft}$$

$$\text{Elev}_B = \text{elev}_A + \text{elev}_{A\text{-to-}B} = 622.9 + (-12.8) = 610.1 \text{ ft above MSL}$$

For stadia work, it is usually convenient to make the value of R the same as the value of the HI, thus reducing Eq. (10.4) to

$$\Delta \text{Elev}_{A\text{-to-}B} = \frac{KI \sin 2\alpha}{2}$$

R is made equal to HI by setting a rod target at the HI height, then aligning the telescope center hair with the target center *before* taking the top and bottom hair readings and the slope-angle reading. If a standard rod target is unavailable, substitute a strip of surveyor's flagging tape that has been twisted into a thin strand and tied onto the rod. The target height on the rod is established by holding the rod vertical next to the instrument and sliding the target (or plastic tape strip) up or down until it is at the same height as the telescope's horizontal axis.

10.3 PRISM POLES

With the increased use of EDMs, it occurred to surveyors that stadia sights could be replaced by observations on an EDM reflector attached to a vertically held pole. Some two-piece range poles incorporate the same thread as that used on a reflector prism base, making it easy to attach a standard single-prism (or larger) reflector to a range pole, use a rod level to hold the range pole vertical, then measure both distance and vertical angle to the prism center.

Prism poles now on the market incorporate a built-in bull's-eye level, can be telescoped to place the prism at a desired height, and are available in a variety of lengths. Figure 10.4 shows a prism pole with three telescoping sections that can be extended from 4 ft to as high as 12 ft—the bull's-eye level is in the light-colored projection about 10 in. below the rodman's hands.

The bull's-eye level allows the prism pole to be maintained to well within 20 minutes of the true vertical. For a prism set 5 ft above ground, a 20-minute error would amount to just 0.03 ft, or 0.06 ft for a prism set 10 ft high. Thus, a prism pole affords higher precision for rapid distance measurement—and for measurement of elevation differences—than is obtainable from stadia observations. For this reason, and because it affords a faster surveying method than stadia, use of the prism pole is expected to become near-universal in the not-too-distant future.

Computation of prism-pole observations is quite simple: using EDM-measured slope distances, horizontal distances are determined as discussed in Sec. 2.14 and elevation differences are computed as shown in Sec. 3.3. Because of the lower precision of the prism-pole method—compared with that obtainable with trigonometric leveling—the value of $\Delta\alpha$ in the EDM slope reduction (Sec. 2.14) may be ignored.

FIGURE 10.4
Prism pole. (*Courtesy of Hixon Manufacturing Co.*)

The distance between the telescope center and the EDM center will be small and the effect of this slight offset on either the horizontal or vertical distance computation will be insignificant. With today's total-station instruments, in fact, the EDM line of sight coincides with the telescope line of sight, eliminating $\Delta\alpha$ as a relevant factor in any case.

With advanced-design EDMs, computation comes down to keying in the vertical angle and prism constant, then reading the H and V values off the instrument. With an electronic total-station instrument, the vertical angle is entered automatically, making for even quicker results from prism-pole observations. Built-in computers in most total stations—and in some advanced EDMs—even provide *xyz*-coordinate output once the starting azimuth, the instrument elevation, and the instrument-point coordinates are keyed in. This makes for very rapid acquisition of coordinate points for topographic mapping, as discussed in the next chapter.

Prism-pole observations can be used on reconnaissance traverses and similar surveys on which relatively precise results are not required; it must be understood that the precision of such work will be on the order of a tenth of a foot, both horizontally and vertically.

10.4 LOW-ORDER-OF-PRECISION LEVELS

CONCEPT. A trigonometric leveling method was illustrated in Sec. 3.3 that used a single setup and two EDM reflector positions (see Fig. 3.6). Stadia and prism-pole methods are used in a similar way to even more quickly determine elevation differences with alternate backsights and foresights. For either method, the operation

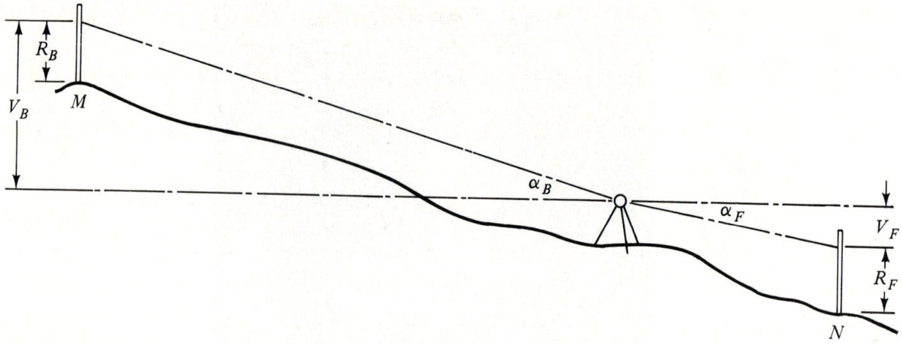

FIGURE 10.5
Low-precision levels with stadia or prism pole.

is diagrammed in Fig. 10.5, in which vertical angles are designated α, center hair readings R, and vertical elevation differences V, with backsight data indicated by subscript B and foresight data by subscript F.

In Fig. 10.5, stadia or prism-pole levels are carried forward from point M to point N using a single instrument setup. From the diagram, it is apparent that

$$\mathbf{\Delta Elev}_{M\text{-to-}N} = \boldsymbol{R_B - V_B + V_F - R_F} \tag{10.5}$$

where $V = (KI \sin 2\alpha)/2$ [Eq. (10.3)] for stadia work; for the prism-pole method, $V = s \sin \alpha$, where s is the corrected EDM slope distance. In either case, the algebraic sign of α must be observed: positive for an elevation angle, negative for a depression angle.

STADIA RECORDING AND COMPUTATION. Field-book recording is done in the five-column format shown in Example 10.5, with top, middle, and bottom hair readings entered in that order in the backsight and foresight "Rod" columns.

Example 10.5. Stadia levels were run from point M (elev. 820.7 ft) to point O, using two instrument setups, each similar to Fig. 10.5. From the recorded field data below, compute the elevation of point O.

	Backsight		**Foresight**	
Sta.	**Rod, ft**	α	**Rod,ft**	α
	9.78			
M	6.40	$+2°13'$		
	3.02			
	6.77		4.98	
N	4.58	$-2°38'$	3.63	$-1°21'$
	2.39		2.28	
			7.74	
O			4.52	$+3°03'$
			1.30	

Solution. The following is an 11-column format that applies the recorded data to Eq. (10.5), as shown by the column headings. I_B and I_F are, respectively, backsight and foresight stadia intercepts; that is, bottom hair readings subtracted from top hair readings. R_B and R_F are the center-hair readings for the backsight and foresight, respectively. $V_B = (KI_B \sin 2\alpha_B)/2$ and $V_F = (KI_F \sin 2\alpha_F)/2$, which become 50.0 $I_B \sin\alpha_B$ and 50.0 $I_F \sin\alpha_F$ in this case, since K is not specified and is thus assumed to be 100.0.

	Backsight readings					Foresight readings				
Sta.	I_B, ft	α_B	V_B, ft	R_B ft	$R_B - V_B$, ft	I_F, ft	α_F	V_F, ft	R_F, ft	$V_F - R_F$, ft
M	6.76	+2°13′	+26.1	6.40	−19.7					
N	4.38	−2°38′	−20.1	4.58	+24.7	2.70	−1°21′	−6.4	3.63	−10.0
O						6.44	+3°03′	+34.2	4.52	+29.7
					ΣBS = + 5.0					ΣFS = +19.7

$$\Delta \text{Elev}_{M\text{-to-}O} = \Sigma(R_B - V_B) + \Sigma(V_F - R_F) = \Sigma\text{BS} + \Sigma\text{FS} = +5.0 + 19.7 = +24.7 \text{ ft}$$

$$\text{Elev}_O = \text{elev}_M + \Delta\text{elev}_{M\text{-to-}O} = 820.7 + 24.7 = 845.4 \text{ ft}$$

PRISM-POLE RECORDING AND COMPUTATION. Stadia levels often involve long stadia sights and large associated stadia intercepts. For this reason, a single stadia sight takes up much of the length of the rod, particularly when underbrush makes it impossible to see the lower portion of the rod face. That being the case, the observer has little control over where the center hair falls on the rod face; thus, it is not ordinarily feasible to set the center hair on the same rod point for every sight. With prism poles, on the other hand, the prism may be readily set at a convenient, fixed height. With the prism height the same for both the backsight and the foresight, Eq. (10.5) reduces to $\Delta\text{elev}_{M\text{-to-}N} = -V_B + V_F$, where $V = s \sin\alpha$, as before. Thus, for prism-pole leveling with prism height fixed:

$$\mathbf{\Delta Elev}_{M\text{-to-}N} = -s_B \sin\alpha_B + s_F \sin\alpha_F \tag{10.6}$$

Prism-pole levels may be recorded and computed in the seven-column format in the next example. The EDM s_B and s_F values are EDM slope distances that have been corrected for atmospheric effects and prism constant. The other data-entry columns are self-explanatory. The $-s_B \sin\alpha_B$ and $+s_F \sin\alpha_F$ columns are part of the computing process, reflecting Eq. (10.6).

Example 10.6. A three-setup line of prism-pole levels was run from point R, elevation 1437.6 ft above mean sea level, to point U, with data recorded below. Compute the MSL elevation of point U.

Sta	s_B, ft	α_B	$-s_B \sin\alpha_B$, ft	s_F, ft	α_F	$+s_F \sin\alpha_F$, ft
R	944.61	+1°18′33″				
S	866.97	−2°31′04″		1011.91	+0°51′12″	
T	746.34	+3°44′52″		852.05	−2°27′38″	
U				804.13	+1°36′07″	

Solution. The backsight and foresight values of $s \sin \alpha$ are computed and entered into the fourth and seventh columns, the column sums are taken, and final results obtained as shown below:

Sta	s_B, ft	α_B	$-s_B \sin \alpha_B$, ft	s_F, ft	α_F	$+s_F \sin \alpha_F$, ft
R	944.61	+1°18′33″	−21.58			
S	866.97	−2°31′04″	+38.09	1011.91	+0°51′12″	+15.07
T	746.34	+3°44′52″	−48.78	852.05	−2°27′38″	−36.58
U				804.13	+1°36′07″	+22.48
			ΣBS = −32.27			ΣFS = +0.97

$$\text{Elev}_U = \text{elev}_R + \Delta\text{elev}_{R\text{-to-}U} = \text{elev}_R + \Sigma\text{BS}_{R\text{-to-}U} + \Sigma\text{FS}_{R\text{-to-}U}$$

$$= 1437.6 - 32.27 + 0.97 = 1406.3 \text{ ft}$$

10.5 PROBLEMS

In Probs. 10.1 through 10.5, calculate H and V from the listed stadia observation values. Assume $K = 100.0$.

10.1

Stadia hair readings top to bottom, ft	Vertical angle
12.37	
7.59	+5°12′
2.81	

10.2

Stadia hair readings top to bottom, ft	Vertical angle
7.86	
5.45	−3°29.3′
3.04	

10.3

Stadia hair readings top to bottom, ft	Vertical angle
10.89	
6.51	−6°48′
2.13	

10.4

Stadia hair readings top to bottom, ft	Vertical angle
8.38	
5.17	+2°33.6′
1.96	

10.5

Stadia hair readings top to bottom, ft	Vertical angle
11.43	
7.57	+4°16′30″
3.71	

In Probs. 10.6 through 10.10, the elevation of the instrument setup point is 389.12 ft above MSL and the height-of-instrument over the point is 5.11 ft.

10.6. Compute the elevation of the stadia rod point in Prob. 10.1.

10.7. Compute the elevation of the stadia rod point in Prob. 10.2.

10.8. Compute the elevation of the stadia rod point in Prob. 10.3.

10.9. Compute the elevation of the stadia rod point in Prob. 10.4.

10.10. Compute the elevation of the stadia rod point in Prob. 10.5.

10.11. An instrument's stadia interval factor K was not known. To determine this value, horizontal distances of 147.33 ft, 221.15 ft, and 406.59 ft were accurately measured with a steel tape from the instrument position to points A, B, and C. With telescope horizontal, top and bottom stadia hair readings for each rod position were as follows: point A, 6.06 and 4.24; point B, 6.54 and 3.76; point C, 7.69 and 2.61. Compute an average value for K.

10.12. An observer, wishing to determine the horizontal distance to a point by stadia methods, assumed that K for her instrument was 100 and took the following observations: upper hair 6.91, lower hair 3.65, and vertical angle −5°11′. The height of the instrument was 5.28 ft and the known elevation difference between the occupied point and the observed point was −29.33 ft. Compute the horizontal distance and verify the assumed value of K.

10.13. An obstruction obscured the lower hair reading on a rod held over a stadia point. The upper hair read 12.96 ft, the center hair read 8.41 ft, and the vertical angle was +3°46′. Assuming K = 100.0, determine the values of H and V.

10.14. Stadia levels were run from BM J, elev. 469.37 ft, to point Z via two turning points. Observed data, with stadia hair readings in feet, are stated in the order top hair, center hair, bottom hair, vertical angle. BS to BM J: 8.96, 5.69, 2.42, −1°17′. FS to TP 1: 12.47, 8.06, 3.65, −2°43′. BS to TP 1:11.42, 7.09, 2.76, +1° 29′. FS to TP 2: 9.39, 5.28, 1.17, −2°36′. BS to TP 2:10.86, 6.81, 2.74, +4° 09′. FS to point Z: 12.32, 8.09, 3.86, +3°51′. Set up the stadia level notes and compute the elevation of point Z.

10.15. Stadia levels were run from point Q to point R and then to point S. If the elevation of point Q is 529.12 ft, determine the elevation of points R and S from the following rod readings (ft) and vertical angle readings. BS to point Q: rod reads 9.02, 5.89, 2.76; α = +4°29′. FS to point R: rod reads 11.38, 8.00, 4.62, α = −2°47′. BS to point R: rod reads 10.44, 5.75, 1.06, α = +2°26′. FS to point S: rod reads 8.71, 5.52, 2.33, α = +3°58′.

10.16. Stadia observations were made from both ends of line CD. With the instrument set over point C and sighting on a rod held over point D, top-to-bottom hair readings (ft) were 12.88, 8.31 and 3.74, while the vertical angle read +5°39′. The instrument height at C was 5.21 ft. With the instrument set over point D and sighting on a rod held over point C, top-to-bottom hair readings (ft) were 11.49, 6.91 and 2.34, while the vertical angle read −5°19′. The instrument height at D was 5.11 ft. Calculate the mean horizontal distance and difference in elevation between C and D.

10.17. Referring to Prob. 10.16, if the elevation of point D is 622.73 ft above MSL, what is the elevation of point C?

10.18. When sighting through a perfectly leveled telescope with a stadia interval factor of 100.00, what is the *grade*, in percent, from the instrument center to the bottom hair position?

10.19. A theodolite-EDM combination was used with a prism pole to run low-order levels over rough terrain from point E to point Z. Slope distances and vertical angles are listed as follows in the order in which they were observed along the E-to-Z route. BS to E: 539.61 ft, $\alpha = +2°15.6'$. FS to first TP: 647.33 ft. $\alpha = -1°37.3'$. BS to first TP: 476.80 ft, $\alpha = -3°41.9'$. FS to second TP: 801.93 ft, $\alpha = +1°31.9'$. BS to second TP: 739.63 ft, $\alpha = -4°38.7'$. FS to Z: 582.78 ft, $\alpha = +0°44.6'$. All readings were taken with the prism set at the instrument height. The elevation of point E is 1140.39 ft above MSL. Set up the observations in field-record form and compute the elevation of point Z.

10.20. Referring to Prob. 10.19, it was discovered later that the prism height was set exactly 1 ft below the instrument height on the backsight and foresight for the first setup only. What effect did this have on the leveling results?

10.21. Referring to Prob. 10.19, when taking the final foresight, it was necessary to raise the prism 4.00 ft above the instrument height because of an obstruction along the line of sight, all other data remaining the same. What is the elevation of point Z?

10.22. Low-order levels were run from point A to point B to point C with a total-station instrument using a single setup between consecutive points. At each setup the prism height was set at the instrument height. The elevation at point A was 2540.86 ft above MSL and the observations were as follows: BS to A, 682.39 ft, $\alpha = -2°29'42''$; FS to B, 809.83 ft, $\alpha = -1°47'18''$; BS to B, 719.66 ft, $\alpha = +3°16'42''$; FS to C, 572.84 ft, $\alpha = -1°26'22''$. Compute the elevations of points B and C to the nearest 0.1 ft.

10.23. Refer to Prob. 10.22. From a single setup between point C and point A, a backsight slope distance of 1268.52 ft and vertical angle of $-0°55'35''$ were observed to point C and a foresight slope distance of 1641.77 ft and vertical angle of $+1°05'21''$ were observed to point A. Determine the closure error in the loop and adjust the elevations of points B and C, distributing closure error in proportion to the number of setups.

10.24. In Prob. 10.23, considering both the slope distances involved and the 0.1-ft precision of the results, what is the effect of curvature and refraction? What would be the effect if the results were taken to 0.01 ft rather than 0.1 ft?

CHAPTER
11

TOPOGRAPHY

characteristics of contour lines

11.1 TOPOGRAPHIC PRINCIPLES

MAPS. Proper planning and design of construction projects depend on the availability of good project-area maps that show the locations of existing structures and roads and, more important still, the overall configuration of the terrain. A map that displays man-made objects, streams, shorelines, and similar natural features in their proper horizontal relationship to one another—but not the terrain relief—is called a **planimetric** map. A map that combines the planimetry of an area with a reasonably detailed display of ground-surface elevations is called a **topographic** map.

SCALE. A map is a representation of a fairly small area of the earth's surface on a flat piece of paper. Horizontal distances between points on the ground must be shown on the map in constant proportion. The proportional constant relating map distance to ground distance is **scale**. Scale may be expressed as the map distance that represents a particular ground distance or "ground equivalence"; say, 1 in. = 1000 ft, in which case 1 in. on the map corresponds to 1000 ft on the ground. Scale is also stated as a ratio or representative fraction; say 1:12,000, in which case one unit on the map represents 12,000 of the same units on the ground. Note, in this case, that 1:12,000 scale is another way of saying 1 in. = 1000 ft.

U.S. Geological Survey (USGS) topographic maps at a scale of 1 in. = 2000 ft (1:24,000 scale) are widely sold by establishments that deal in surveying and engineering supplies. A 1:24,000-scale USGS map, called a $7\frac{1}{2}$-minute quadrangle, covers an area whose north-south dimension is $7\frac{1}{2}$ minutes of latitude (about 8.6 miles) and whose east-west dimension is $7\frac{1}{2}$ minutes of longitude (about 6.6 miles at 40° N latitude). USGS maps are well suited to the initial stages of project planning but they rarely provide sufficient detail, nor are they sufficiently up-to-date, for the design and layout of construction projects. Generally, special-purpose topographic maps must be made for each project, at a scale appropriate for the sort of work to be carried out.

The precision with which a project may be laid out is a function of map scale: the larger the scale, the more precisely points may be located, but the greater the number of map sheets needed to cover a given area and the costlier the mapping task. Selection of map scale (or scales) is usually a trade-off between optimum level of detail and the time and cost involved in providing that amount of detail.

TOPOGRAPHIC REPRESENTATION. Another consideration in topographic mapping is how elevations, or differences in elevation, are to be shown. Early maps showed elevation differences by "hachures," which are short lines drawn in the direction of the ground slope; the steeper the slope, the more closely spaced the hachures. Although hachures give a good impression of slope steepness, they do not provide actual elevations. A better method of presentation is the use of **contours**. A contour is a line that passes through points of equal elevation; that is, it shows the line of intersection of a level surface with the ground surface. The shoreline of a calm pond is a contour line for the surface elevation of the pond. Modern topographic maps use contours spaced at a constant **contour interval**; that is, at a constant vertical distance between contours.

Figure 11.1 shows two views of the same coastal area. The top panel is an oblique view showing a river passing between a cliff and a hill before emptying into a small embayment. The bottom panel is a topographic map of the same area. The zero-elevation contour is the sea-level shoreline and other contours are located successively further inland at 20-ft intervals. From the contours, it is seen that the highest cliff elevation showing on the map is somewhat more than 260 ft and that the hill elevation is somewhat more than 280 ft—some topographic maps show an "×" at the summit location and place the summit elevation alongside.

Figure 11.1 also illustrates a number of contour characteristics that are essential for the surveyor to understand. These are listed below in the form of guidelines.

1. Closely spaced contours represent steep slopes; broadly spaced contours indicate gentle slopes.
2. Changing contour spacing shows changing slope; thus, unevenly spaced contours indicate rugged or undulating terrain while evenly spaced contours indicate a constant surface gradient.
3. A contour line is a continuous line that must either close on itself within a map boundary or else originate at one point on the map boundary and pass off the map at another point along the boundary.
4. Concentric contour lines of decreasing perimeter indicate the presence of either a summit or a depression. To avoid confusion, the inner contour line surrounding a depression is usually designated by hachures radiating inward toward the low point.

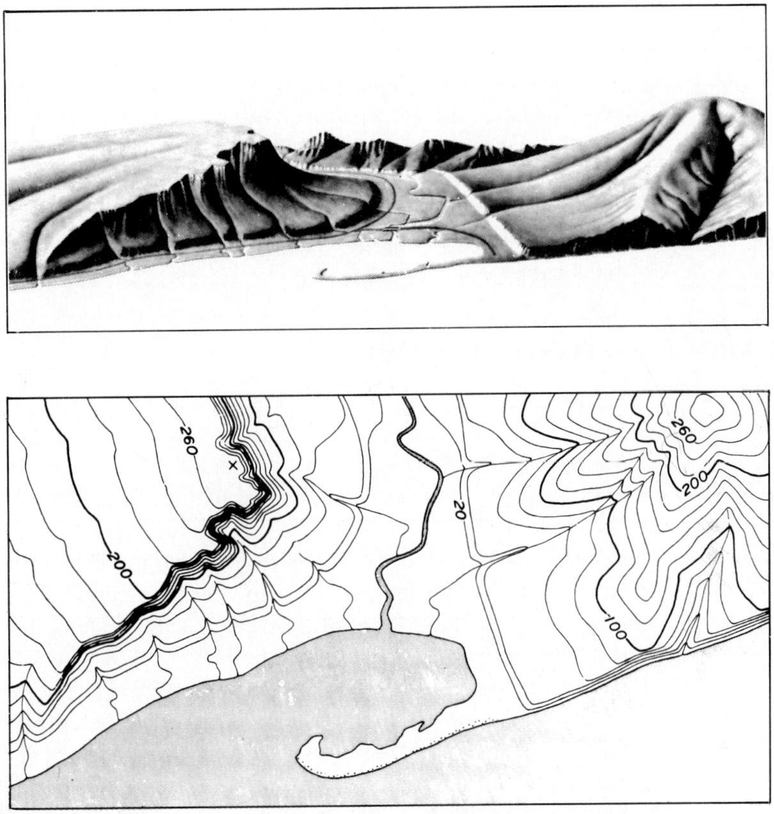

FIGURE 11.1
Topographic representation of a coastal area. (*Courtesy of Department of Interior, U. S. Geological Survey.*)

5. Contours do not cross one another except in the rare case of an overhanging cliff or cavern; they do not merge with one another except in the case of a vertical cliff or wall.
6. The direction of maximum slope is always perpendicular to a contour line.
7. Contours delineating hills or ridges are concave toward higher ground.
8. Contours that cross streams or drainage courses are V-shaped, with the V pointing upstream or upslope.

TOPOGRAPHIC MAP COMPILATION. For large areas, the simplest and least expensive method of compiling topographic maps is by use of **photogrammetry**. Photogrammetric mapping involves stereo aerial photography along with a modest amount of ground surveys to provide photo control for later compilation using specialized instruments, some of them highly automated. Photogrammetric mapping may be done at scales as large as 1 in. = 20 ft and at contour intervals of 1 ft or less, but at considerable expense. For projects no larger than a few acres or even tens of acres, photogrammetric mapping becomes prohibitively costly—it involves not only aircraft and aircrew flight time but also the cost of employing photogrammetric ground-control parties and stereo-instrument operators.

Small areas are more quickly and inexpensively mapped using any of several ground methods. In order of decreasing precision, these are the **controlling-point method**, the **cross-section method**, and the **grid-point method**. The cross-section method is used primarily along transportation rights-of-way to determine the volume of earth to be brought in or taken away. This method is dealt with in some detail in Chap. 14. The controlling-point and grid-point methods are discussed over the balance of this chapter.

11.2 CONTROLLING-POINT METHOD

CONCEPT. Consider again the lower illustration in Fig. 11.1. For the map to have been made in the first place, the surveyor must have established the locations and elevations of a sufficient number of ground points—called **controlling points**—to delineate the key features on the map as well as to determine the proper configuration of the contour lines. Examples of key features are the summit of the round hill, the shoreline, the edges of the cliffs, the banks of the river, the several drainage courses on the flanks of the tall cliff and the round hill, the ridge lines between drainage courses, and low points, including saddles along the ridges that connect high points. To properly delineate contour lines, it is also necessary to locate the lines along which changes in slope occur. In this respect, since changing contour spacing indicates rugged or undulating terrain, there must be sufficient points established to define whatever significant ruggedness or undulation exists. The surveyor should select as many controlling points as are needed to properly define the area, taking neither too many observations nor too few; however, less experienced surveyors should keep in mind that it is better to observe too many points than too few.

DETAILS ABOUT A POINT. Efficiency requires that the surveyor use as few instrument setups as possible when mapping an area. A setup point should itself be a controlling point and it should be located at a place where as many as possible of the other controlling points within the map area are visible. The locations and elevations of the other controlling points are all taken with respect to the setup point, so the setup point must itself have not only a known position and elevation but a known reference azimuth as well.

The setup point is located from a nearby horizontal control point whose elevation is also known. An azimuth, vertical angle, and slope distance are taken *from* the nearby control point *to* the setup point within the map area. Then, with the instrument moved into position over the map-area setup point, its telescope aimed on the control point, the *back* azimuth to the nearby control point is set into the instrument's horizontal circle and the lower motion is clamped. Thus, the horizontal circle will read the correct azimuth to whatever point the telescope cross hair is subsequently pointed. This procedure is called the **azimuth transfer method**.

From the setup position, the surveyor selects controlling points at high and low points within the area, along ridges and drainage paths, at points of abrupt change in grade, and at whatever other points are necessary to define proper spacing of contour lines. Keep in mind that the existing ground configuration has been influenced more by water runoff than by any other factor; thus, establishing the location of all drainage features is of the utmost importance in defining the area's topography.

For each controlling point, the surveyor observes azimuth, vertical angle, and stadia or prism-pole readings, as applicable, recording the results in the field record book. Topographic maps are often compiled by someone other than the on-site surveyor; for that reason, at least, there should be one or more sketches in the field record book showing in some detail the locations of controlling points with respect to the ground features they are intended to delineate. The compiler knows each controlling point is selected because it delineates *something*; the sketch tells the compiler *what*. High and low points are indicated on the sketch with an x, ridges with a dashed line, and drainage paths with solid line segments separated by three dots (a standard map designation for intermittent streams). It is most helpful if a segment of the contour through each point is sketched in as well.

When all controlling points visible from the first setup have been observed, return to the known horizontal position point and establish the next setup point in the same manner as the first. Additional points are then observed, continuing in this way until observations have been taken on all needed controlling points within the map area. For small, open areas, it is usually possible to observe all controlling points from one well-placed setup point.

11.3 DATA FROM STADIA OBSERVATIONS

Figure 11.2 shows the field-book sketch of a small, rectangular area that is to be mapped with stadia methods. The instrument was set up on a ridge line at *B*, a point from which all other controlling points were believed to be visible. Point *B* was established from controlling point *A* (elev. 227.60 ft) at the southwest corner of the survey area. With the instrument at *B*, following a backsight on known point *A*, sights are taken on all the other controlling points in the survey area, labeled *C* through *T* on the sketch. The sketch also shows locations of drainage paths and a ridge line, as well as the estimated orientation of contours that pass through the controlling points; the latter will be of considerable help in the subsequent topographic plot of the area.

Figure 11.3 is the field record of instrument observations (intercept, vertical angle, and azimuth) taken from *B*. The first line shows the starting sight on *A*. The known azimuth from *A* to *B* is 069°22′. The back azimuth to *A* from *B* (069°22′ + 180° = 249°22′) is locked onto the horizontal circle when the vertical cross hair is centered on point *A*. With the horizontal circle locked, all angle readings taken on subsequent controlling points will be correct azimuths. The second observation was on point *C*, next line, and so forth, continuing to the final observation on point *T*. The instrument height at *B* was 5.2 ft. Except as noted, all stadia sights were taken with the center hair on the 5.2-ft mark of the rod. Observed data are recorded in the first four columns of the field record and descriptions of observed points are recorded in the last two columns. For brevity, a property line is designed by PL in the descriptions and a centerline by CL.

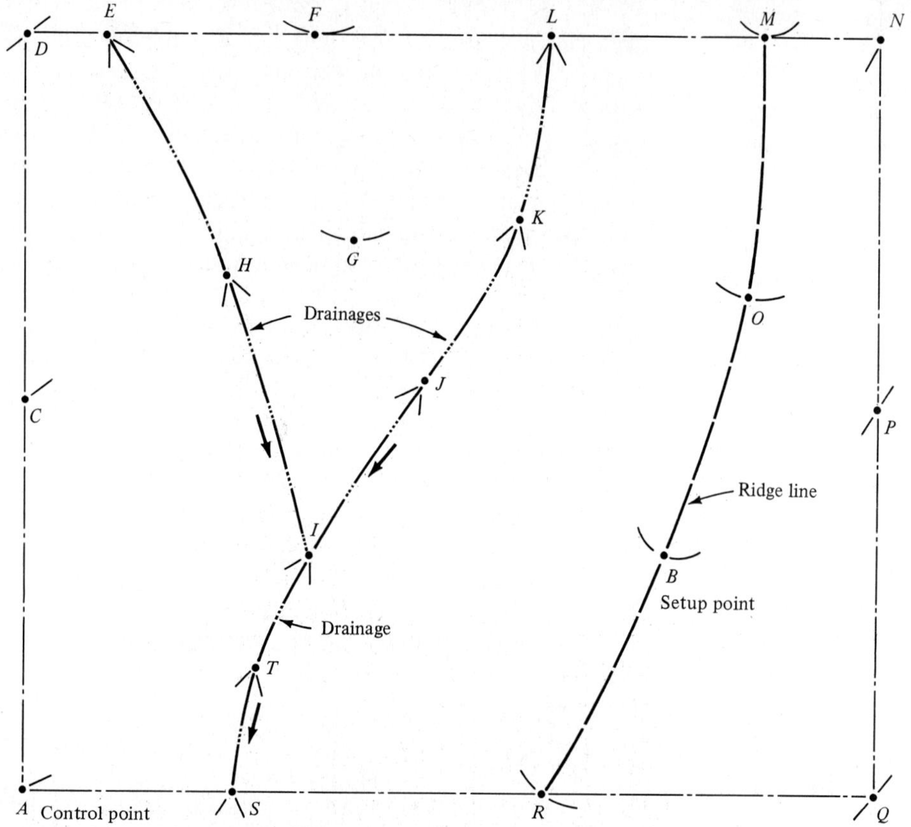

FIGURE 11.2
Field-book sketch of a a small area, showing controlling points and estimated contour paths.

Sta.	I, ft	Vert. angle	Azimuth	Description
A	3.06	−2°49′	249°22′	Southwest corner of area
C	2.94	−0°57′	282°04′	Mid-point of west PL
D	3.62	+0°31′	307°52′	Northwest corner of area
E	3.33	+0°17′	311°25′	West drainage CL @ north PL
F	2.66	+0°41′	324°22′	1st high point on north PL
G	1.90	−0°16′	314°07′	Slope change
H	2.25	−1°44′	301°03′	West drainage CL
I	1.58	−5°37′	268°10′	Drainage junction
J	1.30	−4°15′	303°53′	East drainage CL
K	1.58	−1°45′	336°06′	East drainage CL
L	2.26	−0°05′	346°32′	West drainage CL @ north PL
M	2.26	+3°01′	011°51′	Ridge CL @ north PL
N	2.42	+1°55′ on 8.00′	023°39′	Northeast corner of area
O	1.17	+2°27′	018°22′	Ridge CL

(continued)

Sta.	I, ft	Vert. angle	Azimuth	Description
P	1.18	−1°01′	056°49′	Mid-point of east PL
Q	1.46	−4°26′	137°53′	Southeast corner of area
R	1.19	−3°29′	204°33′	Ridge CL @ south PL
S	2.21	−5°46′	240°05′	Drainage CL @ south PL
T	1.87	−5°40′	253°18′	Drainage CL

FIGURE 11.3
Field-data recording for stadia survey.

11.4 STADIA COMPUTATIONS

Stadia computation equations were derived in the previous chapter. Horizontal distance H is $KI\cos^2\alpha$ [Eq. (10.2)] and difference in elevation, Δelev, is $HI + (KI\sin 2\alpha)/2 - R$ [Eq. (10.4)], where I is the stadia intercept, α is the vertical angle, R is the center hair rod reading, and K, the stadia interval factor, is 100.0 if not otherwise stated. For the data in Fig. 11.3, $H = 100I\cos^2\alpha$ and $\Delta\text{elev} = 5.2 + 50I\sin 2\alpha - R$. But R was 5.2 ft for every observation but one; in those cases, *HI* and *R* cancel, leaving Δelev simply $50I\sin 2\alpha$. In the case of point *N*, the value of *R* was 8.00 ft, so Δelev to *N* was $5.2 + 50(2.42)\sin[2(+1°55')] - 8.0 = 5.2 + 8.1 - 8.0 = +5.3$ ft.

Figure 11.4 shows the computed results of the Fig. 11.3 data. The elevation of the setup point *B* is obtained from the computed Δelev value of −15.0

Sta.	I, ft	Vert. angle	Azimuth	H, ft	ΔElev., ft	Elev., ft
A	3.06	−2°49′	249°22′	305	−15.0	227.60
B						242.6
C	2.94	−0°57′	282°04′	294	−4.9	237.7
D	3.62	+0°31′	307°52′	362	+3.3	245.9
E	3.33	+0°17′	311°25′	333	+1.6	244.2
F	2.66	+0°41′	324°22′	266	+3.2	245.8
G	1.90	−0°16′	314°07′	190	−0.9	241.7
H	2.25	−1°44′	301°03′	225	−6.8	235.8
I	1.58	−5°37′	268°10′	156	−15.4	227.2
J	1.30	−4°15′	303°53′	129	−9.6	233.0
K	1.58	−1°45′	336°06′	158	−4.8	237.8
L	2.26	−0°05′	346°32′	226	−0.3	242.3
M	2.26	+3°01′	011°51′	225	+11.9	254.5
N	2.42	+1°55′ on 8.00′	023°39′	242	+5.3	247.9
O	1.17	+2°27′	018°22′	117	+5.0	247.6
P	1.18	−1°01′	056°49′	118	−2.1	240.5
Q	1.46	−4°26′	137°53′	145	−11.3	231.3
R	1.19	−3°29′	204°33′	119	−7.2	235.4
S	2.21	−5°46′	240°05′	219	−22.1	220.5
T	1.87	−5°40′	253°18′	185	−18.4	224.2

FIGURE 11.4
Computation of stadia survey.

ft *from* B to A. Since A is 15.0 ft lower than B, B is 15.0 ft higher than A, or 227.60 ft + 15.0 ft = 242.6 ft. Δelev values to all subsequent controlling points are added to the 242.6-ft elevation of B to obtain individual controlling-point elevations.

11.5 DATA AND COMPUTATIONS FOR PRISM-POLE OBSERVATIONS

Stadia data and computations were presented first because they represent the traditional method of acquisition of topographic data. Because of its significantly greater speed and precision, prism-pole observation is fast becoming the method of choice for today's surveyors. As discussed in Chap. 10, the horizontal positional error of a prism-pole reflector set at *HI* height is probably no greater than 0.03 ft with respect to a ground point on which the prism-pole base is set. Given the small vertical angles that are typical for topographic observations, elevation-difference precision on the order of 0.01 ft is attainable. On the other hand, the horizontal and vertical positions of ordinary ground points are not well defined. Thus, even though prism-pole observations allow horizontal-distance precisions on the order of 0.1 ft or better and elevation-difference precisions approaching 0.01 ft, that level of accuracy, though comforting, is rarely needed for topographic work.

Figure 11.5 shows equivalent prism-pole data for the Fig. 11.2 survey area, taken from the same instrument setup position and to the same controlling points. The prism-pole reflector was set at the instrument *HI* in every case. s designates slope distance, corrected for prism constant.

Prism-pole computations use the basic slope angle and slope distance equations developed in Chap. 2. Thus, $H = s\cos\alpha$ and $V = s\sin\alpha$. For a prism-pole reflector

Sta	s, ft	Vert. angle	Azimuth	Description
A	305.68	−2°49′	249°22′	Southwest corner of area
C	294.33	−0°57′	282°04′	Midpoint of west PL
D	361.90	+0°31′	307°52′	Northwest corner of area
E	332.79	+0°17′	311°25′	West drainage CL @ north PL
F	266.13	+0°41′	324°22′	First high point on north PL
G	189.74	−0°16′	314°07′	Slope change
H	225.29	−1°44′	301°03′	West drainage CL
I	157.31	−5°37′	268°10′	Drainage junction
J	129.37	−4°15′	303°53′	East drainage CL
K	158.41	−1°45′	336°06′	East drainage CL
L	225.85	−0°05′	346°32′	West drainage CL @ north PL
M	225.43	+3°01′	011°51′	Ridge CL @ north PL
N	242.52	+1°15′	023°39′	Northeast corner of area
O	116.96	+2°27′	018°22′	Ridge CL
P	118.22	−1°01′	056°49′	Midpoint of east PL
Q	145.48	−4°26′	137°53′	Southeast corner of area
R	118.82	−3°29′	204°33′	Ridge CL @ south PL
S	220.45	−5°46′	240°05′	Drainage CL @ south PL
T	186.23	−5°40′	253°18′	Drainage CL

FIGURE 11.5
Field-data recording for prism-pole survey.

Sta.	s, ft	Vert. angle	Azimuth	H, ft	ΔElev., ft	Elev., ft
A	305.68	−2°49′	249°22′	305.3	−15.02	227.60
B						242.62
C	294.33	−0°57′	282°04′	294.3	− 4.88	237.74
D	361.90	+0°31′	307°52′	361.9	+ 3.26	245.88
E	332.79	+0°17′	311°25′	332.8	+ 1.65	244.27
F	266.13	+0°41′	324°22′	266.1	+ 3.17	245.79
G	189.74	−0°16′	314°07′	189.7	− 0.88	241.74
H	225.29	−1°44′	301°03′	225.2	− 6.81	235.81
I	157.31	−5°37′	268°10′	156.6	−15.40	227.22
J	129.37	−4°15′	303°53′	129.0	− 9.59	233.03
K	158.41	−1°45′	336°06′	158.3	− 4.84	237.78
L	225.85	−0°05′	346°32′	225.8	− 0.33	242.29
M	225.43	+3°01′	011°51′	225.1	+11.86	254.48
N	242.52	+1°15′	023°39′	242.5	+ 5.29	247.91
O	116.96	+2°27′	018°22′	116.9	+ 5.00	247.62
P	118.22	−1°01′	056°49′	118.2	− 2.10	240.52
Q	145.48	−4°26′	137°53′	145.0	−11.25	231.37
R	118.82	−3°29′	204°33′	118.6	− 7.22	235.40
S	220.45	−5°46′	240°05′	219.3	−22.15	220.47
T	186.23	−5°40′	253°18′	185.3	−18.39	224.23

FIGURE 11.6
Computation of prism-pole survey.

set at the height of the instrument center, Δelev is identical to *V*, or $\Delta\text{elev} = s \sin\alpha$. These two equations for *H* and Δelev are applied to the data in Fig. 11.5 to compute the prism-pole results shown in Fig. 11.6.

The results shown in Fig. 11.6, though simple enough to compute, are provided automatically on the better EDMs currently on the market. With an EDM of this sort, the prism constant is keyed in, along with the atmospheric factors, to convert measured slope distance to corrected slope distance. Once the vertical angle is keyed in, *H* and *V* appear on the EDM display in whichever units are desired. With total-station instruments that provide digital readout of angles, the value of the vertical angle is automatically read into the equations, along with the corrected slope distance; the instrument then computes and displays *H* and *V*. Many total-station instruments use azimuth to compute and display *x* and *y* components of *H*. This feature is of particular value if an automated plotter is to be used to make a contour map.

11.6 CONTROLLING-POINT PLOTTING AND CONTOURING

MANUAL PLOT. Figures 11.4 and 11.6 show the azimuths and the computed horizontal distances and elevation differences from point *B* to all of the other controlling points in the survey. The next step is to plot the positions of the points on a sheet of paper at a convenient scale, labeling each plotted position with its elevation.

First, latitudes and departures are computed from *B* to corner points *A, D, N,* and *Q*. From this, it is determined that the lot is a rectangle that runs 330 ft north-

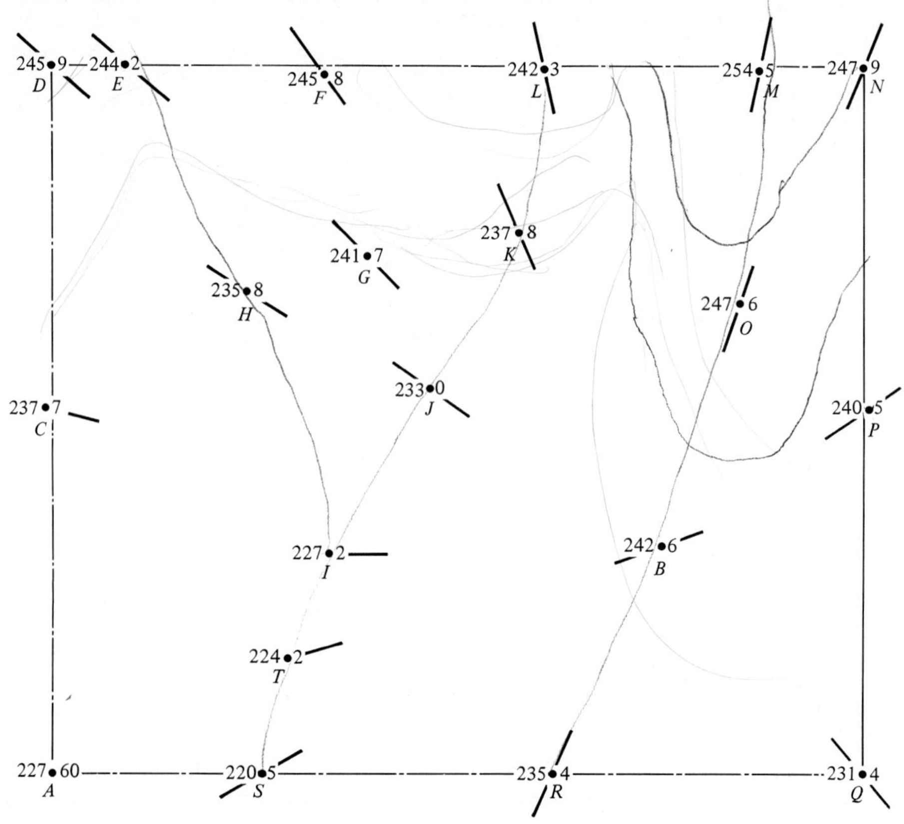

FIGURE 11.7
Point plot of Fig. 11.6 data, showing radials from point *B* and computed elevations.

south and 383 ft east-west and that point *B* lies about 108 ft north and 97 ft west of corner point *Q*. Choosing a convenient scale, points *A*, *B*, *D*, *N*, and *Q* are plotted on an appropriately sized sheet of drafting paper.

With *B* located, radials are run from *B* to the azimuths of all the other controlling points. The correct horizontal scale distances from *B* are laid out along the respective radials and each point is marked with a dot. Figure 11.7 shows the resultant plot of each point, including its letter designation and elevation, the position dot marking the decimal point of the elevation.

CONTOURING. Referring to the field book descriptions (Fig. 11.3 or 11.5), the drainage paths and the ridge line are sketched onto the Fig. 11.7 plot. Referring to the field book sketch of the survey area (Fig. 11.2), segments of the estimated contours through each point are also lightly sketched in.

The controlling-point method is based on the premise that there is a constant slope between adjacent controlling points—were this not so, additional controlling points would have been observed. Using this assumption, construction lines are lightly

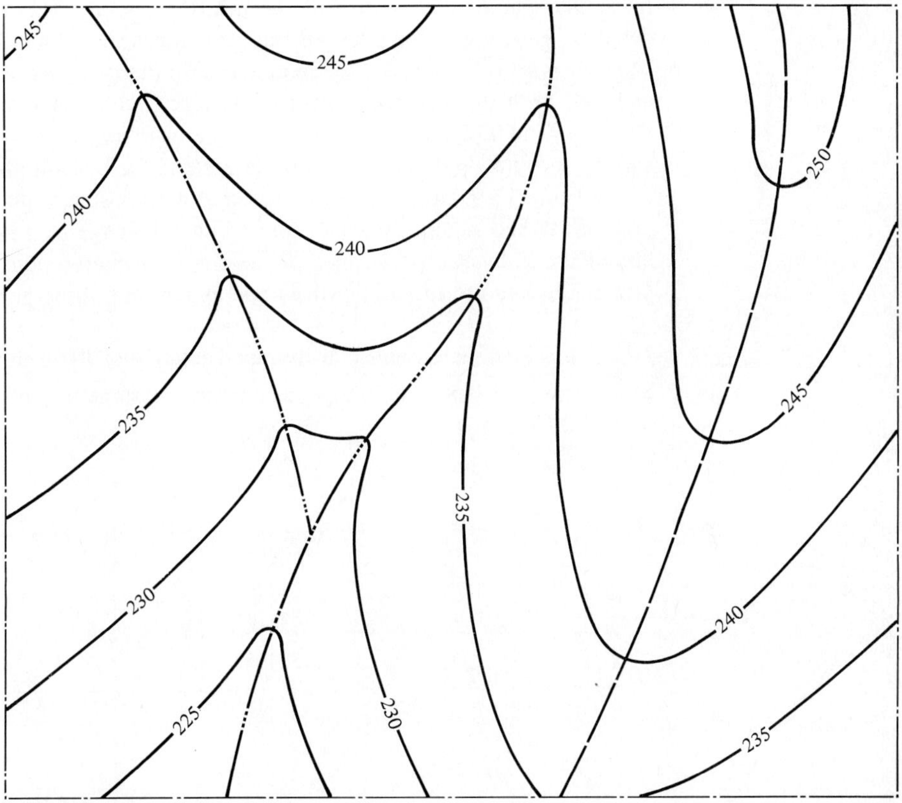

FIGURE 11.8
Topography of Fig. 11.7 area, with 5-ft. contours.

drawn between adjacent controlling points, then the construction lines are marked at points where designated-interval contours cross. In the example, integer 5-ft contours are used. Do not draw construction lines between points on opposite sides of stream beds or ridge lines, no matter how close such points appear.

Referring to the guidelines in Sec. 11.1, connect same-elevation contour crossing points with smooth, logical, lightly drawn curves. This can be a messy business requiring a good deal of erasing and redrawing. When the contour lines have achieved a satisfactory appearance, trace them onto a clean sheet of paper, omitting the controlling-point locations and elevations. Broaden every fifth contour and label it with its correct elevation. Figure 11.8 shows a completed contour map for the previously determined data.

11.7 GRID-POINT METHOD

Where topography is complex, it is better to use controlling points; where it is relatively even, it may be simpler to lay out a grid of intersecting, evenly spaced parallel lines, the spacing chosen to suit the needs of the site. If the spacing is too broad, detail is lost and contour lines are too generalized; if too narrow, more data points are taken than actually needed. Elevations at grid intersections are determined with level and rod. With the grid-point method, assume a constant spacing between grid-point elevations unless there is information to the contrary; drainage paths, ridge lines, and summit locations provide such information and these should be carefully sketched in.

Figure 11.9 shows an area gridded at a 50-ft-square interval with corner elevations, a marked summit, and the approximate drainage path sketched in. The latter is quite easy to locate with respect to nearby grid intersections. The general rules for grid contouring are essentially the same as for controlling-point contouring:

1. Assume each data point is truly representative of the nearby topography.
2. If there is no information to the contrary, assume a constant slope between data points.

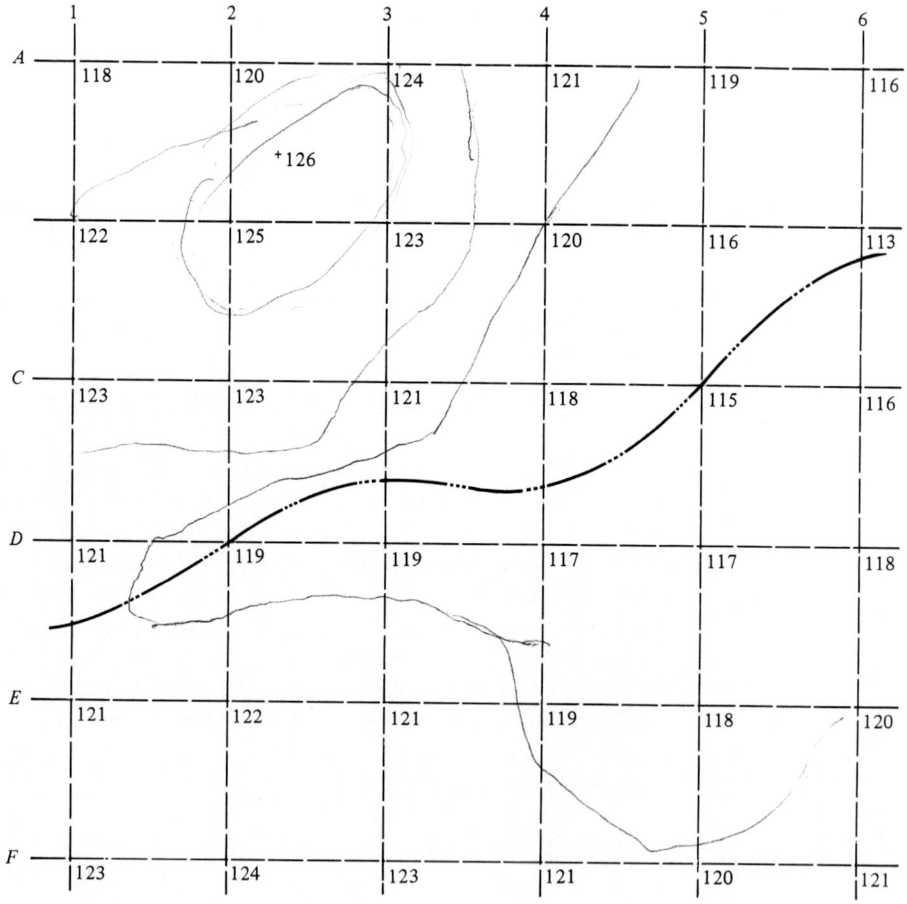

FIGURE 11.9
Grid-point plot, showing drainage and elevations.

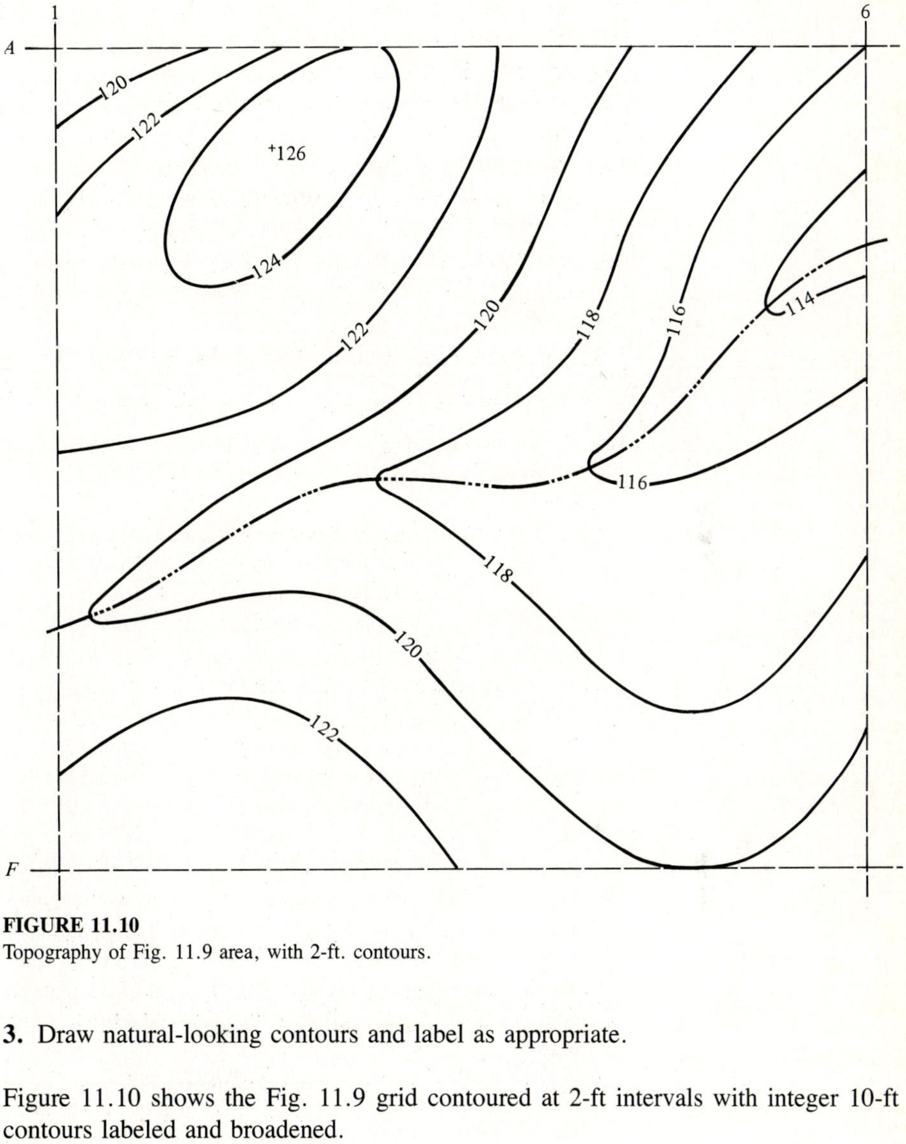

FIGURE 11.10
Topography of Fig. 11.9 area, with 2-ft. contours.

3. Draw natural-looking contours and label as appropriate.

Figure 11.10 shows the Fig. 11.9 grid contoured at 2-ft intervals with integer 10-ft contours labeled and broadened.

11.8 PROBLEMS

11.1. Express the following scales as ratios:
(*a*) 1 in. = 20 ft
(*b*) 1 cm = 400 m
(*c*) 1 in. = $\frac{1}{4}$ mile

11.2. Convert the following ratio scales to ground equivalence; that is, the number of feet on the ground equivalent to 1 in. on the map.

(*a*) 1:3600
(*b*) 1:6000
(*c*) 1:63,360

11.3. When drawing a contour line, would it be correct to allow it to branch into two contours of the same elevation?

11.4. On a 1:4800-scale map, a line is drawn perpendicular to two contours whose elevation difference is 10 ft. If the map distance along the line between the two contours is $1\frac{5}{8}$ in., what is the slope of the ground surface along that line, expressed in percent?

11.5. The scaled distance between two adjacent 2-ft contours is $3\frac{1}{16}$ in. on a map whose scale is 1 in. = 20 ft. What is the ground-surface slope along the scaled line, to the nearest minute?

Referring to Probs. 11.6 and 11.7 below, a map is to be drawn on a sheet of 24 in. × 36 in. drafting paper whose long axis is to lie east-west. Map scale must be a multiple of ten; that is, 1 in. = 10 ft or 1 in. = 20 ft or 1 in. = 30 ft, and so forth.

11.6. Determine the largest scale that can be used to map a parcel of land with the following corner coordinates in feet, maintaining a 1-in. minimum border on all sides of the figure.

Property corner A : $x = 3456.12; y = 1498.63$
Property corner B : $x = 3644.71; y = 2071.32$
Property corner C : $x = 5019.37; y = 1671.63$
Property corner D : $x = 3718.41; y = 843.98$

11.7. Determine the largest scale that can be used to map a parcel of land with the following corner coordinates in feet, maintaining a 1-in. minimum border on all sides of the figure.

Property corner A : $x = 8844.19; y = 4618.73$
Property corner B : $x = 9113.62; y = 4237.46$
Property corner C : $x = 9746.49; y = 4084.88$
Property corner D : $x = 8817.88; y = 3962.71$
Property corner E : $x = 8619.32; y = 4333.01$

11.8. Why is ground control needed in photogrammetric mapping?

11.9. The stadia field notes below were obtained from an instrument setup at station *E*, elevation 555.7 ft, to each listed controlling point. The stadia interval factor was 100.0 and the center hair was placed on the *HI* height for each observation. For each controlling point, compute distances from station *E* to the nearest foot and elevations to the nearest 0.1 ft.

Sta.	*I*,ft	Vert. angle	Azimuth	Description
A	6.23	−5°53′	149°52′	SE property corner
B	7.58	−1°36′	227°00′	SW property corner
C	~~6.38~~	~~−0°46′~~	278°57′	NW property corner
D	3.34	−6°45′	074°43′	NE property corner
1	5.33	−6°58′	177°28′	Drainage CL on PL *A–B*
2	3.69	−6°39′	223°55′	Drainage CL; direction change
3	5.57	−2°25′	256°34′	Drainage CL on PL *B–C*
4	1.21	+4°57′	314°42′	Ridge high point on PL *C–D*
5	2.18	−5°39′	140°30′	Ridge CL
6	5.29	−5°52′	142°24′	Ridge CL on PL *A–D*

11.10. The prism-pole field notes below were obtained from an instrument setup at station E, elevation 555.7 ft, to each listed controlling point. Instrument height and prism height were the same for all observations. For each controlling point, compute distances from E to the nearest foot and elevations to the nearest 0.1 ft.

Sta.	s,ft	Vert. angle	Azimuth	Description
A	619.67	−5°53′	149°52′	SE property corner
B	757.55	−1°36′	227°00′	SW property corner
C	638.51	−0°46′	278°57′	NW property corner
D	331.52	−6°45′	074°43′	NE property corner
1	528.92	−6°58′	177°28′	Drainage CL on PL A–B
2	366.52	−6°39′	223°55′	Drainage CL; direction change
3	556.81	−2°25′	256°34′	Drainage CL on PL B–C
4	120.16	+4°57′	314°42′	Ridge high point on PL C–D
5	217.32	−5°39′	140°30′	Ridge CL
6	526.11	−5°52′	142°24′	Ridge CL on PL A–D

11.11. Using the field and computed data from Prob. 11.9 or 11.10, prepare a topographic map with a scale of 1 in. = 200 ft and a 10-ft contour interval.

11.12. The stadia field notes below were obtained from a theodolite setup at station E (elev. 380.7 ft), the high point on the northwest end of an L-shaped ridge, to each listed controlling point. The stadia interval factor was 100.0 and the center hair was placed on the *HI* height for each observation. For each controlling point, compute distances from station E to the nearest foot and elevations to the nearest 0.1 ft.

Sta.	I,ft	Vert. angle	Azimuth	Description
A	1.38	−2°22′	304°40′	NW property corner
B	3.30	−0°54′	074°51′	NE property corner
C	3.92	−0°10′	125°32′	SE property corner
D	2.61	−1°26′	205°35′	SW property corner
1	1.10	+0°03′	169°24′	High point at bend of ridge
2	1.92	+0°02′	136°12′	High point at SE end of ridge
3	1.14	−2°19′	256°00′	High point on PL A–D
4	0.88	−1°38′	018°04′	High point on PL A–B
5	1.85	−0°56′	089°01′	Center of NE slope

11.13. The prism-pole field notes below were obtained from a theodolite setup at station E (elev. 380.7 ft), the high point on the northwest end of an L-shaped ridge, to each listed controlling point. Instrument height and prism height were the same for all observations. For each controlling point, compute distances from E to the nearest foot and elevations to the nearest 0.1 ft.

Sta.	s,ft	Vert. angle	Azimuth	Description
A	138.31	−2°22′	304°40′	NW property corner
B	330.19	−0°54′	074°51′	NE property corner
C	391.84	−0°10′	125°32′	SE property corner

(continued)

Sta.	s,ft	Vert. angle	Azimuth	Description
D	261.27	−1°26′	205°35′	SW property corner
1	109.92	+0°03′	169°24′	High point at bend of ridge
2	192.33	+0°02′	136°12′	High point at SE end of ridge
3	114.39	−2°19′	256°00′	High point on PL *A–D*
4	88.10	−1°38′	018°04′	High point on PL *A–B*
5	185.22	−0°56′	089°01′	Center of NE slope

11.14. Using the field and computed data from Prob. 11.12 or 11.13, prepare a topographic map with a scale of 1 in. = 100 ft and a 1-ft contour interval.

11.15. The stadia field notes below were obtained from a theodolite setup at occupied property corner *A*, elevation 239.8 ft. The stadia interval factor was 100.0. The center hair was placed on the *HI* height (5.10 ft) for each observation except points 4 and 10. For each controlling point, compute distances from station *A* to the nearest foot and elevations to the nearest 0.1 ft.

Sta.	*I*,ft	Vert. angle	Azimuth	Description
B	2.14	+4°08′	294°03′	SW property corner
C	3.81	+3°00′	320°28′	NW property corner
D	3.32	+3°34′	000°47′	N property corner
E	3.07	+1°50′	059°32′	E property corner
1	0.50	+3°12′	294°03′	W drainage CL on PL *A–B*
2	2.00	+3°09′	328°32′	W drainage CL
3	3.59	+2°44′	328°06′	W drainage CL on PL *C–D*
4	3.29	+4°58′ on 9.18′	349°46′	W ridge CL on PL *C–D*
5	2.23	+4°15′	349°46′	W ridge CL
6	1.07	+3°48′	347°08′	W ridge CL
7	0.68	+0°10′	059°52′	E drainage CL on PL *A–E*
8	1.78	+2°27′	026°10′	E drainage CL
9	2.88	+2°51′	015°14′	E drainage CL on PL *D–E*
10	2.80	+4°03′ on 8.57′	027°00′	E ridge CL on PL *D–E*
11	2.25	+3°08′	038°35′	E ridge CL
12	1.89	+2°31′	059°52′	E ridge CL on PL *A–E*

11.16. The prism-pole field notes below were obtained from a theodolite setup at property corner *A* (elev. 239.8 ft) to each listed controlling point. Instrument height and prism height were the same for all observations. For each controlling point, compute distances from *A* to the nearest foot and elevations to the nearest 0.1 ft.

Sta.	s,ft	Vert. angle	Azimuth	Description
B	213.36	+4°08′	294°03′	SW property corner
C	380.82	+3°00′	320°28′	NW property corner
D	331.24	+3°34′	000°47′	N property corner

(continued)

Sta.	s,ft	Vert. angle	Azimuth	Description
E	306.86	+1°50′	059°32′	E property corner
1	50.58	+3°12′	294°03′	W drainage CL on PL *A–B*
2	199.70	+3°09′	328°32′	W drainage CL
3	358.71	+2°44′	328°06′	W drainage CL on PL *C–D*
4	328.30	+4°15′	349°46′	W ridge CL on PL *C–D*
5	222.21	+4°15′	349°46′	W ridge CL
6	106.94	+3°48′	347°08′	W ridge CL
7	68.00	+0°10′	059°52′	E drainage CL on PL *A–E*
8	178.26	+2°27′	026°10′	E drainage CL
9	287.81	+2°51′	015°14′	E drainage CL on PL *D–E*
10	279.63	+3°20′	027°00′	E ridge CL on PL *D–E*
11	224.84	+3°08′	038°35′	E ridge CL
12	189.18	+2°31′	059°52′	E ridge CL on PL *A–E*

11.17. Using the field and computed data from Prob. 11.15 or 11.16, prepare a topographic map with a scale of 1 in. = 100 ft and a 5-ft contour interval.

11.18. The stadia field notes below were obtained from a theodolite setup at occupied control station *E*, the top center of a dam, elevation 326.8 ft. The stadia interval factor was 100.0. The center hair was placed on the *HI* height for all observations. Compute distances from station *E* to the nearest foot and elevations to the nearest 0.1 ft.

Sta.	I,ft	Vert. angle	Azimuth	Description
A	7.46	+0°06′	243°31′	SW property corner
B	7.17	+0°10′	290°15′	NW property corner
C	3.93	−1°54′	049°02′	NE property corner
D	4.36	−0°17′	137°30′	SE property corner
1	6.71	−0°07′	265°33′	W drainage CL on PL *A–B*
2	4.68	−0°33′	247°25′	W drainage CL
3	2.65	−1°28′	247°25′	W drainage CL at pond
4	2.38	−1°38′	277°01′	NW pond shoreline
5	1.31	−2°58′	304°58′	N pond shoreline at dam base
6	1.27	−3°04′	177°57′	S pond shoreline at dam base
7	3.98	−0°21′	286°30′	General terrain between pond and property corner *B*
8	3.87	−0°07′	310°31′	General terrain on PL *B–C*
9	2.89	±0°00′	331°47′	N end dam CL on PL *B–C*
10	2.53	−1°31′	007°05′	Downstream dam toe on PL *B–C*
11	0.69	−6°58′	049°02′	Downstream dam toe on CL of drainage
12	3.23	−2°23′	065°30′	E drainage CL on PL *C–D*
13	3.60	+0°01′	153°29′	Downstream dam toe on PL *D–A*
14	3.62	−0°50′	205°26′	S end dam CL on PL *D–A*

11.19. The prism-pole field notes below were obtained from a theodolite setup at control station *E* at the top center of a dam (elev. 326.8 ft) to each listed controlling point. Instrument height and prism height were the same for all observations. For each controlling point, compute distances from E to the nearest foot and elevations to the nearest 0.1 ft.

Sta.	s,ft	Vert. angle	Azimuth	Description
A	746.34	+0°06′	243°31′	SW property corner
B	717.41	+0°10′	290°15′	NW property corner
C	393.04	−1°54′	049°02′	NE property corner
D	436.11	−0°17′	137°30′	SE property corner
1	671.09	−0°07′	265°33′	W drainage CL on PL A–B
2	468.02	−0°33′	247°25′	W drainage CL
3	265.02	−1°28′	247°25′	W drainage CL at pond
4	238.27	−1°38′	277°01′	NW pond shoreline
5	131.05	−2°58′	304°58′	N pond shoreline at dam base
6	127.44	−3°04′	177°57′	S pond shoreline at dam base
7	398.01	−0°21′	286°30′	General terrain between pond and property corner B
8	386.88	−0°07′	310°31′	General terrain on PL B–C
9	289.46	±0°00′	331°47′	N end dam CL on PL B–C
10	252.95	−1°31′	007°05′	Downstream dam toe on PL B–C
11	68.83	−6°58′	049°02′	Downstream dam toe on CL of drainage
12	322.74	−2°23′	065°30′	E drainage CL on PL C–D
13	359.86	+0°01′	153°29′	Downstream dam toe on PL D–A
14	362.34	−0°50′	205°26′	S end dam CL on PL D–A

11.20. Using the field and computed data from Prob. 11.18 or 11.19, prepare a topographic map with a scale of 1 in. = 200 ft and a 2-ft contour interval.

Problems 11.21 through 11.24 each involve a 300-ft by 400-ft tract divided into 100-ft square grids. Grid-corner elevations are listed in their proper orientation, the NW corner elevation at top left, the NE corner elevation at top right, and so forth. Each problem requires preparation of a map at a scale of 1 in. = 100 ft, using the indicated contour interval.

11.21. Using the following grid-corner elevations, prepare a topographic map with a 2-ft contour interval.

101.5	103.3	104.6	105.8	106.8
104.4	106.1	106.9	108.0	107.7
107.2	109.0	110.5	111.6	109.0
110.0	111.8	114.9	111.5	109.2

11.22. Using the following grid-corner elevations, prepare a topographic map with a 1-ft contour interval.

229.7	229.1	227.6	225.5	226.8
230.6	229.0	227.1	226.6	227.5
231.2	230.2	230.7	230.1	229.9
232.3	232.5	233.0	233.1	232.9

11.23. Using the following grid-corner elevations, prepare a topographic map with a 5-ft contour interval.

323	321	316	314	313
313	323	325	327	324
322	327	332	339	332
319	321	323	324	326

11.24. Using the following grid-corner elevations, prepare a topographic map with a 2-ft contour interval.

170.9	172.0	175.1	176.5	174.6
175.7	176.8	178.4	182.2	176.4
176.5	180.8	178.2	177.0	175.2
174.6	177.0	177.2	173.8	171.7

CHAPTER 12

CONSTRUCTION SURVEYS

12.1 INTRODUCTION

The phrase "construction surveys" can be confusing because much of what is done by licensed surveyors (or by contractors acting as their own, on-site surveyors) involves either preconstruction layout or verification that ongoing construction is taking place where it should be. To a greater or lesser extent, the surveying activities discussed to this point are construction-related. Examples are right-of-way leveling, transit field operations, traverse, stadia and prism-pole operations, and topographic surveys. The final two chapters have an even stronger construction flavor: horizontal and vertical curves are covered in Chap. 13; cross sections and volumes in Chap. 14.

Since all of these construction-related topics have broad application, it would have been inappropriate to confine them to a single chapter labeled "construction." Instead, this chapter is limited to those surveying concepts and tasks that are not only construction-specific, but are ordinarily carried out on-site by contractors, both before and during construction.

12.2 BUILDING SITING

BUILDABLE AREA. From the land developer's standpoint, the shape and size of building lots are the result of a series of decisions concerning subdivision of a much larger tract of land: where streets and utilities will go, what part of the development will be used for residential construction, what part will be set aside for commercial buildings, schools and parks, and so forth. The contractor, on the other hand, is concerned with an individual lot and how to go about placing a building on it.

The contractor must locate a building on the right property, clear of **easements** and **setbacks**. An easement is that portion of the property in which the community or a utility has a reasonable and definable interest, to the extent that the owner's full use of the property is restricted. A **street easement** provides for future maintenance, or even enlargement, of the roadway. A street easement is described in terms of distance from a street centerline, normally extending well onto the adjacent property. A **utility easement** ensures access by utility people to that portion of the property lying adjacent to sewer, water, power, telephone, and cable lines. When all utility lines are at the front of the lot, the utility easements usually fall within the street easement; when utility lines are placed along the rear property line, the lot will have both a street easement in front and a utility easement at the rear. A rear utility easement is normally specified in terms of a distance in from the rear property line. A **drainage easement**, often situated along a rear or side property line, is for the purpose of maintaining a drainage path for rainwater runoff.

A setback defines an area adjacent to each property line on which no building construction is allowed. Setback distances are ordinarily written into the zoning code for a municipality. Although setbacks are usually more restrictive than easements, it often occurs that setback lines and easement lines coincide.

Figure 12.1 shows a rectangular building lot 90 ft wide and 100 ft deep situated along a residential street. Distances from the street centerline to the front property line and to the front easement are 20 ft and 40 ft, respectively; thus, no construction is allowed over the front 20 ft of the property. The 25-ft rear utility easement and 10-ft side setbacks prohibit construction in those areas. **Buildable area**, shaded in Fig. 12.1, is 55 ft deep (100-ft lot depth less the 20-ft front easement and the 25-ft rear easement) and 70 ft wide (90-ft lot width less two 10-ft side setbacks).

SITING WITHIN THE BUILDABLE AREA. Because construction profits are generally a function of house size, the builder will often place a relatively large house on a given lot, or at least a relatively wide house. However, a subdivision consisting of maximum-width houses squeezed onto each lot will have an undesirably crowded, regimented appearance that could detract from the prices asked. A little leeway, then, is good. Consider the case of a 60-ft-wide by 40-ft-deep house to be situated, in some manner, within the buildable area (shaded) of the Fig. 12.1 lot. If the builder wishes to place the sides of the building parallel to the property lines, there will be a total leeway of 10 ft, side to side, and 15 feet, front to back. The usual situation in many older subdivisions is to place the house as close to the street as possible (to increase back yard size) and center the house between the side property lines to maximize the distance from adjacent houses. The end result—hardly pleasing to the eye—is row upon row of evenly spaced houses facing primly toward the street.

A welcome change from this regimented appearance takes place when some or most of the houses are angled with respect to the street. The problem for the individual builder is to ensure that one or more of the foundation corners do *not* cross the boundary of the buildable area.

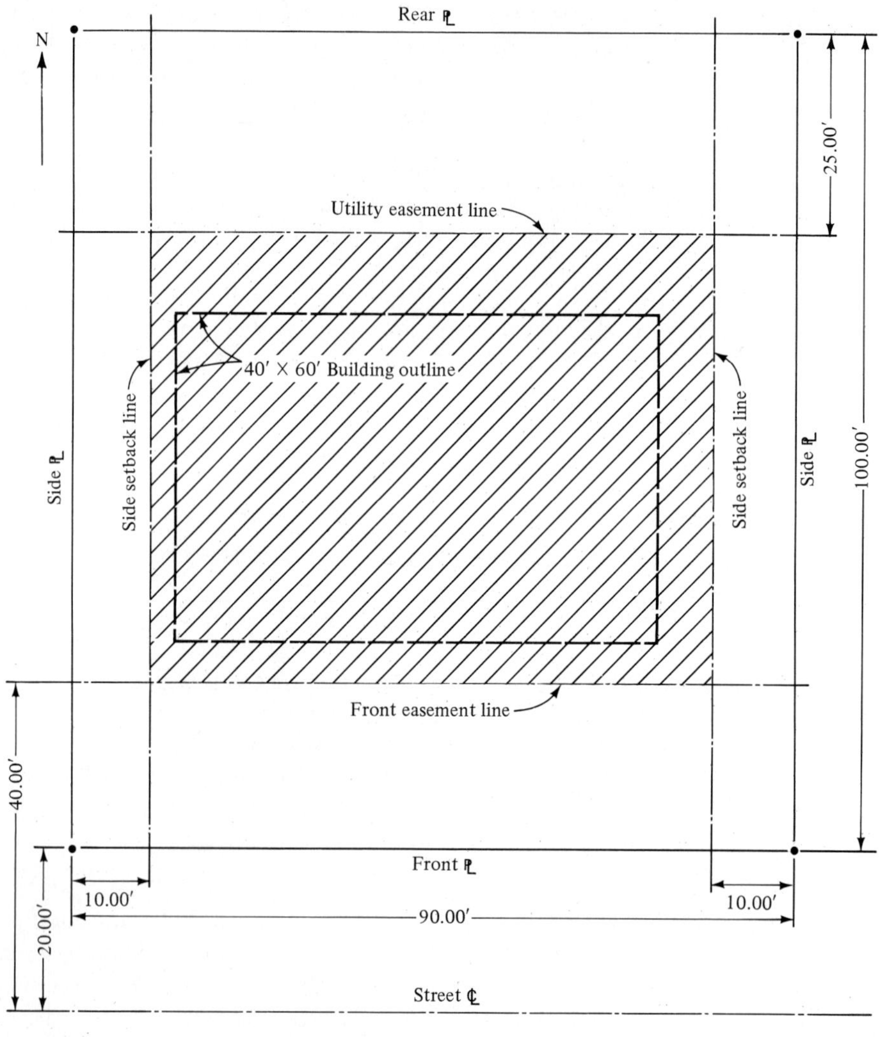

FIGURE 12.1
Buildable area within easement lines and setback lines.

Figure 12.2 shows a rectangular foundation that has been turned through an angle ϕ, at which point the front and back corners touch the front and rear easement lines. The angle made by the foundation diagonal with the foundation front face is β. The foundation swing is limited by the depth D, of the buildable area. From the angles and dimensions of Fig. 12.2, it is seen that $\sin(\phi + \beta) = D/\text{diagonal}$, or $\phi + \beta = \sin^{-1}(D/\text{diagonal})$. This leads to

$$\boldsymbol{\phi_D = \sin^{-1}\left(\frac{D}{\text{diagonal}}\right) - \beta} \qquad \textbf{(12.1)}$$

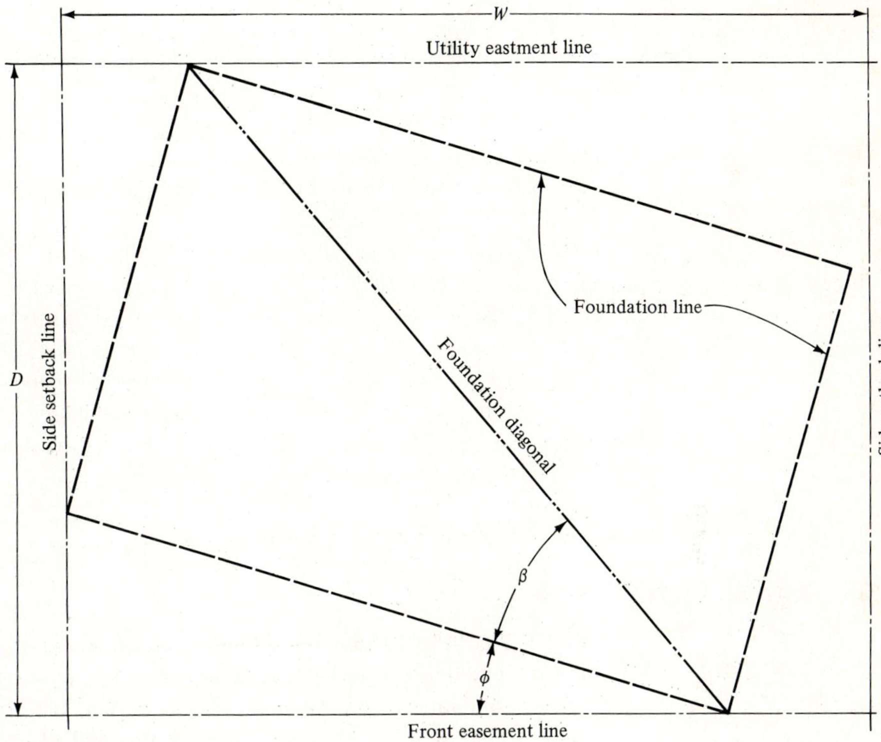

FIGURE 12.2
Permissible foundation rotation within buildable area.

On the other hand, if the amount of swing is restricted by the width, W, of the buildable area, the following equation applies:

$$\boldsymbol{\phi}_W = \sin^{-1}\left(\frac{\mathbf{W}}{\mathbf{diagonal}}\right) - \boldsymbol{\beta}' \tag{12.2}$$

where β' is the *complement* of β, or $(90° - \beta)$. The maximum swing angle is the smaller of ϕ_D or ϕ_W.

Example 12.1. What is the maximum value of the foundation swing angle for a 60.00-ft-wide by 40.00-ft-deep rectangular foundation that is to be placed within the Fig. 12.1 buildable area?

Solution. The foundation diagonal is $\sqrt{60^2 + 40^2} = 72.111$ ft. $\beta = \tan^{-1}\frac{40}{60} = 33°41'24''$ and $\beta' = 90° - \beta = 56°18'36''$. Dimensions W and D are 70.00 ft and 55.00 ft, respectively. Substituting into Eqs. (12.1) and (12.2),

$$\phi_D = \sin^{-1}\left(\frac{55.00}{72.111}\right) - 33°41'24''$$

$$= 49°42'14'' - 33°41'24'' = 16°00'50''$$

$$\phi_W = \sin^{-1}\left(\frac{70.00}{72.1111}\right) - 56°18'36''$$

$$= 76°06'08'' - 56°18'36'' = 19°47'32''$$

From this, it is seen that the *depth* of the buildable area is the more restrictive dimension for this foundation size and that the maximum swing angle is therefore 16°00′50″.

To be on the safe side, it is better to allow a small additional leeway in foundation placement. In the Example 12.1 case, it is probably better to limit swing angle to 15°00′ or 15°30′.

Computation of tilt angle ϕ is simpler with rectangular foundations. Nevertheless, the principle remains the same no matter what the shape; that is, there is always at least one diagonal that defines the maximum allowable swing of the foundation. Diagonal length may be determined from a quick examination of the building's foundation plan.

12.3 FOUNDATION LAYOUT

STAKEOUT BY BASE LINE AND OFFSET. Consider the simple problem of setting a foundation parallel to a setback or easement line. Assume that the 60-ft by 40-ft rectangular foundation in Example 12.1 is to be laid out with its west side 2.50 ft into the buildable area from the west setback line and its street side 4.00 ft into the buildable area from the front easement line, as shown in Fig. 12.3. Thus, the west side will be 12.50 ft in from the west property line (10.00-ft side setback plus 2.50 ft) and the street side will be 24.00 ft from the front property line (20.00 ft from the front property line to the front easement line plus 4.00 ft). The front and back sides of the foundation will thus lie 24.00 ft and 64.00 ft, respectively, from the front property line and the west and east sides of the foundation will lie 12.50 ft and 72.50 ft, respectively, from the west property line.

From an instrument setup at *A*, the southwest property corner, point *B* is established 24.00 ft from *A* along the line of sight from *A* to the northwest property corner. Point *C* is established 64.00 ft from *A* along the same line of sight. The instrument is moved to *B* and, after taking a foresight on the northwest property corner, a 90° angle is turned to the right. Foundation corners 1 and 4 are then staked out 12.50 ft and 72.50 ft from *B*, respectively, along the 90° line of sight. The instrument is then set up over point *C*, a backsight is taken on property corner *A*, a 90° angle is turned to the left, and foundation corners 2 and 3 are staked out. After staking, both diagonals are measured as a check on the work.

The base-line–and–offset method is easy to understand and, for simple foundations laid parallel to property lines, the layout procedure is fairly easy to perform. On the other hand, when the shape of the foundation is fairly complex and/or when foundations are angled from the property line, the angle-and-distance method is often better.

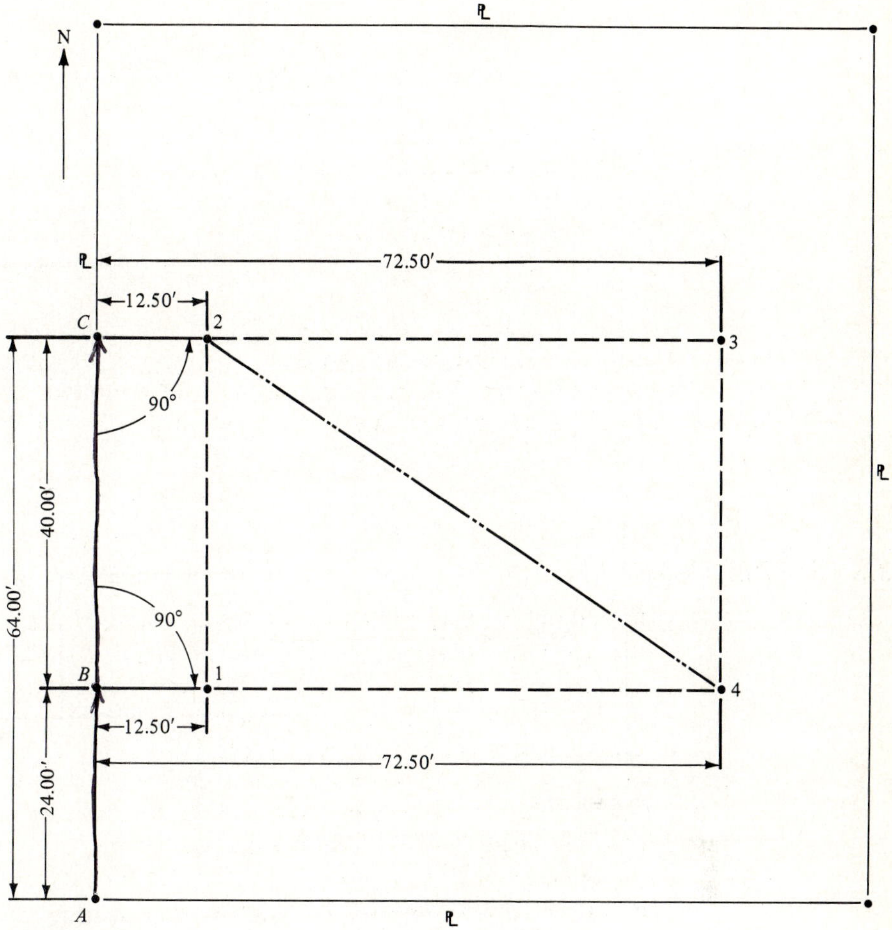

FIGURE 12.3
Stakeout plan, base line and offset method.

STAKEOUT BY ANGLE AND DISTANCE. In Fig. 12.4, a foundation is to be built parallel to the setback and easement lines. Each of its 10 foundation corners (numbered 0 through 9) is to be staked out. Corner 0 is the origin of a coordinate system whose x axis parallels the front easement line and whose y axis parallels the side-setback line. The instrument is set over point 0 and a zero-azimuth point is offset the same distance from the side setback line as point 0. Corner points 1 through 9 are then staked out by turning angles β_1 through β_9 and measuring distances d_{0-1} through d_{0-9}.

Angles β_i and distances d_i are computed from foundation dimensions as follows:

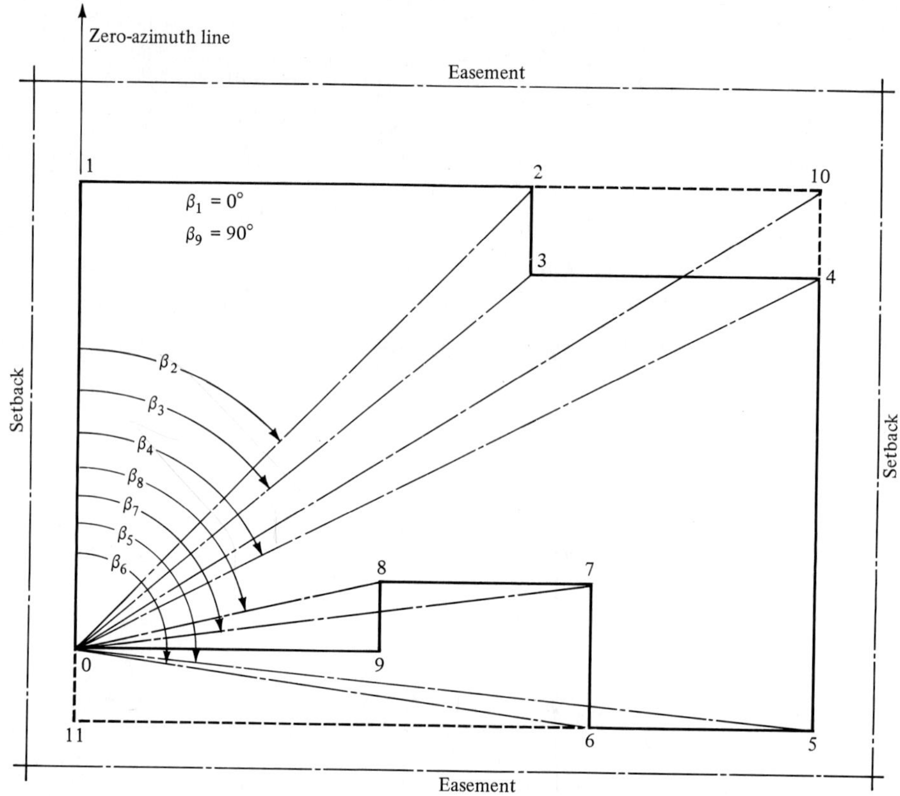

FIGURE 12.4
Foundation layout by angles and distances.

$$\beta_2 = \tan^{-1}\frac{x \text{ distance}_{1-2}}{y \text{ distance}_{0-1}}; \quad d_{0-2} = \sqrt{(x \text{ distance}_{1-2})^2 + (y \text{ distance}_{0-1})^2}$$

$$\beta_3 = \tan^{-1}\frac{x \text{ distance}_{1-2}}{y \text{ distance}_{0-1} + y \text{ distance}_{2-3}};$$

$$d_{0-3} = \sqrt{(x \text{ distance}_{1-2})^2 + (y \text{ distance}_{0-1} + y \text{ distance}_{2-3})^2}$$

and so forth, noting that the algebraic sign for distances conforms to the usual *xy*-coordinate convention, with *y* positive up and negative down, *x* positive right and negative left. It is somewhat simpler to consider the angle and distance equations in the following terms:

$$\boldsymbol{\beta}_i = \mathbf{tan}^{-1}\frac{\Sigma \mathbf{x}_{0-i}}{\Sigma \mathbf{y}_{0-i}} \tag{12.3}$$

$$\mathbf{d}_{0-i} = \sqrt{\Sigma \mathbf{x}_{0-i}^2 + \Sigma \mathbf{y}_{0-i}^2} \tag{12.4}$$

where β_i designates the angle, taken at the origin foundation corner 0, from the zero-azimuth direction to the *i*th foundation corner, d_{0-i} designates the distance from the

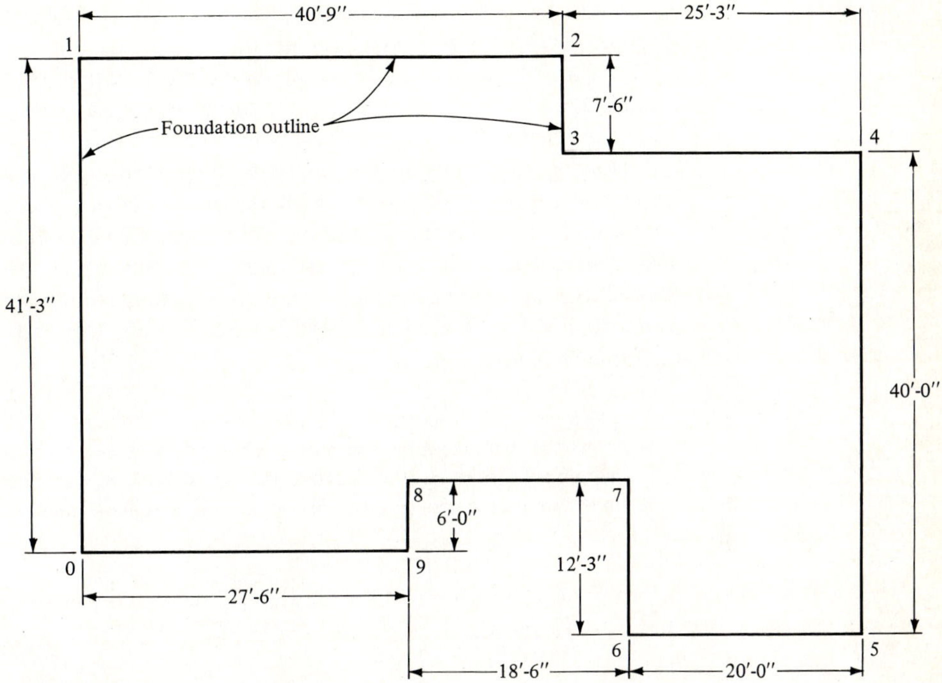

FIGURE 12.5
Side dimensions of the Fig. 12.4 foundation.

foundation corner at 0 to the ith foundation corner, Σx_{0-i} is the *cumulative* sum of the x distances from the foundation corner at 0 to the ith foundation corner, and Σy_{0-i} is the *cumulative* sum of the y distances from the foundation corner at 0 to the ith foundation corner.

Example 12.2. Figure 12.5 shows side dimensions for the Fig. 12.4 foundation layout. For an instrument set up at 0, and taking a common backsight on corner 1, compute angles and distances to foundation corners 1 through 9.

Solution. The format below provides for the systematic solution of Eq. (12.3) and (12.4).

		***x* distance, ft**		***y* distance, ft**			
Side	**Cumul.**	**Side**	**Σx**	**Side**	**Σy**	β	d_{0-i}**, ft**
0−1	0−1	±0.00	±0.00	+41.25	+41.25	00°00.0′	41.25
1−2	0−2	+40.75	+40.75	±0.00	+41.25	44°39.0′	57.98
2−3	0−3	±0.00	+40.75	−7.50	+33.75	50°22.1′	52.91
3−4	0−4	+25.25	+66.00	±0.00	+33.75	62°55.0′	74.13
4−5	0−5	±0.00	+66.00	−40.00	−6.25	95°24.6′	66.30
5−6	0−6	−20.00	+46.00	±0.00	−6.25	97°44.2′	46.42
6−7	0−7	±0.00	+46.00	+12.25	+6.00	82°34.1′	46.39
7−8	0−8	−18.50	+27.50	±0.00	+6.00	77°41.5′	28.15
8−9	0−9	±0.00	+27.50	−6.00	±0.00	90°00.0′	27.50
9−0	0−0	−27.50	±0.00	±0.00	±0.00		

As before, the completed stakeout is checked by measuring one or more computed diagonals between nonorthogonal corners.

Another approach "squares out" the building outline by setting points 1, 10, 5, and 11 (Fig. 12.4) using the angle-and-distance method, stretching strings along the perimeter, then setting the other points by taped distances from the string lines.

Laying out an angled foundation by angle and distance is almost as simple as laying out a parallel foundation with respect to setback and easement lines. First, the location of the instrument-setup point is determined from a previously computed swing angle ϕ and from appropriate foundation dimensions. Then, swing angle ϕ is applied to each computed β value, adding if the foundation is to be swung clockwise, subtracting if it is to be swung counterclockwise. In either event, all d values remain the same as before.

Example 12.3. The Fig. 12.5 residence is to be rotated 12°00.0′ clockwise to align the street face with the ground contours. Corner 5 is to lie 0.50 ft inside the front easement and corner 0 is to lie 0.50 ft inside the side setback. Locate points 0 and 5 with respect to the front easement line and the adjacent side setback line.

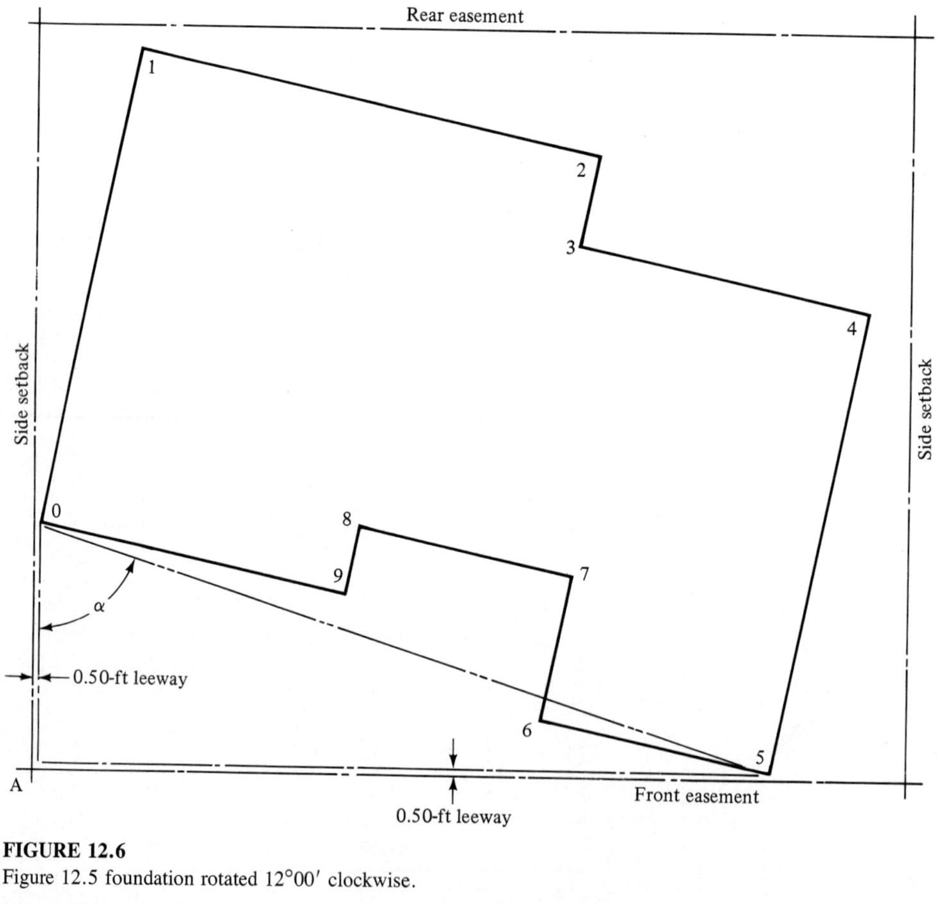

FIGURE 12.6
Figure 12.5 foundation rotated 12°00′ clockwise.

Solution. Figure 12.6 shows the solution geometry, the angle at 0 being α. The β angle to corner point 5 is 107°25.6′ (the 95°25.6′ value determined in Example 12.2 plus the 12°00.0′ clockwise swing); thus, the value of α is 72°34.4′ (180° − 107°25.6′). The side leg of the solution triangle is $d_{0-5} \cos \alpha$, or 66.30 cos 72°34.4′ = 19.86 ft. Thus, the y distance to point 0 from point A is 19.86 ft plus the 0.50 ft leeway = 20.36 ft and the x distance is 0.50 ft.

The street leg of the triangle is $d_{0-5} \sin \alpha$, or 66.30 sin 72°34.4′ = 63.26 ft. Thus, the x distance to point 5 from point A is 63.26 ft plus the 0.50 ft leeway = 63.76 ft and the y distance is 0.50 ft.

Note that the location of point 5 is not needed except as a partial check of the subsequent stakeout.

One method for staking out the Example 12.3 foundation corners is to establish corner 0, set the instrument up over it, establish a zero-azimuth line parallel to the side-setback line, turn the 12°00.0′ clockwise angle to the new line from 0 to 1, and establish the location of corner point 1. The horizontal circle is then reset to 00°00.0′, the vertical cross hair is set on point 1, and the angles on all subsequently points are taken to the values computed in Example 12.3.

Corner points may also be staked out by leaving the original azimuth on the instrument's horizontal circle and adding 12°00.0′ to *each* of the angles, β_i, to corner points 2 through 9. Distances d_{0-1} through d_{0-9} are of course the same as computed in Example 12.2.

12.4 SETTING AND LEVELING FORMS

The purpose of setting foundation corners is to locate the foundation forms, ordinarily built on-site using dimension lumber and plywood. Unfortunately, foundation forms hide the corner points as soon as construction begins. But corner-point locations must be known so foundation position and alignment may be checked during all phases of foundation construction. To do this, corner-point references are established out of the immediate construction area, usually with batter boards, illustrated in Fig. 12.7.

Batter boards are built from 2″-by-4″ vertical pieces and a 1″-by-6″ or 2″-by-6″ horizontal crosspiece. The 2-by-4's are driven into the ground a short distance from

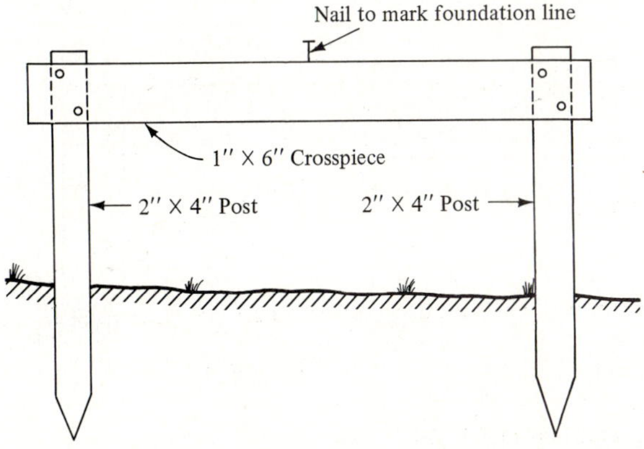

FIGURE 12.7
Batter board.

a corner point, straddling a foundation-wall line. The horizontal crosspiece is nailed between the 2-by-4's with its top at the elevation of the top surface of the completed foundation. A nail is then driven partway home into the crosspiece, on line with the foundation-wall line.

Figure 12.8 shows the Fig. 12.5 foundation layout with batter boards in place, one on each side of each foundation-wall line. With batter boards in place, both the position and height of any foundation-wall form may be checked by stretching a string taut between the batter-board nails at the two line ends—no need for an instrument to be on hand. This may be done whenever it is convenient, and by whatever crew happens to be working on foundation-form setup.

Batter boards are quite satisfactory for residential foundation construction. For large building sites, involving heavy equipment, the vulnerability of batter-board arrays might preclude their use. In this case, foundation-form alignment and elevation may be periodically reestablished from offset control points, using either a theodolite or a laser-alignment instrument such as the one shown in Fig. 12.9. For elevation reference alone, laser levels (Chap.3) are easy to use and are finding increasing popularity on construction sites.

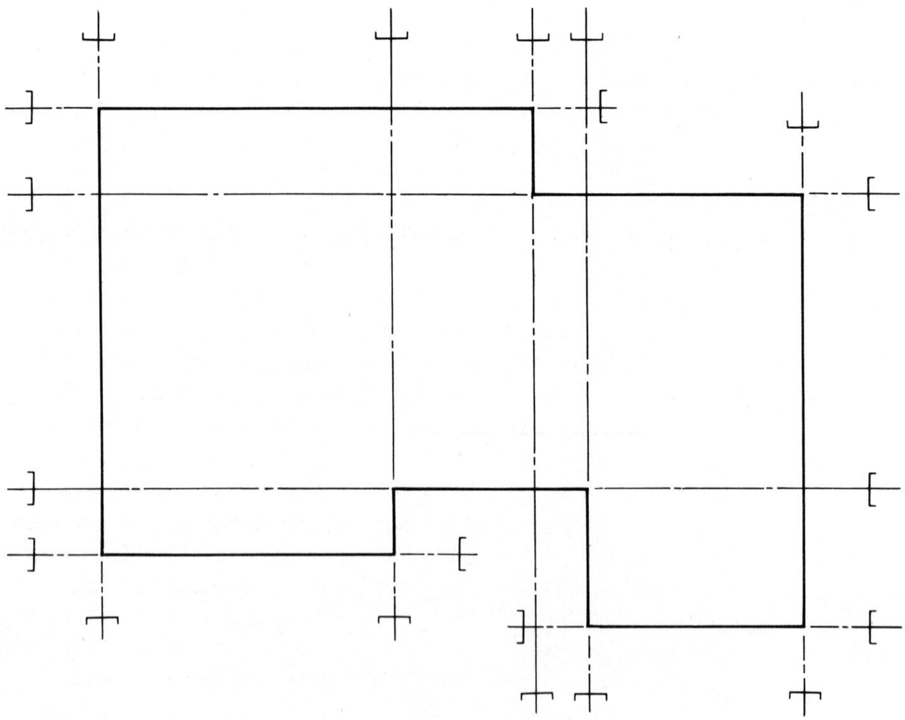

FIGURE 12.8
Figure 12.5 foundation, showing batter board positions.

FIGURE 12.9
Beam Aligner 4700. (*Courtesy of Laser Alignment, Inc.*)

12.5 PIPELINES

PIPELINE GRADES. Storm sewers and sanitary sewers are put in place with sufficient grade to make the lines self-cleaning under gravity flow. Pipes of average roughness require a constant, minimum grade of 0.50% for storm sewers and 0.67% for sanitary sewers; with smooth-walled pipe, grades can be less. Grade is not critical for pipelines under pressure, such as water mains.

TRADITIONAL PROCEDURE. The older method of establishing a gravity-flow pipeline is with batter boards straddling the pipe trench. Figure 12.10, left, shows a batter board whose crosspiece is nailed a fixed distance above the desired elevation of the pipe "invert"; that is, above the inside bottom of the pipe. The grade of the batter boards is thus the same as the grade of the invert. A nail on the crosspiece marks the centerline of the pipeline and a string held taut between successive nails provides a running elevation against which a fixed-height mark on a vertically held measuring rod may be set. The right side of Fig. 12.10 is a side view of a simple measuring rod used to maintain the string-line–to–invert distance.

The method for setting the elevation of the batter-board crosspieces is somewhat similar to profile leveling. The pipeline trench is dug to somewhat below its final depth and the batter-board vertical pieces driven at regular intervals—say, 50 ft—along the pipeline right of way. A level is set up and its *HI* established from nearby vertical control. The surveyor, having determined beforehand the desired elevations of the batter-board crosspieces, computes intermediate foresight values (IFSs) needed to establish those elevations from the instrument *HI*. Each crosspiece is raised or lowered until the proper IFS value is achieved, then nailed in place.

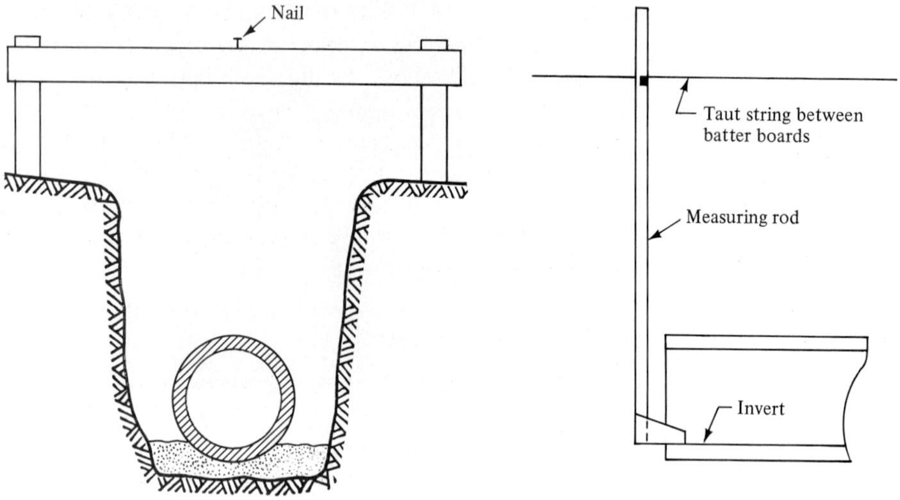

FIGURE 12.10
Batter board set over pipeline trench. Inset: Simple measuring rod set on pipe invert.

FIGURE 12.11
Setting pipeline grade with Fig. 12.9 laser aligner. (*Courtesy of Laser Alignment, Inc.*)

LASER ALIGNMENT. Batter boards can be awkward to set up and they can get in the way of construction. A newer procedure involves the use of a visible-light laser.

Figure 12.11 shows the Fig. 12.9 laser aligner propped on a piece of concrete block to place it at the desired elevation. The pipe grade has been set into the instrument (0.400%, according to the digital readout) and the laser line of sight is being projected along the intended centerline at that grade. Sand in the trench bottom (called "bedding") is shoveled under the pipe section to bring the two ends to their proper elevations. As additional pipe lengths are added, only the far end need be positioned with respect to the laser beam.

12.6 TUNNELING

It is becoming apparent that the professional surveyor is ordinarily not needed on a construction site once work is underway. Given the shortage of surveyors, this is just as well. Most contractors with a basic understanding of surveying and a working knowledge of today's easy-to-use instruments are quite capable of handling about all the surveying tasks likely to occur on a given construction project.

Tunneling is an activity that once required frequent monitoring by surveyors to insure that work was progressing in the proper direction and along the appropriate grade. Now, however, automated tunneling machines are available that are guided by a laser beam. About the only task the surveyor must perform is to establish the location and direction of the laser-alignment device with respect to the tunnel centerline, then provide reference points for repositioning the laser if it should be moved or jarred out of position.

12.7 PROBLEMS

12.1. Though many construction surveying operations may be done by the contractor, what particular associated function must almost always be handled by a licensed surveyor?

12.2. A title search is normally conducted prior to issuance of title insurance on a piece of property. What information should be obtained during the title search that may be relevant to the positioning of a structure on the property?

12.3. What factors, other than required setbacks and the existence of easements, might bear upon the positioning of a structure on a piece of property?

The buildings in Probs. 12.4 through 12.7 are to be positioned on rectangular building lots 70.00 ft wide and 100.00 ft deep. The minimum setback requirements are 25.00 ft front and rear, 8.00 ft on the sides.

12.4. What is the maximum value of the foundation swing angle for a 47.50-ft-wide by 40.00-ft-deep rectangular foundation?

12.5. What is the maximum value of the foundation swing angle for a 44.00-ft-wide by 40.00-ft-deep rectangular foundation?

12.6. What is the maximum value of the foundation swing angle for a 48.00-ft-wide by 45.00-ft-deep rectangular foundation?

12.7. What is the maximum value of the foundation swing angle for a 50.00-ft-wide by 40.00-ft-deep rectangular foundation?

In solving Probs. 12.8 through 12.15, assume (1) all properties are rectangular and large enough to accommodate the specified buildings, (2) buildings are to lie parallel to the property lines with foundation walls lying along the specified setback lines, and (3) the instrument will be set up over the left, front property corner with a zero initial setting on the left, rear property corner.

12.8. Calculate the angles and distances required to lay out a rectangular building 50.75 ft wide by 32.00 ft deep if the front setback is 25.00 ft and the side setback is 7.50 ft.

12.9. Calculate the angles and distances required to lay out a rectangular building 67.25 ft wide by 43.33 ft deep if the front setback is 30.00 ft and the side setback is 7.00 ft.

12.10. Calculate the angles and distances required to lay out a rectangular building 55.00 ft wide by 40.00 ft deep if the front setback is 27.00 ft and the side setback is 8.00 ft.

12.11. Calculate the angles and distances required to lay out a rectangular building 72.00 ft wide by 35.00 ft deep if the front setback is 20.00 ft and the side setback is 8.50 ft.

12.12. Referring to Prob. 12.4, compute the angles and distances required to lay out the described building at the calculated maximum swing angle, turned counterclockwise.

12.13. Referring to Prob. 12.5, compute the angles and distances required to lay out the described building at the calculated maximum swing angle, turned clockwise.

12.14. Referring to Prob. 12.6, compute the angles and distances required to lay out the described building at the calculated maximum swing angle, turned counterclockwise.

12.15. Referring to Prob. 12.7, compute the angles and distances required to lay out the described building at the calculated maximum swing angle, turned clockwise.

In solving Probs. 12.16 through 12.19, refer to the sketch below showing the side property-line bearings and the front property-line bearing for a cul-de-sac building lot. The instrument is to be set over the left front property corner with a zero initial setting along the westerly property boundary. *A* and *B* refer to the front corners of a rectangular building. *Hint:* As a first step, use the numbered and lettered corners in the sketch to aid in solving missing-data traverse 0-1–*A*-*B*-2-3-0.

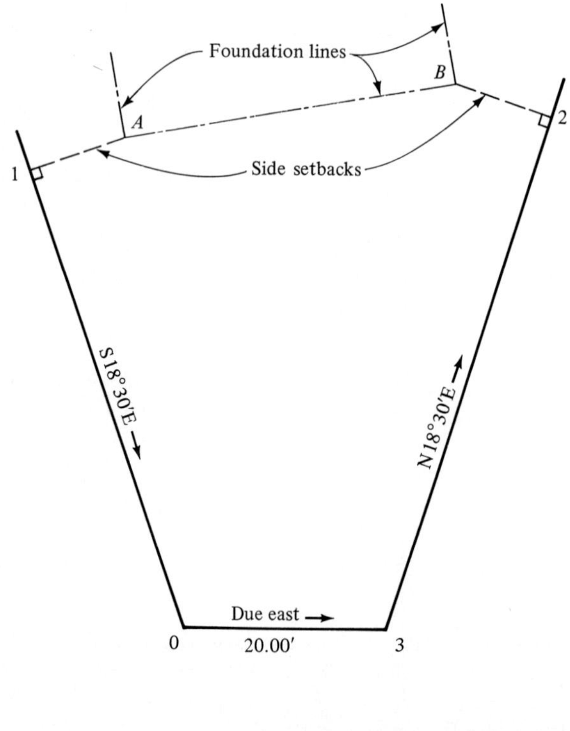

12.16. Dimension *AB* is 34.00 ft and *AB* bears N80°.00′E. Side setback (perpendicular to the side property lines) is 10.00 ft. Calculate the angles and distances to lay out building corners *A* and *B* if the two corners are to just touch their respective side setback lines.

12.17. Dimension *AB* is 36.50 ft and *AB* bears S80°00′E. Side setback (perpendicular to the side property lines) is 7.50 ft. Calculate the angles and distances to lay out building corners *A* and *B* if the two corners are to just touch their respective side setback lines.

12.18. Dimension *AB* is 35.67 ft and *AB* bears due east. Side setback (perpendicular to the side property lines) is 9.00 ft. Calculate the angles and distances to lay out building corners *A* and *B* if the two corners are to just touch their respective side setback lines.

12.19. Dimension *AB* is 37.25 ft and *AB* bears N85°00′E. Side setback (perpendicular to the side property lines) is 8.00 ft. calculate the angles and distances to lay out building corners *A* and *B* if the two corners are to just touch their respective side setback lines.

12.20. A level with a stadia interval factor of 100.0 was set up over the centerline of a storm sewer trench, leveled, and pointed along the sewer centerline. If the end of each pipe section was set so the lower stadia hair fell on a constant reading for a rod held on the pipe invert, what would be the grade of the sewer line in percent?

12.21. A transit was set up exactly on line between two consecutive manhole covers over a storm sewer line. Stadia shots, shown below with readings in feet, were then taken to rods held vertically on the centers of the manhole covers. If $K = 100.0$, what is the difference in elevation between the two manhole covers?

	Intercept	Vertical angle	Center hair
Manhole *A*	1.76	−2°16′	4.17
Manhole *B*	2.39	+3°42′	5.89

12.22. In Prob. 12.21 above, if the vertical distance from the top of the manhole cover to the sewer invert was 5.35 ft at *A* and 8.76 ft at *B*, what is the percent grade of the sewer line between *A* and *B*?

CHAPTER 13

CURVES

13.1 INTRODUCTION TO CIRCULAR AND PARABOLIC CURVES

CIRCULAR CURVES. The horizontal layout of a transportation right-of-way is characterized by straight lines, called **tangents**, that are connected by horizontal curves. For the most part, such curves have constant radius; that is, they are **circular curves**. Figure 13.1 shows a series of tangents that intersect at deflection angles Δ_i and are connected by circular curves of radius R_i.

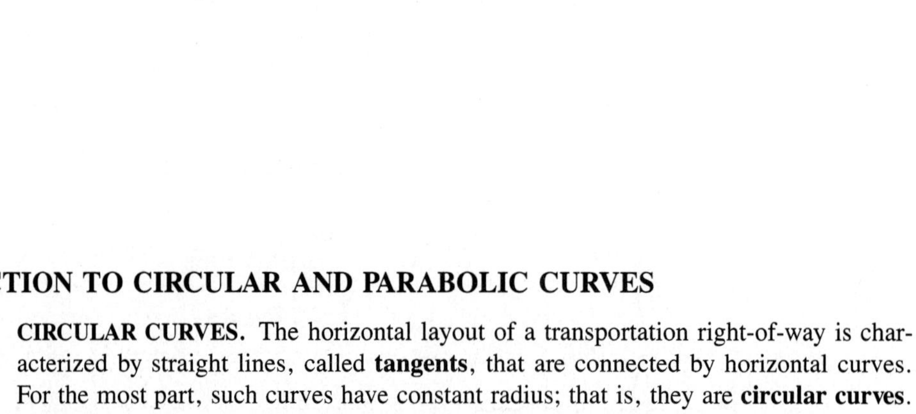

FIGURE 13.1
Intersecting tangents connected by horizontal circular curves.

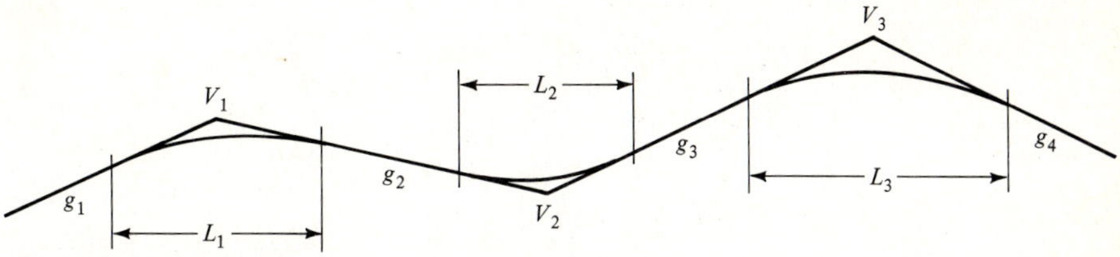

FIGURE 13.2
Intersecting grades connected by vertical curves.

On occasion, particularly on railroads, a circular curve is joined to a tangent with a transition curve to eliminate sudden, violent changes in lateral acceleration going into or out of a circular curve. This is because railroad rolling stock must follow the course of the rails; highway vehicles, on the other hand, are afforded a good deal of steering latitude entering or leaving horizontal curves. Except for speed-change ramps leading on and off freeways, highway curves are for the most part circular curves. Transition segments on horizontal curves may consist of multiple-radius circular arcs or, more often, they may be spirals. In either event, analysis of transition segments is left to route surveying texts—discussion of horizontal curves in this text will be limited to circular curves.

PARABOLIC CURVES. Just as intersecting tangents are connected by horizontal curves, intersecting transportation *grades* are connected by vertical curves. Because of the acceleration problem, circular curves make inappropriate vertical curves, so grades are connected by **parabolic curves**. A parabolic curve affords constant vertical acceleration for vehicles moving at constant speed. Figure 13.2 shows a profile of a series of grades g_i that intersect at V_i and are connected by parabolic vertical curves of *horizontal* length L_i.

13.2 CIRCULAR CURVE TERMINOLOGY

CURVATURE. A circular curve may be defined by **degree of curve**, or D, with D_c the angle subtended by a 100-ft chord ("chord definition") and D_a the angle subtended by a 100-ft arc ("arc definition").

Chord-definition curves. The circular-curve segment in Fig. 13.3 has a 100-ft chord that subtends angle D_c. If D_c were 2°, the circular curve would be called a 2° curve, by chord definition.

From Fig. 13.3, sin $(D_c/2) = 50/R$. Transposing this expression leads to two equations that relate R to D_c:

$$D_c = 2\sin^{-1}\frac{50}{R} \tag{13.1a}$$

and

$$R = \frac{50}{\sin(D_c/2)} \tag{13.1b}$$

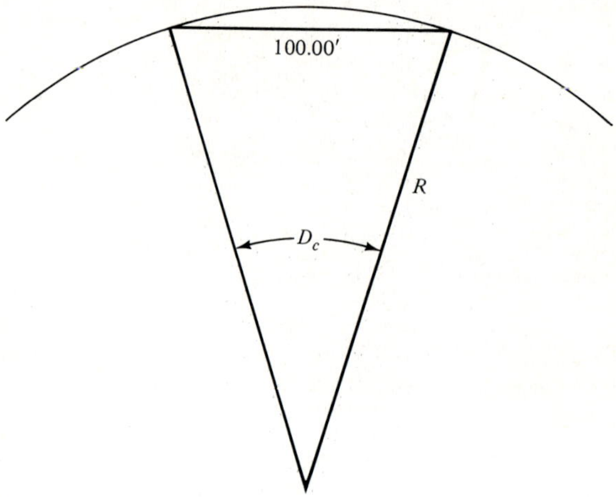

FIGURE 13.3
Chord-definition circular curve.

Current practice limits use of the chord definition to railroads, with curves defined by their degree of curve rather than radius. For this reason, Eq. (13.1*b*) has greater application than Eq. (13.1*a*).

Example 13.1. D_c for a railroad curve is 2°30′ (exactly). What is the curve radius?

Solution. Substituting into Eq. (13.1*b*),

$$R = \frac{50}{\sin\,(2°30'/2)} = 2292.01 \text{ ft}$$

Arc-definition curves. Highway surveys are characterized by stations at integer 100-ft intervals along the centerline of the right-of-way. Thus, highway curves are defined by arc-definition degree of curve—or by radius. Figure 13.4 shows a full circle of radius R whose 100-ft arc subtends angle D_a. The full circle has a total angle of 360° and a circumference of $2\pi R$. By proportion, the 100-ft arc is to subtended angle D_a as the full circumference is to 360°. That is,

$$\frac{100.00}{D_a} = \frac{2\pi R}{360°}$$

This expression leads to the companion equations

$$\mathbf{R = \frac{5729.578}{D_a}} \tag{13.2a}$$

and

$$\mathbf{D_a = \frac{5729.578}{R}} \tag{13.2b}$$

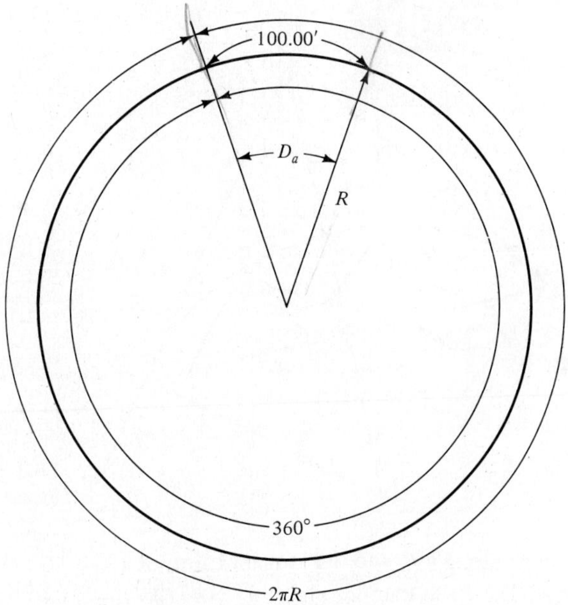

FIGURE 13.4
Arc-definition circular curve.

Example 13.2. D_a for a highway curve is 2°50′, exactly. What must the radius be?

Solution. Substituting into Eq. (13.2*a*),

$$R = \frac{5729.578}{2°50'} \Rightarrow\Rightarrow \frac{5729.578}{2.83333\ldots} = 2022.20 \text{ ft}$$

Example 13.3. A highway curve has a radius of 1500.00 ft. What is its degree of curve, to the nearest second?

Solution. Substituting into Eq. (13.2*b*),

$$D_a = \frac{5729.578}{R} = \frac{5729.578}{1500.00} = 3°49'11''$$

GEOMETRY OF CIRCULAR CURVES. Figure 13.1 shows a series of tangents that intersect at different deflection angles Δ. For any pair of intersecting tangents, there is an infinite number of circular arcs that could connect them. Curve selection requires an understanding of the various elements that constitute a single curve. Figure 13.5 defines those elements for a circular curve of radius *R* connecting two tangents that intersect at deflection angle Δ.

The intersection point of the two tangents is called the **point of intersection**, or **PI**. Assuming the survey is progressing from left to right, the beginning point of the curve, called the **point of curvature**, or **PC**, is on the left. The end point of the curve, called the **point of tangency** or **PT**, is on the right. The **tangent distance *T***

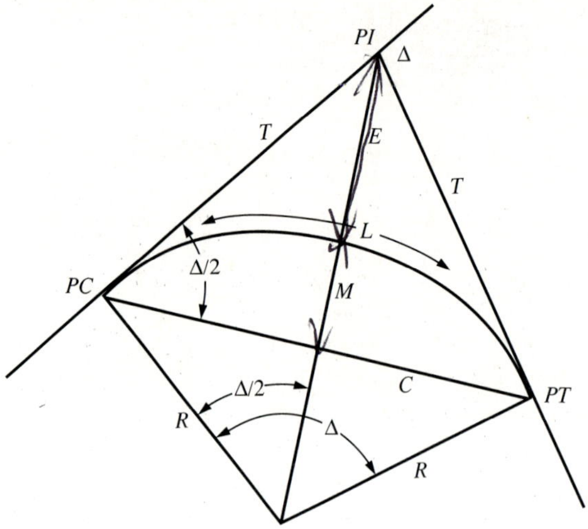

FIGURE 13.5
Elements of a circular curve.

is the distance along the entering tangent from the PC to the PI, or along the departing tangent from the PI to the PT. The full chord distance from the PC to the PT is ***C***. A line from the radius center to the PI defines two more elements: the **middle ordinate** ***M*** is the distance along that line from the full chord to the curve; the **external distance** ***E*** is the distance along that line from the curve to the PI. ***L*** is the length of the full arc of the curve, from PC to PT. The angle subtended by the full arc of length L is Δ, which, by geometry, is the same as the deflection angle of intersection, Δ. The half-angle on either side of the bisector of Δ is $\Delta/2$, which, by geometry, is the same as the angle between the full chord and the entering or departing tangent.

Inspection of Fig. 13.5 shows that the following equations apply:

$$\mathbf{T = R\tan\frac{\Delta}{2}} \tag{13.3}$$

$$\mathbf{C = 2R\sin\frac{\Delta}{2}} \tag{13.4}$$

$\text{Cos}\,(\Delta/2) = R/(E + R) \Rightarrow\Rightarrow \quad E = R/\cos\,(\Delta/2) - R$, which leads to

$$\mathbf{E = R\left[\frac{1}{\cos\,(\Delta/2)} - 1\right]} \tag{13.5}$$

Similarly, $M = R - R\cos\,(\Delta/2)$, which leads to

$$\mathbf{M = R\left(1 - \cos\frac{\Delta}{2}\right)} \tag{13.6}$$

D_a is the angle subtended by a 100-ft arc on the curve and Δ is the angle subtended by the full arc, L. Thus, 100 ft is to D_a as the full arc is to Δ; that is, $100/D_a = L/\Delta$. This leads to the expression

$$\mathbf{L = \frac{100\Delta}{D_a}} \tag{13.7}$$

But highway curves are defined in terms of R more often than D_a. Substituting D_a from Eq. (13.2*b*) into Eq. (13.7),

$$L = \frac{\Delta R}{57.29578} \tag{13.8}$$

13.3. CIRCULAR CURVE SELECTION AND COMPUTATION

PRELIMINARY CALCULATION OF CIRCULAR CURVE. A curve is specified initially by Δ, the deflection angle of intersection of the approach tangent with the departing tangent, and by the preliminary station of the PI. The circular curve that connects the tangents is defined by selecting any one of the elements R, D, T, L, E, or M. Equations (13.3) through (13.8), in turn, define the remainder of the elements.

Example 13.4. Approach and departing tangents for a highway right-of-way intersect at an angle $\Delta = 10°36'$ (exactly). If the curve is specified by $D_a = 2°30'$ (exactly), compute R, T, L, E, and M.

Solution. Substituting into Eqs. (13.2*a*) and (13.7),

$$R = \frac{5729.578}{2°30'} \Rrightarrow \frac{5729.578}{2.5} = 2291.83 \text{ ft}$$

$$L = \frac{100(10°36')}{2°30'} \Rrightarrow \frac{100(10.6)}{2.5} = 424.00 \text{ ft}$$

Substituting the value of R into Eqs. (13.3), (13.5), and (13.6),

$$T = 2291.83 \tan 5°18' = 212.61 \text{ ft}$$

$$E = 2291.83\left(\frac{1}{\cos 5°18'} - 1\right) = 9.84 \text{ ft}$$

$$M = 2291.83(1 - \cos 5°18') = 9.80 \text{ ft}$$

The other original specification for a curve is the initial station of the PI. The word "initial" is used because, once the curve is in place, the stationing of the right-of-way proceeds along the arc of the curve, which is a shorter path than by way of the PI. The station of the PC is determined by subtracting tangent distance T, in stations, from the original value of the PI; then the station of the PT is determined by adding the full arc length of the curve, L, in stations, to the station value of the PC.

Example 13.5. The station of the PI in Example 13.4 is 87+14.66 (that is, the PI lies 87 full, 100-ft stations plus 14.66 ft from the project's initial point, station 0+00). Compute the stations of the PC and the PT.

Solution. The station of the PC is the station of the PI minus the tangent distance T, in stations. From Example 13.4, $T = 212.61$ ft, or 2+12.61 stations. The format on the following page is used to solve for the PC:

$$\text{Station of PC} = \text{station of PI} - T(\text{stations})$$

$$\begin{array}{rr} \text{Station of PI} = & 87+14.66 \\ -T(\text{stations}) & \underline{-\ 2+12.61} \\ \text{Station of PC} = & 85+02.05 \end{array}$$

The station of the PT is the just-computed station of the PC plus the length of the curve computed earlier, in stations.

$$\text{Station of PT} = \text{station of PC} + L$$

$$\begin{array}{rr} \text{Station of PC} = & 85+02.05 \\ +L(\text{stations}) & \underline{+\ 4+24.00} \\ \text{Station of PT} = & 89+26.05 \end{array}$$

FINAL SELECTION AND CALCULATION OF CIRCULAR CURVE. As indicated earlier, two intersecting tangents may be connected by an infinite number of circular arcs of differing radii. Selection of a particular radius value (or selection of a particular value of any of the other elements, such as *D, T, L, E,* or *M*) to satisfy a particular design criterion will change the values of all the other elements because their defining equations are interdependent.

Example 13.6. After completing the preliminary curve calculations in Examples 13.4 and 13.5, it was decided that *L* should not exceed 400 ft, that *R* is to be an integer number of hundreds of feet, and that *R* is otherwise to be as large a value as possible. Compute the final values of *R, T, L, E, M*, and the stations of the PC and PT.

Solution. The maximum value of *R* is related to the maximum value of *L* by Eq. (13.8): $L = \Delta R/57.29578$. Transposing Eq. (13.8) and substituting the maximum value of *L*, 400.00 ft,

$$R_{\max} = \frac{57.29578L_{\max}}{\Delta} = \frac{57.29578(400.00)}{10.6^\circ} = 2161.10 \text{ ft}$$

But *R* is to be an integer 100-ft value, so the next *smaller* integer value is chosen to maintain *L* less than 400 ft; that is, $R = 2100$ ft.

The final value of *R* is substituted into Eqs. (13.8), (13.3), (13.5), and (13.6) to produce

$$L = \frac{\Delta R}{57.29578} = \frac{10.6(2100.00)}{57.29578} = 388.51\text{ft}$$

$$T = R \tan \frac{\Delta}{2} = 2100.00 \tan 5^\circ 18' = 194.81 \text{ ft}$$

$$E = R\left(\frac{1}{\cos(\Delta/2)} - 1\right) = 2100.00\left(\frac{1}{\cos 5^\circ 18'} - 1\right) = 9.02 \text{ ft}$$

$$M = R\left(1 - \cos\frac{\Delta}{2}\right) = 2100.00(1 - \cos 5^\circ 18') = 8.98 \text{ ft}$$

The stations of the PC and PT are computed as before:

$$\begin{aligned} \text{PC station} &= \text{PI station} - T \\ &= 87+14.66 \\ &\underline{-1+94.81} \\ &85+19.85 \end{aligned} \qquad \begin{aligned} \text{PT station} &= \text{PC station} + L \\ &= 85+19.85 \\ &\underline{+3+88.51} \\ &89+08.36 \end{aligned}$$

13.4 RIGHT-OF-WAY STATIONING FOR CIRCULAR CURVES

During the roadway planning stage, the right-of-way consists of a series of tangents that follow a path from PI to PI, the PIs carrying station values appropriate to that initial-design route—this is known as "old" stationing. As roadway design advances, curves are inserted, resulting in shorter distances by way of the curves than by way of the PIs. The shorter route distance requires "new" stationing for points along the right-of-way.

Figure 13.6 shows the initial roadway survey, with two PI stations and the station for a pipeline crossing carrying the old stationing values. The distances from PI_1 to PI_2 and from PI_2 to the pipeline crossing, in stations, are equal to the differences in station values between those specified points.

Figure 13.7 shows the same right-of-way with curves inserted. Distances along the right-of-way now follow the path of the curves. Connecting straight section s_1 lies between the PT of the first curve and the PC of the second curve; its length must be the old-station difference between PI_1 and PI_2 minus the sum of T_1 and T_2. Similarly, connecting straight section s_2 lies between the PT of the second curve and the pipeline; its length must be the old-station difference between PI_2 and the pipeline crossing minus T_2. The new station of the pipeline crossing must equal the sum of the following distances, in stations: station 0+00 to PC_1, L_1, s_1, L_2 and s_2.

Example 13.7. In Fig. 13.7, R_1 is 1000.00 ft and R_2 is 800.00 ft. Compute the new stationing values for the PCs, PTs, and the pipeline crossing.

Solution.

$$T_1 = R_1 \tan \frac{\Delta_1}{2} = 1000.00 \tan 25°00' = 466.31 \text{ ft}$$

$$PC_1 = \text{old } PI_1 - T_1 = 27{+}16.00 - 4{+}66.31 = 22{+}49.69$$

$$L_1 = \frac{R_1 \Delta_1}{57.29578} = \frac{1000.00(50.0000)}{57.29578} = 872.66 \text{ ft}$$

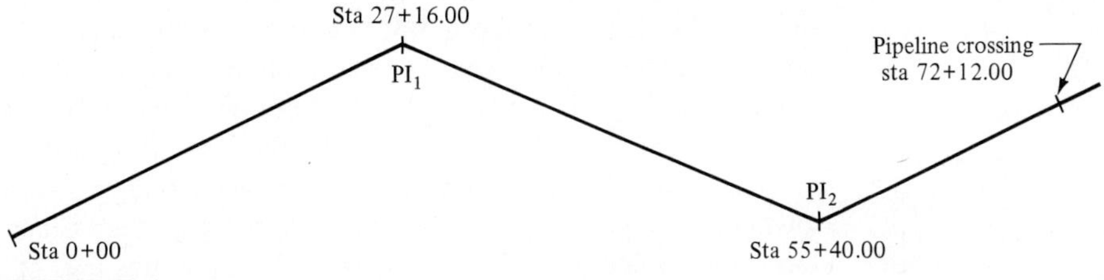

FIGURE 13.6
Initial roadway layout without curves, showing "old stationing."

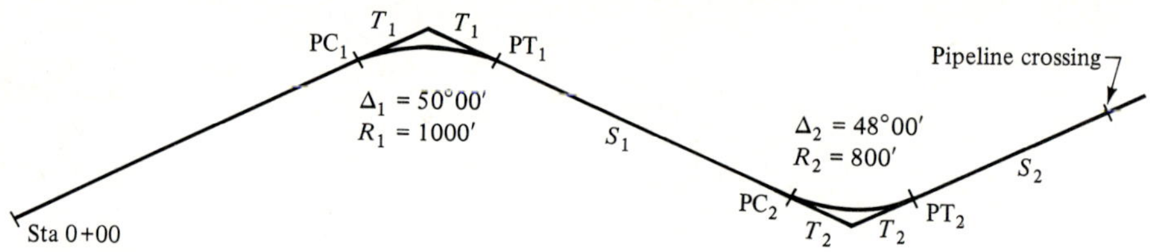

FIGURE 13.7
Figure 13.6 roadway layout, curves added.

$$PT_1 = PC_1 + L_1 = 22+49.69 + 8+72.66 = 31+22.35$$

$$T_2 = R_2 \tan \frac{\Delta_2}{2} = 800.00 \tan 24°00' = 356.18 \text{ ft}$$

$$s_1 = \text{old } PI_2 - \text{old } PI_1 - (T_1 + T_2)$$

$$= 55+40.00 - 27+16.00 - (4+46.31 + 3+56.18) = 20+01.51$$

$$PC_2 = PT_1 + s_1 = 31+22.35 + 20+01.51 = 51+23.86$$

$$L_2 = \frac{R_2\Delta_2}{57.29578} = \frac{800.00(48.0000°)}{57.29578} = 670.21 \text{ ft}$$

$$PT_2 = PC_2 + L_2 = 51+23.86 + 6+70.21 = 57+94.07$$

$$s_2 = \text{old pipeline–crossing station} - \text{old } PI_2 - T_2$$

$$= 72+12.00 - 55+40.00 - 3+56.18 = 13+15.82$$

$$\text{New pipeline-crossing station} = PT_2 + s_2$$

$$= 57+94.07 + 13+15.82 = 71+09.89$$

The distance saved by way of the curves must equal the difference in the old and new stationing at the pipeline crossing. As a check:

$$\text{Distance saved} = 2T_1 - L_1 + 2T_2 - L_2$$

$$= 2(466.31) - 872.66 + 2(356.18) - 670.21 + 102.11 \text{ ft}$$

$$72+12.00 - 71+09.89 = 1+02.11 \text{ stations} \checkmark$$

13.5 CIRCULAR CURVE LAYOUT

CURVE LAYOUT BY DEFLECTION ANGLES. Stations are staked out along a straight right-of-way by extending a line of sight along the right-of-way centerline and establishing survey points at station intervals along that line of sight. Stations on a curve, on the other hand, are located by setting up on the curve PC, then taking deflection angles to integer curve stations. The traditional layout method uses a steel tape for measuring successive chord distances from one curve station to the next. Figure 13.8 shows the geometry for laying out full 100-ft stations on the curve.

Because the PC is almost never situated on an integer station, there is almost always a less-than-100-ft arc distance, c_1', between the PC and the first full station on the curve. Subsequent full-station–to–full-station arcs are, of course, 100.00 ft. The final arc distance on the curve, from the last full station to the PT, is c_2'.

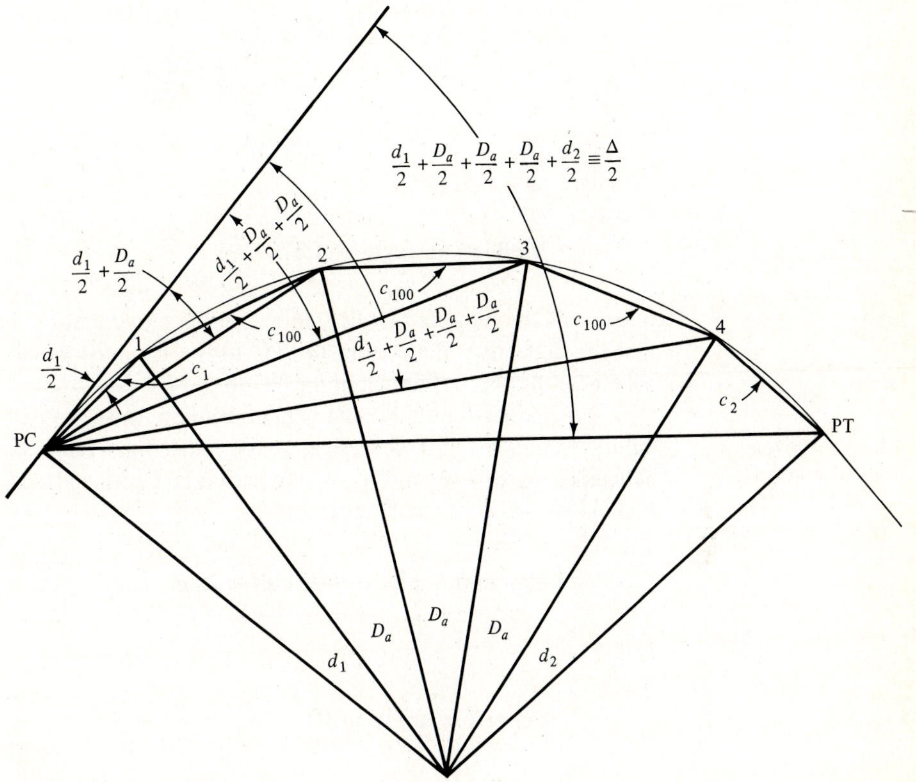

FIGURE 13.8
Deflection angles on circular curve, with station-to-station chords.

Recall that D_a is the angle subtended by a 100-ft arc; that is, by the distance between full stations along the curve. The angles subtended by arcs c_1' and c_2' are d_1 and d_2, respectively. It can be said that c_1' is to its subtended angle d_1 as a 100-ft arc on the curve is to its subtended angle D_a; that is, $c_1'/d_1 = 100/D_a$. Transposition produces the expression $d_1 = c_1'D_a/100$. From Fig. 13.8, the deflection angle from the PC to the first curve station is $d_1/2$, or

$$\frac{d_1}{2} = \frac{c_1'D_a}{200} \tag{13.9}$$

The angle subtended by the arc from the PC to the second full station must be $d_1 + D_a$; thus, the deflection angle to the second full station must be half that, or $d_1/2 + D_a/2$. From this, we see that deflection angles to full stations beyond the second station are obtained by adding successive values of $D_a/2$ to previous deflection angles, or

$$\textbf{Deflection angle to } \boldsymbol{m}\textbf{th full station} = \frac{d_1}{2} + \frac{(m-1)D_a}{2} \tag{13.10}$$

Consider now that subtended angle $d_2 = c_2' D_a/100$, leading to

$$\frac{d_2}{2} = \frac{c_2' D_a}{200} \tag{13.11}$$

The deflection angle to the PT must be the deflection angle to the last full station on the curve plus $d_2/2$, or

$$\textbf{Deflection angle to PT} = \frac{d_1}{2} + \frac{(n-1)D_a}{2} + \frac{d_2}{2} \equiv \frac{\Delta}{2} \tag{13.12}$$

where n is the number of full stations on the curve and the identity sign $\equiv$ indicates that the deflection angle from the PC to the PT *must* equal half the total angle, Δ, subtended by the full arc from the PC to the PT.

For short-radius curves, it is often better to stake out half-stations (that is, station values ending in +50 as well as +00) rather than full-station intervals. The angle subtended by a 50-ft arc along the curve is $D_a/2$, and successive deflection-angle increments are $D_a/4$, leading to

$$\textbf{Deflection angle to } m\textbf{th half-station} = \frac{d_1}{2} + \frac{(m_{50}-1)D_a}{4} \tag{13.13}$$

and

$$\textbf{Deflection angle to PT} = \frac{d_1}{2} + \frac{(n_{50}-1)D_a}{4} + \frac{d_2}{2} \equiv \frac{\Delta}{2} \tag{13.14}$$

where m_{50} is the mth half station on the curve, n_{50} is the number of half-stations on the curve, with $d_1/2$ and $d_2/2$ computed from appropriate c_1 and c_2 values—appropriate in the sense that those arc values cannot exceed 50 ft.

COMPUTATION OF BETWEEN-STATION CHORDS. With this method, illustrated in Fig. 13.8, chords are measured successively from the PC to the first station on the curve, from the first station to the second, from the second to the third, and so forth, with the final chord measured from the last station to the PT. Chord-distance equations are as follows:

$$\textbf{Chord from PC to first station, } c_1 = 2R \sin \frac{d_1}{2} \tag{13.15}$$

$$\textbf{Chord between full stations, } c_{100} = 2R \sin \frac{D_a}{2} \tag{13.16}$$

$$\textbf{Chord between half-stations, } c_{50} = 2R \sin \frac{D_a}{4} \tag{13.17}$$

$$\textbf{Chord from last station to PT, } c_2 = 2R \sin \frac{d_2}{2} \tag{13.18}$$

Example 13.8. Compute full-station deflection angles and lengths of successive chords for the circular curve in Example 13.6.

Solution. From Example 13.6, $\Delta = 10°36'$, $R = 2100.00$, the station of the PC is 85+19.85, and the station of the PT is 89+08.36. Using Eq. (13.2*b*) to find D_a,

$$D_a = \frac{5729.578}{R} = \frac{5729.578}{2100.00} = 2.728370° = 2°43'42''$$

Thus, $D_a/2 = 2.728370°/2 = 1.364185° = 1°21'51''$. Substituting into Eq. 13.16,

$$c_{100} = 2R \sin \frac{D_a}{2} = 2(2100.00) \sin 1.364185° = 99.99 \text{ ft}$$

The arc distance to the first full station from the PC, c_1', is the first full-station value minus the last station of the PC, or 86+00 − 85+19.85 = 80.15 ft. Substituting into Eq. (13.9),

$$\frac{d_1}{2} = \frac{c_1' D_a}{200} = \frac{80.15(2.728370)}{200} = 1.093394° = 1°05'36''$$

Substituting into Eq. (13.15),

$$c_1 = 2R \sin \frac{d_1}{2} = 2(2100.00) \sin 1.093394° = 80.15 \text{ ft}$$

The arc distance from the last full station to the PT, c_2', is the PT station value minus the last full-station value, or 89+08.36 − 89+00 = 8.36 ft. Substituting into Eq. (13.11),

$$\frac{d_2}{2} = \frac{c_2' D_a}{200} = \frac{8.36(2.728370)}{200} = 0.114046° = 0°06'51''$$

Substituting into Eq. (13.18),

$$c_2 = 2R \sin \frac{d_2}{2} = 2(2100.00) \sin 0.114046° = 8.36 \text{ ft}$$

Successive deflection angles are found by application of Eq. (13.10). For the deflection angle to the first station, $m = 1$ and $m - 1 = 0$; thus the deflection angle is $d_1/2$, or 1°05′36″. For the deflection angle to the second station, $m = 2$ and $m - 1 = 1$; thus, the deflection angle is $d_1/2 + D_a/2$, or 1°05′36″ + 1°21′51″ = 2°27′27″. Deflection angles to subsequent stations, obtained by adding $D_a/2$ to the previous sum, are tabulated as follows, beginning at the PC station:

Station	Deflection angle from PC
PC @ 85+19.85	0°00′00″
86+00	1°05′36″
87+00	2°27′27″
88+00	3°49′18″
89+00	5°11′09″
PT @ 89+08.36	5°18′00″

The deflection angles tabulated above are the accumulation of $d_1/2$, $d_2/2$, and $D_a/2$ values, each of which had been rounded to the nearest second. By good luck, the above deflection angle to the PT is half the value of Δ, to the second. However, it is often the case that the deflection angle to the PT, as determined from Eq. (13.12), differs from $\Delta/2$ by several seconds due to cumulative round-off error.

Example 13.9. Compute all deflection angles and chords for the Example 13.6 curve using half-stations rather than full stations.

Solution. The values of Δ, R, and D_a, as well as the stations of the PC and PT, are the same as before. The value of $D_a/4$ is needed to compute deflection angles and c_{50}, the chord for a 50-ft arc.

$$\frac{D_a}{4} = \frac{2.728370^\circ}{4} = 0.682093^\circ = 0^\circ 40' 56''$$

Substituting into Eq. (13.17),

$$c_{50} = 2R \sin \frac{D_a}{4} = 2(2100.00) \sin 0.682093^\circ = 50.00 \text{ ft}$$

The station of the PC is 85+19.85; thus, the first half-station on the curve is 85+50. In this case, $c_1' = 85{+}50 - 85{+}19.85 = 30.15$ ft. Substituting into Eq. (13.9),

$$\frac{d_1}{2} = \frac{c_1' D_a}{200} = \frac{30.15(2.728370^\circ)}{200} = 0.411302^\circ = 0^\circ 24' 41''$$

Substituting into Eq. (13.15),

$$c_1 = 2R \sin \frac{d_1}{2} = Z(2100.00) \sin 0.411302^\circ = 30.15 \text{ ft}$$

The last half-station on the curve is the same as the last full station; that is 89+00. Thus, the arc distance from the last half-station to the PT is still 8.36 ft and the values of $d_2/2$ and C_2 are the same as before: 0°06′51″ and 8.36 ft, respectively.

Deflection angles are computed from Eqs. (13.13) and (13.14), using the same format as before.

Station	Deflection angle from PC
PC @ 85+19.85	0°00′00″
85+50	0°24′41″
86+00	1°05′37″
86+50	1°46′33″
87+00	2°27′29″
87+50	3°08′25″
88+00	3°49′21″
88+50	4°30′17″
89+00	5°11′13″
PT @ 89+08.36	5°18′04″

This time, because of cumulative round-off error in the nearest-second values of deflection angles, the tabulated sum of the deflection angles from Eq. (13.14) is a few seconds different from $\Delta/2$.

COMPUTATION OF PC-TO-STATION CHORDS. With an EDM and a prism pole, it is simpler to locate stations on the curve by chord distances to each station *from the PC*, as indicated in Fig. 13.9.

Example 13.10. For the Example 13.8 deflection-angle data, compute chord distances from the PC to each station on the curve and to the PT.

Solution. The chord distance from the PC to any point is $2R$ multiplied by the sine of the deflection angle from the PC to that point. Computations are tabulated on the next page from the deflection angles determined in Example 13.8.

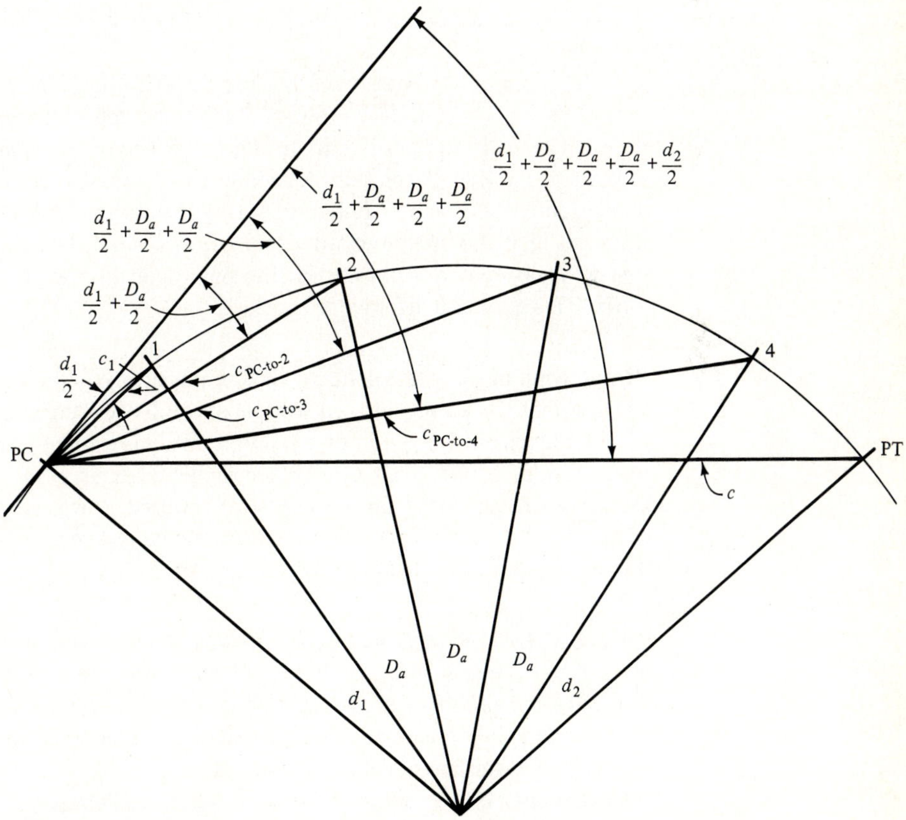

FIGURE 13.9
Deflection angles on circular curve, with PC-to-station chords.

Station	Deflection angle from PC	Chord from PC, ft, 2*R* sin (def. angle)
PC @ 85+19.85	0°00′00″	0.00
86+00	1°05′36″	80.14
87+00	2°27′27″	180.09
88+00	3°49′18″	279.94
89+00	5°11′09″	379.62
PT @ 89+08.36	5°18′00″	387.96

CURVE STAKEOUT

PI available. Once the curve is computed, each curve station, from PC to PT, will have to be staked out in the field. If the PI is available, the instrument is set up over the PI, the telescope is sighted back down the approach tangent, and the PC is staked out a distance T along that line from the PI. Deflection angle Δ is then turned to the direction of the departing tangent and the PT is staked out a distance T along that line from the PI.

The instrument is next set up over the PC, a backsight is taken in the direction of the approach tangent (i.e., toward the PI), the deflection angle to the first station is turned, chord distance c_1 is laid off along the line of sight, and a tack is set in a hub to mark the point. Thereafter, deflection angles are turned to the remaining stations on the curve and station hubs are set by measuring *either* station-to-station chord distances or PC-to-station chord distances, as desired. Because the PT was staked out first by measurement from the PI, the final deflection angle and chord measurement to the PT station are afforded a check.

PI not available. Occasionally, it will be physically impossible to set up over the PI, because the PI will fall in a canyon or in a hillside—after all, the curve was placed there for a reason, and avoiding a hill or a ravine is a sound reason. In that case, locate the station of the PC on the approach tangent, set up over the PC, backsight along the approach tangent (with the telescope inverted,) turn successive deflection angles (with the telescope direct), and proceed with the stakeout as before. The difference is that there is no final check on the location of the PT.

"MOVING UP" ON A CIRCULAR CURVE. Another real-world problem is the presence of an obstacle on a line of sight from the PC to a curve station. Figure 13.10 illustrates the problem in a simple situation in which the line of sight to the third station on the curve is obstructed by a tree. To get around this problem, the surveyor simply moves the instrument to the last station visible from the PC—in this case, the second station on the curve. From that point, a backsight is taken on the PC, with the initial set at zero; then the *same* deflection angles are taken to subsequent curve stations as would have been taken from the PC.

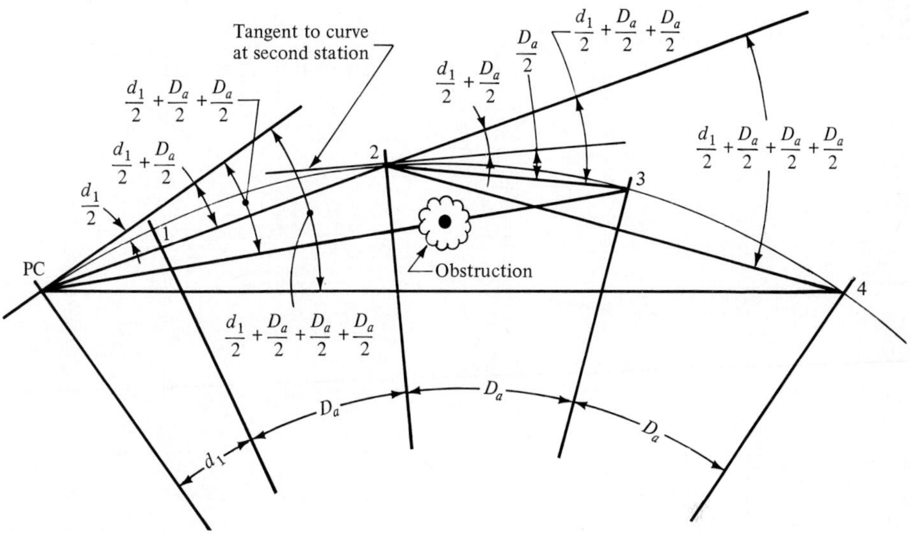

FIGURE 13.10
Obstruction interfering with curve layout.

Figure 13.10 also illustrates the geometric explanation for why this is possible. In the figure, a short line has been drawn tangent to the curve at the second station from the PC. The angle between that tangent and the chord from the PC is identical to the deflection angle from the PC to the second station: $d_1/2 + D_a/2$. The angle between the tangent and the projection of the chord from the PC *through* the second station is also $d_1/2 + D_a/2$. The angle between the tangent and the chord running from the second station to the third station is $D_a/2$. Thus, for an instrument set up on the second curve station and backsighted on the PC, the deflection angle to the third station is $d_1/2 + D_a/2 + D_a/2$, which is the same as the original deflection angle from the PC to the third station. Similarly, deflection angles from the second-curve-station setup to all subsequent stations on the curve are identical to those originally computed for the PC.

13.6 PARABOLIC CURVE TERMINOLOGY

CURVE DEFINITIONS. As discussed earlier, a parabolic curve provides a transition from one grade to the next without the sudden changes in acceleration that would occur if a circular curve were used. Figure 13.11 is a profile view of a parabolic curve that joins entering grade $\mathbf{g}_1$ to departing grade $\mathbf{g}_2$, where the units of g_1 and g_2 are *feet per station*, which is the same as percent grade. The beginning point of the vertical curve is designated as the **BVC** while the end point of the vertical curve is designated as the **EVC**. Curve length ***L*** is the *horizontal* distance, in *stations*, between the BVC and the EVC. The point of intersection of grade g_1 with grade g_2 is ***V***, which lies halfway, horizontally, between the BVC and the EVC.

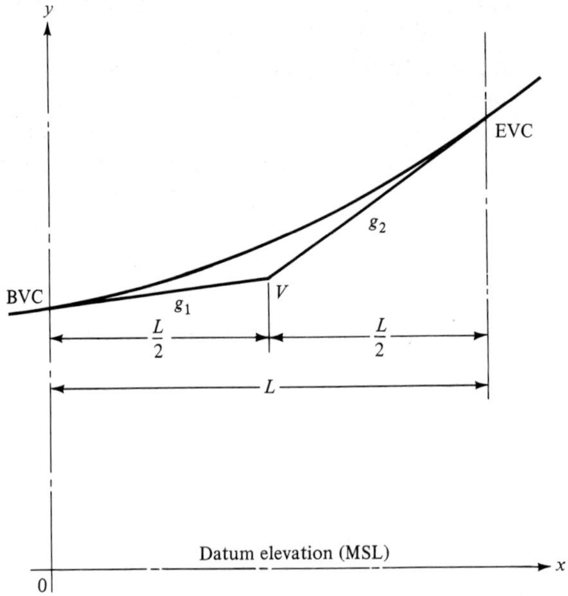

FIGURE 13.11
Parabolic curve joining two grades.

Parabolic-curve equations are further defined by the *xy*-coordinate system shown in Fig. 13.11, in which the y axis passes through the BVC and the x axis lies at the elevation of the vertical datum, usually taken as mean sea level, or MSL.

The constant-vertical-acceleration property of a parabolic curve may also be expressed as a constant rate of change of vertical grade, $\boldsymbol{r}$, from the entering grade g_1 to the departing grade g_2 over the length of the curve, L. That is,

$$r = \frac{g_2 - g_1}{L} \tag{13.19}$$

Equation (13.19) may also be expressed:

$$L = \frac{g_2 - g_1}{r} \tag{13.20}$$

Thus, for any parabolic curve connecting two grades, either r or L may be specified and the other computed from Eq. (13.19) or (13.20). To reconcile the units of g (feet per station) and L (stations), the units of r are *feet per station per station.*

The general equation for a parabola is $y = ax^2 + bx + c$. The specific equation for a parabolic curve, developed from the calculus in App. Sec. A.4, is

$$y = \frac{r}{2}x^2 + g_1 x + \text{elev}_{\text{BVC}} \tag{13.21}$$

where $\boldsymbol{y}$ is the elevation, in feet, of a point on the curve lying a distance x, in stations, from the BVC. This is in keeping with the vertical-curve convention that horizontal dimensions are in stations and vertical dimensions are in feet.

Although Fig. 13.11 shows a parabolic curve that connects two positive (upward sloping) grades, it is often the case that a vertical curve will connect a positive entering grade with a negative departing grade, in which case there will be a *high* point on the curve; or it will connect a negative entering grade with a positive departing grade, in which case there will be a *low* point on the curve. Appendix Sec. A.4 also develops, from Eq. (13.21), the following equation for determining the horizontal distance, in stations, from the BVC to a high or low point on the curve:

$$x = -\frac{g_1}{r} \tag{13.22}$$

The x distance may also be determined, without computing the value of r beforehand, by substituting $(g_2 - g_1)/L$ for r, leading to

$$x = \frac{g_1 L}{g_1 - g_2} \tag{13.23}$$

In Eq. (13.23), the positions of g_2 and g_1 in the denominator have been reversed to eliminate the negative sign that occurs in the Eq. (13.22) version.

13.7 EQUAL-TANGENT VERTICAL CURVE COMPUTATIONS

STATION ELEVATIONS ALONG A CURVE. Figure 13.11 shows an equal-tangent vertical curve; one in which the horizontal distance from the BVC to point V is the same as the horizontal distance from V to the EVC. Points on equal-tangent vertical curves are computed from the parabolic curve equations as demonstrated in the following example.

Example 13.11. For the equal-tangent parabolic curve shown in Fig. 13.12, compute the elevations of the BVC, the EVC, all full stations, and the station and elevation of the summit.

Solution. The entering grade g_1, from Fig. 13.12, is +1.50% (assume an exact value). The curve length L is the horizontal distance, in stations, from the BVC to the EVC, or eight stations (again, an exact value is assumed). The elevation of V must equal the elevation of the BVC plus half the curve length multiplied by entering grade g_1; that is, $\text{elev}_V = \text{elev}_{\text{BVC}} + g_1L/2$. Thus, $\text{elev}_{\text{BVC}} = \text{elev}_V - g_1L/2$. Substituting for L, g_1, and the elevation of point V,

$$\text{Elev}_{\text{BVC}} = 247.90 - 1.50\left(\frac{8.00}{2}\right) = 241.90 \text{ ft}$$

Similarly, $\text{elev}_{\text{EVC}} = \text{elev}_V + g_2L/2$. Substituting,

$$\text{Elev}_{\text{EVC}} = 247.90 + (-2.50)\left(\frac{8.00}{2}\right) = 237.90 \text{ ft}$$

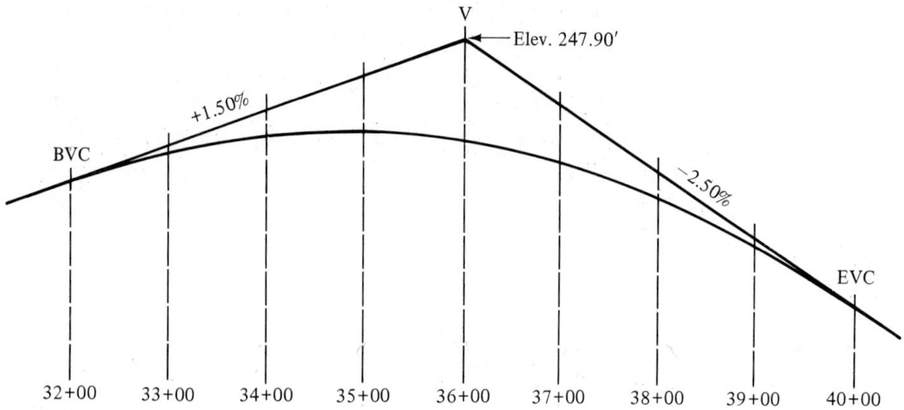

FIGURE 13.12
Example 13.11 parabolic curve.

The elevations of stations on the curve are obtained from Eq. (13.21) after computing r from Eq. (13.19). Substituting into Eq. (13.19),

$$r = \frac{g_2 - g_1}{L} = \frac{-2.50 - 1.50}{8.00} = -0.5000 \text{ ft} \cdot \text{station}^{-2}$$

In the tabular format below, the fourth, fifth, and sixth columns represent the three terms to the right of the equals sign in Eq. (13.21). The sum of those three terms (y, or elevation of x) is shown in the last column.

Station	x, sta	x^2, sta^2	$(r/2)x^2$,ft	g_1x, ft	Elev$_{BVC}$, ft	Elev$_x$, ft
32+00	0	0	0	0	241.90	241.90
33+00	1	1	−0.25	+1.50	241.90	243.15
34+00	2	4	−1.00	+3.00	241.90	243.90
35+00	3	9	−2.25	+4.50	241.90	244.15
36+00	4	16	−4.00	+6.00	241.90	243.90
37+00	5	25	−6.25	+7.50	241.90	243.15
38+00	6	36	−9.00	+9.00	241.90	241.90
39+00	7	49	−12.25	+10.50	241.90	240.15
40+00	8	64	−16.00	+12.00	241.90	237.90

Substituting into Eq. (13.22) for the value of x at the summit,

$$x_{\text{summit}} = -\frac{g_1}{r} = \frac{-(+1.50)}{(-0.500)} = 3.00 \qquad \text{(stations from the BVC)}$$

The summit station is 32+00 + 3+00 = 35+00 (elevation shown above).

INTERMEDIATE POINTS ON A PARABOLIC CURVE. To find the elevation of a point that does not lie on a full station, simply substitute into Eq. (13.21) with whatever x value is involved.

Example 13.12. Using curve data from Example 13.11, find the elevation at station 34+68.50.

Solution. Station 34+68.50 is 2.6850 stations from the BVC; thus, the value of x is 2.6850. Substituting into Eq. (13.21),

$$y = \frac{r}{2}x^2 + g_1 x + \text{elev}_{\text{BVC}}$$

$$= -\left(\frac{0.500}{2}\right)(2.6850)^2 + 1.50(2.6850) + 241.90 = 244.13 \text{ ft}$$

LOCATING A GIVEN ELEVATION ON A PARABOLIC CURVE. To find the curve station corresponding to a particular elevation, begin with the same equation, $y = (r/2)x^2 + g_1 x + \text{elev}_{\text{BVC}}$. In this case, y, the elevation of the point whose x value is desired, is already known. Thus,

$$\text{Elev}_x = \frac{r}{2}x^2 + g_1 x + \text{elev}_{\text{BVC}}$$

$$\frac{r}{2}x^2 + g_1 x + (\text{elev}_{\text{EVC}} - \text{elev}_x) = 0$$

The second equation is in the form $ax^2 + bx + c = 0$, where a is $r/2$, b is g_1, and c is $(\text{elev}_{\text{BVC}} - \text{elev}_x)$. From Chap. 1, an equation of this form may be solved for the value of x using the quadratic equation theorem [Eq. (1.2) from Chap. 1].

Substituting into Eq. (1.2) with $r/2$ for a, g_1 for b, and $(\text{elev}_{\text{BVC}} - \text{elev}_x)$ for c,

$$\boldsymbol{x = \frac{-g_1 \pm \sqrt{g_1^2 - 2r(\text{elev}_{\text{BVC}} - \text{elev}_x)}}{r}} \qquad \textbf{(13.24)}$$

Example 13.13. Using data from Example 13.11, find the curve station whose elevation is 240.00 ft.

Solution. In this case, elev_x is 240.00 ft; thus, $(\text{elev}_{\text{BVC}} - \text{elev}_x) = 241.90 - 240.00 = 1.90$ ft. Substituting into Eq. (13.24),

$$x = \frac{-1.50 \pm \sqrt{1.50^2 - 2(-0.500)(1.90)}}{-0.500} = 3.0000 \pm 4.0743$$

The value $3.0000 - 4.0743$ is negative; thus, it falls off the curve to the left. The correct value of x is $3.000 + 4.0743 = 7.0743$. The station of x is therefore 32+00 + 7+07.43, or 39+07.43.

ODD-LENGTH, ODD-STATION CURVES. The Fig. 13.12 curve is a simple illustration in which V is situated directly on a full station and L is an exact, even number of stations in length. Often, however, V occurs between stations and L is not an integer number of hundreds of feet in length, let alone an even number of hundreds. In this case, computations are just the same as in Example 13.11, though the numerical values are slightly more awkward to deal with.

Example 13.14. An entering grade of −3.60% meets a departing grade of +2.20% at station 119+60, the two grades intersecting at an elevation of 714.10 ft above MSL. If the two grades are connected by a 650-ft equal-tangent curve, compute the elevations of the BVC, EVC, and all full stations, as well as the location and elevation of the low point on the curve.

Solution. Substituting into Eq. (13.19) to find the value of r,

$$r = \frac{g_2 - g_1}{L} = \frac{2.20 - (-3.60)}{6.50} = +0.8923 \text{ ft·station}^{-2}$$

$$\text{Elev}_{\text{BVC}} = \text{elev}_V - \frac{g_1 L}{2} = 714.10 - \frac{(-3.60)(6.50)}{2} = 725.80 \text{ ft}$$

$$\text{Elev}_{\text{EVC}} = \text{elev}_V + \frac{g_2 L}{2} = 714.10 + \frac{(+2.20)(6.50)}{2} = 721.25 \text{ ft}$$

The BVC station is the station of $V - L/2$ = 119+60 − 3+25 = 116+35. The EVC station is the station of $V + L/2$ = 119+60 + 3+25 = 122+85.

The first station on the curve must be station 117+00, so distance x to the first station is 0.65 station. All other stations on the curve are at one-station intervals from the first station; that is, x = 1.65, 2.65, 3.65 and so forth. The elevations of the curve stations are tabulated below in the same manner as in Example 13.11.

Station	x, sta	x^2, sta^2	$(r/2)x^2$, ft	$g_1 x$, ft	Elev_{BVC}, ft	Elev_x, ft
116+35	0	0	0	0	725.80	725.80
117+00	0.65	0.4225	0.188	−2.340	725.80	723.65
118+00	1.65	2.7225	1.215	−5.940	725.80	721.08
119+00	2.65	7.0225	3.133	−9.540	725.80	719.39
120+00	3.65	13.3225	5.944	−13.140	725.80	718.60
121+00	4.65	21.6225	9.647	−16.740	725.80	718.71
122+00	5.65	31.9225	14.242	−20.340	725.80	719.70
122+85	6.50	42.2500	18.850	−23.400	725.80	721.25

From Eq. (13.22), the low point occurs at $x = -g_1/r$. Substituting,

$$x = -\frac{-3.60}{0.8923} = 4.0345$$

Low point at station 116+35 + 4+03.45 = station 120+38.45

Because the low point doesn't fall on a known station, its elevation must be computed by substituting the x value of the low point into Eq. (13.21):

$$\text{Elev}_{\text{low}} = \frac{0.8923(4.0345)^2}{2} + (-3.60)4.0345 + 725.80 = 718.54 \text{ ft}$$

13.8 INTERSECTION OF AN EQUAL-TANGENT CURVE WITH A POINT

To maintain vertical clearance, the road surface at a particular station may have to be held to a specific elevation. In Fig. 13.13, g_1 and g_2 are known, as is the station and elevation of V. The point to be held is P, whose elevation is y_P and whose horizontal distance from V is h_P. The unknown curve length L must be whatever value satisfies the condition that the parabolic curve connecting g_1 and g_2 must also pass through point P.

From Eq. (13.21), $y_P = (r/2)x_P^2 + g_1 x_P + \text{elev}_{\text{BVC}}$. Transposing terms,

$$\frac{r}{2}x_P^2 + g_1 x_P + \text{elev}_{\text{BVC}} - y_P = 0$$

But $r/2 = (g_2 - g_1)/2L$, $\text{elev}_{\text{BVC}} = \text{elev}_V - g_1 L/2$ and $x_P = L/2 + h_P$, where h_P is positive if x_P is greater than x_V, negative if x_P is less than x_V. Substituting,

$$\left(\frac{g_2 - g_1}{2L}\right)\left(\frac{L}{2} + h_P\right)^2 + g_1\left(\frac{L}{2} + h_P\right) + \text{elev}_V - \frac{g_1 L}{2} - \text{elev}_P = 0$$

Squaring $(L/2 + h_P)$,

$$\left(\frac{g_2 - g_1}{2L}\right)\left(\frac{L^2}{4} + Lh_P + h_P^2\right) + \frac{g_1 L}{2} + g_1 h_P + \text{elev}_V - \frac{g_1 L}{2} - \text{elev}_P = 0$$

Canceling the $g_1 L/2$ terms and multiplying by $2L$,

$$(g_2 - g_1)\left(\frac{L^2}{4} + Lh_P + h_P^2\right) + 2g_1 L h_P + (2L)\text{elev}_V - (2L)\text{elev}_P = 0$$

Collecting terms,

$$\left(\frac{g_2 - g_1}{4}\right)L^2 + [(g_2 + g_1)h_P + 2(\text{elev}_V) - 2(\text{elev}_P)]L + (g_2 - g_1)h_P^2 = 0 \qquad \textbf{(13.25)}$$

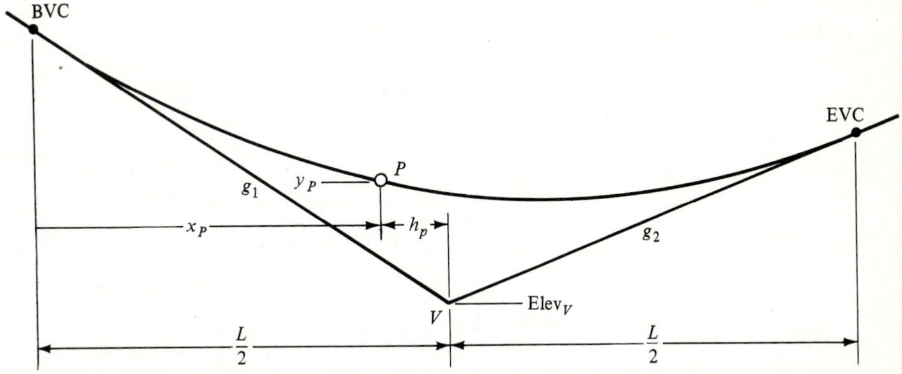

FIGURE 13.13
Intersection of equal-tangent curve with a point.

Example 13.15. In Fig. 13.13, $g_1 = -2.50\%$, $g_2 = +1.50\%$, $h_P = -1.0000$ stations (negative, because x_P falls before x_V), elev$_V$ is 1220.00 ft, and elev$_P$ is 1227.00 ft. Compute the length of the curve.

Solution. The value of $g_2 - g_1 = 1.50 - (-2.50) = +4.00$. The value of $g_2 + g_1 = 1.50 + (-2.50) = -1.00$. Substituting into Eq. (13.25),

$$L^2 + [(-1.00)(-1.00) + 2(1220.00) - 2(1227.00)]L + 4.00(-1.00)^2 = 0$$

$$L^2 - 13.00L + 4.00 = 0$$

Solving with the quadratic equation theorem [Eq. (1.2)],

$$L = 6.50 \pm 6.1847 \text{ stations}$$

But if $L = 6.50 - 6.1847$, the length of the curve will be only 0.3153 station (31.53 ft) and point P will fall before the start of the curve, which is not feasible. Thus, L is $6.50+6.1847 = 12.6847$ stations.

With L known, the elevations and locations of the BVC and the EVC, as well as the elevations of stations on the curve, may be computed as demonstrated in earlier examples.

13.9 UNEQUAL-TANGENT PARABOLIC CURVES

So far, discussion has centered on curves whose horizontal distances from the BVC to V and from V to the EVC are equal. An unequal-tangent curve, shown in Fig. 13.14, is one in which the distance from the BVC to V is *not* the same as the distance from V to the EVC, resulting in a compound curve composed of two equal-tangent curves that meet, at the station of V, at the point of **compound vertical curvature**, or CVC. In effect, the CVC is the EVC for the first equal-tangent curve and the BVC for the second equal-tangent curve.

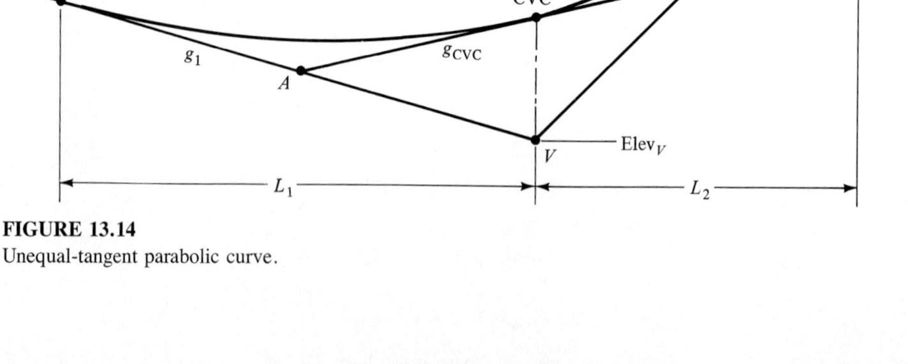

FIGURE 13.14
Unequal-tangent parabolic curve.

At the CVC, the two equal-tangent curves are tangent to a line AB that joins the midpoint A of the approach tangent with the midpoint B of the departing tangent. Thus, the grade of line AB is both the departing grade g_2 for the first equal-tangent curve and the entering grade g_1 for the second equal-tangent curve. The two equal-tangent curves are computed and tabulated separately, after first computing the elements they hold in common:

$$\text{Elev}_A = \text{elev}_V - \frac{g_1 L_1}{2} \qquad \text{and} \qquad \text{Elev}_B = \text{elev}_V + \frac{g_2 L_2}{2}$$

The grade of the CVC, which is the same as the grade of line AB, is the elevation difference from A to B divided by the horizontal distance from A to B. That is,

$$g_{\text{CVC}} = \text{grade}_{AB} = \frac{\text{elev}_B - \text{elev}_A}{L_1/2 + L_2/2}$$

$$= \frac{g_2 L_2/2 + g_1 L_1/2}{L_1/2 + L_2/2}$$

which becomes

$$g_{\text{CVC}} = \frac{g_2 L_2 + g_1 L_1}{L_1 + L_2} \qquad \textbf{(13.26)}$$

The elevation of the CVC is the elevation of point A plus half the horizontal distance L_1 multiplied by the grade of the CVC. That is,

$$\text{Elev}_{\text{CVC}} = \text{elev}_A + \frac{g_{\text{CVC}} L_1}{2} = \text{elev}_V - \frac{g_1 L_1}{2} + \frac{g_{\text{CVC}} L_1}{2}$$

which becomes

$$\textbf{Elev}_{\text{CVC}} = \textbf{elev}_V + \frac{(g_{\text{CVC}} - g_1) L_1}{2} \qquad \textbf{(13.27)}$$

Example 13.16. In Fig. 13.14, the BVC is at station 21+00, the EVC is at station 31+00, and V, whose elevation is 1260.00 ft, is at station 27+00. The entering grade g_1 is -1.50% and the departing grade g_2 is $+5.00\%$. For this unequal-tangent curve, compute the elevations of the BVC, the EVC, the CVC, and the other full stations, as well as the elevation and station of the low point.

Solution. The grade of the CVC is obtained first by substituting into Eq. (13.26)

$$g_{\text{CVC}} = \frac{+5.00(4.00) + [-1.50(6.00)]}{6.00 + 4.00} = +1.10\%$$

Substituting into Eq. (13.27),

$$\text{Elev}_{\text{CVC}} = 1260.00 + [1.10 - (-1.50)]3.00 = 1260.00 + 7.80 = 1267.80 \text{ ft}$$

Then,

$$\text{Elev}_{\text{BVC}} = \text{elev}_V - g_1 L_1 = 1260.00 - (-1.50)(6.00) = 1269.00 \text{ ft}$$

and

$$\text{Elev}_{\text{EVC}} = \text{elev}_V + g_2 L_2 = 1260.00 + (5.00)(4.00) = 1280.00 \text{ ft}$$

Substituting into Eq. (13.19) for the two curves,

$$r_1 = \frac{g_{CVC} - g_1}{L_1} = \frac{+1.1 - (-1.50)}{6.00} = +0.4333 \text{ ft} \cdot \text{station}^{-2}$$

$$\frac{r_1}{2} = +0.2167 \text{ ft} \cdot \text{station}^{-2}$$

and

$$r_2 = \frac{g_2 - g_{CVC}}{L_2} = \frac{+5.00 - (+1.10)}{4.00} = +0.9750 \text{ ft} \cdot \text{station}^{-2}$$

$$\frac{r_2}{2} = +0.4875 \text{ ft} \cdot \text{station}^{-2}$$

The elevations of points on both curves are tabulated from Eq. (13.21). For the first curve:

Station	x	x^2	$(r/2)x^2$	g_1x	**Elev$_{BVC}$**	**Elev$_x$**
21+00	0	0	0	0	1269.00	1269.00
22+00	1	1	0.2167	−1.50	1269.00	1267.72
23+00	2	4	0.8667	−3.00	1269.00	1266.87
24+00	3	9	1.9500	−4.50	1269.00	1266.45
25+00	4	16	3.4667	−6.00	1269.00	1266.47
26+00	5	25	5.4167	−7.50	1269.00	1266.92
27+00	6	36	7.8000	−9.00	1269.00	1267.80 ✓

and for the second curve:

Station	x	x^2	$(r/2)x^2$	$g_{CVC}x$	**Elev$_{CVC}$**	**Elev$_x$**
27+00	0	0	0	0	1267.80	1267.80
28+00	1	1	0.4875	1.10	1267.80	1269.39
29+00	2	4	1.9500	2.20	1267.80	1271.95
30+00	3	9	4.3875	3.30	1267.80	1274.49
31+00	4	16	7.8000	4.40	1267.80	1280.00 ✓

From the above tabulations, the low point of the combined curve is in the first section: Substituting into Eq. (13.22),

$$x_{\text{low}} = -\frac{g_1}{r_1} = \frac{-(-1.50)}{0.4333} = 3.4615 \Rightarrow\Rightarrow \text{station } 24+46.15$$

Substituting into Eq. (13.21),

$$\text{Elev}_{\text{low}} = \left(\frac{r_1}{2}\right) x_{\text{low}}^2 + g_1 x_{\text{low}} + \text{elev}_{\text{BVC}}$$

$$= +0.2167(3.4615)^2 + (-1.5)(3.4615) + 1269.00 = 1266.40 \text{ ft}$$

13.10 PROBLEMS

13.1. Calculate degree of curvature by the arc and chord definitions for circular curves of radius:
(*a*) 225.00 ft (*b*) 816.00 ft (*c*) 2800.00 ft

13.2. Calculate degree of curvature by the arc and chord definitions for circular curves of radius:
(*a*) 350.00 ft (*b*) 1015.00 ft (*c*) 3125.00 ft

13.3. Determine, from the given degree of curvature, the radius of each of the following arc-definition curves:
(*a*) $D = 1°35'30''$ (*b*) $D = 2°12'00''$ (*c*) $D = 4°28'15''$

13.4. Referring to Prob. 13.3, determine the radius of each curve assuming the given degree of curvature is based on chord definition.

13.5. Determine, from the given degree of curvature, the radius of each of the following chord-definition curves:
(*a*) $D = 0°54'00''$ (*b*) $D = 1°42'30''$ (*c*) $D = 3°15'22''$

13.6. Referring to Prob. 13.5, determine the radius of each curve assuming the given degree of curvature is based on arc definition.

In Probs. 13.7 through 13.11, values of Δ, D_a, R, T, L, C, E, and M are needed for the following arc-definition curves. From the given exact values, compute the values of the missing elements.

13.7. For a highway curve, $D_a = 2°48'$ and $\Delta = 14°52'51''$.

13.8. For a highway curve, $\Delta = 29°31'12''$ and $L = 641.74$ ft.

13.9. For a highway curve, $\Delta = 16°24'$ and $T = 235.90$ ft.

13.10. For a highway curve, $R = 2546.48$ ft and $M = 5.77$ ft.

13.11. $\Delta = 30°00'$ and E cannot exceed 40 ft, while D_a must be the smallest possible value to the nearest integer half-degree.

13.12. In Prob. 13.7, if the station of the PI is 27+56.02, compute the stations of the PC and the PT.

13.13. In Prob. 13.8, if the station of the PI is 41+78.39, compute the stations of the PC and the PT.

13.14. In Prob. 13.9, if the station of the PI is 37+68.46, compute the stations of the PC and the PT.

13.15. In Prob. 13.10, if the station of the PI is 15+09.21, compute the stations of the PC and the PT.

13.16. In Prob. 13.11, if the station of the PI is 22+39.86, compute the stations of the PC and the PT.

13.17. Referring to Probs. 13.7 and 13.12, compute full-station deflection angles and successive chords required to lay out the curve.

13.18. Referring to Probs. 13.8 and 13.13, compute full-station deflection angles and successive chords required to lay out the curve.

13.19. Referring to Probs. 13.9 and 13.14, compute full-station deflection angles and successive chords required to lay out the curve.

13.20. Referring to Probs. 13.10 and 13.15, compute half-station deflection angles and successive chords required to lay out the curve.

13.21. Referring to Probs. 13.11 and 13.16, compute half-station deflection angles and successive chords required to lay out the curve.

13.22. Using Prob. 13.17 deflection-angle data, compute chord distances *from* the PC *to each station* on the curve and to the PT.

13.23. Using Prob. 13.18 deflection-angle data, compute chord distances *from* the PC *to each station* on the curve and to the PT.

13.24. Using Prob. 13.19 deflection-angle data, compute chord distances *from* the PC *to each station* on the curve and to the PT.

13.25. Using Prob. 13.20 deflection-angle data, compute chord distances *from* the PC *to each station* on the curve and to the PT.

13.26. Using Prob. 13.21 deflection-angle data, compute chord distances *from* the PC *to each station* on the curve and to the PT.

13.27. Compute the "new stationing" values for PCs and PTs based on the arc-definition curve data shown below and provide a check.

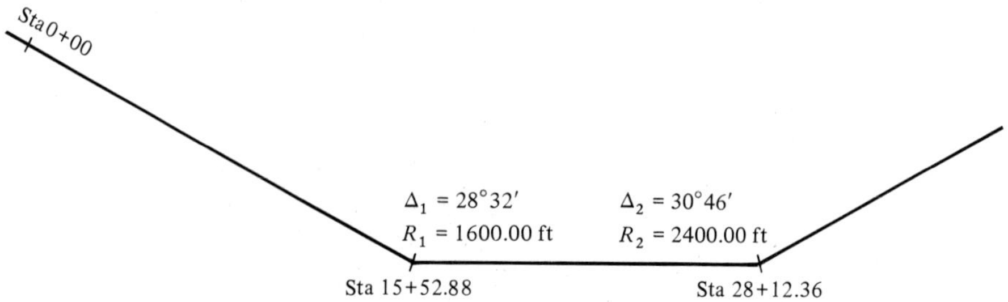

13.28. Referring to the sketch for Prob. 13.27, and assuming arc-definition curves, compute the "new stationing" values for the PCs and PTs if the station of the first PI is 21+39.47, $\Delta_1 = 15°42'$, $R_1 = 900.00$ft., the station of the second PI is 38+72.63, $\Delta_2 = 24°16'$, and $R_2 = 1800.00$ ft (assume Δ values are exact).

13.29. The point of intersection of two tangents on a highway curve could not be set due to the nature of the terrain. As shown in the sketch below, a 563.72-ft cutoff line was measured to the departing tangent from station 50+18.32 on the approach tangent and the two deflection angles were observed. A curve whose exact D_a is 8°45′ is to connect the approach tangent with the departing tangent. From the given information, compute Δ, the station of the PI, curve elements R, T, L, C, E, and M, and the stations of the PC and PT.

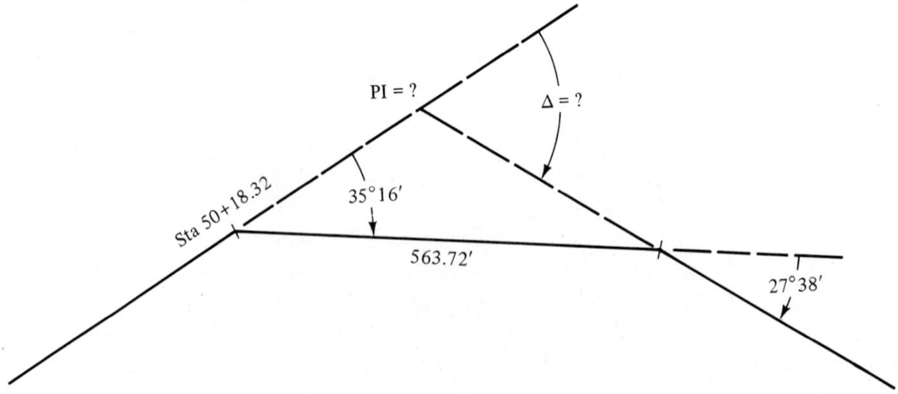

13.30. Referring to Prob. 13.29, compute full-station deflection angles and successive chords required to lay out the curve.

13.31. Three segments of the centerline of a proposed road intersect as shown below. Compute the radius of the single circular curve that is tangent to all three segments.

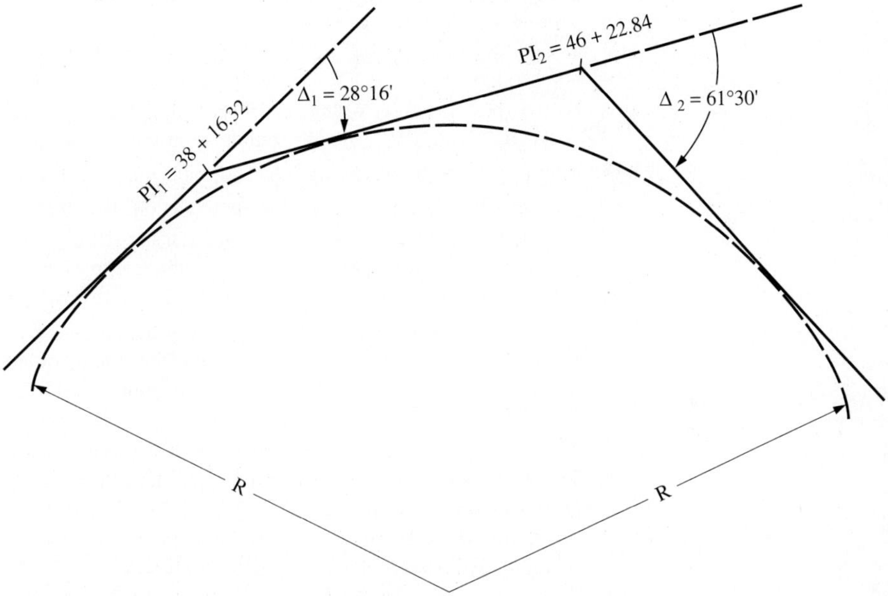

13.32. Do the same as Prob. 13.31 except the station of PI_1 = 27+62.34, Δ_1 = 32°44′, the station of PI_2 = 39+28.66, and Δ_2 = 56°20′.

13.33. The approach grade to a 600-ft equal-tangent vertical curve, g_1, is −2.25% and the departing grade g_2 is +4.50%. *V* is at station 41+00.00, elevation 343.82 ft. Compute the stations and elevations of the BVC, EVC, and all full stations.

13.34. The approach grade to an 800-ft equal-tangent vertical curve, g_1, is +3.00% and the departing grade g_2 is −1.75%. *V* is at station 23+00.00, elevation 561.16 ft. Compute the stations and elevations of the BVC, EVC, and all full stations.

13.35. The approach grade to a 500-ft equal-tangent vertical curve, g_1, is +4.50% and the departing grade g_2 is +2.75%. *V* is at station 36+25.50, elevation 643.89 ft. Compute the stations and elevations of the BVC, EVC, and all full stations.

13.36. The approach grade to a 700-ft equal-tangent vertical curve, g_1, is −6.00% and the departing grade g_2 is −3.50%. *V* is at station 47+42.00, elevation 712.32 ft. Compute the stations and elevations of the BVC, EVC, and all full stations.

13.37. Referring to Prob. 13.33, determine the station and elevation of the low point on the curve.

13.38. Referring to Prob. 13.34, determine the station and elevation of the high point on the curve.

13.39. Referring to Prob. 13.35, determine the station and elevation of the high point on the curve.

13.40. Referring to Prob. 13.36, determine the station and elevation of the low point on the curve.

13.41. An equal-tangent vertical curve is to connect an entering grade g_1 of +3.25% with a departing grade g_2 of −1.75%, with the rate of change of grade, r, not to exceed −0.4500 ft·station^{-2}. V is at station 55+36.00, elevation 369.84 ft. Determine the shortest possible curve of integer 100-ft length, then compute the stations and elevations of the BVC, the EVC, all full stations, and the high point on the curve.

13.42. An equal-tangent vertical curve is to connect an entering grade g_1 of −6.50% with a departing grade g_2 of +2.20%, with the rate of change of grade, r, not to exceed +0.7500 ft·station^{-2}. V is at station 47+00.00, elevation 821.93 ft. Determine the shortest possible curve of integer 100-ft length, then compute the stations and elevations of the BVC, the EVC, all full stations, and the low point on the curve.

13.43. Referring to Prob. 13.34, find the station(s) on the curve with elevation of 552.00 ft.

13.44. Referring to Prob. 13.35, find the station(s) on the curve with elevation of 636.00 ft.

13.45. An entering grade of −4.00% intersects a departing grade of +2.50% at V, station 36+00.00, elevation 630.00 ft. Compute the length of an equal-tangent curve connecting these two grades that intersects a point of elevation of 635.00 ft at station 37+50.00.

13.46. An entering grade of +8.00% intersects a departing grade of −1.50% at V, station 64+00.00, elevation 450.00 ft. Compute the length of an equal-tangent curve connecting these two grades that intersects a point of elevation of 427.00 ft at station 61+40.00.

13.47. An entering grade of +2.00% intersects a departing grade of −3.50% at V, station 52+00.00, elevation 355.40 ft. An unequal-tangent vertical curve, with BVC at station 48+00.00 and EVC at station 55+00.000, is to connect the two grades. Compute the elevations of the BVC, EVC, CVC, and all full stations on the curve, as well as the station and elevation of the high point on the curve.

13.48. An entering grade of −4.25% intersects a departing grade of +2.40% at V, station 31+00.00, elevation 629.06 ft. An unequal-tangent vertical curve, with BVC at station 29+00.00 and EVC at station 35+00.000, is to connect the two grades. Compute the elevations of the BVC, EVC, CVC, and all full stations on the curve, as well as the station and elevation of the low point on the curve.

13.49. Referring to Fig. 13.14, prove algebraically that line AB is parallel to a line connecting the BVC with the EVC.

CHAPTER 14

EARTHWORK

14.1 INTRODUCTION

Construction projects require, as a first step, the movement of earth to provide a predetermined surface upon which construction may proceed. In this respect, the surveyor must first establish the topography of the site in order to compute the volume of earth to be brought in (called **fill**) or the volume to be taken away (called **cut**) in order to bring the surface of the site to its required elevation(s). The movement of earth into or out of the project area, along with additional operations required to establish a firm surface at a predetermined grade, is called **earthwork**. Although earthwork is involved, to a greater or lesser extent, in activities as disparate as homebuilding and dam construction, its most prevalent use is in highway work.

In laying out highway rights-of-way, every effort is made to follow the natural terrain in order to minimize the cost of earthwork. Other considerations—unexpectedly high land-acquisition costs over the preferred route, difficult terrain, or specifications that limit grade or curvature along a right-of-way—can often result in much higher volumes of cut and fill within a highway project area than would have been the case if earthwork cost were the only consideration.

Because of irregularities in the natural terrain, it is not feasible for a surveyor to determine exactly the configuration of the ground surface along the project right-of-way. Similarly, the moving and compacting of earth is carried out with large, ungainly equipment, so the finished surface is bound to deviate somewhat from the intended surface. For these reasons, the earthwork volume unit is the **cubic yard** (yd^3) rather than the more precise-seeming cubic foot.

14.2 CROSS SECTION TERMINOLOGY

The basis for computation of highway cut and fill volumes is cross sections taken at appropriate intervals along the right-of-way. Figure 14.1 shows a somewhat complex section of highway right-of-way that passes from cut (bottom) to fill (top). The plan

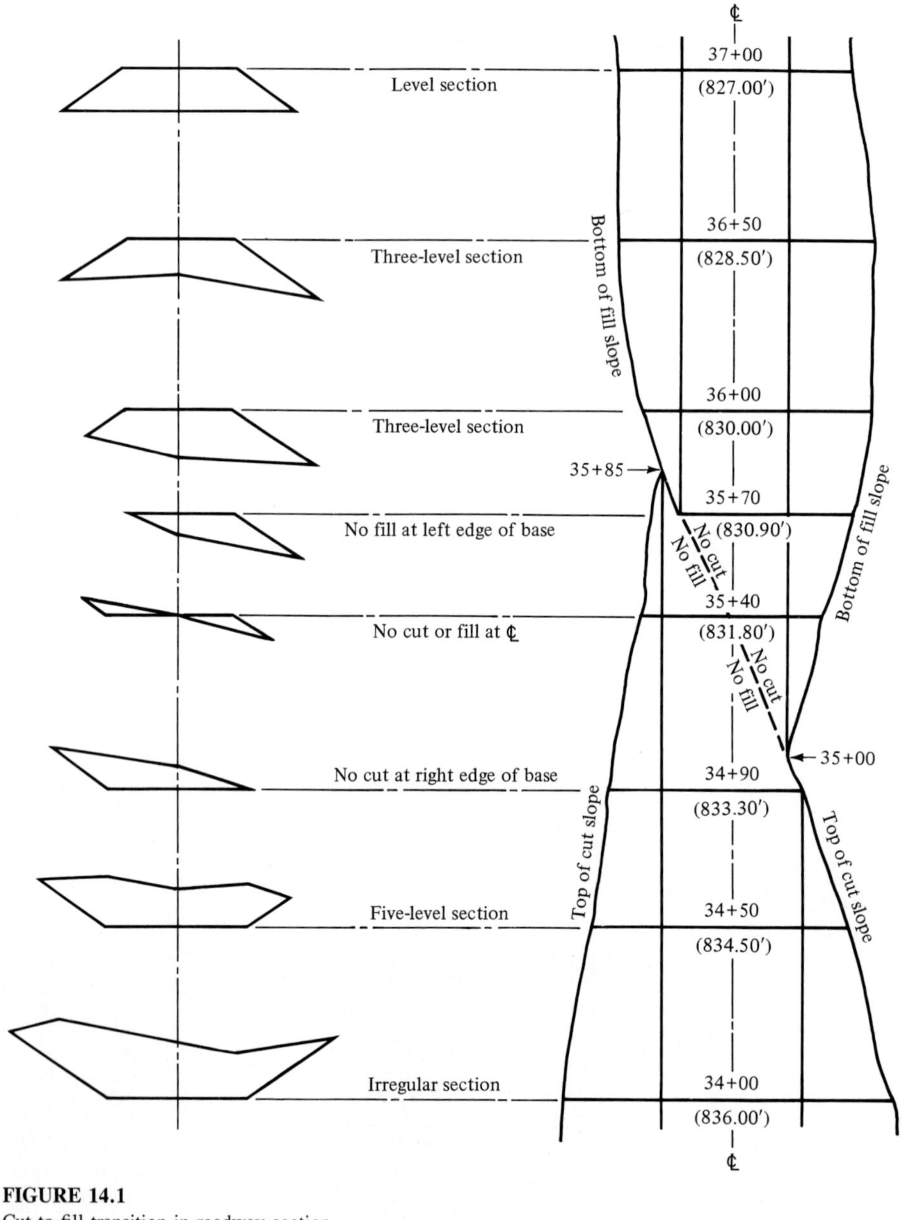

FIGURE 14.1
Cut-to-fill transition in roadway section.

view, at right, shows the right-of-way centerline with lines drawn transversely at half-stations or at key points in the transition from cut to fill. The cross sections are shown at left in Fig. 14.1, opposite the corresponding stations. Each cross section is bounded by the natural ground surface, by the **base** (the level surface, lying either above or below the ground surface, on which the road will be built) and by two sloping lines, called **side slopes**, leading from the edges of the base to the ground surface. To make room for drainage ditches, the base is wider in cut; in fill, the runoff simply passes down the side slope.

The numbers in parentheses at right in Fig. 14.1 are the elevations of the road base, established from the design specifications for this roadway (note the constant −3.00% grade along this section).

The Fig. 14.1 cross sections do not show the undulations of natural terrain. The ground surface has been generalized to illustrate four cross-section categories: the **level** section, the **three-level** section, the **five-level** section, and the **irregular section**, the designation depending on the number and location of the level-rod points determined on the ground for a particular cross section.

In a level section (sta 37+00), the ground surface is considered level throughout the section. The ground surfaces in the 2 three-level sections (sta 36+00 and sta 36+50) are considered to slope evenly from the centerline to the left edge of the section and from the centerline to the right edge. A five-level section (sta 34+50) is assumed to have four even slopes between these five elevation points: left edge of section, ground point half the base width left of the centerline, centerline, ground point half the base width right of the centerline, and right edge of section. An irregular section (sta 34+00) is one that may not be reasonably placed in one of the other categories. Methods and equations for computing the areas of these cross sections are covered in Sec. 14.4.

Most cross sections that cover the **transition zone** from cut to fill (sta 34+90 to sta 35+70 in Fig. 14.1) involve rather simple figures with four or three sides. Treatment of transition-zone cross sections is covered in Sec. 14.7

14.3 FIELD SURVEY AND STAKEOUT OF CROSS SECTIONS

CUT AND FILL DETERMINATION. The cross section for station 36+00 in Fig. 14.1 is shown in greater detail in Fig. 14.2, with a level set up adjacent to the right-of-way. The height-of-instrument (HI) of 831.1 ft was established from profile leveling procedures (Chap. 3). The rod reading at a ground point on the cross section is called the **ground rod**. From specifications for this section of roadway (see Fig. 14.1), the **grade elevation** at station 36+00 is to be 830.0 ft. If the level rod were to be held with its base at the level of the grade elevation, its reading would be the HI minus the grade elevation, a value called the **grade rod**. The grade rod can be either positive or negative; in this case, it is 831.1 − 830.0 = +1.1 ft.

The vertical distance to the grade elevation from a ground point on the cross section is called fill if the grade lies above ground; cut if the grade lies below ground. Grade-rod values may be expressed as follows in terms of the amount of cut or fill at the right-of-way centerline, where **ground rod**$_c$ is the centerline ground-rod value:

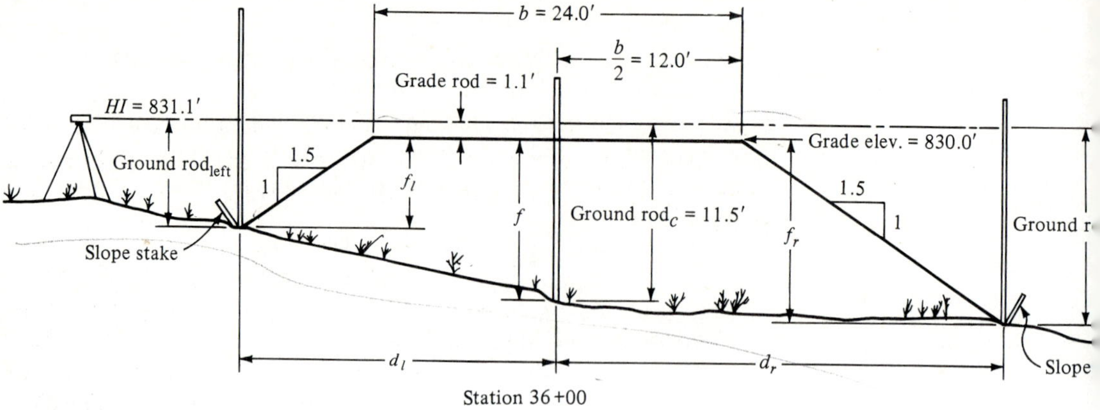

FIGURE 14.2
Cross-section levels at station 36+00 in Fig. 14.1.

$$\textbf{Grade rod} = \textbf{ground rod}_c + \textbf{center cut} \tag{14.1a}$$

and

$$\textbf{Grade rod} = \textbf{ground rod}_c - \textbf{center fill} \tag{14.1b}$$

Once the grade-rod value is established for a given cross section, the amount of cut or fill at any point in that cross section is determined from the ground rod reading at that point and from the equation:

$$\textbf{Cut or fill} = \textbf{ground rod} - \textbf{grade rod} \tag{14.2}$$

where the numerical result will be positive for fill, negative for cut. The fill at the left edge of the embankment in Fig. 14.2, f_l, is thus the ground-rod reading at the left edge minus 1.1 ft; similarly, the fill at the right edge of the embankment, f_r, must be the ground-rod reading at the right edge minus 1.1 ft.

The small triangles on either side slope in Fig. 14.2 indicate the manner in which side slope is designated; that is, for every linear unit of vertical *rise* along the slope, there are **s** units of horizontal *run*. Slope is designated in the form s:1. In the Fig. 14.2 example, for section 36+00 s is $1\frac{1}{2}$, so the side slope is a $1\frac{1}{2}$-to-1 slope, noted fractionally as $1\frac{1}{2}:1$ or decimally as 1.5:1. The decimal notation will be used in this text.

SETTING SLOPE STAKES. Figure 14.2 shows slope stakes marking the intersection of the side slopes with the natural grade, with d_l the distance from centerline to the left slope stake and d_r the distance from centerline to the right slope stake. Unfortunately, the surveyor must determine the values of both distances by trial. This is because the ground-surface profile along that section is not exactly known to start with and the distances are determined by the equations

$$d = \frac{b}{2} + cs \tag{14.3a}$$

or

$$d = \frac{b}{2} + fs \tag{14.3b}$$

where $\boldsymbol{b}$ is the width of the road base and **s** is side slope, while $\boldsymbol{c}$ or $\boldsymbol{f}$ is the value of cut or fill (both are assumed positive in this case) at a distance $\boldsymbol{d}$ from the centerline. Because the c or f values are not exactly known, Eq. (14.3*a*) or (14.3*b*) must be satisfied by trial.

Example 14.1. By trial, find d_l and f_l at section 36+00 in Fig. 14.2.

Solution procedure. If section 36+00 were a level section, f_l would be the same as the centerline fill. From Eq. (14.2), this is 11.5 − 1.1 = 10.4 ft. Substituting into Eq. (14.3*b*), d_l = 12.0 + (10.4)(1.5) = 27.6 ft. However, the ground slopes up significantly from the section centerline to the left slope–stake location; thus, d_l must be somewhat less than 27.6 ft. As a first trial, indicated d_A in Fig. 14.3, the rod is placed about 5 ft closer to the centerline, at 23.0 ft, and a ground-rod reading of 6.2 ft is taken. From Eq. (14.2), f_A = 6.2 − 1.1 = 5.1 ft. Substituting into Eq. (14.3*b*), d_l = 12.0 + (5.1)(1.5) = 19.6 ft, which is different from the 23.0 trial distance d_A. The 19.6-ft computed distance is used for the second trial distance d_B, and a new ground-rod reading of 6.9 ft is obtained for a rod held at that point. From Eq. (14.2), f_l = 6.9 − 1.1 = 5.8 ft. Substituting again into Eq. (14.3*b*), d_l = 12.0 + (5.8)(1.5) = 20.7 ft. Considering the results so far, the next trial rod position should lie between 19.6 ft and 20.7 ft, and somewhat closer to the latter. A third trial distance, d_C, is chosen at 20.4 ft, which is three quarters of the way between 19.6 ft and 20.7 ft. With the rod now held 20.4 ft from the centerline, a ground-rod reading of 6.7 ft is obtained. From Eq. (14.2), f_l = 6.7 − 1.1 = 5.6 ft. Substituting into Eq. (14.3*b*), d_l = 12.0 + (5.6)(1.5) = 20.4 ft, which is the same value as d_C. The accepted values of f_l and d_l are therefore 5.6 ft and 20.4 ft, respectively.

By chance, the third trial value for d_l is the same as the corresponding computed value of d_l. In practice, agreement within 0.1 ft is quite satisfactory and experienced surveyors usually achieve that level of agreement as quickly as in the above example.

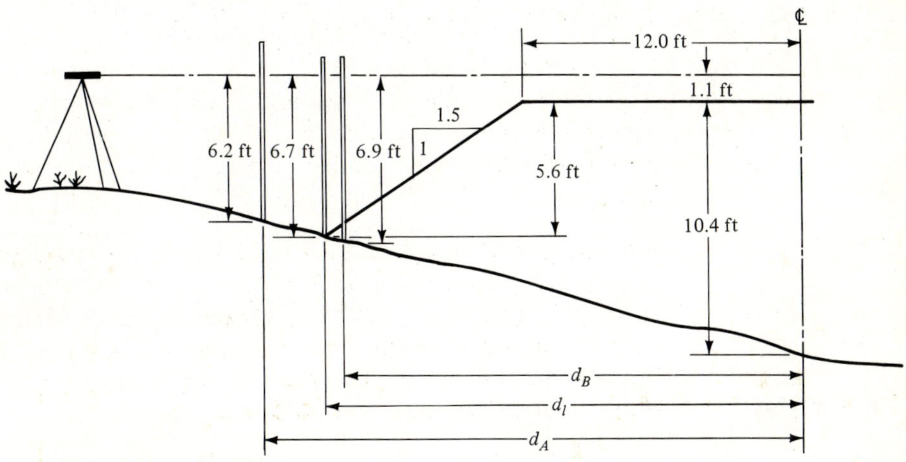

FIGURE 14.3
Slope-stake-location detail, station 36+00.

Station	Grade Elev., ft	Ground Elev., ft	Cross sections		
36+00	830.0	819.6	$\frac{F5.6}{20.4}$	$F10.4$	$\frac{F11.8}{29.7}$

FIGURE 14.4
Cross-section field notes, station 36+00.

FIELD-RECORD NOTATION. By a similar field-survey procedure, the values of f_r and d_r are found to be 11.8 ft and 29.7 ft, respectively. Only the final results are recorded in the field record. Figure 14.4 illustrates the field-record format for station 36+00.

Slope stakes are set with one side facing the centerline and they are driven on the slant. Most surveyors slant these stakes *outward* from the centerline in fill, as indicated in Fig. 14.2, but *inward* toward the centerline in cut. Reflecting field-book notation, slope stakes are marked (using crayon or felt-tip marker) on the side facing

Cross Sections on Sasquatch Road,
Thurston County, Washington

Station	Grade elev., ft	Ground ℄ elev., ft	Cross sections (Bases: Cut 32.0′; Fill 24.0′; s = 1.5:1)				
37+00	827.0	817.6		$\frac{F9.4}{26.1}$	F9.4	$\frac{F9.4}{26.1}$	
36+50	828.5	821.2		$\frac{F9.2}{25.8}$	F7.3	$\frac{F12.8}{31.2}$	
36+00	830.0	819.6		$\frac{F5.6}{20.4}$	F10.4	$\frac{F11.8}{29.7}$	
35+70	830.9	826.4		$\frac{0}{12.0}$	F4.5	$\frac{F9.7}{26.6}$	
35+40	831.8	831.8		$\frac{C2.8}{20.2}$	0	$\frac{F4.8}{19.2}$	
34+90	833.3	837.9		$\frac{C7.9}{27.8}$	C4.6	$\frac{0}{16.0}$	
34+50	834.5	842.4	$\frac{C10.1}{31.2}$	$\frac{C10.9}{16.0}$	C7.9	$\frac{C9.2}{16.0}$	$\frac{C6.0}{25.0}$
34+00	836.0	848.4	$\frac{C14.4}{37.6}$	$\frac{C17.2}{26.6}$	C12.4	$\frac{C9.8}{12.1}$	$\frac{C13.3}{36.0}$

FIGURE 14.5
Cross-section field notes, stations 34+00 to 37+00.

the centerline with the amount of cut or fill, along with the distance out from center. The station is marked on the other side of the stake.

Figure 14.5 shows the completed cross-section field notes for the stations in Fig. 14.1. Note that the stations are written sequentially *up* the page, in keeping with the convention that forward progress is designated upward. The bottom-to-top notation provides the added advantage of placing the left rod and distance values to the left of the center values and the right rod and distance values to the right of center, eliminating the need for "left" and "right" labels in the field record book.

14.4 COMPUTING AREAS OF CROSS SECTIONS

As stated earlier, standardized cross sections include level sections, three-level sections, five-level sections, and irregular sections. Each requires a different method for computing cross-sectional area, depending on section geometry.

LEVEL SECTION. In the level section shown in Fig. 14.6, the key dimensions are the roadway base width $\boldsymbol{b}$, the depth of fill $\boldsymbol{f}$, which is constant throughout the section, and s, the side-slope designator. Two vertical dashed lines divide the cross section into a rectangle and two triangles. The ratio of triangle base to triangle height is s:1, so a height of f must have a corresponding base width of sf. The area of the rectangle is bf and the area of each triangle (base multiplied by half the height) is $sf(f/2)$. Thus, the sum of the three areas is $bf + 2[sf(f/2)]$. Collecting terms, the area of a level section is

$$A_L = f(b + sf) \tag{14.4}$$

If the level section in Fig. 14.6 had been in cut, rather than fill, the dimension $\boldsymbol{c}$, from the ground surface down to the road base, would have been substituted for f in Eq. (14.4).

Example 14.2. From the cross-section field notes in Fig. 14.5, compute the cross-sectional area at station 37+00.

Solution. From cross-section data, $f = 9.4$ ft. From additional data listed at top, right, the side-slope factor s is 1.5, and base width b is 24.0 ft for a fill section. Substituting into Eq. (14.4),

$$A_L = f(b + sf) = 9.4[24.0 + (1.5)(9.4)] = 358.14 \Rightarrow\Rightarrow 358.1 \text{ ft}^2$$

THREE-LEVEL SECTION. In Fig. 14.7, the roadway base is again b, the distances from the centerline to the left and right slope stakes are d_l and d_r, respectively, and

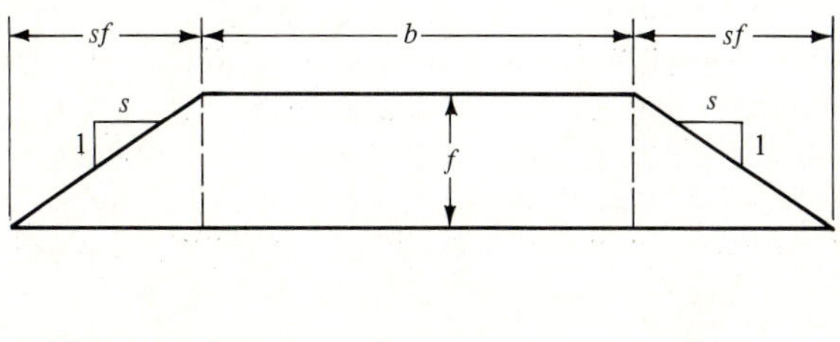

FIGURE 14.6
Level section in fill.

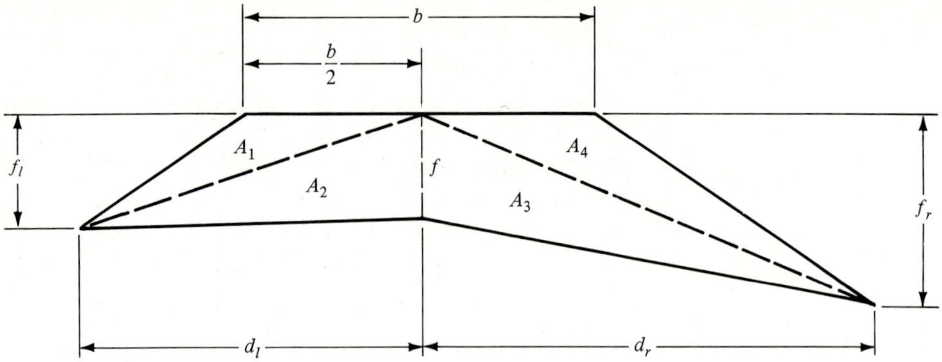

FIGURE 14.7
Three-level section in fill.

the fill depths are f on the centerline, f_l at the left slope stake and f_r at the right slope stake. Dashed lines divide the cross section into triangles A_1, A_2, A_3, and A_4. The area of a triangle is base width multiplied by height divided by 2. Triangles A_1 and A_4 have bases of $b/2$ and heights of f_l and f_r, respectively; thus, areas of A_1 and A_4 are $bf_l/4$ and $bf_r/4$, respectively. Triangles A_2 and A_3 have base widths f and heights d_l and d_r, respectively; thus, the areas of A_2 and A_3 are $f\,d_l/2$ and $f\,d_r/2$, respectively. The area of the entire figure is the sum of the areas of its triangles: $A_1 + A_2 + A_3 + A_4$. Collecting terms, the area of a three-level section is

$$\mathbf{A_{3-L} = \frac{b}{4}(f_l + f_r) + \frac{f}{2}(d_l + d_r)} \qquad \mathbf{(14.5)}$$

Had the section in Fig. 14.7 been in cut rather than fill, dimensions c, c_l and c_r would be substituted in Eq. (14.5) for the corresponding fill dimensions.

Example 14.3. From the cross-section field notes in Fig. 14.5, compute the cross-sectional area at station 36+50.

Solution. From Fig. 14.5, the station 36+50 dimensions are $b = 24.0$ ft, $f = 7.3$ ft, $f_l = 9.2$ ft, $f_r = 12.8$ ft, $d_l = 25.8$ ft, and $d_r = 31.2$ ft. Substituting into Eq. (14.5),

$$A_{3-L} = \frac{b}{4}(f_l + f_r) + \frac{f}{2}(d_l + d_r)$$

$$= 6.0(9.2 + 12.8) + 3.65(25.8 + 31.2) = 340.05 \Rightarrow\Rightarrow 340.0 \text{ ft}^2$$

FIVE-LEVEL SECTION. Figure 14.8 shows a five-level section in cut, with key dimensions $\boldsymbol{b}$, $\boldsymbol{d_l}$, $\boldsymbol{d_r}$, $\boldsymbol{c}$, $\boldsymbol{c_l}$, and $\boldsymbol{c_r}$. Note that the c_l and c_r dimensions (or f_l and f_r in fill) are taken at the left and right edges of the roadway base, rather than at the left and right slope stakes, as is the case with a three-level section.

The dashed lines in Fig. 14.8 divide the section into triangles A_1 through A_6. The base of triangle A_1 is c_l and its height is $d_l - b/2$; thus, its area is $c_l(d_l - b/2)/2$. Similarly, the area of triangle A_6 is $c_r(d_r - b/2)/2$. The base of triangle A_2 is c_l and its height is $b/2$; thus, its area is $c_l(b/2)/2$, or $c_l b/4$. Similarly, the area of triangle A_5

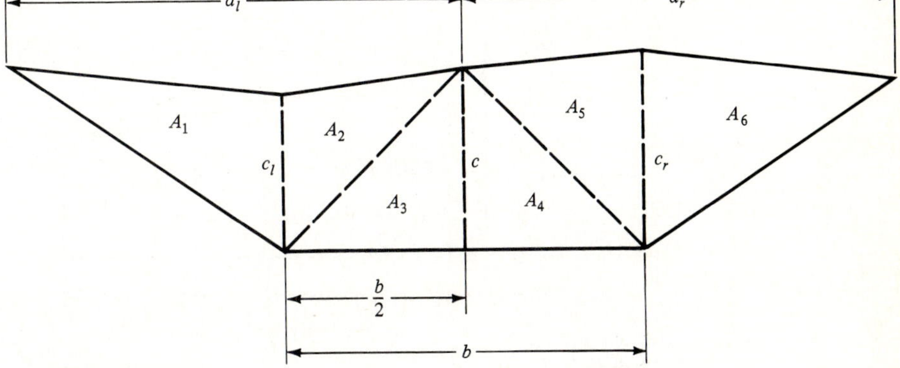

FIGURE 14.8
Five-level section in cut.

is $c_r b/4$. The common base of triangles A_3 and A_4 is c and the height is $b/2$ in both cases; thus, the areas of both A_3 and A_4 are $cb/4$. The area of the entire cross section is the sum of areas A_1 through A_6. Collecting terms and canceling where feasible, the area of a five-level section is expressed as

$$\mathbf{A_{5-L} = \frac{c_l d_l + cb + c_r d_r}{2}} \tag{14.6}$$

Note that f_l, f, and f_r would be substituted for c_l, c, and c_r in Eq. (14.6) if the section were in fill rather than cut.

Example 14.4. From the data in Fig. 14.5, compute the area of the cross section at station 34+50.

Solution. Since this section is in cut rather than fill, the width of the road base, b, is 32.0 ft rather than 24.0 ft, as indicated at top, right in Fig. 14.5. From the cross-section data for station 34+50: c_l = 10.9 ft, c = 7.9 ft, c_r = 9.2 ft, d_l = 31.2 ft, and d_r = 25.0 ft. Substituting into Eq. (14.6),

$$A_{5-L} = \frac{c_l d_l + cb + c_r d_r}{2}$$

$$= \frac{(10.9)(31.2) + (7.9)(32.0) + (9.2)(25.0)}{2} = 411.44 \Rightarrow\Rightarrow 411.4 \text{ ft}^2$$

IRREGULAR SECTION. Irregular sections do not use standardized area equations; they must be considered individually as a combination of trapezoids and triangles. Figure 14.9 is an enlarged, dimensioned view of the section at station 34+00, in Fig. 14.1. There are four trapezoidal shapes (*ABCL, LCDJ, JDEI,* and *IEFG*) whose dimensions are shown and two triangles (*ABK* and *HFG*) whose base dimensions are the left and right centerline-to-slope–stake distances minus half the base width. The area of the entire irregular section is the sum of the four trapezoidal areas minus the sum of the two triangular areas.

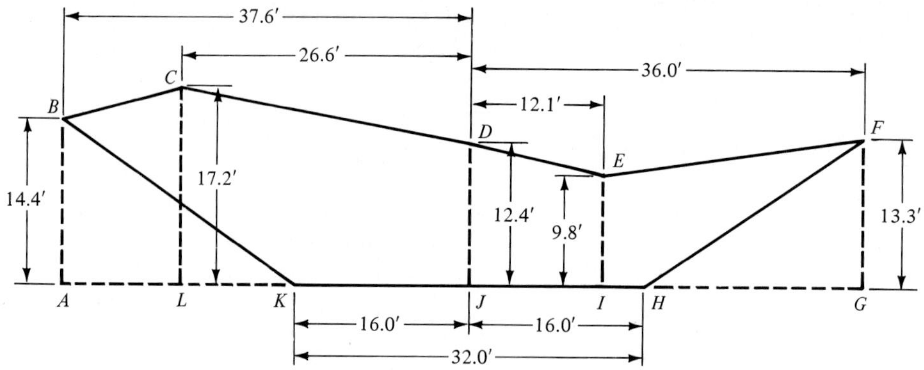

FIGURE 14.9
Irregular section in cut at station 34+00 in Fig. 14.1.

Example 14.5. Compute the area of the irregular section at station 34+00 from dimensions given in Fig. 14.9.

Solution. The area of a trapezoid is the base width multiplied by the average of the two sides that lie perpendicular to the base. The base, *AL*, of trapezoid *ABCL* is $37.6 - 26.6 = 11.0$ ft and the heights of sides *AB* and *LC* are 14.4 ft and 17.2 ft; thus, $\text{area}_{ABCL} = 11.0(14.4 + 17.2)/2 = 173.80\ \text{ft}^2$. Similarly, $\text{area}_{LCDJ} = 26.6(17.2 + 12.4)/2 = 393.68\ \text{ft}^2$; $\text{area}_{JDEI} = 12.1(12.4 + 9.8)/2 = 134.31\ \text{ft}^2$; $\text{area}_{IEFG} = (36.0 - 12.1)(9.8 + 13.3)/2 = 276.04\ \text{ft}^2$. The area of a triangle is the base width multiplied by half the height; thus, $\text{area}_{ABK} = (37.6 - 16.0)14.4/2 = 155.52\ \text{ft}^2$ and $\text{area}_{HFG} = (36.0 - 16.0)13.3/2 = 133.00\ \text{ft}^2$. $\text{Area}_{\text{total}} = \text{area}_{ABCL} + \text{area}_{LCDJ} + \text{area}_{JDEI} + \text{area}_{IEFG} - \text{area}_{ABK} - \text{area}_{HFG} = 173.80 + 393.68 + 134.31 + 276.04 - 155.52 - 133.00 = 689.31 \Rightarrow\Rightarrow 689.3\ \text{ft}^2$.

An alternate method for calculating areas of irregular cross sections is to define the coordinates of the labeled points in Fig. 14.9, then solve using the area-by-coordinates method demonstrated in Sec. 8.9 for land areas.

Example 14.6. Compute the area of the irregular section at station 34+00 using the area-by-coordinates method.

Solution. To keep all coordinate values positive and as small as possible, the origin is set at point *A*. Point *K*, then, has coordinates of $x = 21.6\ (37.6' - 16.0')$, $y = 0$, that is, (21.6,0). In the same manner, working around the figure clockwise from the *K*, the corner points are as follows: *K* (21.6, 0.0), *B* (0.0, 14.4), *C* (11.0, 17.2), *D* (37.6, 12.4), *E* (49.7, 9.8), *F* (73.6, 13.3), and *H* (53.6, 0.0). Substituting into Eq. (8.7),

$$\text{Area} = [0.0(37.6 - 0.0) + 14.4(21.6 - 11.0) + 17.2(0.0 - 37.6) + 12.4(11.0 - 49.7)$$

$$+ 9.8(37.6 - 73.6) + 13.3(49.7 - 53.6) + 0.0(73.6 - 21.6)]/2$$

$$= -689.315 \Rightarrow\Rightarrow 689.3\ \text{ft}^2 \quad \checkmark \qquad \text{(negative sign is ignored)}$$

In the case of hand-drawn or computer-generated cross sections, it is often simpler and more efficient to determine cross-sectional areas with a mechanical or electronic planimeter.

14.5 VOLUME BY AVERAGE END AREA

Cross-sectional areas are used to compute the volumes of earthwork lying between those cross sections. One method of computing such volumes is based on the simple, slightly erroneous assumption that the volume between two parallel cross sections is equal to the distance between the sections multiplied by the average of the two cross-sectional areas. In equation form: $V(\text{ft}^3) = L(A_1 + A_2)/2$, where L is distance between sections in feet while A_1 and A_2 are the cross-sectional areas of the two ends in square feet. To find volume in standard earthwork units of cubic yards, volume in cubic feet is divided by 27 cubic feet per cubic yard; that is, $V(\text{yd}^3) = L(A_1 + A_2)/[(2)(27)]$, leading to

$$\boldsymbol{V_e = \frac{L(A_1 + A_2)}{54}} \qquad \textbf{(14.7)}$$

where $\boldsymbol{V_e}$ is average-end-area volume in cubic yards, $\boldsymbol{L}$ is in feet, $\boldsymbol{A_1}$ and $\boldsymbol{A_2}$ are in square feet.

Example 14.7. Compute the average-end-area volume between stations 34+00 and 34+50, Figs. 14.1 and 14.5.

Solution. The cross-sectional areas at stations 34+00 and 34+50, determined in Examples 14.5 and 14.4, are 689.3 ft^2 and 411.4 ft^2. The distance between sections, from station designations, is 50 ft. Substituting into Eq. (14.7),

$$V_e = \frac{50(689.3 + 411.4)}{54} = 1019.17 \Rightarrow\Rightarrow 1019.2 \text{ yd}^3$$

14.6 VOLUME BY PRISMOIDAL FORMULA

The average-end-area equation is widely used because it is simple; however, it is not theoretically correct. A more precise approach assumes that the volume between two stations is a **prismoid**; that is, a solid bounded by planes, the end faces being parallel and having the same number of sides. The equation for the volume of a prismoid is $V = L(A_1 + 4A_m + A_2)/6$, where L is distance between end-areas A_1 and A_2, but A_m is the area of a midsection between A_1 and A_2, whose *dimensions* are the *average* of the end-area dimensions. Note that A_m is *not* the mean of areas A_1 and A_2; if it were, the prismoidal and average-end-area equations would be identical.

Consider the volume of a square-based pyramid as determined from the average-end-area and prismoidal formulas. The pyramid's base dimension is b and its height is h. The correct equation for its volume is $b^2h/3$. From the average-end-area formula, the volume is $h(0+b^2)/2 = b^2h/2$; however, from the prismoidal formula, the volume is $h[0+4(b/2)^2+b^2]/6 = b^2h/3$, the correct value. In this extreme example, average-end-area volume exceeds prismoidal volume by 50%—a clear indication that, in most cases, the average-end-area equation will give a larger volume than the prismoidal formula.

As before, the volume obtained from the prismoidal formula is converted to cubic yards by dividing by 27 $ft^3 \cdot yd^{-3}$, leading to

$$\mathbf{V_p = \frac{L(A_1 + 4A_m + A_2)}{162}} \qquad \textbf{(14.8)}$$

where $\mathbf{V}_p$ is prismoidal volume in cubic yards, $\boldsymbol{L}$ is in feet, and $\boldsymbol{A_1}, \boldsymbol{A_2}$ and $\boldsymbol{A_m}$ are in square feet.

Example 14.8. Compute the prismoidal volume between stations 36+00 and 36+50 in Fig. 14.5.

Solution. These sections are in fill, with a roadway base width of 24.0 ft. Figure 14.10 shows a format for computing cross-sectional areas, with station designations at left, cross-section notes in the middle, Eq. (14.5) computations of cross-sectional area at right, and the computation of the prismoidal-formula volume below the area computations. Note that the cross-section dimensions for the midsection at 36+25 are the *computed* mean of the corresponding dimensions of the cross sections for stations 36+00 and 36+50.

Equation (14.8) is based on the assumption that the two end sections have the same number of corner points, permitting the computation of average center-section dimensions as demonstrated in Example 14.8. Note that the two end sections need not necessarily be precisely the same figure. A three-level section on one end will work with a level section on the other end, since the value of the center height for a level section is known. Similarly, a five-level section on one end will work with an irregular section on the other end that has the same number of ground-level points. Even in cases where the end sections are wholly incompatible, however, Eq. (14.8) may be used if A_m is determined from field measurements taken of the midsection to establish its figure type and its dimensions.

Station	**Cross-section notes**			**Area** $= \frac{b}{4}(f_l + f_r) + \frac{f}{2}(d_l + d_r)$
36+50	$\frac{F9.2}{25.8}$	F7.3	$\frac{F12.8}{31.2}$	$A_1 = 6.0(9.2 + 12.8) + 3.65(25.8 + 31.2) = 340.0\ ft^2$
36+25	$\frac{F7.4}{23.1}$	F8.85	$\frac{F12.3}{30.45}$	$A_m = 6.0(7.4 + 12.3) + 4.425(23.1 + 30.45) = 335.2\ ft^2$
36+00	$\frac{F5.6}{20.4}$	F10.4	$\frac{F11.8}{29.7}$	$A_2 = 6.0(5.6 + 11.8) + 5.2(20.4 + 29.7) = 364.9\ ft^2$

$$Y_p = \frac{L(A_1 + 4A_m + A_2)}{162} = 50.0[340.0 + 4(355.2) + 364.9] = \underline{656.1\ yd^3}$$

FIGURE 14.10
Prismoidal volume computation.

14.7 VOLUME THROUGH TRANSITION

In Fig. 14.1, the right-of-way between stations 34+00 and 35+00 consists exclusively of cut while the right-of-way between stations 35+85 and 37+00 consists exclusively of fill. A transition zone within a right-of-way contains volumes of both cut and fill. In Fig. 14.1, the transition zone lies entirely between stations 35+00 (beginning of fill) and 35+85 (end of cut). For computation purposes, however, the beginning of the fill section is assumed to lie at the nearby station where the cut slope intersects the edge of the cut base (station 34+90 in Fig. 14.1), while the end of the cut section is assumed to lie at the nearby station where the fill slope intersects the edge of the fill base (station 35+70 in Fig. 14.1).

Consider the cut portion of the transition zone in Fig. 14.1. It comprises two separate volumes. The first, bounded by a four-sided figure at station 34+90 and a triangle at station 35+40, is computed using the average-end-area equation. The second, a pyramid whose base is the station 35+40 triangle and whose top is assumed to lie at station 35+70, is computed from the pyramidal-volume equation: volume equals base area multiplied by one-third of the height.

Example 14.9. From data in Fig. 14.5, compute the volume of the cut transition zone between stations 34+90 and 35+70.

Solution. Consider first the portion between station 34+90 and station 35+40. Station 34+90 is a form of three-level section with $b = 32.0$ ft, $c_l = 7.9$ ft, $d_l = 27.8$ ft, and $c = 4.6$ ft, but c_r is zero and d_r is half the base width in cut, or 16.0 ft. Substituting into Eq. (14.5), $A_{34+90} = (32.0/4)(7.9 + 0) + (4.6/2)(27.8 + 16.0) = 163.9$ ft^2. The triangular section at station 35+40 has a base width equal to half the roadway base width in cut, 16.0 ft, and a height equal to c_l, or 2.8 ft. The area of a triangle is base width multiplied by half the height. Substituting, $A_{35+40} = 16.0(2.8)/2 = 22.4$ ft^2. The distance between stations 34+90 and 35+40 is 50.0 ft. Substituting the length and area values into the average-end-area equation, $V_e = 50(163.9 + 22.4)/54 = 172.5$ yd^3.

The pyramidal portion between stations 35+40 and 35+70 has as its base the station 35+40 triangle whose area is 22.4 ft^2. The tip of the pyramid (by convention) is at station 35+70; thus, its height is 30.0 ft. $V_{\text{pyramid}} = (A_{\text{base}})(\text{height}/3)$. Substituting, $V = 22.4(30.0/3) = 224$ ft^3 $\Rightarrow\Rightarrow$ 8.3 yd^3.

The total cut volume in the transition zone is the sum of the volumes of the two cut segments: $V_{\text{total cut}} = 172.5 + 8.3 = 180.8$ yd^3.

The fill portion of the transition zone in Fig. 14.1 also comprises two segments: from the four-sided figure at station 35+70 to the triangle at station 35+40 and from the station 35+40 triangular base to the tip of the pyramid, assumed to lie at station 34+90.

Example 14.10. From data in Fig. 14.5, compute the volume of the fill transition zone between stations 35+70 and 34+90.

Solution. Begin with the larger volume lying between stations 35+70 and 35+40; the cross section of the former is a modified three-level section with $b = 24.0$ ft,

$f = 4.5$ ft, $f_r = 9.7$ ft, and $d_r = 26.6$ ft, but f_l is zero and d_l is 12.0 ft, which is half the base width in fill. Substituting these values into Eq. (14.5), $A_{35+70} = [(24.0/4) \times (0 + 9.7) + (4.5/2)(12.0 + 26.6)] = 145.0\ \text{ft}^2$. The triangular section at station 35+40 has a base width equal to half the roadway base width in fill, 12.0 ft, and a height equal to f_r, or 4.8 ft. Substituting into the triangle-area equation, $A_{35+40} = 12.0(4.8)/2 = 28.8\ \text{ft}^2$. The distance between stations 35+70 and 35+40 is 30.0 ft. Substituting the length and area values into the average-end-area equation, $V_e = 30(145.0 + 28.8)/54 = 96.6\ \text{yd}^3$.

The pyramidal section between stations 35+40 and 34+90 has as its base the station 35+40 triangle whose area is 28.8 ft² and a height of 50 ft. Substituting into the equation for pyramidal volume, $V_{\text{pyramid}} = 28.8(50/3) = 480.0\ \text{ft}^3 \Rightarrow\Rightarrow 17.8\ \text{yd}^3$.

The total volume of fill in the transition zone is the sum of the two volumes just computed: $96.6 + 17.8 = 114.4\ \text{yd}^3$.

For the transition section in Examples 14.9 and 14.10, the computed volume of cut and fill combined is $180.8 + 114.4 = 295.2\ \text{yd}^3$. Had cut and fill volumes been computed more precisely, taking the ends of the cut and fill slopes as stations 35+85 and 35+00, respectively, the combined cut-fill volume would have been 295.7 yd³. In this case, then, use of the computational convention made little difference in the total computed earthwork volume, underestimation of cut being reasonably balanced by overestimation of fill. In fact, the normal tendency in using the simplified convention is for the combined-volume error to fall well within the larger errors inherent in using simplified cross sections and in using either of the two simplified volume equations. Only in cases where there is a very significant difference between the costs of cut and fill might it be worthwhile to use the more precise computation method in transition sections.

14.8 BORROW-PIT VOLUMES

Engineering projects typically require that fill material be obtained off-site, from one or more **borrow pits**, whose excavation rights have been purchased from project funds. Excavated materials are priced primarily by volume; thus, one must determine the amount of material that has been taken away. This is done by dividing the surface of the borrow pit into squares of appropriate size, then taking level observations to the grid corners before and after excavation. The volume of removed material is the sum of the volumes removed from each square, taken as the average drop at the four corners of the square multiplied by the area of the square. The method is equally effective for computing fill volume.

On a construction site, the grid is staked out with transit and tape or EDM. Corners are designated by letters and numbers as shown in Fig. 14.11. With a level set up at an appropriate point, a backsight is taken on a known elevation to establish the HI, then foresights are observed on all grid corners and their elevations are tabulated. The cut or fill depth at each grid corner is the difference between the original ground elevation at that corner and the final grade elevation the project requires.

Although the total cut or fill volume may be computed by determining the mean cut or fill height for each square and multiplying by the area of the square, there is a quicker method. Consider the fact that the cut or fill height at a given corner is used in the volume computation from one to four times, depending on the location of the corner. In Fig. 14.11, the cut or fill height at corner A,0 is used only once; at corner A,1 it is used twice; at C,3, it is used three times; and at B,1 it is used four times. As a simplification, multiply cut or fill height at each corner by the number of times, n, the corner appears in the volume computation. The sum of the multiples is the cumulative sum of the cut or fill heights used in computing the volumes within all grid squares. Dividing by four produces the cumulative sum of the *average* cut or fill heights for all squares. Multiplying by the area of a *single* square and dividing by 27 provides the total volume in cubic yards. The resulting equation for borrow-pit volume is

$$V_{BP} = \frac{[\Sigma n(\text{corner cut or fill heights})]A_{sq}}{108} \tag{14.9}$$

where Σ indicates a summation of n multiples of corner cut or fill values and A_{sq} is the area of a single grid square.

Example 14.11. The Fig. 14.11 area is to be excavated to an elevation of 358.0 ft for a building site. The grid spacing for the figure is 30.0 ft, horizontally and vertically. Profile-level values for the grid corners are shown on the next page. Compute the volume of earth to be excavated.

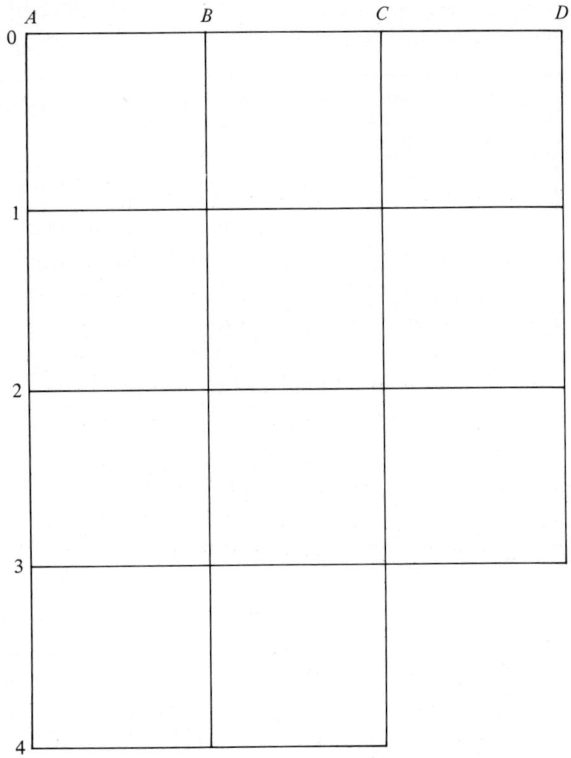

FIGURE 14.11
Layout of borrow pit.

Point	BS	HI	FS	Elev, ft
BM2	4.22	364.70		360.48
A,0			5.2	359.5
B,0			5.4	359.3
C,0			5.7	359.0
D,0			5.9	358.8
A,1			4.7	360.0
B,1			4.8	359.9
C,1			5.2	359.5
D,1			5.5	359.2
A,2			4.2	360.5
B,2			4.7	360.0
C,2			4.8	359.9
D,2			5.0	359.7
A,3			3.8	360.9
B,3			4.0	360.7
C,3			4.6	360.1
D,3			4.6	360.1
A,4			3.4	361.3
B,4			3.7	361.0
C,4			4.2	360.5

Solution. Two columns are added to the above notes (see below): the cut at each corner (difference between the corner elevation and the 358.0-ft final grade), and the corner cut multiplied by n. The sum of the values in the final column is the bracketed summation in Eq. (14.9).

Point	BS	HI	FS	Elev, ft	Cut	n(Cut)
BM2	4.22	364.70		360.48		
A,0			5.2	359.5	1.5	1.5
B,0			5.4	359.3	1.3	2.6
C,0			5.7	359.0	1.0	2.0
D,0			5.9	358.8	0.8	0.8
A,1			4.7	360.0	2.0	4.0
B,1			4.8	359.9	1.9	7.6
C,1			5.2	359.5	1.5	6.0
D,1			5.5	359.2	1.2	2.4
A,2			4.2	360.5	2.5	5.0
B,2			4.7	360.0	2.0	8.0
C,2			4.8	359.9	1.9	7.6
D,2			5.0	359.7	1.7	3.4
A,3			3.8	360.9	2.9	5.8
B,3			4.0	360.7	2.7	10.8
C,3			4.6	360.1	2.1	6.3
D,3			4.6	360.1	2.1	2.1
A,4			3.4	361.3	3.3	3.3
B,4			3.7	361.0	3.0	6.0
C,4			4.2	360.5	2.5	2.5
					Σ =	87.7 ft

Each square is 30 ft by 30 ft, or 900 ft^2. Substituting the corner-height summation and the area of each square into Eq. (14.9),

$$V_{BP} = \frac{87.7(900)}{108} = 731 \text{ yd}^3$$

14.9 PROBLEMS

14.1. A slope stake is to be set for a fill section with a 32.0-ft roadbed width and side slopes of 2:1. Based on the design elevation of the road, the grade rod is computed to be 3.60 ft. For an estimated fill of 8.4 ft, compute the trial ground rod and distance from the centerline.

14.2. The centerline elevation at a full station on a route survey was 563.17 ft, a point where design drawings indicate a cut of 4.72 ft will be required. Roadbed width, including ditches, is to be 40.0 ft and side slopes are to be 1.5:1. If the HI of the level instrument is 570.95 ft and the estimated cut on the right side for that station is 10.2 ft, compute the grade rod, the trial ground rod, and the trial distance from the centerline for the right slope stake.

14.3. Referring to Prob. 14.2, if the estimated cut on the left side is 1.4 ft, compute the grade rod, the trial ground rod, and the trial distance from the centerline for the left slope stake.

14.4. The roadbed width, including shoulders, for a level fill section is 52.0 ft. Side slopes are 2:1. If the grade rod is 3.90 ft and the fill at the centerline is 6.3 ft, compute the ground rod and distance from the centerline required to set the left and right slope stakes.

Refer to the field-record notes below in solving Probs. 14.5 through 14.8.

Station	Cross sections				
	Roadbed width: 32.0 ft; Side slope: 2:1				
29+00		$\frac{F8.6}{32.9}$	F6.2	$\frac{F11.4}{38.8}$	
28+00		$\frac{F5.5}{27.0}$	F5.5	$\frac{F5.5}{27.0}$	
27+00	$\frac{F10.2}{36.4}$	$\frac{F6.7}{16.0}$	F8.3	$\frac{F4.1}{16.0}$	$\frac{F6.4}{28.8}$
26+00	$\frac{F9.2}{34.4}$	$\frac{F10.4}{24.3}$	F6.5	$\frac{F5.2}{11.4}$	$\frac{F11.2}{38.4}$

14.5. Compute the area of the cross section at station 29+00.

14.6. Compute the area of the cross section at station 28+00.

14.7. Compute the area of the cross section at station 27+00.

14.8. Compute the area of the cross section at station 26+00.

Refer to the field-record notes on the next page in solving Probs. 14.9 through 14.12.

Station	Cross sections				
	Roadbed width: 40.0 ft; Side slope: 1.5:1				
20+00		$\frac{C8.6}{32.9}$	C4.6	$\frac{C6.2}{29.3}$	
19+00		$\frac{C6.4}{29.6}$	C6.4	$\frac{C6.4}{29.6}$	
18+00	$\frac{C6.8}{30.2}$	$\frac{C7.5}{20.0}$	C5.2	$\frac{C5.4}{20.0}$	$\frac{C4.2}{26.3}$
17+00	$\frac{C4.0}{26.0}$	$\frac{C7.8}{15.0}$	C4.4	$\frac{C4.1}{6.0}$	$\frac{C11.2}{36.8}$

14.9. Compute the area of the cross section at station 20+00.

14.10. Compute the area of the cross section at station 19+00.

14.11. Compute the area of the cross section at station 18+00.

14.12. Compute the area of the cross section at station 17+00.

14.13. Using the results from Probs. 14.5 and 14.6, compute the average-end-area volume between stations 28+00 and 29+00.

14.14. Using the results from Probs. 14.6 and 14.7, compute the average-end-area volume between stations 27+00 and 28+00.

14.15. Using the results from Probs. 14.7 and 14.8, compute the average-end-area volume between stations 26+00 and 27+00.

14.16. Using the results from Probs. 14.9 and 14.10, compute the average-end-area volume between stations 19+00 and 20+00.

14.17. Using the results from Probs. 14.10 and 14.11, compute the average-end-area volume between stations 18+00 and 19+00.

14.18. Using the results from Probs. 14.11 and 14.12, compute the average-end-area volume between stations 17+00 and 18+00.

14.19. Using the results from Probs. 14.5 and 14.6, compute the prismoidal volume between stations 28+00 and 29+00.

14.20. Using the results from Probs. 14.7 and 14.8, compute the prismoidal volume between stations 26+00 and 27+00.

14.21. Using the results from Probs. 14.9 and 14.10, compute the prismoidal volume between stations 19+00 and 20+00.

14.22. Using the results from Probs. 14.11 and 14.12, compute the prismoidal volume between stations 17+00 and 18+00.

14.23. Using the field-record notes shown below, compute cut and fill volumes in the station 41+80–to–station 43+00 transition zone.

Station	Cross sections		
	Roadbed: 32.0 ft in fill, 40.0 ft in cut Side slope: 2:1 in fill, 1.5:1 in cut		
43+00	$\frac{F6.3}{28.6}$	F4.0	$\frac{0.0}{16.0}$
42+64	$\frac{F3.4}{22.8}$	0.0	$\frac{C4.2}{26.3}$
41+80	$\frac{0.0}{20.0}$	C6.2	$\frac{C8.0}{32.0}$

14.24. Using the field-record notes shown below, compute cut and fill volumes in the station 63+37–to–station 65+22 transition zone.

Station	Cross sections		
	Roadbed: 28.0 ft in fill, 36.0 ft in cut Side slope: 2:1 in fill, 1.5:1 in cut		
65+22	$\frac{0.0}{18.0}$	C4.8	$\frac{C6.4}{27.6}$
64+43	$\frac{F4.4}{22.8}$	0.0	$\frac{C4.0}{24.0}$
63+37	$\frac{F9.2}{32.4}$	F5.3	$\frac{0.0}{14.0}$

14.25. A borrow-pit site was laid out using a 50-ft square grid, with corner elevations in feet as shown below. When use of the pit was terminated, its floor had a constant elevation of 563.8 ft. Compute the amount of material removed from the site.

	A	*B*	*C*	*D*
0	563.9	566.2	569.7	568.3
1	564.1	565.8	567.4	571.5
2	566.0	571.1	573.7	577.4
3	565.3	568.6	571.6	575.6
4	565.2	567.7	569.2	573.1

14.26. Do the same as Prob. 14.25 except the site was expanded to include an additional row of grid figures as shown below:

	A	*B*	*C*	*D*
5	564.3	567.1	568.0	570.2

14.27. Using data from Probs. 14.25 and 14.26, what would be the final level elevation of the floor of the borrow pit if a total of 10,000 yd^3 of material were taken out before abandoning the site?

14.28. During site preparation for a shopping-mall parking lot, a grid was staked out on a 100-ft square. Preconstruction elevations were obtained as shown below, alphabetic designations running from west to east and numeric designations from north to south. If the desired constant elevation along the north edge of the parking lot is 352.50 ft and the lot is to have a −1.00% north-to-south slope, what is the net volume of earth that must be removed or brought in to achieve the desired grade?

	A	*B*	*C*	*D*	*E*
0	351.7	354.9	357.4	355.6	353.1
1	352.4	356.1	358.3	355.8	354.7
2	355.2	355.9	359.2	351.6	355.3
3	353.1	352.0	354.6	347.3	357.2
4	349.3	350.6	348.5	344.8	344.4
5	346.0	348.9	343.6		
6	344.2	347.4	341.7		

14.29. Do the same as Prob. 14.28 except the north-to-south slope is to be −0.70%.

APPENDIX A

DERIVATIONS

INTRODUCTION

This text is intended for students who, as a group, either have not taken calculus or have never become comfortable with the subject. For that reason, equations whose derivation rely on the calculus are not presented in the text narrative. The authors feel strongly that most equations that are important to surveying should not be simply pulled out of the air. This appendix is provided for those derivations considered inappropriate for inclusion in the chapter narrative.

A-1 EQUATION (2.3)

An inaccuracy in measuring slope angle will result in an error in the final computation of horizontal distance, even if the slope-distance measurement itself is precise. Consider Eq. (2.1) in Chap. 2:

$$\boldsymbol{H = s\cos\alpha} \qquad \textbf{(2.1)}$$

where H is horizontal distance, s is slope distance, and α is slope angle. Assuming the inaccuracy in slope-distance measurement to be negligible, Eq. (2.1) is differentiated to produce

$$dH = -s\sin\alpha\ d\alpha$$

where dH is the error in horizontal distance due to an infinitesimal error $d\alpha$ in the measurement of slope angle α. But $d\alpha$, in the calculus context, is an impractically small value for use in a surveying error equation. Consider instead a value, $\Delta\alpha$, of the magnitude expected for ordinary errors in slope-angle measurement. Substituting $\Delta\alpha$ for $d\alpha$ and ΔH for dH,

$$\Delta H = -s \sin \alpha \Delta \alpha \tag{2.3}$$

A-2 EQUATION (2.7)

Inaccuracies in measuring difference in elevation, d, between the two ends of a precisely measured slope distance will result in an error in the final computed horizontal distance. In this case, horizontal distance H is computed from Eq. (2.6):

$$H = \sqrt{s^2 - d^2} \tag{2.6}$$

Assuming negligible error in s, Eq. (2.6) is differentiated to produce

$$dH = -\left(\frac{d}{\sqrt{s^2 - d^2}}\right) dd$$

$$= -\left(\frac{d}{H}\right) dd$$

where dH is the error in horizontal distance due to an infinitesimal error dd in the measurement of elevation difference d. As before, Δd and ΔH are substituted for dH and dd to produce

$$\Delta H = -\left(\frac{d}{H}\right) \Delta d \tag{2.7}$$

A-3 EQUATIONS (5.8) and (5.11)

Consider a value V that is a function of a number of measured values $K, L, M, \ldots, U$; that is,

$$V = f(K, L, M, \ldots, U)$$

From this designation of V, numbers of surveying texts have developed the following simplified version of the **error propagation equation**:

$$\sigma_V^2 = \left(\frac{\partial V}{\partial K}\right)^2 \sigma_K^2 + \left(\frac{\partial V}{\partial L}\right)^2 \sigma_L^2 + \left(\frac{\partial V}{\partial M}\right)^2 \sigma_M^2 + \cdots + \left(\frac{\partial V}{\partial U}\right)^2 \sigma_U^2$$

where the designation $\partial V/\partial K$ indicates a partial derivative of the overall function V with respect to K, $\partial V/\partial L$ is a partial derivative of V with respect to L, and so forth, while σ_K through σ_V are the standard errors in values K through V.

Consider V to be the sum of a number of measured values $K, L, M, \ldots, U$, such that $V = f(K, L, M, \ldots, U) = K + L + M + \cdots + U$. Applying the simplified error propagation equation,

$$\sigma_V^2 = \sigma_K^2 + \sigma_L^2 + \sigma_M^2 + \cdots + \sigma_U^2$$

Substituting σ_s, the standard error of a series of measurements, for σ_V, and the numerals 1 through n for the subscripts K through U, then taking the square root of both sides,

$$\boldsymbol{\sigma}_s = \sqrt{\boldsymbol{\sigma}_1^2 + \boldsymbol{\sigma}_2^2 + \boldsymbol{\sigma}_3^2 + \cdots + \boldsymbol{\sigma}_n^2} \qquad \textbf{(5.8)}$$

Consider now a different situation in which V is the product of two values K and L such that $V = f(K, L) = KL$. Again applying the simplified error propagation equation,

$$\boldsymbol{\sigma}_V^2 = L^2\boldsymbol{\sigma}_K^2 + K^2\boldsymbol{\sigma}_L^2$$

Substituting σ_A, the standard error of a rectangular area, for σ_V and X and Y for K and L, then taking the square root, produces

$$\boldsymbol{\sigma}_A = \sqrt{Y^2\boldsymbol{\sigma}_X^2 + X^2\boldsymbol{\sigma}_Y^2} \qquad \textbf{(5.11)}$$

In the same manner, the simplified error propagation equation may be used to derive quickly the standard error of a circular area, an elliptical area, the volumes of a cube, sphere, pyramid, rectangular parallelopiped, or any other two- or three-dimensional figures that can be expressed in algebraic form.

A-4 EQUATIONS (13.21), (13.22), and (13.23)

A parabolic vertical curve connecting two grades provides a constant vertical acceleration for a vehicle moving at constant horizontal velocity from one grade to another. The property of constant vertical acceleration may also be defined as constant rate of change of grade. Consider Fig. A.1, with an entering grade g_1, connected to a departing grade g_2 by a parabolic vertical curve of length L. The rate of change of grade, r, is defined by

$$\boldsymbol{r} = \frac{\boldsymbol{g}_2 - \boldsymbol{g}_1}{\boldsymbol{L}} \qquad \textbf{(13.20)}$$

Where L is in stations and g is in percent grade, which is the same as *feet per station*. The dimensions of r are therefore *feet per station per station*.

In Fig. A.1 the y axis passes through the beginning of the vertical curve (BVC) and the x axis passes through the elevation datum (most often mean sea level, or MSL). The general equation for a parabola is

$$y = ax^2 + bx + c$$

Coefficients a, b, and c will now be redefined in terms that relate to the xy-coordinate system in Fig. A.1 and to the dimensions of parabolic vertical curves.

Substituting $x = 0$ into the general equation for a parabola, it is seen that $y = c$. But at the point on the curve in Fig. A.1 defined by $x = 0$, the value of y is the elevation of the BVC. Thus, $c = \text{elev}_{\text{BVC}}$.

Differentiating the general equation for a parabola,

$$\frac{dy}{dx} = 2ax + b$$

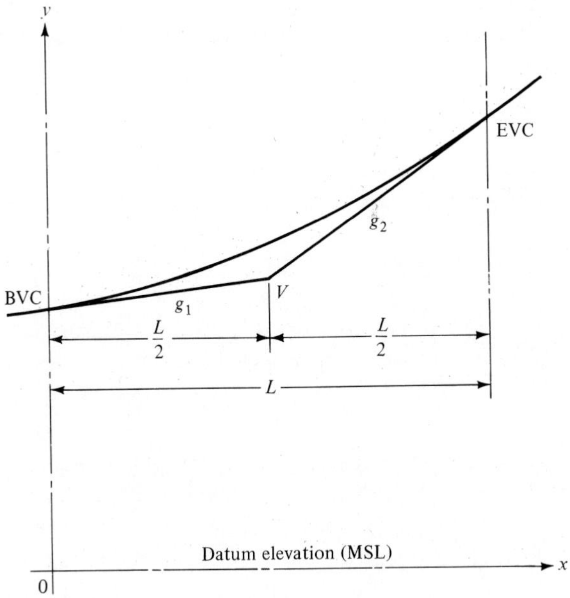

FIGURE A.1

At $x = 0$, $dy/dx = b$. But dy/dx is the slope of the curve, or grade, and the grade at $x = 0$ is the entering grade g_1. Thus, $b = g_1$.

Differentiating again the general equation for a parabola,

$$\frac{d^2y}{dx^2} = 2a$$

But d^2y/dx^2 is the rate of change of grade, which has already been defined as r. Thus, $r = 2a$ and $a = r/2$.

Substituting the derived values of the coefficients back into the general equation for a parabola,

$$\mathbf{y = \frac{r}{2}x^2 + g_1x + elev_{BVC}} \tag{13.21}$$

Remember that the units of r are feet per station per station and the units of g are feet per station. Thus, x must be expressed in stations and units of y are feet.

The high or low point of a curve is defined as a point where the slope, or grade, of the curve is zero. Taking the derivative of Eq. (13.21) and setting it to zero,

$$\frac{dy}{dx} = rx + g_1 = 0$$

This leads to the following expressions for x at the high point or low point of a parabolic curve:

$$\mathbf{x = -\frac{g_1}{r}} \tag{13.22}$$

But $r = (g_2 - g_1)/L$. Substituting for r, the value of x at the high or low point of a parabolic curve may also be expressed as

$$x = \frac{g_1 L}{g_1 - g_2} \tag{13.23}$$

APPENDIX B

NOTEKEEPING IN FIELD BOOKS

Example 1 Measuring Distance with a Steel Tape and Determining Pace

Example 2 Measuring Traverse Sides with a Steel Tape

Example 3 Differential Levels around Traverse

Example 4 Three-Wire Leveling

Example 5 Determination of Collimation Correction ("C" Factor)

Example 6 Measuring Angles by Repetition with a Transit and Closing the Horizon

Example 7 Measuring Angles by Repetition with a Repeating Theodolite

Example 8 Stadia Observations for Topographic Mapping

Example 9 Foundation Layout by Angles and Distances

Example 10 Circular Curve Layout by Deflection Angles

MEASURING DISTANCE WITH A STEEL TAPE & DETERMINING PACE

1. Measuring distance

Point	Direction	Measured Distance (ft.)	Difference	Mean Distance (ft.)	Precision
A					
	NE	352.432			
B			0.069	352.466	1/5100
	SW	352.501			
A					

2. Determining pace - S. Jones

	Natural Paces	Direction
	123	A → B
	121	B → A
	124	A → B
	122	B → A
Mn.	122.5	

$$\text{Mean pace length} = \frac{\text{Mn. dist.}}{\text{Mn. No. of paces}} = \frac{352.466}{122.5} = 2.877 = 2.88 \text{ ft.}$$

$$\text{Mean paces}/100' = \frac{100}{\text{Mn. pace length}} = \frac{100}{2.877} = 34.755 = 34.8$$

Example 1

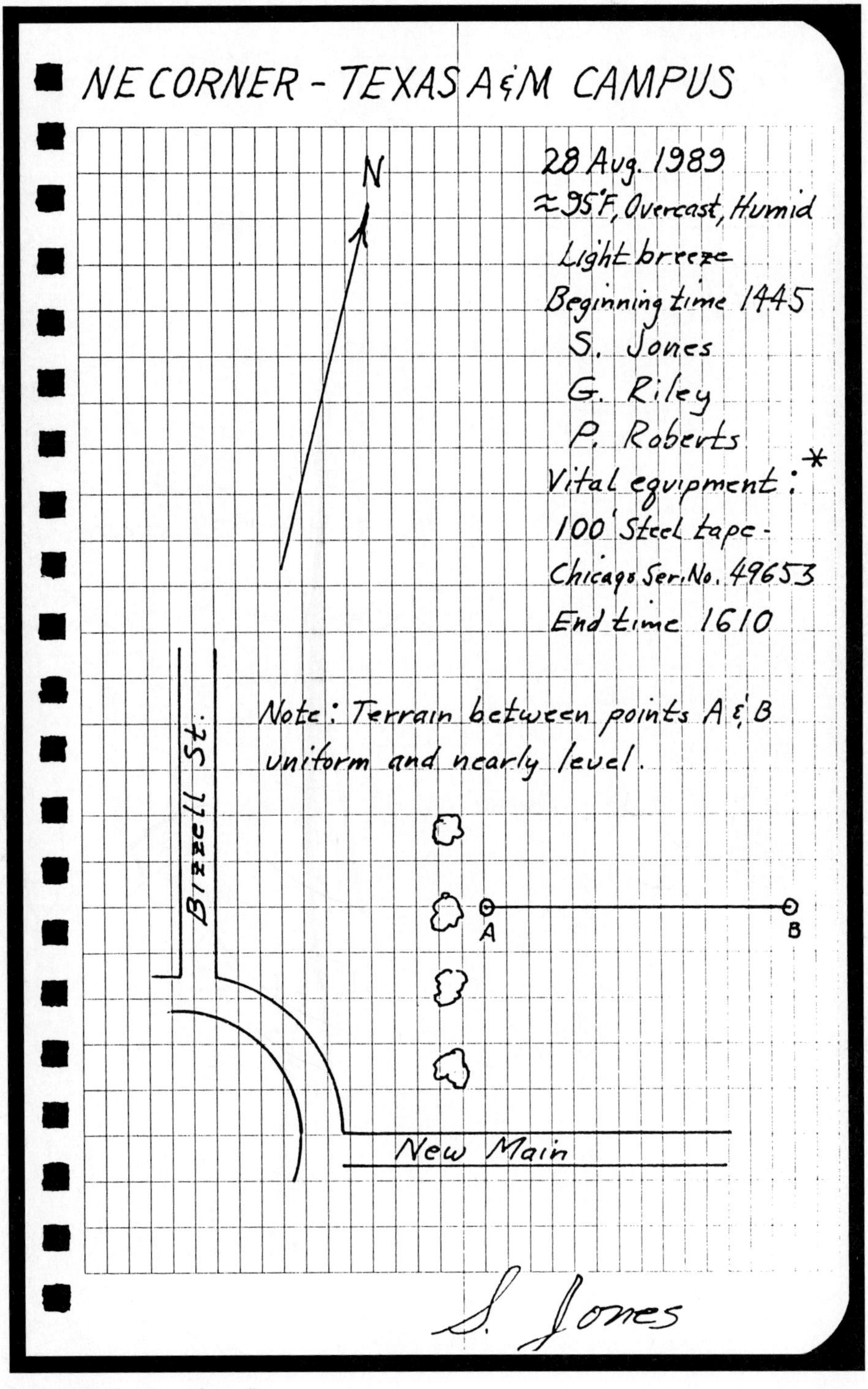

Example 1 (continued).

* Equipment that can affect surveying accuracy.

MEASURING TRAVERSE SIDES WITH A STEEL TAPE

Point	c/w Measurement (ft.)	cc/w Measurement (ft.)	Mean (ft.)	Difference	Precision
CE-4					
	125.902	125.879	125.890	0.023	1/5500
A					
	212.451	212.416	212.434	0.035	1/6100
B					
	100.568	100.585	100.576	0.017	1/5900
C					
	162.104	162.073	162.088	0.031	1/5200
D					
	121.072	121.088	121.080	0.016	1/7600
E					
	124.028	124.~~002~~ 011	124.~~015~~ 020	~~0.026~~ 0.017	1/~~4800~~ 7300
CE-4					

Note: Specifications require that precision must not be less than 1/5000.

Example 2

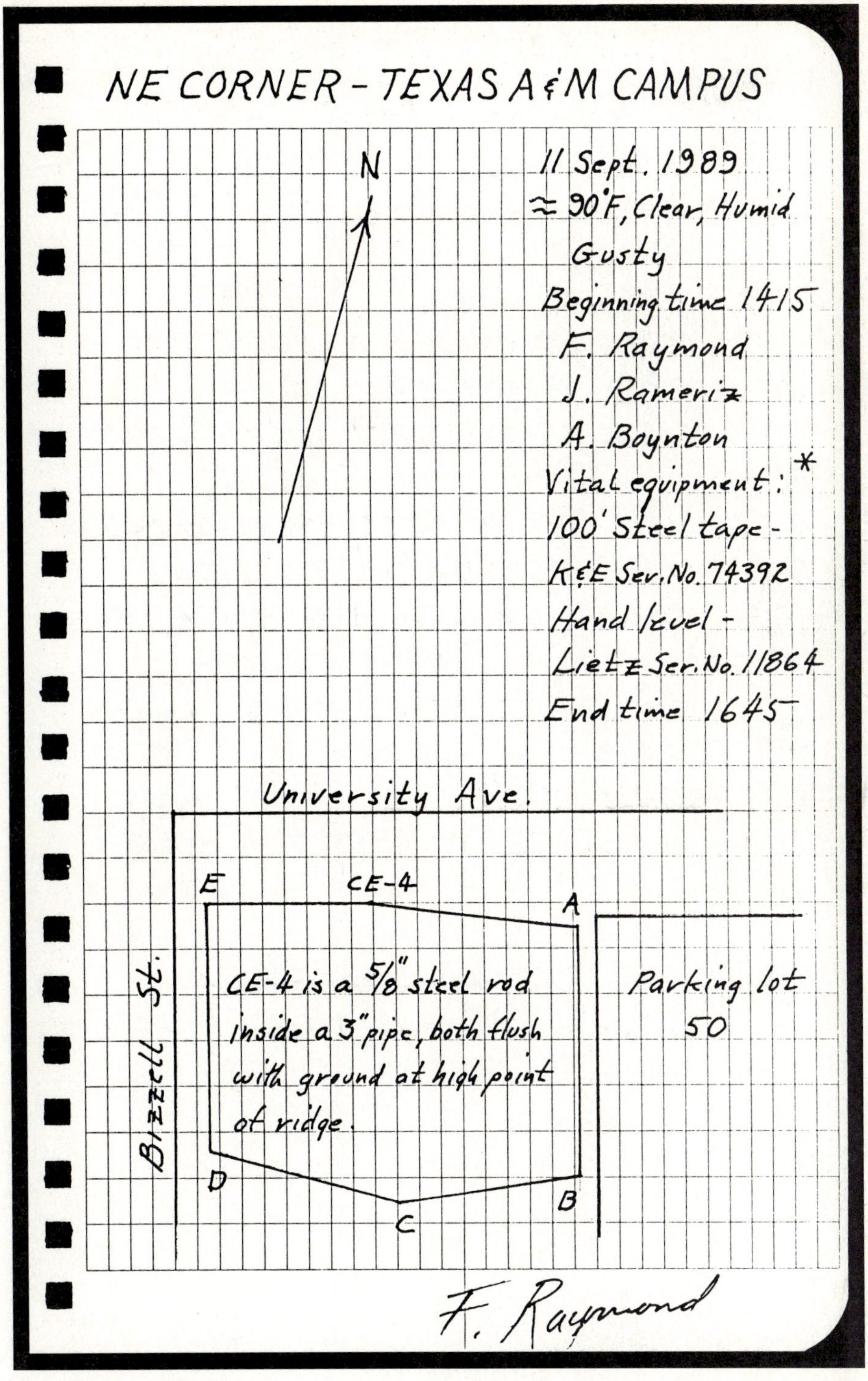

Example 2 (continued).

DIFFERENTIAL LEVELS AROUND TRAVERSE

Point	B.S. (ft.)	H.I. (ft.)	F.S. (ft.)	Elev. (ft.)	Dist. B.S./F.S. (Paces)
BMCE-4	1.126	340.234		339.108	24
BM A	2.384	331.980	10.638	329.596	39/24
BM B	7.019	333.537	5.462	326.518	21/38
BM C	3.662	334.353	2.846	330.691	30/20
IFS #1			(6.118)	(328.235)	(10)
BM D	6.501	336.570	4.284	330.069	22/31
BM E	10.191	343.372	3.389	333.181	34/21
T.P. 1	5.336	347.300	1.408	341.964	6/35
BMCE-4			8.173	339.127	6

ΣBS + 36.219 $\quad$ ΣFS − 36.200 $\quad$ 176/175

ΣFS − 36.200 $\quad$ = 351

0.019 + 339.108 = 339.127 ✓ $\quad$ × 2.88 ft/P.

= 1011 ft.

Allowable closure (3rd Order) $= \pm 0.050\sqrt{\frac{1011}{5280}}$

$= \pm 0.022$ ft.

Actual closure = 0.019 ft. ✓

Example 3 Differential Levels around Traverse

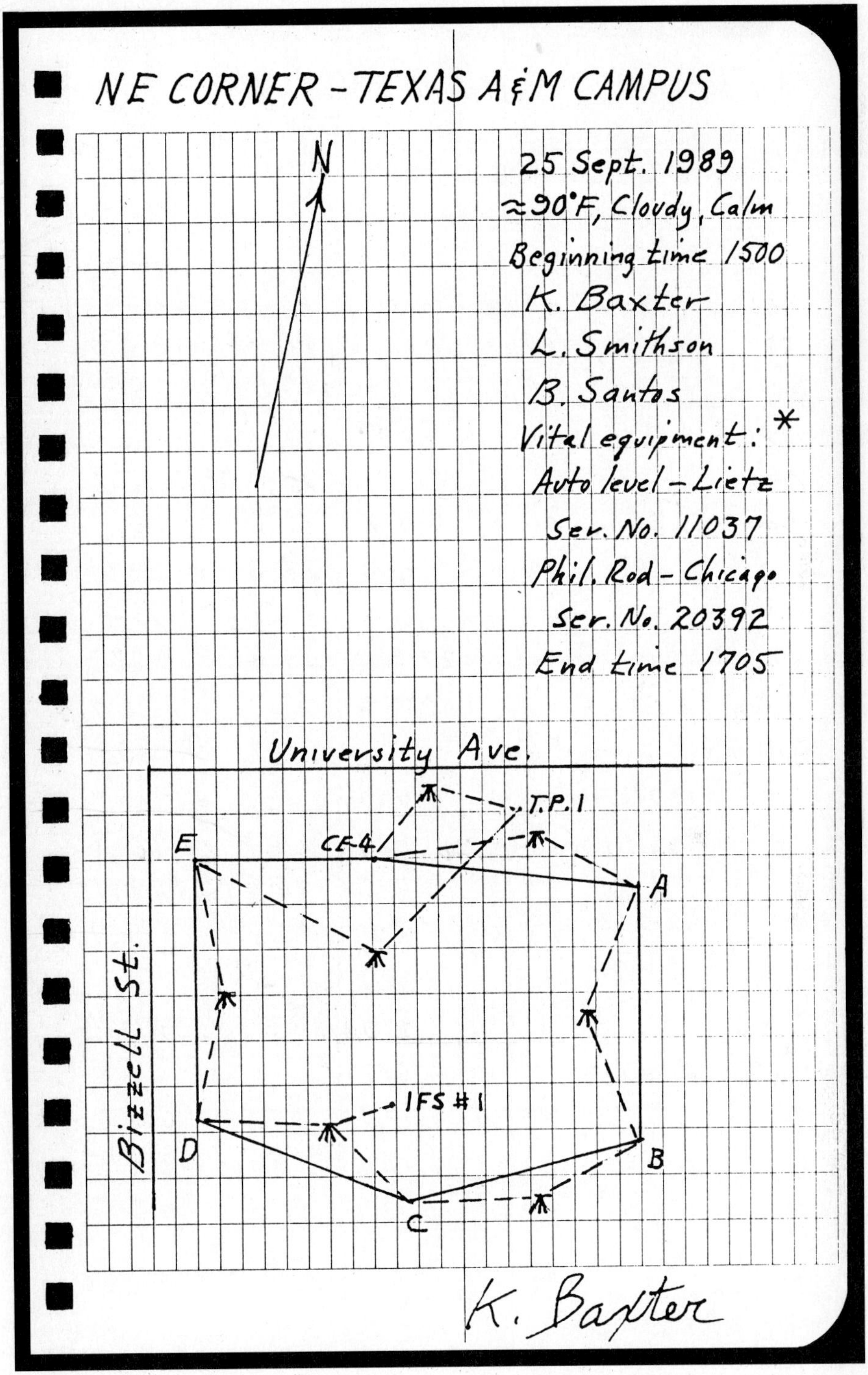

Example 3 (continued).

THREE WIRE LEVELING

Sta.	Backsight Data Rdgs.	Backsight Data Stadia	Foresight Data Rdgs.	Foresight Data Stadia	Elev. (ft)
BM CE-4					339.108
	5.397		7.809		
	5.171	22.6	7.568	24.1	+ 5.170
	4.942	22.9	7.331	23.7	344.278
Sums	15.510	45.5	22.708	47.8	− 7.569
Mn	5.170		7.569		
T.P. 1					336.709
	3.881		9.476		
	3.632	24.9	9.238	23.8	+ 3.631
	3.380	25.2	9.002	23.6	340.340
Sums	10.893	50.1	27.716	47.4	− 9.239
Mn	3.631		9.239		
TBM 1		95.6		95.2	331.101
	9.922		4.069		
	9.669	25.3	3.828	24.1	+9.670
	9.419	25.0	3.584	24.4	340.771
Sums	29.010	50.3	11.481	48.5	−3.827
Mn	9.670		3.827		
T.P. 2		145.9		143.7	336.944
	6.772		4.632		
	6.533	23.9	4.376	25.6	+6.532
	6.290	24.3	4.124	25.2	343.476
Sums	19.595	48.2	13.132	50.8	− 4.377
Mn	6.532		4.377		
BM CE-4		194.1		194.5	339.099

Example 4

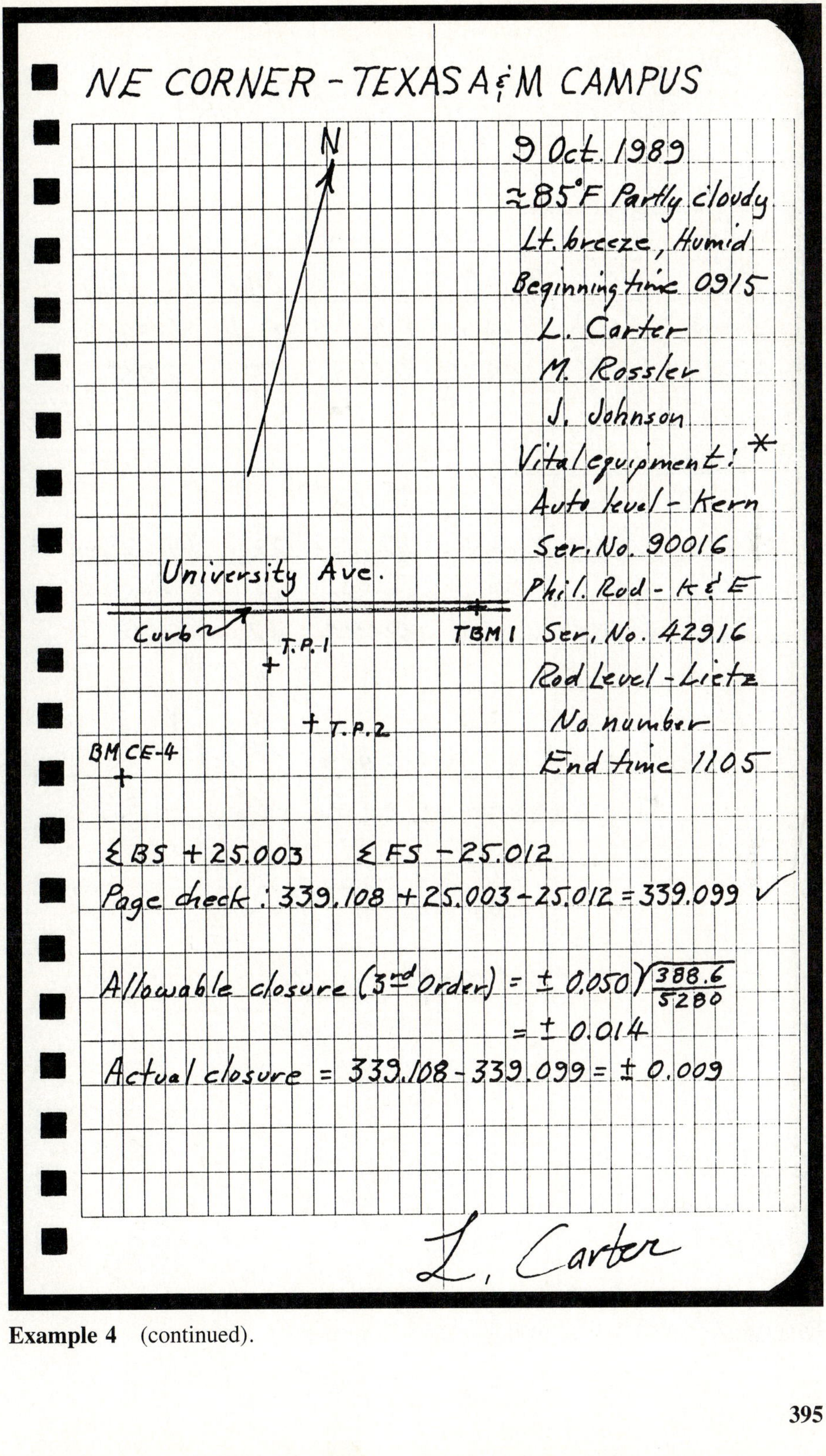
NE CORNER - TEXAS A&M CAMPUS

N

University Ave.

Curb

TBM 1

T.P. 1

T.P. 2

BM CE-4

9 Oct. 1989

≈85°F Partly cloudy

Lt. breeze, Humid

Beginning time 0915

L. Carter

M. Rossler

J. Johnson

Vital equipment: *

Auto level - Kern

Ser. No. 90016

Phil. Rod - K&E

Ser. No. 42916

Rod Level - Lietz

No number

End time 1105

Σ BS +25.003 Σ FS −25.012

Page check: 339.108 + 25.003 − 25.012 = 339.099 ✓

Allowable closure (3rd Order) = $\pm 0.050\sqrt{\frac{388.6}{5280}}$

= ± 0.014

Actual closure = 339.108 − 339.099 = ± 0.009

L. Carter

Example 4 (continued).

DETERMINATION OF COLLIMATION CORRECTION ("C" FACTOR)

Set-up 1

R_{BS} (ft.)	R_{FS} (ft.)	$R_{BS}-R_{FS}$ (ft.)	D_{FS} (ft.)	D_{BS} (ft.)	$D_{FS}-D_{BS}$ (ft.)
4.363	3.786	0.577	230	25	205

Set-up 2

R'_{BS} (ft.)	R'_{FS} (ft.)	$R'_{BS}-R'_{FS}$ (ft.)	D'_{BS} (ft.)	D'_{FS} (ft.)	$D'_{BS}-D'_{FS}$ (ft.)
5.066	4.531	0.535	239	32	207

$$\text{"C" factor} = \frac{(R_{BS}-R_{FS})-(R'_{BS}-R'_{FS})}{(D_{FS}-D_{BS})+(D'_{BS}-D'_{FS})}$$

$$= \frac{0.577-0.535}{205+207} = \frac{0.042}{412}$$

$$= +0.000102 \text{ ft./ft.}$$

$$\approx 1/8'' \text{ in } 100 \text{ ft.}$$

Example 5 Determination of Collimation Correction ("C" Factor)

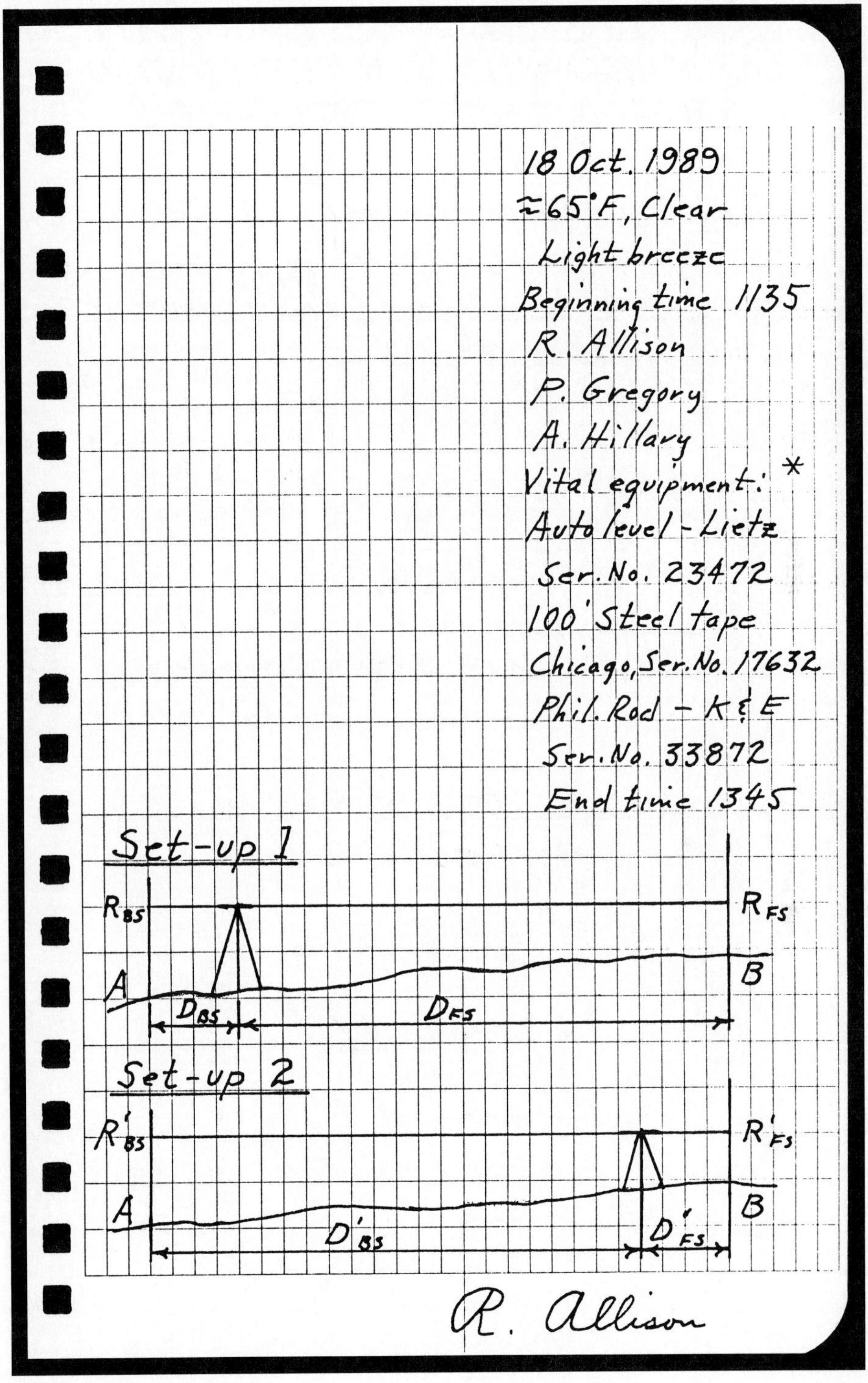

Example 5 (continued).

MEASURING ANGLES BY REPETITION WITH A TRANSIT & CLOSING THE HORIZON

⊼ at Pt. B, All Angles Turned Clockwise

4 Repetitions / Angle

Pt.	Tel.	Vern. A	Vern. B	Mean Vern.	Unadj. Mn. Angle	Adj. Mn. Angle
		Pt. C	to	Pt. A		
C	D	0°00'00"	00'20"	0°00'10"		
A	D	147°22'20"				
A	P	229°29'40"	30'20"	229°30'00"	147°22'27.5"	147°22'25"
		Pt. A	to	Pt. C		
A	P	229°29'40"	30'20"	229°30'00"		
C	P	82°07'00"				
C	D	0°00'20"	00'40"	0°00'30"	212°37'37.5"	212°37'35"
					360°00'05"	360°00'00"

Hor. circle least count $= \frac{20'}{60} = \frac{1}{3}' = 20''$

Vert. circle least count $= \frac{30'}{60} = \frac{1}{2}' = 30''$

Horizon closure $= 360° - 360°00'05'' = -05''$

Adjustment $= -05''/2 = -2.5''$ each ∡

Vert. circle	(D)	+5°12'00"
	(P)	+5°13'00"
	Mn. =	+5°12'30" = Correct vert. ∡

Vert. circle index error = Direct − Mean

$= 5°12'00'' - 5°12'30'' = -30''$

Example 6 Measuring Angles by Repetition with a Transit and Closing the Horizon

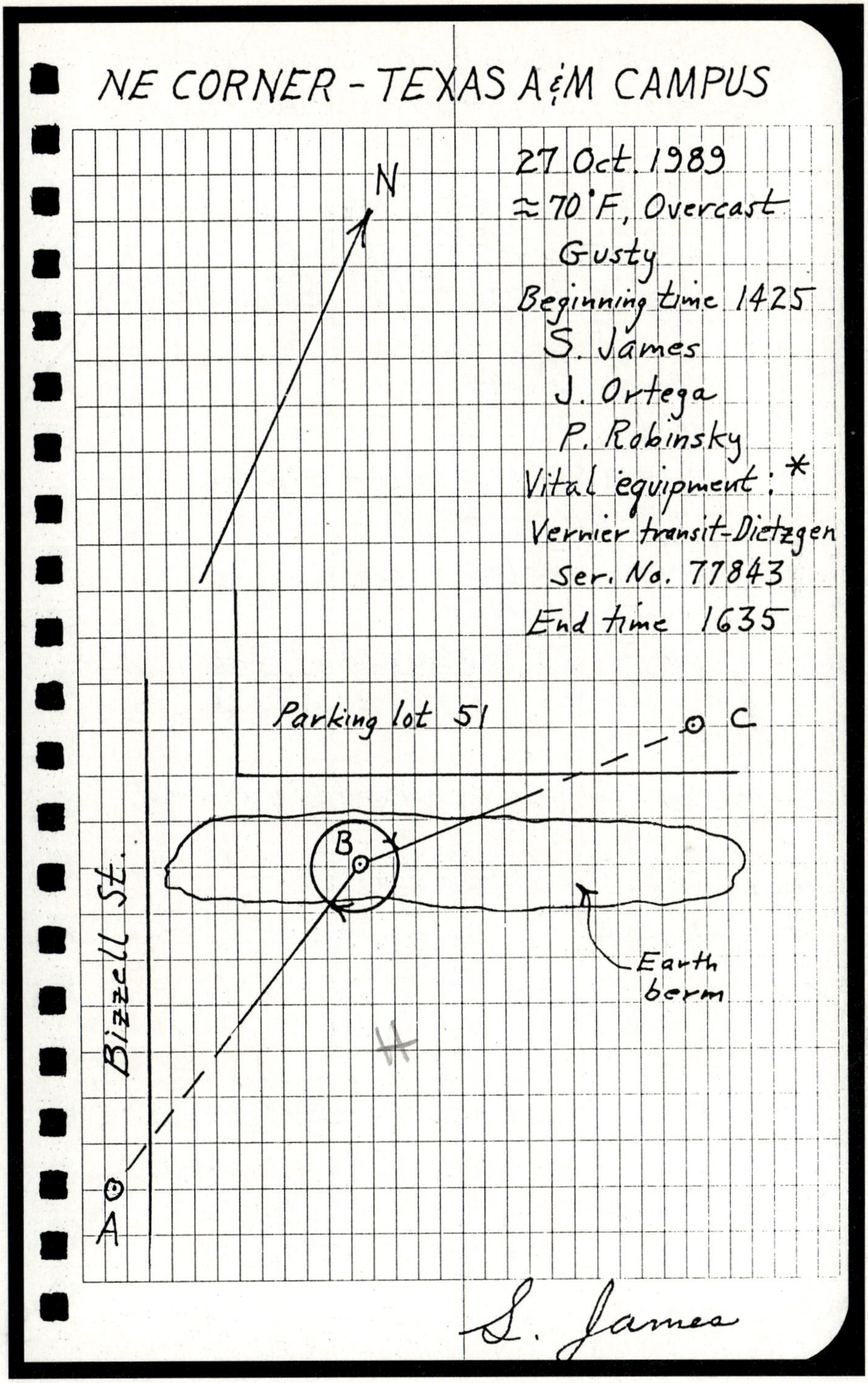

Example 6 (continued).

MEASURING ANGLES BY REPETITION WITH A REPEATING THEODOLITE

⊼ at Pt. B, All Angles Turned Clockwise
4 Repetitions / Angle

Pt.	Telescope	Scale Reading	Unadj. Mn. Angle	Adj. Mn. Angle
		Pt. C to Pt. A		
C	Direct	0° 00.0′		
A	Direct	147° 22.3′		
A	Plunged	229° 30.2′	147° 22.55′	147° 22.58′
		Pt. A to Pt. C		
A	Plunged	229° 30.2′		
C	Plunged	82° 07.0′		
C	Direct	359° 59.8′	212° 37.40′	212° 37.42′
			359° 59.95′	360° 00.00′

Hor. and Vert. circle least count = 1′ (0.1′ estimated)

Horizon closure = 360° − 359° 59.95′ = 0.05′

Angle adjustment = + 0.03′ ∡CBA
+ 0.02′ ∡ABC

Vert. circle (D) 91° 12.6′
(P) 268° 47.6 (360 − 268° 47.6 = 91° 12.4′)
Mn. = 91° 12.5′ = Correct zenith angle

Vert. circle index error = Direct − Mean
= 91° 12.6′ − 91° 12.5′ = + 0.1′

Example 7

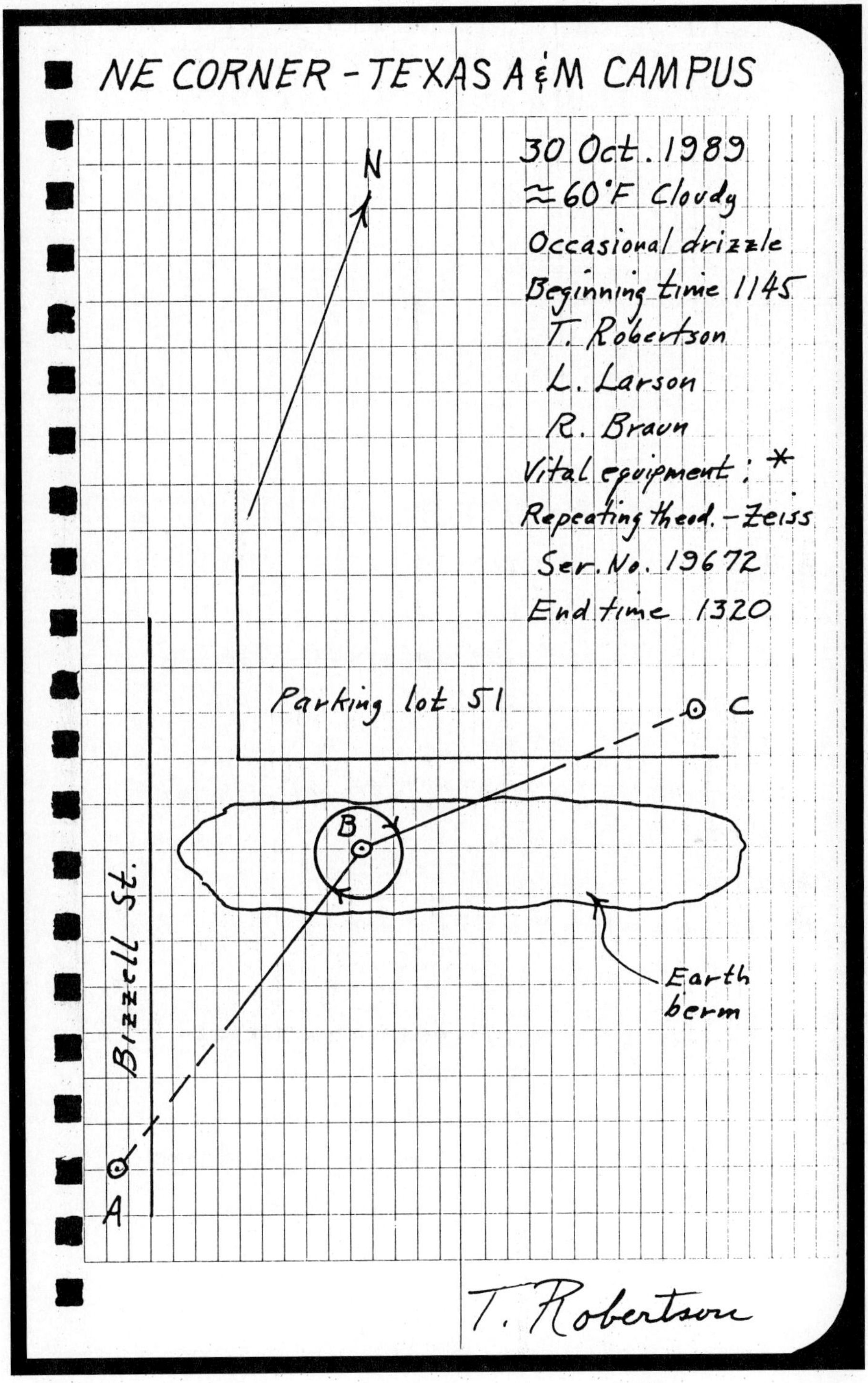

Example 7 (continued).

STADIA OBSERVATIONS FOR TOPOGRAPHIC MAPPING

⊼ at Sta. B, Elev. 462.48 ft., h.i. = 5.05', K = 100

All angles turned clockwise

Sta.	Rod Rdgs.	I	Zenith Angle (Z)	Azimuth	H (ft.)	V (ft.)	Elev. (ft.)
C	5.490		87°17.9'				
	5.050	0.878	(87°18.1')	135°18.1'	88	4.13	466.6
	4.612						
1	6.485		90°00.9'				
	5.710	1.550	(90°01.1')	162°29.4'	155	−0.05	461.8
	4.935		on 5.710				
2	6.109		89°26.0'				
	5.203	1.816	(89°26.2')	204°35.9'	182	1.79	464.1
	4.293		on 5.203				
3	2.798		90°18.2'				
	2.161	1.273	(90°18.4')	257°05.9'	127	−0.68	464.7
	1.525		on 2.161				
4	10.031		91°41.7'				
	7.869	4.320	(91°41.9')	310°55.9'	432	−12.80	446.9
	5.711		on 7.869				

Horiz. & Vert. circle least count = 1' (est. 0.1')

Vert. circle index error determination

83°08.8' (D) 276°50.8' (R) ⇒ 360 − 276°50.8' = 83°09.2'

Mn. = 83°09.0' V.C. Index Error = D − Mn = −0.2'

Note: ( ) = Z corrected for index error

Example 8

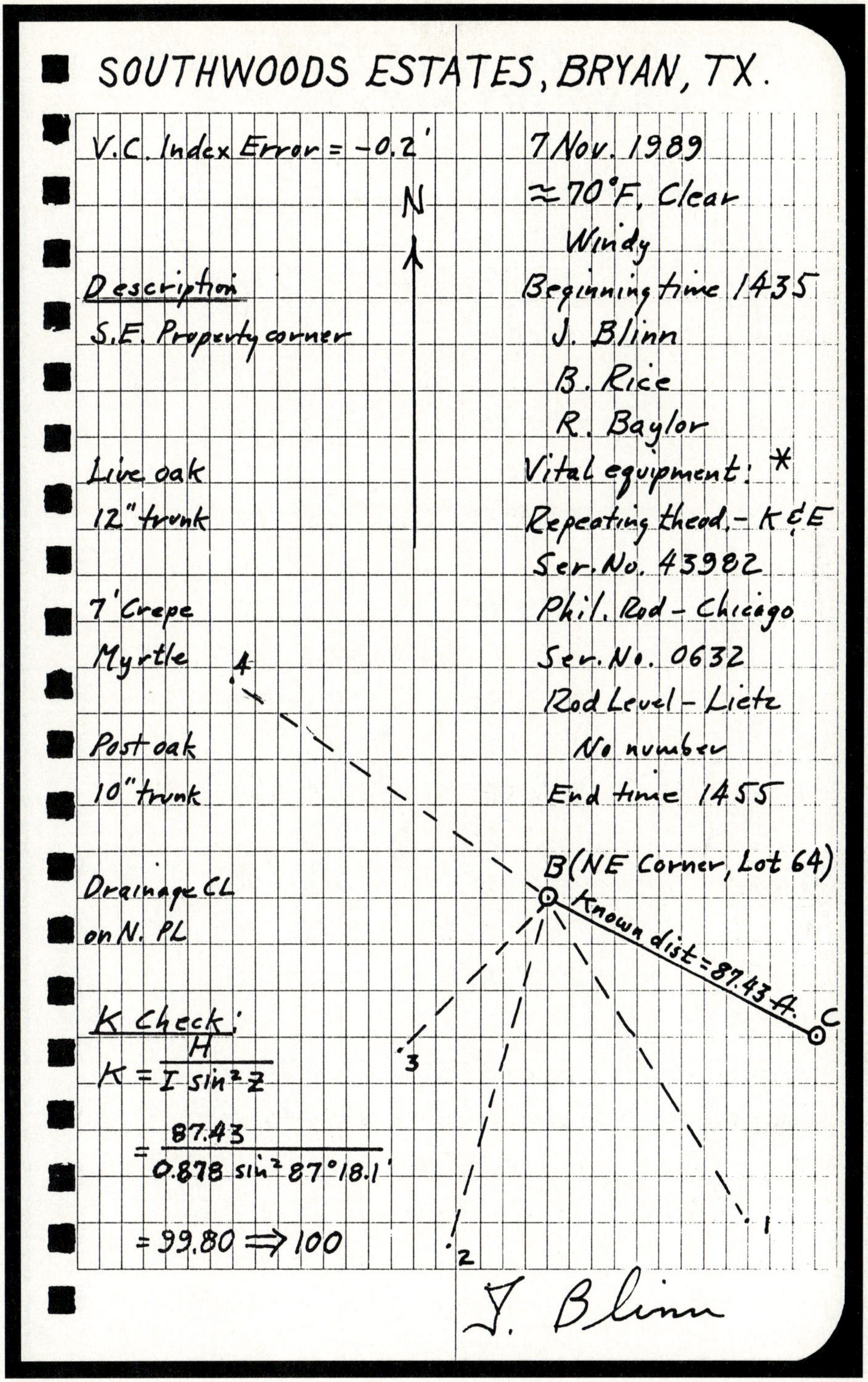

Example 8 (continued).

FOUNDATION LAYOUT BY ANGLES AND DISTANCES

⊼ at Pt. "0", All Angles Turned Clockwise

Line		Angle (β)	Dist. (ft.)
0 – NW	Prop. Cor.	0°00′00″	
0 – 1		13°37′37″	33.956
0 – 2		38°39′35″	12.806
0 – 3		55°12′39″	57.838
0 – 4		66°56′55″	25.539
0 – 5		90°00′00″	8.000
0 – 6		90°00′00″	23.500
0 – 7		90°00′00″	47.500

Diagonal 1-7 and 3-5 = 51.471 ft.
Diagonal 2-6 and 4-5 = 18.446 ft.

Example 9

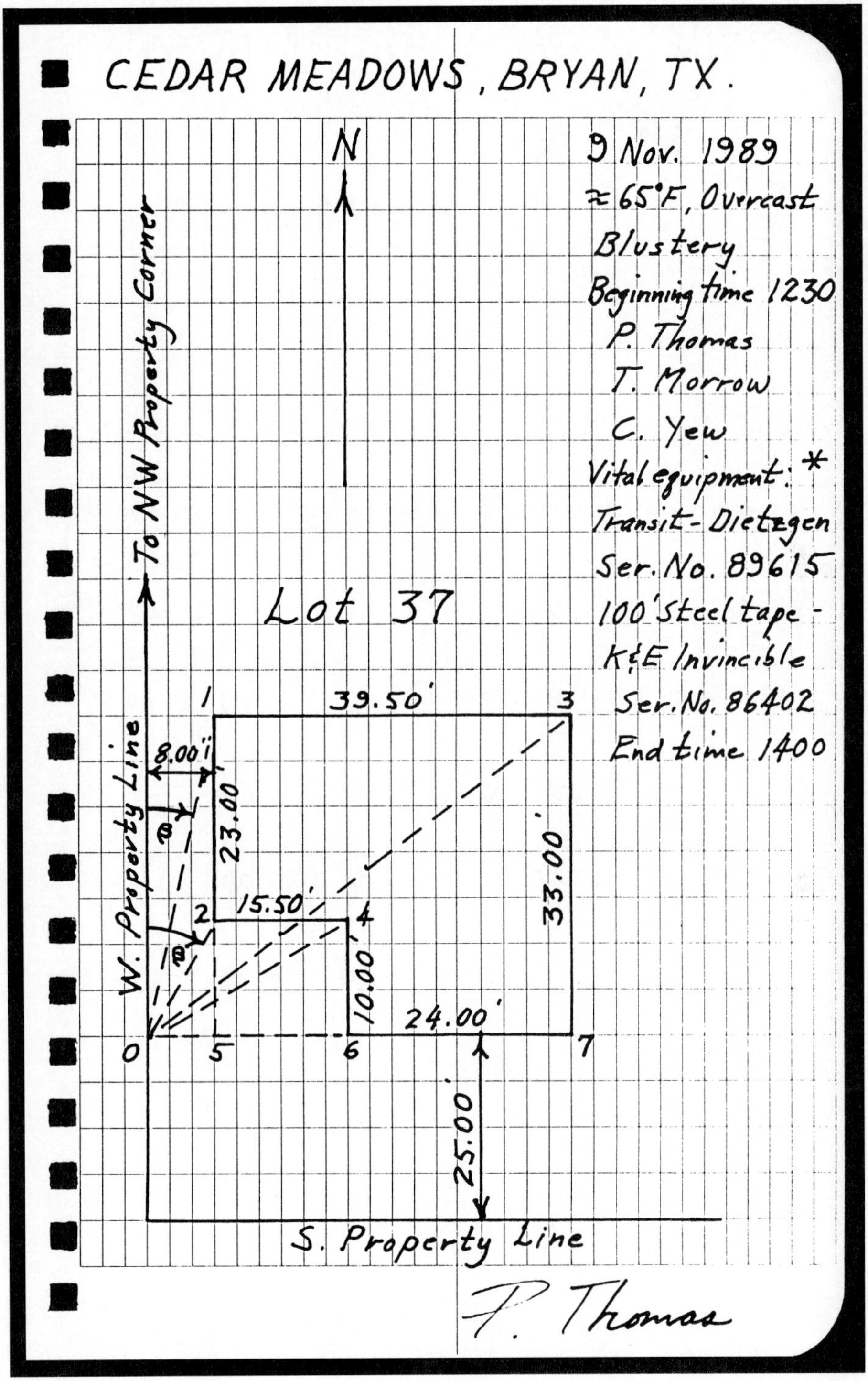

Example 9 (continued.)

CIRCULAR CURVE LAYOUT BY DEFLECTION ANGLES

⊼ at Sta. PC, All Angles Turned Clockwise

Sta.	c′ (arc)	c (chord)	d/2	Cumulative d/2	Curve Parameters
51		65.00			PI = 48+90.14
PT					D_a = 12°00′
50+35.00	35.00	34.99	2°06′00″	18°00′00″ = Δ/2	Δ = 36°00′R
					L = 300.00′
50+00	50.00	49.98	3°00′00″	15°54′00″	R = 477.46′
					T = 155.14′
49+50	50.00	49.98	3°00′00″	12°54′00″	LC = 295.09
					E = 24.57′
49+00	50.00	49.98	3°00′00″	9°54′00″	M = 23.37
48+50	50.00	49.98	3°00′00″	6°54′00″	
48+00	50.00	49.98	3°00′00″	3°54′00″	
47+50	15.00	15.00	0°54′00″	0°54′00″	
PC 47+35.00		35.00			
47					

Example 10

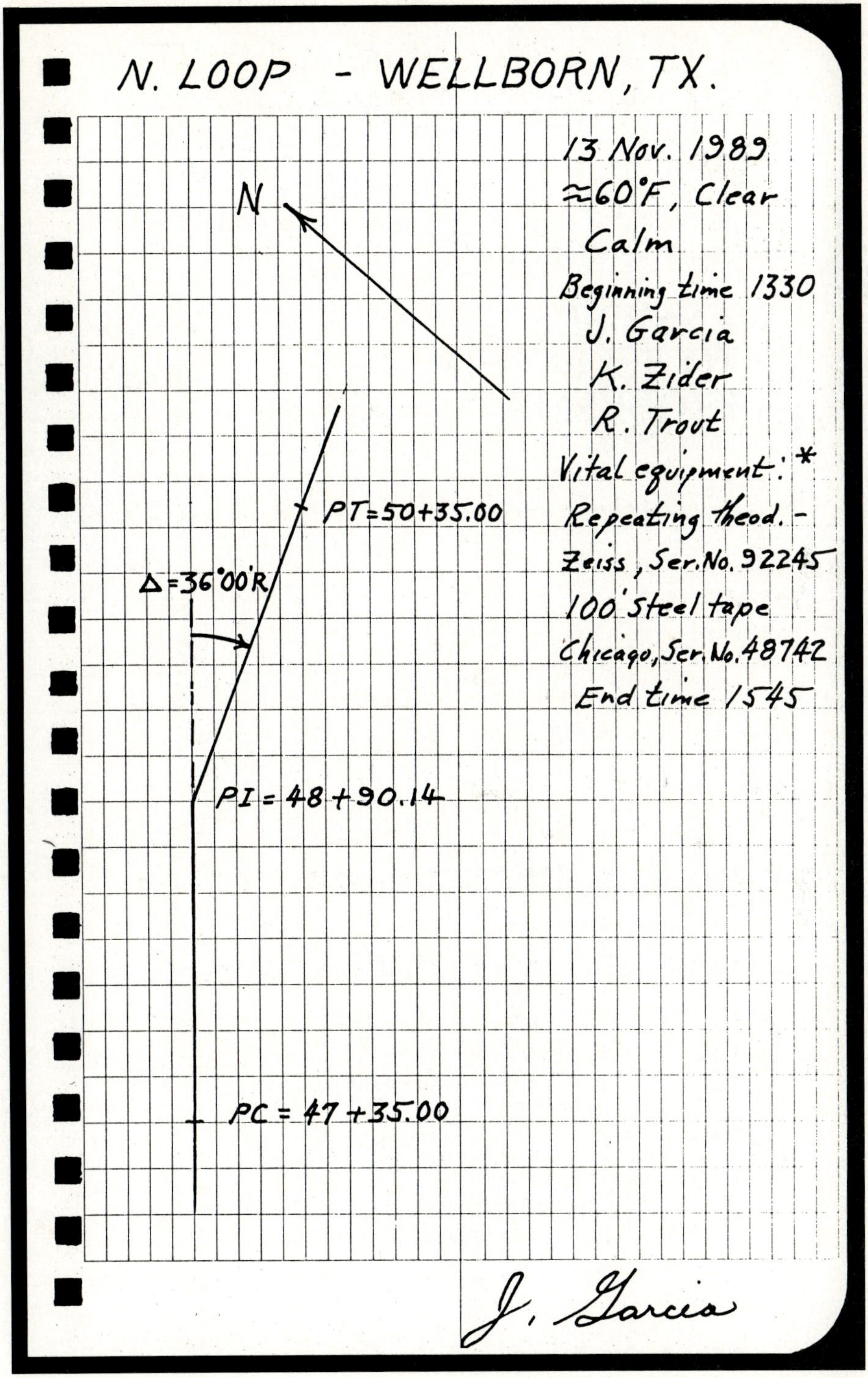

Example 10 (continued).

ANSWERS TO SELECTED PROBLEMS

CHAPTER 1

1.2 (*a*)	18.23 m
1.6	1.0025 ha
1.7 (*c*)	123.1131°
1.10 (*c*)	2.738303 rad
1.12 (*c*)	0.04118
1.16 (*c*)	$y = -13.5$
1.18 (*d*)	$x = +0.971$ or -0.562
1.22	$m = +0.526$, $b = -4.42$
1.25	$x = +8.000$, $y = -4.000$
1.30 (*c*)	20.56879473
1.33	48.1 ft
1.38	3d angle = 32°20′ Side opposite 84°28′ angle = 421.98 ft Side opposite 32°20′ angle = 226.75 ft

CHAPTER 2

2.1	1.77%
2.3 (*c*)	812.17 ft
2.4 (*f*)	373.51 ft

2.6	0.008 ft
2.11	$\pm 00°01'17.3''$
2.15	0.03 ft
2.18 (*a*)	535.34 ft
2.19 (*e*)	650.33 ft
2.23	25.0 lb
2.26	$n_g = 1.0003002$
2.30 (*a*)	E = ±0.039 ft
2.34	2169.65 ft

CHAPTER 3

3.2	23.6 miles
3.5	1348.44 ft
3.10	1348.48 ft
3.14	7.826 ft
3.19	Closure error = ±0.03 ft
3.24	Second order, Class II
3.27	367.82 ft
3.33	39.000 ft
3.34	2.565 ft
3.36	$\text{Elev}_{\text{BM}\,K}$ = 677.437 ft; closure error = 0.0118 ft
3.39	Adjusted $\text{elev}_{\text{BM}\,K}$ = 677.441 ft
3.43	Elevation-difference misclosure = 0.034 ft
3.45	Elev_{1+25} = 429.1 ft; $\text{elev}_{\text{BM Stop}}$ = 424.49 ft

CHAPTER 4

4.1	$\angle A = 25°12'02''$
4.5	$21°31'10''$
4.9	$20''$
4.14	$69°21'50''$
4.16	$72°22'07.5''$
4.21	$E = 1.83''$
4.23	$V = +4°13''$
4.27 (*b*)	$V_{set} = -5°31'$
4.29	$V = +3°45.9'$
4.31	$Z_{\text{set}} = 93°21.9'$

CHAPTER 5

5.1	$\sigma = \pm 0.051$ ft
5.4	$M = 77°15'07.5''$
5.10	$M = 423.9$ ft
5.12	$\sigma_s = \pm 0.08$ ft

5.15	$\sigma_s = \pm 0.005$ ft
5.18	$\sigma = 21$ ft^2
5.22	$M_w = 426.48$ ft
5.28	$C_{adj} = 121°39'50''$
5.30	$C_{adj} = 121°39'39.4''$

CHAPTER 6

6.1	$C_{adj} = 37°30'$L
6.6	Offset = 0.004 ft.
6.8	Angular error = 52″
6.11	Error = 3′26.3″
6.14	Offset = 1.07 ft
6.17	$B'B = 37.487$ ft

CHAPTER 7

7.2 (*c*)	243°12′32″
7.5 (*c*)	S38°29.2′W
7.7 (*c*)	N63°12′32″E
7.9 (*c*)	038°29.2′
7.11	$Az_{BC} = 140°30'$
7.15	$Az_{BC} = 031°22'$
7.21	$\angle B = 92°59'$
7.24	$O°35'$W (annual change approx. 7.7′W)
7.26	Brg_{mag}N42°31′W = Brg_{true} N28°51′W
7.30	Brg_{mag}N55°10′E = Brg_{true} N49°34′E
7.34 (*c*)	2°13′

CHAPTER 8

8.5	Adj. Brg_{BC} = N67°21′04″W
8.9	Adj. Brg_{LA} = N64°11′18″E
8.13	$Lat_{BC} = +127.11$ ft; $Dep_{BC} = -304.63$ ft; Closure = 0.15 ft
8.17	$Lat_{LA} = +162.17$ ft; $Dep_{LA} = +335.30$ ft; Closure = 0.22 ft
8.21	$y_C = 377.00$ ft; $x_C = 411.72$ ft
8.25	$y_A = 9583.75$ ft; $x_A = 16,214.74$ ft
8.29	Third order, Class II
8.33	Third order, Class II
8.37	Area = 1.4472 acres

CHAPTER 9

9.1	$AB = 441.55$ ft
9.5	$Az_{Tower-Stack} = 104°37'22''$

9.7	Brg_{AQ} = N57°34′26″W
9.9	Y_I = 664.65 ft
9.11	X_I = 10,222.71 ft
9.13	AB = 229.31 ft; Brg_{AB} = S38°22′33″W
9.15	BC = 386.43 ft; Brg_{BC} = S33°04′35″W
9.17	PI = 207.05 ft
9.20	Az_{MI} = 227°32′28″
9.23	BI = 1815.73 ft
9.25	LI = 213.46 ft
9.27	CD = 217.94 ft
9.29	EA = 241.18 ft
9.31	Az_{CD} = 348°23′16″
9.33	CD = 265.39 ft
9.36	Az_{BC} = 253°18′49″

CHAPTER 10

10.1	H = 948 ft
10.3	V = −103.0 ft
10.7	Elev = 359.5 ft
10.11	K_{ave} = 80.2
10.15	$Elev_R$ = 445.4 ft
10.16	$\Delta Elev_{C \text{ to } D}$ = +86.3 ft
10.19	$Elev_Z$ = 1220.4 ft
10.22	$Elev_B$ = 2545.3 ft

CHAPTER 11

11.1 (*b*)	1:40,000
11.4	1.5%
11.6	1 in. = 60 ft
11.9	Station *D*: H = 329 ft; elev = 516.7 ft
11.12	Station *A*: H = 138 ft; elev = 375.0 ft
11.15	Station *B*: H = 213 ft; elev = 255.2 ft
11.18	Station *C*: H = 393 ft; elev = 313.8 ft

CHAPTER 12

12.4	10°30′40″
12.8	To corner 1: angle = 16°41′57″; distance = 26.10 ft
12.10	To corner 4: angle = 66°48′05″; distance = 68.54 ft
12.12	To corner 3: angle = 36°50′55″; distance = 91.22 ft
12.16	To corner *A*: angle = 11°45′48″; distance = 49.05 ft
12.20	−0.50%

CHAPTER 13

13.1 (*a*)	$D_a = 25°27'53''$
13.3 (*b*)	$R = 2604.35$ ft
13.5 (*c*)	$R = 1759.88$ ft
13.7	$R = 2046.28$ ft
13.10	$\Delta = 7°42'56''$
13.15	PC @ sta 13+37.49
13.20	Deflection angle to first half-station = 00°08′27″
13.25	Chord to second half-station = 62.51 ft
13.27	PC_2 @ Sta 21+35.16
13.29	$\Delta = 62°54'$
13.33	BVC @ sta 38+00; elev of sta 39+00 = 348.88 ft
13.35	BVC @ sta 33+75.50; elev of first full station = 633.73 ft
13.37	Station of low point = 40+00
13.41	$L = 1200.00$ ft
13.44	34+52.47
13.47	$\text{Elev}_{\text{CVC}} = 350.69$ ft

CHAPTER 14

14.1	Trial ground rod = 12.0 ft
14.4	Centerline distance = 38.6 ft
14.6	$A_L = 236.5\ \text{ft}^2$
14.9	$A_{3-L} = 291.1\ \text{ft}^2$
14.11	$A_{5-L} = 288.3\ \text{ft}^2$
14.14	1019.1 yd^3
14.19	1142.1 yd^3
14.23	Fill = 139.4 yd^3
14.25	6164.4 yd^3

REFERENCES

CHAPTER 1

Burington, R. S., *Handbook of Mathematical Tables and Formulas,* Handbook Publishers, Inc., Sandusky, Ohio, 1940.

Davis, R. E., Foote, F. S., Anderson, J. M., and Mikhail, E. M., *Surveying: Theory and Practice,* 6th ed., McGraw-Hill, Inc., New York, 1981.

Evett, J. B., *Surveying,* John Wiley & Sons, New York, 1979.

Federal Register, "Proposal to Retain the Unit Known as the U.S. Survey Foot," FR Document 88-16174, Washington, D.C., July 18, 1988.

Moffitt, F. H., and Bouchard, H., *Surveying,* 8th ed., Harper & Row, New York, 1987.

Stem, J. E., "Request for Comments: International vs. U.S. Survey," *P.O.B.,* vol. 14, no. 1, October-November 1988.

Whitten, C. A., "Which Foot to Use? International or U.S. Survey," *P.O.B.,* vol. 10, no. 3, February-March 1985.

CHAPTER 2

Davis, R. E., Foote, F. S., Anderson, J. M., and Mikhail, E. M., *Surveying: Theory and Practice,* 6th ed., McGraw-Hill, Inc., New York, 1981.

Handbook of Chemistry and Physics, 43d ed., The Chemical Rubber Publishing Company, Cleveland, Ohio, 1961.

McDonnell, P. W., "P.O.B. 1987 EDMI Survey," *P.O.B.,* vol. 13, no. 1, October-November 1987.

Moffitt, F. H., and Bouchard, H., *Surveying,* 8th ed., Harper & Row, New York, 1987.

Schmidt, M. O., and Wong, K. W., *Fundamentals of Surveying,* 3d ed., PWS Engineering, Boston, Mass., 1985.

Sears, F. W., *Principles of Physics II: Electricity and Magnetism.* Addison-Wesley Press, Inc., Cambridge, Mass., 1945.

Sears, F. W., *Principles of Physics III: Optics,* Addison-Wesley Press, Inc., Cambridge, Mass., 1945.

Wolf, P. R., and Brinker, R. C., *Elementary Surveying,* 8th ed., Harper & Row, New York, 1989.

CHAPTER 3

Davis, R. E., Foote, F. S., Anderson, J. M., and Mikhail, E. M., *Surveying: Theory and Practice,* 6th ed., McGraw-Hill, Inc., New York, 1981.

Kavanagh, B. F., *Surveying with Construction Applications,* Prentice-Hall, Inc., Englewood Cliffs, N.J., 1989.

McDonnell, P. W., "P.O.B. 1987 Level Survey," *P.O.B.,* vol. 13, no. 2, December 1987–January 1988.

Moffitt, F. H., and Bouchard, H., *Surveying,* 8th ed., Harper & Row, New York, 1987.

Ross, A., and Gonzalez, I, "Rotating Laser Beam Systems Applications for Surveyors," *P.O.B.,* vol. 14, no. 1, October-November 1988.

Schmidt, M. O., and Wong, K. W., *Fundamentals of Surveying,* 3d ed., PWS Engineering, Boston, Mass., 1985.

Wolf, P. R., and Brinker, R. C., *Elementary Surveying,* 8th ed., Harper & Row, New York, 1989.

Zilkoski, D. B., "Impact and Benefits of the North American Vertical Datum of 1988," National Geodetic Survey, NOAA, Rockville, Md., 1985.

CHAPTER 4

Davis, R. E., Foote, F. S., Anderson, J. M., and Mikhail, E. M., *Surveying: Theory and Practice,* 6th ed., McGraw-Hill, Inc., New York, 1981.

McDonnell, P. W., "P.O.B. 1988 Transit/Theodolite Survey," *P.O.B.,* vol. 13, no. 6, August-September 1988.

McDonnell, P. W., "P.O.B. 1989 Total Station Survey," *P.O.B.,* vol. 14, no. 4, April-May 1989.

Moffitt, F. H., and Bouchard, H., *Surveying,* 8th ed., Harper & Row, New York, 1987.

Ross, A., and Gonzalez, I., "Rotating Laser Beam Systems Applications for Surveyors," *P.O.B.,* vol. 14, no. 1, October-November 1988.

Schmidt, M. O., and Wong, K. W., *Fundamentals of Surveying,* 3d ed., PWS Engineering, Boston, Mass., 1985.

Wolf, P. R., and Brinker, R. C., *Elementary Surveying,* 8th ed., Harper & Row, New York, 1989.

CHAPTER 5

Davis, R. E., Foote, F. S., Anderson, J. M., and Mikhail, E. M., *Surveying: Theory and Practice,* 6th ed., McGraw-Hill, Inc., New York, 1981.

Freund, J. E., *Mathematical Statistics,* Prentice-Hall, Inc., Englewood Cliffs, N.J., 1962.

Moffitt, F. H., and Bouchard, H., *Surveying,* 8th ed., Harper & Row, New York, 1987.

Reddick, H. W., and Miller, F. H., *Advanced Mathematics for Engineers,* 2d ed., John Wiley & Sons, Inc., New York, 1947.

Schmidt, M. O., and Wong, K. W., *Fundamentals of Surveying,* 3d ed., PWS Engineering, Boston, Mass., 1985.

CHAPTER 6

Davis, R. E., Foote, F. S., Anderson, J. M., and Mikhail, E. M., *Surveying: Theory and Practice,* 6th ed., McGraw-Hill, Inc., New York, 1981.

Moffitt, F. H., and Bouchard, H., *Surveying,* 8th ed., Harper & Row, New York, 1987.

Wolf, P. R., and Brinker, R. C., *Elementary Surveying,* 8th ed., Harper & Row, New York, 1989.

CHAPTER 7

Davis, R. E., Foote, F. S., Anderson, J. M., and Mikhail, E. M., *Surveying: Theory and Practice,* 6th ed., McGraw-Hill, Inc., New York, 1981.

Moffitt, F. H., and Bouchard, H., *Surveying,* 8th ed., Harper & Row, New York, 1987.

Peddie, N. W., "1985 Magnetic Models for the United States," *EOS Transactions,* American Geophysical Union, vol. 68, no. 20, 1987.

Wolf, P. R., and Brinker, R. C., *Elementary Surveying,* 8th ed., Harper & Row, New York, 1989.

CHAPTER 8

Davis, R. E., Foote, F. S., Anderson, J. M., and Mikhail, E. M., *Surveying: Theory and Practice,* 6th ed., McGraw-Hill, Inc., New York, 1981.

Moffitt, F. H., and Bouchard, H., *Surveying,* 8th ed., Harper & Row, New York, 1987.

Schmidt, M. O., and Wong, K. W., *Fundamentals of Surveying,* 3d ed., PWS Engineering, Boston, Mass., 1985.

Sprinsky, W. H., "Traverse Adjustments for Observations from Modern Survey Instruments," *Journal of Surveying Engineering,* vol. 113, no. 2, 1987.

Wolf, P. R., and Brinker R. C., *Elementary Surveying,* 8th ed., Harper & Row, New York, 1989.

CHAPTER 9

Brand, M. J. D., "Beyond Digital Mapping," *Civil Engineering,* vol. 57, no. 12, 1987.

Breed, C. B., *Surveying,* 2d ed., John Wiley & Sons, New York, 1942.

Collins, J., "GPS Surveying Techniques," *P.O.B.,* vol. 11, no. 4, April-May 1986.

Collins, J., "GPS Hardware/Software/Satellites," seminar paper from Surveying Automation in the Field, presented at P.O.B. Conference "Surveying Automation '89," Anaheim, Calif., June 7 1989.

Collins, J., "Shooting for a Full GPS Constellation," *Professional Surveyor,* vol. 9, no. 4, July-August 1989.

Davis, R. E., Foote, F. S., Anderson, J. M., and Mikhail, E. M., *Surveying: Theory and Practice,* 6th ed., McGraw-Hill, Inc., New York, 1981.

Donahue, J. G., "Land Base Accuracy, Is It Worth the Cost?" *ACSM Bulletin,* December 1988.

Ethridge, M. M., "The National Geodetic Reference System of the Future," *P.O.B.,* vol. 14, no. 2, December 1988–January 1989.

Goad, C. C., "On the Move with GPS," *P.O.B.,* vol. 14, no. 4, April-May 1989.

Harlow, C., "Trends in GIS," *Professional Surveyor,* vol. 8, no. 6, July-August 1988.

Kavanagh, B. F., *Surveying with Construction Applications,* Prentice-Hall, Inc., Englewood Cliffs, N.J., 1989.

Manual of Remote Sensing, 2d, ed., American Society of Photogrammetry, Falls Church, Va., 1983.

Moffitt, F. H., and Bouchard, H., *Surveying,* 8th ed., Harper & Row, New York, 1987.

Robinson, R., "GIS Goes Public," *Civil Engineering,* vol. 58, no. 6, 1988.

Schmidt, M. O., and Wong, K. W., *Fundamentals of Surveying,* 3d ed., PWS Engineering, Boston, Mass., 1985.

Simkowitz, H. J., "GIS: Technology for Transportation," *Civil Engineering,* vol. 59, no. 6, 1989.

Stem, J. E., *State Plane Coordinate System of 1983,* NOAA Manual NOS NGS 5, National Oceanic and Atmospheric Administration, Rockville, Md., 1989.

Wolf, P. R., and Brinker, R. C., *Elementary Surveying,* 8th ed., Harper & Row, New York, 1989.

CHAPTER 10

Davis, R. E., Foote, F. S., Anderson, J. M., and Mikhail, E. M., *Surveying: Theory and Practice,* 6th ed., McGraw-Hill, Inc., New York, 1981.

Moffitt, F. H., and Bouchard, H., *Surveying,* 8th ed., Harper, & Row, New York, 1987.

Schmidt, M. O.., and Wong, K. W., *Fundamentals of Surveying,* 3d ed., PWS Engineering, Boston, Mass., 1985.

Wolf, P. R., and Brinker, R. C., *Elementary Surveying,* 8th ed., Harper & Row, New York, 1989.

CHAPTER 11

Davis, R. E., Foote, F. S., Anderson, J. M., and Mikhail, E. M., *Surveying: Theory and Practice,* 6th ed., McGraw-Hill, Inc., New York, 1981.

Moffitt, F. H., and Bouchard, H., *Surveying,* 8th ed., Harper & Row, New York, 1987.
Schmidt, M. O., and Wong, K. W., *Fundamentals of Surveying,* 3d ed., PWS Engineering, Boston, Mass., 1985.
Wolf, P. R., and Brinker, R. C., *Elementary Surveying,* 8th ed., Harper & Row, New York, 1989.

CHAPTER 12

Davis, R. E., Foote, F. S., Anderson, J. M., and Mikhail, E. M., *Surveying: Theory and Practice,* 6th ed., McGraw-Hill, Inc., New York, 1981.
Kavanagh, B. F., *Surveying with Construction Applications,* Prentice-Hall, Inc., Englewood Cliffs, N.J., 1989.
Wolf, P. R., and Brinker, R. C., *Elementary Surveying,* 8th ed., Harper & Row, New York, 1989.

CHAPTER 13

Davis, R. E., Foote, F. S., Anderson, J. M., and Mikhail, E. M., *Surveying: Theory and Practice,* 6th ed., McGraw-Hill, Inc., New York, 1981.
Moffitt, F. H., and Bouchard, H., *Surveying,* 8th ed., Harper & Row, New York, 1987.
Schmidt, M. O., and Wong, K. W., *Fundamentals of Surveying,* 3d ed., PWS Engineering, Boston, Mass,., 1985.
Wolf, P. R., and Brinker, R. C., *Elementary Surveying,* 8th ed., Harper & Row, New York, 1989.

CHAPTER 14

Breed, C. B., *Surveying,* 2d ed., John Wiley & Sons, New York, 1942.
Davis, R. E., Foote, F. S., Anderson, J. M., and Mikhail, E. M., *Surveying: Theory and Practice,* 6th ed., McGraw-Hill, Inc., New York, 1981.
Moffitt, F. H., and Bouchard, H., *Surveying,* 8th ed., Harper & Row, New York, 1987.
Schmidt, M. O., and Wong, K. W., *Fundamentals of Surveying,* 3d ed., PWS Engineering, Boston, Mass., 1985.
Wolf, P. R., and Brinker, R. C., *Elementary Surveying,* 8th ed., Harper & Row, New York, 1989.

INDEX